BOB SCHOL

#17A - 25 W

WATERLOO, ONT

884-4649

$$\nu = \frac{\mu}{\rho}$$

$$\gamma = \rho g$$

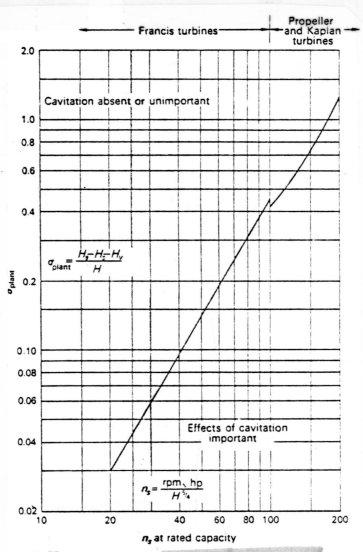

Francis turbines ← Propeller and Kaplan turbines →

Cavitation absent or unimportant

$$\sigma_{plant} = \frac{H_s - H_{\bar{z}} - H_v}{H}$$

Effects of cavitation important

$$n_s = \frac{rpm \sqrt{hp}}{H^{5/4}}$$

n_s at rated capacity

FIG. 7-5 Experience limits of plant sigma versus specific speed for hydraulic turbines.

Table III. PROPERTIES OF WATER*

Temperature °Fahrenheit	Specific Weight γ lb/ft³	Mass Density ρ lb-sec²/ft⁴	Dynamic Viscosity μ x 10⁵ lb-sec/ft²	Kinematic Viscosity ν x 10⁵ sq ft/sec	Surface Energy† σ x 10³ lb/ft	Vapor Pressure p_v lb/sq in	Bulk Modulus E x 10⁻³ lb/sq in
32	62.42	1.940	3.746	1.931	5.18	0.09	290
40	62.43	1.940	3.229	1.664	5.14	0.12	295
50	62.41	1.940	2.735	1.410	5.09	0.18	300
60	62.37	1.938	2.359	−1.217	5.04	0.26	312
70	62.30	1.936	2.050	-1.059	5.00	0.36	320
80	62.22	1.934	1.799	− 0.930	4.92	0.51	323
90	62.11	1.931	1.595	0.826	4.86	0.70	326
100	62.00	1.927	1.424	0.739	4.80	0.95	329
110	61.86	1.923	1.284	0.667	4.73	1.24	331
120	61.71	1.918	1.168	0.609	4.65	1.69	333
130	61.55	1.913	1.069	0.558	4.60	2.22	332
140	61.38	1.908	0.981	0.514	4.54	2.89	330
150	61.20	1.902	0.905	0.476	4.47	3.72	328
160	61.00	1.896	0.838	0.442	4.41	4.74	326
170	60.80	1.890	0.780	0.413	4.33	5.99	322
180	60.58	1.883	0.726	0.385	4.26	7.51	318
190	60.36	1.876	0.678	0.362	4.19	9.34	313
200	60.12	1.868	0.637	0.341	4.12	11.52	308
212	59.83	1.860	0.593	0.319	4.04	14.7	300

* Most of this table was taken from "Hydraulic Models," A.S.C.E. Manual of Engineering Practice, No. 25, A.S.C.E., 1942.

† In contact with air.

Table IV. PROPERTIES OF AIR*

Temperature T °Fahrenheit	Specific Weight γ lb/ft³	Mass Density ρ lb-sec²/ft⁴	Dynamic Viscosity μ lb-sec/ft²	Kinematic Viscosity ν ft²/sec.
−40	0.0946	0.00294	3.12×10^{-7}	1.06×10^{-4}
−20	0.0903	0.00280	3.25	1.16
0	0.0863	0.00268	3.38	1.26
20	0.0827	0.00257	3.50	1.36
40	0.0795	0.00247	3.62×10^{-7}	1.46×10^{-4}
60	0.0764	0.00237	3.74	1.58
80	0.0735	0.00228	3.85	1.69
100	0.0709	0.00220	3.96	1.80
150	0.0651	0.00202	4.23	2.07
200	0.0602	0.00187	4.49×10^{-7}	2.40×10^{-4}

* Approximate values at standard atmospheric pressure.

SIXTH EDITION

FLUID MECHANICS

VICTOR L. STREETER

Professor of Hydraulics
University of Michigan

E. BENJAMIN WYLIE

Professor of Civil Engineering
University of Michigan

McGRAW-HILL BOOK COMPANY

New York St. Louis San Francisco Düsseldorf
Johannesburg Kuala Lumpur London Mexico Montreal
New Delhi Panama Paris São Paulo
Singapore Sydney Tokyo Toronto

Library of Congress Cataloging in Publication Data
Streeter, Victor Lyle, date
    Fluid mechanics.
    Includes bibliographical references.
    1. Fluid mechanics.   I. Wylie, E. Benjamin, joint author.
TA357.S8 1975   532   74-9930
ISBN 0-07-062193-4

FLUID MECHANICS

23456789 KPKP 798765

This book was set in Modern 8A by Mono of Maryland Incorporated.
The editors were B. J. Clark and J. W. Maisel,
the designer was Stephen Naab,
the production supervisor was Sam Ratkewitch.
New drawings were done by J & R Services, Inc.
The cover was designed by Rafael Hernandez.

CONTENTS

Preface   xi

PART I   FUNDAMENTALS OF FLUID MECHANICS

Chapter 1   Fluid Properties and Definitions   3
1.1   Definition of a Fluid   4
1.2   Force, Mass, and Length Units   7
1.3   Viscosity   9
1.4   Continuum   13
1.5   Density, Specific Volume, Specific Weight, Specific Gravity,
    Pressure   13
1.6   Perfect Gas   14
1.7   Bulk Modulus of Elasticity   17
1.8   Vapor Pressure   18
1.9   Surface Tension   18

Chapter 2   Fluid Statics   27
2.1   Pressure at a Point   27
2.2   Basic Equation of Fluid Statics   29
2.3   Units and Scales of Pressure Measurement   33
2.4   Manometers   38
2.5   Forces on Plane Areas   44
2.6   Force Components on Curved Surfaces   53
2.7   Buoyant Force   60
2.8   Stability of Floating and Submerged Bodies   64
2.9   Relative Equilibrium   71

Chapter 3 Fluid-Flow Concepts and Basic Equations   109
  3.1 The Concepts of System and Control Volume   109
  3.2 Application of the Control Volume to Continuity, Energy, and Momentum   114
  3.3 Flow Characteristics; Definitions   116
  3.4 Continuity Equation   121
  3.5 Euler's Equation of Motion along a Streamline   127
  3.6 Reversibility, Irreversibility, and Losses   129
  3.7 The Steady-State Energy Equation   130
  3.8 Interrelationships between Euler's Equations and the Thermodynamic Relations   132
  3.9 The Bernoulli Equation   134
  3.10 Application of the Bernoulli and Energy Equations to Steady Fluid-Flow Situations   140
  3.11 Applications of the Linear-Momentum Equation   144
  3.12 The Moment-of-Momentum Equation   173

Chapter 4 Dimensional Analysis and Dynamic Similitude   207
  4.1 Dimensional Homogeneity and Dimensionless Ratios   208
  4.2 Dimensions and Units   210
  4.3 The Π Theorem   211
  4.4 Discussion of Dimensionless Parameters   223
  4.5 Similitude; Model Studies   226

Chapter 5 Viscous Effects: Fluid Resistance   239
  5.1 Laminar, Incompressible, Steady Flow between Parallel Plates   241
  5.2 Laminar Flow through Circular Tubes and Circular Annuli 249
  5.3 The Reynolds Number   254
  5.4 Prandtl Mixing Length; Velocity Distribution in Turbulent Flow   258
  5.5 Rate Processes   262
  5.6 Boundary-Layer Concepts   266
  5.7 Drag on Immersed Bodies   280
  5.8 Resistance to Turbulent Flow in Open and Closed Conduits 286
  5.9 Steady Uniform Flow in Open Channels   287
  5.10 Steady Incompressible Flow through Simple Pipe Systems   291
  5.11 Lubrication Mechanics   309

Chapter 6 Compressible Flow   333
  6.1 Perfect-Gas Relationships   333
  6.2 Speed of a Sound Wave; Mach Number   340

6.3 Isentropic Flow   343
6.4 Shock Waves   350
6.5 Fanno and Rayleigh Lines   355
6.6 Adiabatic Flow with Friction in Conduits   359
6.7 Frictionless Flow through Ducts with Heat Transfer   366
6.8 Steady Isothermal Flow in Long Pipelines   371
6.9 Analogy of Shock Waves to Open-Channel Waves   375

Chapter 7 Ideal-Fluid Flow   389
7.1 Requirements for Ideal-Fluid Flow   389
7.2 The Vector Operator ∇   390
7.3 Euler's Equation of Motion   396
7.4 Irrotational Flow; Velocity Potential   401
7.5 Integration of Euler's Equations; Bernoulli Equation   403
7.6 Stream Functions; Boundary Conditions   406
7.7 The Flow Net   411
7.8 Three-dimensional Flow   414
7.9 Two-dimensional Flow   428

PART 2 APPLICATIONS OF FLUID MECHANICS

Chapter 8 Fluid Measurement   449
8.1 Pressure Measurement   449
8.2 Velocity Measurement   451
8.3 Positive-Displacement Meters   456
8.4 Rate Meters   457
8.5 Electromagnetic Flow Devices   482
8.6 Measurement of River Flow   482
8.7 Measurement of Turbulence   482
8.8 Measurement of Viscosity   483

Chapter 9 Turbomachinery   498
9.1 Homologous Units; Specific Speed   498
9.2 Elementary Cascade Theory   504
9.3 Theory of Turbomachines   506
9.4 Impulse Turbines   511
9.5 Reaction Turbines   517
9.6 Pumps and Blowers   523
9.7 Centrifugal Compressors   530
9.8 Cavitation   534

Chapter 10 Steady Closed-Conduit Flow   544
10.1 Exponential Pipe-Friction Formulas   544

10.2   Hydraulic and Energy Grade Lines   547
10.3   The Siphon   553
10.4   Pipes in Series   556
10.5   Pipes in Parallel   560
10.6   Branching Pipes   563
10.7   Networks of Pipes   565
10.8   Computer Program for Steady-State Hydraulic Systems   569
10.9   Conduits with Noncircular Cross Sections   578
10.10   Aging of Pipes   580

Chapter 11   Steady Flow in Open Channels   590
11.1   Classification of Flow   591
11.2   Best Hydraulic-Channel Cross Sections   592
11.3   Steady Uniform Flow in a Floodway   595
11.4   Hydraulic Jump; Stilling Basins   596
11.5   Specific Energy; Critical Depth   600
11.6   Gradually Varied Flow   604
11.7   Classification of Surface Profiles   611
11.8   Control Sections   614
11.9   Computer Calculation of Gradually Varied Flow   615
11.10   Transitions   619

Chapter 12   Unsteady Flow   629
             Flow in Closed Conduits   630
12.1   Oscillation of Liquid in a U Tube   630
12.2   Establishment of Flow   644
12.3   Surge Control   645
12.4   Description of the Waterhammer Phenomenon   647
12.5   Differential Equations for Calculation of Waterhammer   649
12.6   The Method-of-Characteristics Solution   654
12.7   Boundary Conditions   658

        Open-Channel Flow   673
12.8   Frictionless Positive Surge Wave in a Rectangular Channel   674
12.9   Frictionless Negative Surge Wave in a Rectangular Channel   676
12.10   Flood Routing in Prismatic Channels   680
12.11   Mechanics of Rainfall-Runoff Relations for Sloping Plane
        Areas   687

        Appendixes   701
   A   Force Systems, Moments, and Centroids   701
   B   Partial Derivatives and Total Differentials   706
   C   Physical Properties of Fluids   711

D   Notation   719

E   Computer Programming Aids   723

   E.1   Quadratures; Numerical Integration by Simpson's
        Rule   723

   E.2   Parabolic Interpolation   724

   E.3   Solution of Algebraic or Trancendental Equations by
        the Bisection Method   726

   E.4   Solution of Trancendental or Algebraic Equations by
        the Newton-Raphson Method   727

   E.5   Runge-Kutta Solution of Differential Equations   728

Answers to Even-numbered Problems   732

Index   739

PREFACE

The general pattern of the text is unchanged in this revision in that it is divided into two parts, the first emphasizing fundamentals and the second applications. The chapter contents remain the same as in previous editions although much of the material is updated. The most noticeable change in the revision is the incorporation of the metric system of units (SI) in the text, examples, and problems. The SI and Engilsh units are now given equal emphasis. The generalized control-volume derivation has been improved so that the limiting procedures are more easily visualized. Several examples focusing on environmental issues have been included; these problems generally require some special information or are limited in certain features because of the normal complexity of natural situations.

Some material that was included in the previous edition has been removed in this revision, primarily in Part 2, while the material on computer applications has been strengthened. Chapter 8 on flow measurement no longer carries descriptive material on many devices and the turbomachinery chapter omits fluid torque converters and fluid complings. The graphical and algebraic waterhammer solutions have been removed from Chap. 12.

The use of the digital computer in fluid-flow applications is recommended. The addition of a fairly general program to analyze steady liquid flow in piping systems provides the reader with great flexibility in this field. Pumps, pipelines, and reservoirs can be treated in simple systems or in complex networks. The treatment of turbulent flow in pipelines with empirical (exponential) pipe-flow equations has been added to Chap. 10. A more general program is provided for computation of steady gradually varied flow in open channels.

The unsteady-flow chapter has been reorganized in its treatment of waterhammer problems, with a larger variety of boundary conditions considered. Relevant improvements have also been incorporated in the flood-routing development and accompanying program.

The first six chapters form the basis for a first course in fluids, with selected materials from Chap. 7 and topics from Part 2. Part 2 contains enough material for a second course, including support for laboratory.

The assistance of Mr. Joel Caves in developing some of the examples relating to environmental issues is gratefully acknowledged, as is the advice received from many reviewers.

VICTOR L. STREETER
E. BENJAMIN WYLIE

FLUID
MECHANICS

1

FUNDAMENTALS OF FLUID MECHANICS

In the first three chapters of Part 1, the properties of fluids, fluid statics, and the underlying framework of concepts, definitions, and basic equations for fluid dynamics are discussed. Dimensionless parameters are next introduced, including dimensional analysis and dynamic similitude. Chapter 5 deals with real fluids and the introduction of experimental data into fluid-flow calculations. Compressible flow of both real and frictionless fluids is then treated. The final chapter on fundamentals deals with two- and three-dimensional ideal-fluid flow. The theory has been illustrated with elementary applications throughout Part 1.

FLUID PROPERTIES AND DEFINITIONS

Fluid mechanics is one of the engineering sciences that form the basis for all engineering. The subject branches out into various specialties such as aerodynamics, hydraulic engineering, marine engineering, gas dynamics, and rate processes. It deals with the statics, kinematics, and dynamics of fluids, since the motion of a fluid is caused by unbalanced forces exerted upon it. Available methods of analysis stem from the application of the following principles, concepts, and laws: Newton's laws of motion, the first and second laws of thermodynamics, the principle of conservation of mass, equations of state relating to fluid properties, Newton's law of viscosity, mixing-length concepts, and restrictions caused by the presence of boundaries.

In fluid-flow calculations, viscosity and density are the fluid properties most generally encountered; they play the principal roles in open- and closed-channel flow and in flow around immersed bodies. Surface-tension effects are of importance in the formation of droplets, in flow of small jets, and in situations where liquid-gas-solid or liquid-liquid-solid interfaces occur, as well as in the formation of capillary waves. The property of vapor pressure, accounting for changes of phase from liquid to gas, becomes important when reduced pressures are encountered.

A liquid-fuel-injection system is an example of an engineering problem in which the performance of the product is significantly affected by the properties of the fluid being handled. Fuel is pumped from a storage tank through a series of fuel lines and spray nozzles. The process is intermittent and occurs at high speed. It appears reasonable to expect that less force, and less power, are needed to pump a light-grade, or "thin," oil than a heavy-grade, or "thick," oil. The terms "light" and "heavy" are qualitative terms which describe how easily the fuel flows. There is a quantitative way of specifying this fluidity property, and it will be described later in this chapter. Indeed it will be necessary to define a fluid in a rigorous manner and to see how our fuel fits this definition.

How the fuel sprays from the nozzle will be affected by how surface tension determines the drop formation. The actual design of the nozzle passages will be influenced by other properties of the liquid.

The fluid flow in the lines is intermittent because fuel is supplied to the spray nozzles only at specific times during the operating cycle of an engine. The duration of fuel delivery is carefully regulated. Consequently there are pulsations of pressure in the system. These pressures can be very high and, surprisingly, also very low. It is possible that when the pressure gets low enough, the fuel may momentarily vaporize and interfere with the expected performance of the system. The pressure pulses are transmitted along the column of liquid in the fuel lines similarly to sound waves in air. These pressure waves may be in such a phase relationship with one another that the waves result in momentary pressure peaks, which are many times the expected system pressures. The speed of the pressure waves depends on a property called the bulk modulus.

The sections which follow point up the importance of the physical properties of a liquid or gas. A number of definitions are also included so that one can be specific about the property, quantity, or assumption being considered.

1.1 DEFINITION OF A FLUID

A fluid is a substance that deforms continuously when subjected to a shear stress, no matter how small that shear stress may be. A shear force is the force component tangent to a surface, and this force divided by the area of the surface is the average shear stress over the area. Shear stress at a point is the limiting value of shear force to area as the area is reduced to the point.

In Fig. 1.1 a substance is placed between two closely spaced parallel plates, so large that conditions at their edges may be neglected. The lower plate is fixed, and a force F is applied to the upper plate, which exerts a shear stress F/A on any substance between the plates. A is the area of the upper plate. When the force F causes the upper plate to move with a steady (nonzero) velocity, no matter how small the magnitude of F, one may conclude that the substance between the two plates is a fluid.

The fluid in immediate contact with a solid boundary has the same velocity as the boundary; i.e., there is no slip at the boundary.[1] This is an ex-

---

[1] S. Goldstein, "Modern Developments in Fluid Dynamics," vol. II, pp. 676–680, Oxford University Press, London, 1938.

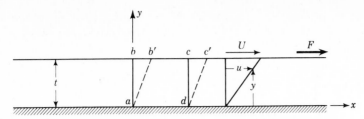

Fig. 1.1 Deformation resulting from application of constant shear force.

perimental fact which has been verified in countless tests with various kinds of fluids and boundary materials. The fluid in the area *abcd* flows to the new position *ab'c'd*, each fluid particle moving parallel to the plate and the velocity *u* varying uniformly from zero at the stationary plate to *U* at the upper plate. Experiments show that other quantities being held constant, *F* is directly proportional to *A* and to *U* and is inversely proportional to thickness *t*. In equation form

$$F = \mu \frac{AU}{t}$$

in which μ is the proportionality factor and includes the effect of the particular fluid. If $\tau = F/A$ for the shear stress,

$$\tau = \mu \frac{U}{t}$$

The ratio U/t is the angular velocity of line *ab*, or it is the *rate of angular deformation* of the fluid, i.e., the rate of decrease of angle *bad*. The angular velocity may also be written du/dy, as both U/t and du/dy express the velocity change divided by the distance over which the change occurs. However, du/dy is more general as it holds for situations in which the angular velocity and shear stress change with *y*. The velocity gradient du/dy may also be visualized as the rate at which one layer moves relative to an adjacent layer. In differential form,

$$\tau = \mu \frac{du}{dy} \tag{1.1.1}$$

is the relation between shear stress and rate of angular deformation for one-dimensional flow of a fluid. The proportionality factor μ is called the *viscosity* of the fluid, and Eq. (1.1.1) is *Newton's law of viscosity.* In the second book of his "Principia," Newton considered the circular motion of fluids as part of his studies of the planets and wrote

Hypothesis

The resistance arising from the want of lubricity in the parts of a fluid, is, other things being equal, proportional to the velocity with which the parts of a fluid are separated from one another.

A plastic substance cannot fulfill the definition of a fluid because it has an initial yield shear stress that must be exceeded to cause a continuous deformation. An elastic substance placed between the two plates would deform a certain *amount* proportional to the force, but not continuously at a definite rate. A complete vacuum between the plates would not result in a constant final rate but in an ever-increasing rate. If sand were placed between the two plates, dry friction would require a *finite* force to cause a continuous motion. Thus sand will not satisfy the definition of a fluid.

Fluids may be classified as Newtonian or non-Newtonian. In Newtonian fluid there is a linear relation between the magnitude of applied shear stress and the resulting rate of deformation [μ constant in Eq. (1.1.1)], as shown in Fig. 1.2. In non-Newtonian fluid there is a nonlinear relation between the magnitude of applied shear stress and the rate of angular deformation. An

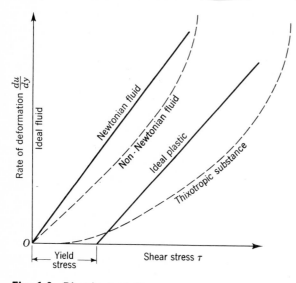

Fig. 1.2 Rheological diagram.

ideal plastic has a definite yield stress and a constant linear relation of τ to du/dy. A *thixotropic* substance, such as printer's ink, has a viscosity that is dependent upon the immediately prior angular deformation of the substance and has a tendency to take a set when at rest. Gases and thin liquids tend to be Newtonian fluids, while thick, long-chained hydrocarbons may be non-Newtonian.

For purposes of analysis, the assumption is frequently made that a fluid is nonviscous. With zero viscosity the shear stress is always zero, regardless of the motion of the fluid. If the fluid is also considered to be incompressible, it is then called an *ideal* fluid and plots as the ordinate in Fig. 1.2.

1.2 FORCE, MASS, AND LENGTH UNITS

In this text units of the English system and the International System (SI) are employed. The basic English units are the *pound* force (lb), the *slug* mass, and the *foot* length (ft). The basic SI units are the *newton* force (N), the *kilogram* mass (kg), and the *meter* length (m). These units are consistent in that the unit of force accelerates the unit of mass 1 unit of length per second per second. The pound mass (lb_m) is used in some tabulations of properties and is related to the slug by 32.174 lb_m = 1 slug. The kilogram force (kg_f) is also used in some countries and is related to the newton by 9.806 N = 1 kg_f.

Abbreviations of SI units are written in lowercase (small) letters for terms like hours (h), meters (m), and seconds (s). When a unit is named after a person, the abbreviation (but not the spelled form) is capitalized; for example, watt (W), pascal (Pa), or newton (N). Multiples and submultiples in powers of 10 are indicated by prefixes, which are also abbreviated, as in such familiar combinations as centimeter (cm), for 10^{-2} m, and kilogram (kg) for 10^3 g. Common prefixes are shown in Table 1.1. Note that prefixes may

Table 1.1 *Selected prefixes for powers of 10 in SI units*

Multiple	SI prefix	Abbre-viation	Multiple	SI prefix	Abbre-viation
10^9	giga	G	10^{-3}	milli	m
10^6	mega	M	10^{-6}	micro	μ
10^3	kilo	k	10^{-9}	nano	n
10^{-2}	centi	c	10^{-12}	pico	p

not be doubled up: The correct form for 10^{-9} is the prefix n-, as in nanometers; combinations like millimicro-, formerly acceptable, are no longer to be used.

The pound of force is defined in terms of the pull of gravity, at a specified (standard) location, on a given mass of platinum. At standard gravitation, $g = 32.174$ ft/s², the body having a pull of one pound has a mass of one pound mass. When Newton's second law of motion is written in the form

$$\mathbf{F} = \frac{m}{g_0}\,\mathbf{a} \tag{1.2.1}$$

and applied to this object falling freely in a vacuum at standard conditions

$$1 \text{ lb} = \frac{1 \text{ lb}_m}{g_0}\,32.174 \text{ ft/s}^2$$

it is clear that

$$g_0 = 32.174 \text{ lb}_m \cdot \text{ft/lb} \cdot \text{s}^2 \tag{1.2.2}$$

Similarly, in SI units

$$1 \text{ kg}_f = \frac{1 \text{ kg}}{g_0}\,9.806 \text{ m/s}^2$$

and

$$g_0 = 9.806 \text{ kg} \cdot \text{m/kg}_f \cdot \text{s}^2 \tag{1.2.3}$$

The number g_0 is a constant, independent of location of application of Newton's law and dependent only on the particular set of units employed. At any other location than standard gravity, the mass M of a body remains constant, but the weight (force or pull of gravity) varies:

$$W = M\,\frac{g}{g_0} \tag{1.2.4}$$

For example, where $g = 31.0$ ft/s²,

$$10 \text{ lb}_m \text{ weighs } \frac{(10 \text{ lb}_m)\,(31.0 \text{ ft/s}^2)}{32.174 \text{ lb}_m \cdot \text{ft/lb} \cdot \text{s}^2} = 9.635 \text{ lb}$$

The slug is a derived unit of mass, defined as the amount of mass that is accelerated one foot per second per second by a force of one pound. For these units the constant g_0 is unity; that is, 1 slug·ft/lb·s². Since fluid mechanics is so closely tied to Newton's second law, the slug may be defined as

$$1 \text{ slug} \equiv 1 \text{ lb·s}^2/\text{ft} \tag{1.2.5}$$

and the consistent set of units slug, pound, foot, second may be used without the dimensional constant g_0. Similarly in the SI the kilogram, newton, meter, and second are related by

$$1 \text{ N} \equiv 1 \text{ kg·m/s}^2 \tag{1.2.6}$$

and g_0 is not needed. If the pound mass or the kilogram force is to be used in dynamical equations, g_0 must be introduced.

On the inside front cover, many conversions for various units are given. As they are presented in the form of dimensionless ratios equal to 1, they can be used on one side of an equation, as a multiplier; or as a divisor, to convert units.

1.3 VISCOSITY

Of all the fluid properties, viscosity requires the greatest consideration in the study of fluid flow. The nature and characteristics of viscosity are discussed in this section as well as dimensions and conversion factors for both absolute and kinematic viscosity. Viscosity is that property of a fluid by virtue of which it offers resistance to shear. Newton's law of viscosity [Eq. (1.1.1)] states that for a given rate of angular deformation of fluid the shear stress is directly proportional to the viscosity. Molasses and tar are examples of highly viscous liquids; water and air have very small viscosities.

The viscosity of a gas increases with temperature, but the viscosity of a liquid decreases with temperature. The variation in temperature trends can by explained by examining the causes of viscosity. The resistance of a fluid to shear depends upon its cohesion and upon its rate of transfer of molecular momentum. A liquid, with molecules much more closely spaced than a gas, has cohesive forces much larger than a gas. Cohesion appears to be the predominant cause of viscosity in a liquid, and since cohesion decreases with temperature, the viscosity does likewise. A gas, on the other hand, has very small cohesive forces. Most of its resistance to shear stress is the result of the transfer of molecular momentum.

As a rough model of the way in which momentum transfer gives rise to an apparent shear stress, consider two idealized railroad cars loaded with

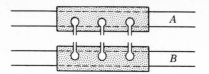

Fig. 1.3   Model illustrating transfer of momentum.

sponges and on parallel tracks, as in Fig. 1.3.   Assume each car has a water
tank and pump, arranged so that the water is directed by nozzles at right
angles to the track.   First, consider A stationary and B moving to the right,
with the water from its nozzles striking A and being absorbed by the sponges.
Car A will be set in motion owing to the component of the momentum of the
jets which is parallel to the tracks, giving rise to an apparent shear stress
between A and B.   Now if A is pumping water back into B at the same rate,
its action tends to slow down B and equal and opposite apparent shear forces
result.   When A and B are both stationary or have the same velocity, the
pumping does not exert an apparent shear stress on either car.

   Within fluid there is always a transfer of molecules back and forth across
any fictitious surface drawn in it.   When one layer moves relative to an adjacent layer, the molecular transfer of momentum brings momentum from
one side to the other so that an apparent shear stress is set up that resists the
relative motion and tends to equalize the velocities of adjacent layers in a
manner analogous to that of Fig. 1.3.   The measure of the motion of one layer
relative to an adjacent layer is du/dy.

   Molecular activity gives rise to an apparent shear stress in gases which
is more important than the cohesive forces, and since molecular activity increases with temperature, the viscosity of a gas also increases with temperature.

   For ordinary pressures viscosity is independent of pressure and depends
upon temperature only.   For very great pressures, gases and most liquids have
shown erratic variations of viscosity with pressure.

   A fluid at rest or in motion so that no layer moves relative to an adjacent
layer will not have apparent shear forces set up, regardless of the viscosity,
because du/dy is zero throughout the fluid.   Hence, in the study of *fluid statics*,
no shear forces can be considered because they do not occur in a static fluid,
and the only stresses remaining are normal stresses, or pressures.   This greatly
simplifies the study of fluid statics, since any free body of fluid can have only
gravity forces and normal surface forces acting on it.

   The dimensions of viscosity are determined from Newton's law of viscosity [Eq. (1.1.1)]. Solving for the viscosity μ

$$\mu = \frac{\tau}{du/dy}$$

and inserting dimensions F, L, T for force, length, and time,

$$\tau: FL^{-2} \qquad u: LT^{-1} \qquad y: L$$

shows that μ has the dimensions $FL^{-2}T$. With the force dimension expressed in terms of mass by use of Newton's second law of motion, $F = MLT^{-2}$, the dimensions of viscosity may be expressed as $ML^{-1}T^{-1}$.

The English unit of viscosity (which has no special name) is 1 lb·s/ft² or 1 slug/ft·s (these are identical). The cgs unit of viscosity, called the *poise* (P), is 1 dyn·s/cm² or 1 g/cm·s. The *centipoise* (cP) is one one-hundredth of a poise. Water at 68°F has a viscosity of 1.002 cP. The SI unit of viscosity in kilograms per meter second or newton-seconds per square meter has no name. The SI unit is 10 times larger than the poise unit.[1]

Kinematic viscosity

The viscosity μ is frequently referred to as the *absolute* viscosity or the *dynamic* viscosity to avoid confusing it with the *kinematic* viscosity ν, which is the ratio of viscosity to mass density,

$$\nu = \frac{\mu}{\rho} \qquad\qquad\qquad (1.3.1)$$

The kinematic viscosity occurs in many applications, e.g., the Reynolds number, which is VD/ν. The dimensions of ν are L^2T^{-1}. The English unit, 1 ft²/s,

---

[1] The conversion from the English unit of viscosity to the SI unit of viscosity is

$$\frac{1 \text{ slug}}{\text{ft·s}} \frac{14.594 \text{ kg}}{\text{slug}} \frac{1 \text{ ft}}{0.3048 \text{ m}} = 47.9 \text{ kg/m·s}$$

or

$$\frac{1 \text{ Engl unit viscosity}}{47.9 \text{ unit SI viscosity}} = 1$$

has no special name; the cgs unit, called the *stoke* (St), is 1 cm²/s. The SI unit[1] of kinematic viscosity is 1 m²/s.

To convert to the English unit of viscosity from the English unit of kinematic viscosity, it is necessary to multiply by the mass density in slugs per cubic foot. To change to the poise from the stoke, it is necessary to multiply by the mass density in grams per cubic centimeter, which is numerically equal to the specific gravity.

EXAMPLE 1.1 A liquid has a viscosity of 0.05 P and a specific gravity of 0.85. Calculate (*a*) the viscosity in SI units; (*b*) the kinematic viscosity in SI units; (*c*) the viscosity in English units; (*d*) the kinematic viscosity in stokes.

$$(a) \quad \mu = \frac{0.05 \text{ g}}{\text{cm} \cdot \text{s}} \frac{1 \text{ kg}}{1000 \text{ g}} \frac{100 \text{ cm}}{1 \text{ m}} = 0.005 \text{ kg/m} \cdot \text{s}$$

$$(b) \quad \nu = \frac{\mu}{\rho} = \frac{0.005 \text{ kg/m} \cdot \text{s}}{0.85 \times 1000 \text{ kg/m}^3} = 5.88 \times 10^{-6} \text{ m}^2/\text{s} = 5.88 \text{ } \mu\text{m}^2/\text{s}$$

$$(c) \quad \mu = 0.005 \text{ kg/m} \cdot \text{s} \times \frac{1 \text{ slug/ft} \cdot \text{s}}{47.9 \text{ kg/m} \cdot \text{s}} = 0.0001044 \text{ slug/ft} \cdot \text{s}$$

$$(d) \quad \nu = \frac{\mu}{\rho} = \frac{0.05 \text{ g/cm} \cdot \text{s}}{0.85 \text{ g/cm}^3} = 0.0588 \text{ cm}^2/\text{s}$$

Viscosity is practically independent of pressure and depends upon temperature only. The kinematic viscosity of liquids, and of gases at a given pressure, is substantially a function of temperature. Charts for the determination of absolute viscosity and kinematic viscosity are given in Appendix C, Figs. C.1 and C.2, respectively.

---

[1] The conversion from the SI unit of kinematic viscosity to the English unit of kinematic viscosity is

$$1 \frac{\text{m}^2}{\text{s}} \left(\frac{1 \text{ ft}}{0.3048 \text{ m}} \right)^2 = 10.764 \text{ ft}^2/\text{s}$$

or

$$\frac{1 \text{ kinematic viscosity unit (SI)}}{10.764 \text{ kinematic viscosity (Engl)}} = 1$$

1.4 CONTINUUM

In dealing with fluid-flow relationships on a mathematical or analytical basis, it is necessary to consider that the actual molecular structure is replaced by a hypothetical continuous medium, called the *continuum*. For example, velocity at a point in space is indefinite in a molecular medium, as it would be zero at all times except when a molecule occupied this exact point, and then it would be the velocity of the molecule and not the mean mass velocity of the particles in the neighborhood. This dilemma is avoided if one considers velocity at a point to be the average or mass velocity of all molecules surrounding the point, say, within a small sphere with radius large compared with the *mean distance between molecules*. With n molecules per cubic centimeter, the mean distance between molecules is of the order $n^{-1/3}$ cm. Molecular theory, however, must be used to calculate fluid properties (e.g., viscosity) which are associated with molecular motions, but continuum equations can be employed with the results of molecular calculations.

In rarefied gases, such as the atmosphere at 50 mi above sea level, the ratio of the mean free path[1] of the gas to a characteristic length for a body or conduit is used to distinguish the type of flow. The flow regime is called *gas dynamics* for very small values of the ratio, the next regime is called *slip flow*, and for large values of the ratio it is *free molecule flow*. In this text only the gas-dynamics regime is studied.

The quantities density, specific volume, pressure, velocity, and acceleration are assumed to vary continuously throughout a fluid (or be constant).

1.5 DENSITY, SPECIFIC VOLUME, SPECIFIC WEIGHT, SPECIFIC GRAVITY, PRESSURE

The *density* ρ of a fluid is defined as its mass per unit volume. To define density at a point the mass Δm of fluid in a small volume $\Delta \mathcal{V}$ surrounding the point is divided by $\Delta \mathcal{V}$ and the limit is taken as $\Delta \mathcal{V}$ becomes a value ϵ^3 in which ϵ is still large compared with the mean distance between molecules,

$$\rho = \lim_{\Delta \mathcal{V} \to \epsilon^3} \frac{\Delta m}{\Delta \mathcal{V}} \tag{1.5.1}$$

For water at standard pressure (14.7 lb/in²) and 39.4°F (4°C), $\rho = 1.94$ slugs/ft³ or 62.43 lb$_m$/ft³ or 1000 kg/m³.

The *specific volume* v_s is the reciprocal of the density ρ; that is, it is the

---

[1] The mean free path is the average distance a molecule travels between collisions.

volume occupied by unit mass of fluid. Hence

$$v_s = \frac{1}{\rho} \qquad\qquad (1.5.2)$$

The *specific weight* γ of a substance is its weight per unit volume. It changes with location,

$$\gamma = \rho g \qquad\qquad (1.5.3)$$

depending upon gravity. It is a convenient property when dealing with fluid statics or with liquids with a free surface.

The *specific gravity* S of a substance is the ratio of its weight to the weight of an equal volume of water at standard conditions.[1] It may also be expressed as a ratio of its density or specific weight to that of water.

The normal force pushing against a plane area divided by the area is the average *pressure*. The pressure at a point is the ratio of normal force to area as the area approaches a small value enclosing the point. If a fluid exerts a pressure against the walls of a container, the container will exert a reaction on the fluid which will be compressive. Liquids can sustain very high compressive pressures, but unless they are extremely pure, they are very weak in tension. It is for this reason that the absolute pressures used in this book are never negative, since this would imply that fluid is sustaining a tensile stress. Pressure has the units force per area and may be pounds per square inch, pounds per square foot, or newtons per square meter, also called pascals (Pa). Pressure may also be expressed in terms of an equivalent length of a fluid column, as shown in Sec. 2.3.

1.6 PERFECT GAS

In this treatment, thermodynamic relationships and compressible-fluid-flow cases have been limited generally to perfect gases. The perfect gas is defined in this section, and its various interrelationships with specific heats are treated in Sec. 6.1.

The perfect gas, as used herein, is defined as a substance that satisfies the *perfect-gas law*

$$pv_s = RT \qquad\qquad (1.6.1)$$

---

[1] These conditions are normally 4°C. In this text standard conditions are considered to be a temperature of 68°F and a pressure of 30 in Hg abs unless specified otherwise.

and that has constant specific heats. p is the absolute pressure, v_s the specific volume, R the gas constant, and T the absolute temperature. The perfect gas must be carefully distinguished from the ideal fluid. An ideal fluid is frictionless and incompressible. The perfect gas has viscosity and can therefore develop shear stresses, and it is compressible according to Eq. (1.6.1).

Equation (1.6.1) is the equation of state for a perfect gas. It may be written

$$p = \rho R T \tag{1.6.2}$$

The units of R can be determined from the equation when the other units are known. For p in pounds per square foot, ρ in slugs per cubic foot, and T (°F + 459.6) in degrees Rankine (°R),

$$R = (\text{lb/ft}^2)(\text{ft}^3/\text{slug} \cdot °R) = \text{ft} \cdot \text{lb/slug} \cdot °R$$

For ρ in pounds mass per cubic foot,

$$R = (\text{lb/ft}^2)(\text{ft}^3/\text{lb}_m \cdot °R) = \text{ft} \cdot \text{lb/lb}_m \cdot °R$$

The magnitude of R in slugs is 32.174 times greater than in pounds mass. Values of R for several common gases are given in Table C.3.

Real gases below critical pressure and above the critical temperature tend to obey the perfect-gas law. As the pressure increases, the discrepancy increases and becomes serious near the critical point. The perfect-gas law encompasses both Charles' law and Boyle's law. Charles' law states that for constant pressure the volume of a given mass of gas varies as its absolute temperature. Boyle's law (isothermal law) states that for constant temperature the density varies directly as the absolute pressure. The volume $\mathcal{V}$ of m mass units of gas is mv_s; hence

$$p\mathcal{V} = mRT \tag{1.6.3}$$

Certain simplifications result from writing the perfect-gas law on a mole basis. A pound mole of gas is the number of pounds mass of gas equal to its molecular weight; e.g., a pound mole of oxygen O_2 is 32 lb_m. With $\bar{v}_s$ being the volume per mole, the perfect-gas law becomes

$$p\bar{v}_s = MRT \tag{1.6.4}$$

if M is the molecular weight. In general, if n is the number of moles of the gas in volume $\mathcal{U}$,

$$p\mathcal{U} = nMRT \tag{1.6.5}$$

since $nM = m$. Now, from Avogadro's law, equal volumes of gases at the same absolute temperature and pressure have the same number of molecules; hence their masses are proportional to the molecular weights. From Eq. (1.6.5) it is seen that MR must be constant, since $p\mathcal{U}/nT$ is the same for any perfect gas. The product MR is called the universal gas constant and has a value depending only upon the units employed. It is

$$MR = 1545 \text{ ft·lb/lb}_m \cdot \text{mole·°R} \tag{1.6.6}$$

The gas constant R can then be determined from

$$R = \frac{1545}{M} \text{ ft·lb/lb}_m \cdot \text{°R} \tag{1.6.7}$$

or in slug units,

$$R = \frac{1545 \times 32.174}{M} \text{ ft·lb/slug·°R} \tag{1.6.8}$$

In SI units[1]

$$R = \frac{8312}{M} \text{ m·N/kg·K} \tag{1.6.9}$$

so that knowledge of molecular weight leads to the value of R. In Table C.3 of Appendix C molecular weights of some common gases are listed.

Additional relationships and definitions used in perfect-gas flow are introduced in Chaps. 3 and 6.

EXAMPLE 1.2 A gas with molecular weight of 44 is at a pressure of 13.0 psia (pounds per square inch absolute) and a temperature of 60°F. Determine its density in slugs per cubic foot.

From Eq. (1.6.8)

$$R = \frac{1545 \times 32.174}{44} = 1129 \text{ ft·lb/slug·°R}$$

---

[1] In 1967 the name of the *degree Kelvin* (°K) was changed to *kelvin* (K).

Then from Eq. (1.6.2)

$$\rho = \frac{p}{RT} = \frac{(13.0 \text{ lb/in}^2)(144 \text{ in}^2/\text{ft}^2)}{(1129 \text{ ft}\cdot\text{lb/slug}\cdot\text{°R})(460 + 60 \text{ °R})} = 0.00319 \text{ slug/ft}^3$$

1.7 BULK MODULUS OF ELASTICITY

In the preceding section the compressibility of a perfect gas is described by the perfect-gas law. For most purposes a liquid may be considered as incompressible, but for situations involving either sudden or great changes in pressure, its compressibility becomes important. Liquid (and gas) compressibility also becomes important when temperature changes are involved, e.g., free convection. The compressibility of a liquid is expressed by its *bulk modulus of elasticity*. If the pressure of a unit volume of liquid is increased by dp, it will cause a volume decrease $- d\mathcal{V}$; the ratio $-dp/d\mathcal{V}$ is the bulk modulus of elasticity K. For any volume $\mathcal{V}$ of liquid,

$$K = -\frac{dp}{d\mathcal{V}/\mathcal{V}} \tag{1.7.1}$$

Since $d\mathcal{V}/\mathcal{V}$ is dimensionless, K is expressed in the units of p. For water at ordinary temperatures and pressures, $K = 300{,}000$ psi $= 2068$ MN/m², where M is the abbreviation for mega ($= 10^6$).

To gain some idea about the compressibility of water, consider the application of 100 psi pressure to 1 ft³ of water. When Eq. (1.7.1) is solved for $- d\mathcal{V}$,

$$-d\mathcal{V} = \frac{\mathcal{V}\, dp}{K} = \frac{1.0 \times 100}{300{,}000} = \frac{1}{3000} \text{ ft}^3$$

Hence, the application of 100 psi to water under ordinary conditions causes its volume to decrease by only 1 part in 3000. As a liquid is compressed, the resistance to further compression increases; therefore K increases with pressure. At 45,000 psi the value of K for water has doubled.

EXAMPLE 1.3 A liquid compressed in a cylinder has a volume of 1 liter (l) (1000 cm³) at 1 MN/m² and a volume of 995 cm³ at 2 MN/m². What is its bulk modulus of elasticity?

$$K = -\frac{\Delta p}{\Delta \mathcal{V}/\mathcal{V}} = -\frac{2 - 1 \text{ MN/m}^2}{(995 - 1000)/1000} = 200 \text{ MN/m}^2$$

1.8 VAPOR PRESSURE

Liquids evaporate because of molecules escaping from the liquid surface. The vapor molecules exert a partial pressure in the space, known as *vapor pressure*. If the space above the liquid is confined, after a sufficient time the number of vapor molecules striking the liquid surface and condensing are just equal to the number escaping in any interval of time, and equilibrium exists. Since this phenomenon depends upon molecular activity, which is a function of temperature, the vapor pressure of a given fluid depends upon temperature and increases with it. When the pressure above a liquid equals the vapor pressure of the liquid, boiling occurs. Boiling of water, for example, may occur at room temperature if the pressure is reduced sufficiently. At 68°F water has a vapor pressure of 0.339 psi, and mercury has a vapor pressure of 0.0000251 psi.

In many situations involving the flow of liquids it is possible that very low pressures are produced at certain locations in the system. Under such circumstances the pressures may be equal to or less than the vapor pressure. When this occurs, the liquid flashes into vapor. This is the phenomenon of cavitation. A rapidly expanding vapor pocket, or cavity, forms, which is usually swept away from its point of origin and enters regions of the flow where the pressure is greater than vapor pressure. The cavity collapses. This growth and decay of the vapor bubbles affect the operating performance of hydraulic pumps and turbines and can result in erosion of the metal parts in the region of cavitation.

1.9 SURFACE TENSION

Capillarity

At the interface between a liquid and a gas, or two immiscible liquids, a film or special layer seems to form on the liquid, apparently owing to attraction of liquid molecules below the surface. It is a simple experiment to place a small needle on a quiet water surface and observe that it is supported there by the film.

The formation of this film may be visualized on the basis of *surface energy* or work per unit area required to bring the molecules to the surface. The surface tension is then the stretching force required to form the film, obtained by dividing the surface-energy term by unit length of the film in equilibrium. The surface tension of water varies from about 0.005 lb/ft at 68°F to 0.004 lb/ft at 212°F. Surface tensions of other liquids are given in Table 1.2.

Table 1.2 *Surface tension of common liquids in contact with air at 68°F (20°C)*

Liquid	Surface tension σ	
	lb/ft	N/m
Alcohol, ethyl	0.00153	0.0223
Benzene	0.00198	0.0289
Carbon tetrachloride	0.00183	0.0267
Kerosene	0.0016–0.0022	0.0233–0.0321
Water	0.00501	0.0731
Mercury:		
In air	0.0352	0.5137
In water	0.0269	0.3926
In vacuum	0.0333	0.4857
Oil:		
Lubricating	0.0024–0.0026	0.0350–0.0379
Crude	0.0016–0.0026	0.0233–0.0379

The action of surface tension is to increase the pressure within a droplet of liquid or within a small liquid jet. For a small spherical droplet of radius r the internal pressure p necessary to balance the tensile force due to the surface tension σ is calculated in terms of the forces which act on a hemispherical free body (see Sec. 2.6),

$$p\pi r^2 = 2\pi r\sigma \qquad \text{or} \qquad p = \frac{2\sigma}{r}$$

For the cylindrical liquid jet of radius r, the pipe-tension equation applies:

$$p = \frac{\sigma}{r}$$

Both equations show that the pressure becomes large for a very small radius of droplet or cylinder.

Capillary attraction is caused by surface tension and by the relative value of adhesion between liquid and solid to cohesion of the liquid. A liquid that *wets* the solid has a greater adhesion than cohesion. The action of surface tension in this case is to cause the liquid to rise within a small vertical tube that is partially immersed in it. For liquids that do not wet the solid, surface tension tends to depress the meniscus in a small vertical tube. When the contact angle between liquid and solid is known, the capillary rise can be com-

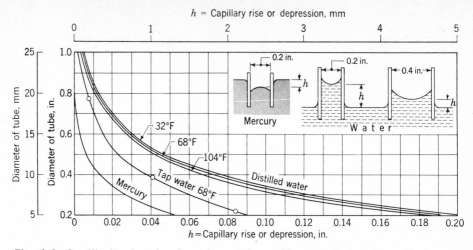

Fig. 1.4 Capillarity in circular glass tubes. (*By permission from R. L. Daugherty, "Hydraulics," copyright 1937, McGraw-Hill Book Company.*)

puted for an assumed shape of the meniscus. Figure 1.4 shows the capillary rise for water and mercury in circular glass tubes in air.

PROBLEMS

1.1 Classify the substance that has the following rates of deformation and corresponding shear stresses:

du/dy, rad/s	0	1	3	5
τ, lb/ft²	15	20	30	40

1.2 Classify the following substances (maintained at constant temperature):

(*a*)
du/dy, rad/s	0	3	4	6	5	4
τ, lb/ft²	2	4	6	8	6	4

(*b*)
du/dy, rad/s	0	0.5	1.1	1.8
τ, N/m²	0	2	4	6

(*c*)
du/dy, rad/s	0	0.3	0.6	0.9	1.2
τ, N/m²	0	2	4	6	8

1.3 A Newtonian liquid flows down an inclined plane in a thin sheet of thickness t (Fig. 1.5). The upper surface is in contact with air, which offers almost no resistance

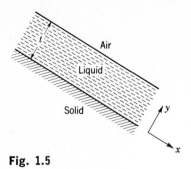

Fig. 1.5

to the flow. Using Newton's law of viscosity, decide what the value of du/dy, y measured normal to the inclined plane, must be at the upper surface. Would a linear variation of u with y be expected?

1.4 What kinds of rheological materials are paint and grease?

1.5 A Newtonian fluid is in the clearance between a shaft and a concentric sleeve. When a force of 500 N is applied to the sleeve parallel to the shaft, the sleeve attains a speed of 1 m/s. If 1500 N force is applied, what speed will the sleeve attain? The temperature of the sleeve remains constant.

1.6 Determine the weight in pounds of 3 slugs mass at a place where $g = 31.7$ ft/s^2.

1.7 When standard scale weights and a balance are used, a body is found to be equivalent in pull of gravity to two of the 1-lb scale weights at a location where $g = 31.5$ ft/s^2. What would the body weigh on a correctly calibrated spring balance (for sea level) at this location?

1.8 Determine the value of proportionality constant g_0 needed for the following set of units: kip (1000 lb), slug, foot, second.

1.9 On another planet, where standard gravity is 3 m/s^2, what would be the value of the proportionality constant g_0 in terms of the kilogram force, gram, millimeter, and second?

1.10 A correctly calibrated spring scale records the weight of a 51-lb$_m$ body as 17.0 lb at a location away from the earth. What is the value of g at this location?

1.11 Does the weight of a 20-N bag of flour denote a force or the mass of the flour? What is the mass of the flour in kilograms? What are the mass and weight of the flour at a location where the gravitational acceleration is one-seventh that of the earth's standard?

1.12 Convert 10.4 SI units of kinematic viscosity to English units of dynamic viscosity if $S = 0.85$.

1.13 A shear stress of 4 dyn/cm^2 causes a Newtonian fluid to have an angular deformation of 1 rad/s. What is its viscosity in centipoises?

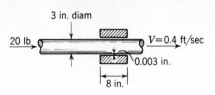

Fig. 1.6

1.14 A plate, 0.5 mm distant from a fixed plate, moves at 0.25 m/s and requires a force per unit area of 2 Pa (N/m²) to maintain this speed. Determine the fluid viscosity of the substance between the plates in SI units.

1.15 Determine the viscosity of fluid between shaft and sleeve in Fig. 1.6.

1.16 A flywheel weighing 500 N has a radius of gyration of 30 cm. When it is rotating 600 rpm, its speed reduces 1 rpm/s owing to fluid viscosity between sleeve and shaft. The sleeve length is 5 cm, shaft diameter 2 cm, and radial clearance 0.05 mm. Determine the fluid viscosity.

1.17 A 1-in-diameter steel cylinder 12 in long falls, because of its own weight, at a uniform rate of 0.5 ft/s inside a tube of slightly larger diameter. A castor-oil film of constant thickness is between the cylinder and the tube. Determine the clearance between the tube and the cylinder. The temperature is 100°F.

1.18 A piston of diameter 5.000 cm moves within a cylinder of 5.010 cm. Determine the percentage decrease in force necessary to move the piston when the lubricant warms up from 0 to 120°C. Use crude-oil viscosity from Fig. C.1, Appendix C.

1.19 How much greater is the viscosity of water at 32°F than at 212°F? How much greater is its kinematic viscosity for the same temperature range?

1.20 A fluid has a viscosity of 4 cP and a density of 50 lb_m/ft^3. Determine its kinematic viscosity in English units and in stokes.

1.21 A fluid has a specific gravity of 0.83 and a kinematic viscosity of 3 St. What is its viscosity in English units and in SI units?

1.22 A body weighing 90 lb with a flat surface area of 2 ft² slides down a lubricated inclined plane making a 30° angle with the horizontal. For viscosity of 1 P and body speed of 3 ft/s, determine the lubricant film thickness.

1.23 What is the viscosity of gasoline at 30°C in poises?

1.24 Determine the kinematic viscosity of benzene at 60°F in stokes.

1.25 Calculate the value of the gas constant R in SI units, starting with $R = 1545/M$ ft·lb/lb_m·°R.

1.26 What is the specific volume in cubic feet per pound mass and cubic feet per slug of a substance of specific gravity 0.75?

1.27 What is the relation between specific volume and specific weight?

1.28 The density of a substance is 2.94 g/cm³. What is its (*a*) specific gravity, (*b*) specific volume, and (*c*) specific weight?

1.29 A force, expressed by $\mathbf{F} = 4\mathbf{i} + 3\mathbf{j} + 9\mathbf{k}$, acts upon a square area, 2 by 2 in, in the *xy* plane. Resolve this force into a normal-force and a shear-force component. What are the pressure and the shear stress? Repeat the calculations for $\mathbf{F} = -4\mathbf{i} + 3\mathbf{j} - 9\mathbf{k}$.

1.30 A gas at 20°C and 2 kg$_f$/cm² has a volume of 40 l and a gas constant $R = 210$ m·N/kg·K. Determine the density and mass of the gas.

1.31 What is the specific weight of air at 60 psia and 90°F?

1.32 What is the density of water vapor at 30 N/cm² abs and 15°C in SI units?

1.33 A gas with molecular weight 44 has a volume of 4.0 ft³ and a pressure and temperature of 2000 psfa (lb/ft² abs) and 600°R, respectively. What are its specific volume and specific weight?

1.34 1.0 kg of hydrogen is confined in a volume of 0.1 m at -40°C. What is the pressure?

1.35 Express the bulk modulus of elasticity in terms of density change rather than volume change.

1.36 For constant bulk modulus of elasticity, how does the density of a liquid vary with the pressure?

1.37 What is the bulk modulus of a liquid that has a density increase of 0.02 percent for a pressure increase of 1000 lb/ft²? For a pressure increase of 50,000 N/m²?

1.38 For $K = 300,000$ psi for bulk modulus of elasticity of water what pressure is required to reduce its volume by 0.5 percent?

1.39 A steel container expands in volume 1 percent when the pressure within it is increased by 10,000 psi. At standard pressure, 14.7 psia, it holds 1000 lb$_m$ water $\rho = 62.4$ lb$_m$/ft³. For $K = 300,000$ psi, when it is filled, how many pounds mass water need be added to increase the pressure to 10,000 psi?

1.40 What is the isothermal bulk modulus for air at 4 kg$_f$/cm² abs?

1.41 At what pressure can cavitation be expected at the inlet of a pump that is handling water at 100°F?

1.42 What is the pressure within a droplet of water of 0.002-in diameter at 68°F if the pressure outside the droplet is standard atmospheric pressure of 14.7 psi?

1.43 A small circular jet of mercury 0.1 mm in diameter issues from an opening. What is the pressure difference between the inside and outside of the jet when at 20°C?

1.44 Determine the capillary rise for distilled water at 104°F in a circular $\frac{1}{4}$-in diameter glass tube.

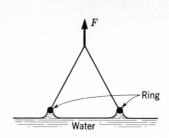

Fig. 1.7

1.45 What diameter of glass tube is required if the capillary effects on the water within are not to exceed 0.02 in?

1.46 Using the data given in Fig. 1.4, estimate the capillary rise of tap water between two parallel glass plates 0.20 in apart.

1.47 A method of determining the surface tension of a liquid is to find the force needed to pull a platinum wire ring from the surface (Fig. 1.7). Estimate the force necessary to remove a 2-cm diameter ring from the surface of water at 20°C. Why is platinum used as the material for the ring?

1.48 A fluid is a substance that

(*a*) always expands until it fills any container
(*b*) is practically incompressible
(*c*) cannot be subjected to shear forces
(*d*) cannot remain at rest under action of any shear force
(*e*) has the same shear stress at a point regardless of its motion

1.49 A 2.0-lb_m object weighs 1.90 lb on a spring balance. The value of g at this location is, in feet per second per second,

(*a*) 30.56   (*b*) 32.07   (*c*) 32.17   (*d*) 33.87   (*e*) none of these answers

1.50 At a location where $g = 30.00$ ft/s², 2.0 slugs is equivalent to how many pounds mass?

(*a*) 60.0   (*b*) 62.4   (*c*) 64.35   (*d*) not equivalent units
(*e*) none of these answers

1.51 The weight, in pounds, of 3 slugs on a planet where $g = 10.00$ ft/s² is

(*a*) 0.30   (*b*) 0.932   (*c*) 30.00   (*d*) 96.53   (*e*) none of these answers

1.52 Newton's law of viscosity relates

(*a*) pressure, velocity, and viscosity
(*b*) shear stress and rate of angular deformation in a fluid
(*c*) shear stress, temperature, viscosity, and velocity
(*d*) pressure, viscosity, and rate of angular deformation
(*e*) yield shear stress, rate of angular deformation, and viscosity

1.53 Viscosity has the dimensions

(a) $FL^{-2}T$ (b) $FL^{-1}T^{-1}$ (c) FLT^{-2} (d) FL^2T (e) FLT^2

1.54 Select the *incorrect* completion. Apparent shear forces

(a) can never occur when the fluid is at rest
(b) may occur owing to cohesion when the liquid is at rest
(c) depend upon molecular interchange of momentum
(d) depend upon cohesive forces
(e) can never occur in a frictionless fluid, regardless of its motion

1.55 Correct units for dynamic viscosity are

(a) dyn·s²/cm   (b) g/cm·s²   (c) g·s/cm   (d) dyn·cm/s²   (e) dyn·s/cm²

1.56 Viscosity, expressed in poises, is converted to the English unit of viscosity by multiplication by

(a) $\frac{1}{479}$ (b) 479   (c) ρ (d) $1/\rho$ (e) none of these answers

1.57 The dimensions for kinematic viscosity are

(a) $FL^{-2}T$ (b) $ML^{-1}T^{-1}$ (c) L^2T^2 (d) L^2T^{-1} (e) L^2T^{-2}

1.58 In converting from the English unit of kinematic viscosity to the stoke, one multiplies by

(a) $\frac{1}{479}$ (b) $1/30.48^2$ (c) 479   (d) 30.48^2 (e) none of these answers

1.59 The kinematic viscosity of kerosene at 90°F in square feet per second is

(a) 2×10^{-5} (b) 3.2×10^{-5} (c) 2×10^{-4} (d) 3.2×10^{-4}
(e) none of these answers

1.60 The kinematic viscosity of dry air at 25°F and 29.4 psia in square feet per second is

(a) 6.89×10^{-5} (b) 1.4×10^{-4} (c) 6.89×10^{-4} (d) 1.4×10^{-3}
(e) none of these answers

1.61 For $\mu = 0.06$ kg/m·s, sp gr $= 0.60$, ν in stokes is

(a) 2.78   (b) 1.0   (c) 0.60   (d) 0.36   (e) none of these answers

1.62 For $\mu = 2.0 \times 10^{-4}$ slug/ft·s, the value of μ in pound-seconds per square foot is

(a) 1.03×10^{-4} (b) 2.0×10^{-4} (c) 6.21×10^{-4} (d) 6.44×10^{-3}
(e) none of these answers

1.63 For $\nu = 3 \times 10^{-4}$ St and $\rho = 0.8$ g/cm³, μ in slugs per foot-second is

(a) 5.02×10^{-7} (b) 6.28×10^{-7} (c) 7.85×10^{-7} (d) 1.62×10^{-6}
(e) none of these answers

1.64   A perfect gas

(*a*) has zero viscosity   (*b*) has constant viscosity   (*c*) is incompressible
(*d*) satisfies $p\rho = RT$ (*e*) fits none of these statements

1.65   The molecular weight of a gas is 28. The value of R in foot-pounds per slug per degree Rankine is

(*a*) 53.3   (*b*) 55.2   (*c*) 1545   (*d*) 1775   (*e*) none of these answers

1.66   The density of air at 10°C and 10 kg_f/cm^2 abs in kilograms per cubic meter is

(*a*) 1.231   (*b*) 12.07   (*c*) 118.4   (*d*) 65.0   (*e*) none of these answers

1.67   How many pounds mass of carbon monoxide gas at 20°F and 30 psia is contained in a volume of 4.0 ft^3?

(*a*) 0.00453   (*b*) 0.0203   (*c*) 0.652   (*d*) 2.175   (*e*) none of these answers

1.68   A container holds 2.0 lb_m air at 120°F and 120 psia. If 3.0 lb_m air is added and the final temperature is 240°F, the final pressure, in pounds per square inch absolute, is

(*a*) 300   (*b*) 362.2   (*c*) 600   (*d*) indeterminable   (*e*) none of these answers

1.69   The bulk modulus of elasticity K for a gas at constant temperature T_0 is given by

(*a*) p/ρ (*b*) RT_0 (*c*) ρp (*d*) ρRT_0 (*e*) none of these answers

1.70   The bulk modulus of elasticity

(*a*) is independent of temperature   (*b*) increases with the pressure
(*c*) has the dimensions of $1/p$ (*d*) is larger when the fluid is more compressible
(*e*) is independent of pressure and viscosity

1.71   For 70 kg_f/cm^2 increase in pressure the density of water has increased, in percent, by about

(*a*) $\frac{1}{300}$ (*b*) $\frac{1}{30}$ (*c*) $\frac{1}{3}$ (*d*) $\frac{1}{2}$ (*e*) none of these answers

1.72   A pressure of 150 psi applied to 10 ft^3 liquid causes a volume reduction of 0.02 ft^3. The bulk modulus of elasticity in pounds per square inch is

(*a*) -750 (*b*) 750   (*c*) 7500   (*d*) 75,000   (*e*) none of these answers

1.73   Surface tension has the dimensions

(*a*) F (*b*) FL^{-1} (*c*) FL^{-2} (*d*) FL^{-3} (*e*) none of these answers

FLUID STATICS

The science of fluid statics will be treated in two parts: the study of pressure and its variation throughout a fluid and the study of pressure forces on finite surfaces. Special cases of fluids moving as solids are included in the treatment of statics because of the similarity of forces involved. Since there is no motion of a fluid layer relative to an adjacent layer, there are no shear stresses in the fluid. Hence, all free bodies in fluid statics have only normal pressure forces acting on their surfaces.

2.1 PRESSURE AT A POINT

The average pressure is calculated by dividing the normal force pushing against a plane area by the area. The pressure at a point is the limit of the ratio of normal force to area as the area approaches zero size at the point. At a point a fluid at rest has the same pressure in all directions. This means that an element δA of a very small area, free to rotate about its center when submerged in a fluid at rest, will have a force of constant magnitude acting on either side of it, regardless of its orientation.

To demonstrate this, a small wedge-shaped free body of unit width is taken at the point (x,y) in a fluid at rest (Fig. 2.1). Since there can be no shear forces, the only forces are the normal surface forces and gravity. So, the equations of equilibrium in the x and y directions are, respectively,

$$\Sigma F_x = p_x\,\delta y - p_s\,\delta s \sin\theta = \frac{\delta x\,\delta y}{2}\,\rho a_x = 0$$

$$\Sigma F_y = p_y\,\delta x - p_s\,\delta s \cos\theta - \gamma\,\frac{\delta x\,\delta y}{2} = \frac{\delta x\,\delta y}{2}\,\rho a_y = 0$$

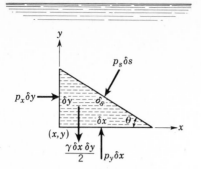

Fig. 2.1 Free-body diagram of wedge-shaped particle.

in which p_x, p_y, p_s are the average pressures on the three faces, γ is the specific weight of the fluid, and ρ is its density. When the limit is taken as the free body is reduced to zero size by allowing the inclined face to approach (x,y) while maintaining the same angle θ and using the geometric relations

$$\delta s \sin \theta = \delta y \qquad \delta s \cos \theta = \delta x$$

the equations simplify to

$$p_x \, \delta y - p_s \, \delta y = 0 \qquad p_y \, \delta x - p_s \, \delta x - \gamma \, \frac{\delta x \, \delta y}{2} = 0$$

The last term of the second equation is an infinitesimal of higher order of smallness and may be neglected. When divided by δy and δx, respectively, the equations can be combined,

$$p_s = p_x = p_y \tag{2.1.1}$$

Since θ is any arbitrary angle, this equation proves that the pressure is the same in all directions at a point in a static fluid. Although the proof was carried out for a two-dimensional case, it may be demonstrated for the three-dimensional case with the equilibrium equations for a small tetrahedron of fluid with three faces in the coordinate planes and the fourth face inclined arbitrarily.

   If the fluid is in motion so that one layer moves relative to an adjacent layer, shear stresses occur and the normal stresses are, in general, no longer the same in all directions at a point. The pressure is then defined as the average of any three mutually perpendicular normal compressive stresses at a point,

$$p = \frac{p_x + p_y + p_z}{3}$$

In a fictitious fluid of zero viscosity, i.e., a frictionless fluid, no shear stresses can occur for any motion of the fluid, and so at a point the pressure is the same in all directions.

2.2 BASIC EQUATION OF FLUID STATICS

Pressure variation in a static fluid

The forces acting on an element of fluid at rest, Fig. 2.2, consist of surface forces and body forces. With gravity the only body force acting, and by taking the y axis vertically upward, it is $-\gamma \, \delta x \, \delta y \, \delta z$ in the y direction. With pressure p at its center (x,y,z), the approximate force exerted on the side normal to the y axis closest to the origin is approximately

$$\left(p - \frac{\partial p}{\partial y} \frac{\delta y}{2} \right) \delta x \, \delta z$$

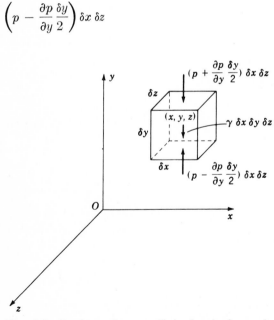

Fig. 2.2 Rectangular parallelepiped element of fluid at rest.

and the force exerted on the opposite side is

$$\left(p + \frac{\partial p}{\partial y} \frac{\delta y}{2} \right) \delta x \, \delta z$$

where $\delta y/2$ is the distance from center to a face normal to y. Summing the forces acting on the element in the y direction gives

$$\delta F_y = -\frac{\partial p}{\partial y} \delta x \, \delta y \, \delta z - \gamma \, \delta x \, \delta y \, \delta z$$

For the x and z directions, since no body forces act,

$$\delta F_x = -\frac{\partial p}{\partial x} \delta x \, \delta y \, \delta z \qquad \delta F_z = -\frac{\partial p}{\partial z} \delta x \, \delta y \, \delta z$$

The elemental force vector $\delta \mathbf{F}$ is given by

$$\delta \mathbf{F} = \mathbf{i}\,\delta F_x + \mathbf{j}\,\delta F_y + \mathbf{k}\,\delta F_z = -\left(\mathbf{i}\frac{\partial p}{\partial x} + \mathbf{j}\frac{\partial p}{\partial y} + \mathbf{k}\frac{\partial p}{\partial z} \right) \delta x \, \delta y \, \delta z - \mathbf{j}\gamma \, \delta x \, \delta y \, \delta z$$

If the element is reduced to zero size, after dividing through by $\delta x \, \delta y \, \delta z = \delta \mathcal{V}$, the expression becomes exact.

$$\frac{\delta \mathbf{F}}{\delta \mathcal{V}} = -\left(\mathbf{i}\frac{\partial}{\partial x} + \mathbf{j}\frac{\partial}{\partial y} + \mathbf{k}\frac{\partial}{\partial z} \right) p - \mathbf{j}\gamma \qquad \lim \delta \mathcal{V} \to 0 \qquad (2.2.1)$$

This is the resultant force per unit volume at a point, which must be equated to zero for a fluid at rest. The quantity in parentheses is the *gradient*, called ∇ (del),

$$\nabla = \mathbf{i}\frac{\partial}{\partial x} + \mathbf{j}\frac{\partial}{\partial y} + \mathbf{k}\frac{\partial}{\partial z} \qquad (2.2.2)$$

and the negative gradient of p, $-\nabla p$, is the vector field $\mathbf{f}$ of the surface pressure force per unit volume,

$$\mathbf{f} = -\nabla p \qquad (2.2.3)$$

The fluid static law of variation of pressure is then

$$\mathbf{f} - \mathbf{j}\gamma = 0 \qquad (2.2.4)$$

In component form,

$$\frac{\partial p}{\partial x} = 0 \qquad \frac{\partial p}{\partial y} = -\gamma \qquad \frac{\partial p}{\partial z} = 0 \qquad\qquad (2.2.5)$$

The partials, for variation in horizontal directions, are one form of Pascal's law, stating that two points at the same elevation in the same continuous mass of fluid at rest have the same pressure.

Since p is a function of y only,

$$dp = -\gamma \, dy \qquad\qquad (2.2.6)$$

This simple differential equation relates the change of pressure to specific weight and change of elevation and holds for both compressible and incompressible fluids.

For fluids that may be considered homogeneous and incompressible, γ is constant, and Eq. (2.2.6), when integrated, becomes

$$p = -\gamma y + c$$

in which c is the constant of integration. The hydrostatic law of variation of pressure is frequently written in the form

$$p = \gamma h \qquad\qquad (2.2.7)$$

in which h is measured vertically downward ($h = -y$) from a free liquid surface and p is the increase in pressure from that at the free surface. Equation (2.2.7) may be derived by taking as fluid free body a vertical column of liquid of finite height h with its upper surface in the free surface. This is left as an exercise for the student.

EXAMPLE 2.1 An oceanographer is to design a sea lab 5 m high to withstand submersion to 100 m, measured from sea level to the top of the sea lab. Find the pressure variation on a side of the container and the pressure on the top if the specific gravity of salt water is 1.020.

$$\gamma = 1.020 \times 9802 \text{ N/m}^3 = 10 \text{ kN/m}^3$$

At the top, $h = 100$ m, and

$$p = \gamma h = 1 \text{ MN/m}^2$$

If y is measured from the top of the sea lab downward, the pressure variation is

$$p = 10(y + 100) \text{ kN/m}^2$$

Pressure variation in a compressible fluid

When the fluid is a perfect gas at rest at constant temperature, from Eq. (1.6.2)

$$\frac{p}{\rho} = \frac{p_0}{\rho_0} \tag{2.2.8}$$

When the value of γ in Eq. (2.2.6) is replaced by ρg and ρ is eliminated between Eqs. (2.2.6) and (2.2.8),

$$dy = \frac{-p_0}{g\rho_0} \frac{dp}{p} \tag{2.2.9}$$

It must be remembered that if ρ is in pounds mass per cubic foot, then $\gamma = g\rho/g_0$ with $g_0 = 32.174 \text{ lb}_m \cdot \text{ft/lb} \cdot \text{s}^2$. If $p = p_0$ when $\rho = \rho_0$, integration between limits

$$\int_{y_0}^{y} dy = -\frac{p_0}{g\rho_0} \int_{p_0}^{p} \frac{dp}{p}$$

yields

$$y - y_0 = -\frac{p_0}{g\rho_0} \ln \frac{p}{p_0} \tag{2.2.10}$$

in which ln is the natural logarithm. Then

$$p = p_0 \exp\left(-\frac{y - y_0}{p_0/g\rho_0}\right) \tag{2.2.11}$$

which is the equation for variation of pressure with elevation in an isothermal gas.

The atmosphere frequently is assumed to have a constant temperature gradient, expressed by

$$T = T_0 + \beta y \tag{2.2.12}$$

For the standard atmosphere, $\beta = -0.00357°F/ft$ ($-0.00651°C/m$) up to the stratosphere. The density may be expressed in terms of pressure and elevation from the perfect-gas law:

$$\rho = \frac{p}{RT} = \frac{p}{R(T_0 + \beta y)} \tag{2.2.13}$$

Substitution into $dp = -\rho g\, dy$ [Eq. (2.2.6)] permits the variables to be separated and p to be found in terms of y by integration.

EXAMPLE 2.2 Assuming isothermal conditions to prevail in the atmosphere, compute the pressure and density at 2000 m elevation if $p = 10^5$ Pa abs, $\rho = 1.24$ kg/m³ at sea level.
    From Eq. (2.2.11)

$$p = 10^5 \text{ N/m}^2 \exp \left\{ - \frac{2000 \text{ m}}{(10^5 \text{ N/m}^2)/[(9.806 \text{ m/s}^2)(1.24 \text{ kg/m}^3)]} \right\}$$

$$= 78,412 \text{ Pa abs}$$

Then, from Eq. (2.2.8)

$$\rho = \frac{\rho_0}{p_0} p = (1.24 \text{ kg/m}^3) \frac{78,412}{100,000} = 0.972 \text{ kg/m}^3$$

    When compressibility of a liquid in static equilibrium is taken into account, Eqs. (2.2.6) and (1.7.1) are utilized.

2.3 UNITS AND SCALES OF PRESSURE MEASUREMENT

Pressure may be expressed with reference to any arbitrary datum. The usual ones are *absolute zero* and *local atmospheric pressure*. When a pressure is expressed as a difference between its value and a complete vacuum, it is called an *absolute pressure*. When it is expressed as a difference between its value and the local atmospheric pressure, it is called a *gage pressure*.
    The *bourdon* gage (Fig. 2.3) is typical of the devices used for measuring

Fig. 2.3 Bourdon gage. (*Crosby Steam Gage and Valve Co.*)

gage pressures. The pressure element is a hollow, curved, flat metallic tube, closed at one end, with the other end connected to the pressure to be measured. When the internal pressure is increased, the tube tends to straighten, pulling on a linkage to which is attached a pointer and causing the pointer to move. The dial reads zero when the inside and outside of the tube are at the same pressure, regardless of its particular value. The dial may be graduated to any convenient units, common ones being pounds per square inch, pounds per square foot, inches of mercury, feet of water, centimeters of mercury, millimeters of mercury, and kilograms force per square centimeter. Owing to the inherent construction of the gage, it measures pressure relative to the pressure of the medium surrounding the tube, which is the local atmosphere.

Figure 2.4 illustrates the data and the relationships of the common units of pressure measurement. Standard atmospheric pressure is the mean pressure at sea level, 29.92 in Hg (rounded to 30 in for slide-rule work). A pressure expressed in terms of a column of liquid refers to the force per unit area at the base of the column. The relation for variation of pressure with altitude in a liquid $p = \gamma h$ [Eq. (2.2.7)] shows the relation between head h, in length of a fluid column of specific weight γ, and the pressure p. In consistent units, p is in pounds per square foot, γ in pounds per cubic foot, and h in feet or p in pascals, γ in newtons per cubic meter, and h in meters. With the specific weight of any liquid expressed as its specific gravity S times the specific weight of water, Eq. (2.2.7) becomes

$$p = \gamma_w \, Sh \qquad\qquad (2.3.1)$$

For water γ_w may be taken as 62.4 lb/ft³ or 9802 N/m³.

When the pressure is desired in pounds per square inch, both sides of the equation are divided by 144,

$$p_{\text{psi}} = \frac{62.4}{144} \, Sh = 0.433 Sh \qquad\qquad (2.3.2)$$

in which h remains in feet.[1]

Local atmospheric pressure is measured by a mercury barometer (Fig. 2.5) or by an *aneroid* barometer which measures the difference in pressure

---

[1] In Eq. (2.3.2) the standard atmospheric pressure may be expressed in pounds per square inch,

$$p_{\text{psi}} = \frac{62.4}{144} \, (13.6) \left(\frac{30}{12}\right) = 14.7$$

when $S = 13.6$ for mercury. When 14.7 is multiplied by 144, the standard atmosphere becomes 2116 lb/ft². Then 2116 divided by 62.4 yields 34 ft H₂O. Any of these designations is for the standard atmosphere and may be called *one atmosphere*, if it is always understood that it is a standard atmosphere and is measured from absolute zero. These various designations of a standard atmosphere (Fig. 2.4) are equivalent and provide a convenient means of converting from one set of units to another. For example, to express 100 ft H₂O in pounds per square inch,

$$\tfrac{100}{34} \, (14.7) = 43.3 \text{ psi}$$

since $\tfrac{100}{34}$ is the number of standard atmospheres and each standard atmosphere is 14.7 psi.

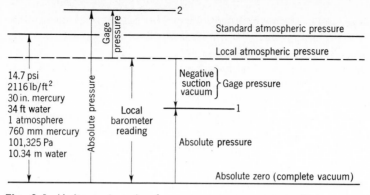

Fig. 2.4 Units and scales for pressure measurement.

between the atmosphere and an evacuated box or tube, in a manner analogous to the bourdon gage except that the tube is evacuated and sealed.

A mercury barometer consists of a glass tube closed at one end, filled with mercury, and inverted so that the open end is submerged in mercury. It has a scale arranged so that the height of column R (Fig. 2.5) can be determined. The space above the mercury contains mercury vapor. If the pressure of the mercury vapor h_v is given in centimeters of mercury and R is measured in the same units, the pressure at A may be expressed as

$$h_v + R = h_A \qquad \text{cm Hg}$$

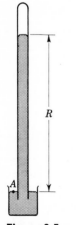

Fig.   2.5
Mercury
barometer.

Although h_v is a function of temperature, it is very small at usual atmospheric temperatures. The barometric pressure varies with location, i.e., elevation, and with weather conditions.

In Fig. 2.4 a pressure may be located vertically on the chart, which indicates its relation to absolute zero and to local atmospheric pressure. If the point is below the local-atmospheric-pressure line and is referred to gage datum, it is called *negative*, *suction*, or *vacuum*. For example, the pressure 18 in Hg abs, as at 1, with barometer reading 29 in, may be expressed as -11 in Hg, 11 in Hg suction, or 11 in Hg vacuum. It should be noted that

$$p_{\text{abs}} = p_{\text{bar}} + p_{\text{gage}}$$

To avoid any confusion, the convention is adopted throughout this text that a *pressure is gage unless specifically marked absolute*, with the exception of the *atmosphere*, which is an absolute pressure unit.

EXAMPLE 2.3   The rate of temperature change in the atmosphere with change in elevation is called its *lapse rate*. The motion of a parcel of air depends on the density of the parcel relative to the density of the surrounding (ambient) air. However, as the parcel ascends through the atmosphere, the air pressure decreases, the parcel expands, and its temperature decreases at a rate known as the *dry adiabatic lapse rate*. A firm wants to burn a large quantity of refuse. It is estimated that the temperature of the smoke plume at 30 ft above the ground will be 20°F greater than that of the ambient air. For the following conditions determine what will happen to the smoke.

    (*a*)   At standard atmospheric lapse rate $\beta = -0.00357°F$ per foot and $t_0 = 70°$.

    (*b*)   At an inverted lapse rate $\beta = 0.002°F$ per foot.

By combining Eqs. (2.2.6) and (2.2.13)

$$\int_{p_0}^{p} \frac{dp}{p} = -\frac{g}{R} \int_{0}^{y} \frac{dy}{T_0 + \beta y} \qquad \text{or} \qquad \frac{p}{p_0} = \left(1 + \frac{\beta y}{T_0}\right)^{-g/R\beta}$$

The relation between pressure and temperature for a mass of gas expanding without heat transfer (isentropic relation, Sec. 6.1) is

$$\frac{T}{T_1} = \left(\frac{p}{p_0}\right)^{(k-1)/k}$$

in which T_1 is the initial smoke absolute temperature and p_0 the initial absolute pressure; k is the specific heat ratio, 1.4 for air and other diatomic gases.

Eliminating p/p_0 in the last two equations gives

$$T = T_1\left(1 + \frac{\beta y}{T_0}\right)^{-[(k-1)/k](g/R\beta)}$$

Since the gas will rise until its temperature is equal to the ambient temperature

$$T = T_0 + \beta y$$

the last two equations may be solved for y. Let

$$a = \frac{-1}{(k-1)g/kR\beta + 1}$$

Then

$$y = \frac{T_0}{\beta}\left[\left(\frac{T_0}{T_1}\right)^a - 1\right]$$

For $\beta = -0.00357°\mathrm{F}$ per foot, $R = 53.3g$ ft·lb/slug·°R, $a = 1.994$, and $y = 10{,}570$ ft. For the atmospheric temperature inversion $\beta = 0.002°\mathrm{F}$ per foot, $a = -0.2717$, and $y = 2680$ ft.

2.4 MANOMETERS

Manometers are devices that employ liquid columns for determining differences in pressure. The most elementary manometer, usually called a *piezometer*, is illustrated in Fig. 2.6a; it measures the pressure in a liquid when it is above zero gage. A glass tube is mounted vertically so that it is connected to the space within the container. Liquid rises in the tube until equilibrium is reached. The pressure is then given by the vertical distance h from the meniscus (liquid surface) to the point where the pressure is to be measured, expressed in units of length of the liquid in the container. It is obvious that the piezometer would not work for negative gage pressures, because air would flow into the container through the tube. It is also impractical for measuring large pressures at A, since the vertical tube would need to be very long. If the specific gravity of the liquid is S, the pressure at A is hS units of length of water.

For measurement of small negative or positive gage pressures in a liquid the tube may take the form shown in Fig. 2.6b. With this arrangement the meniscus may come to rest below A, as shown. Since the pressure at the menis-

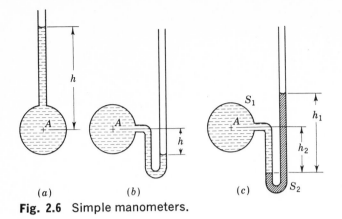

Fig. 2.6 Simple manometers.

cus is zero gage and since pressure *decreases* with elevation,

$$h_A = -hS \quad \text{units of length } H_2O$$

For greater negative or positive gage pressures a second liquid of greater specific gravity is employed (Fig. 2.6c). It must be immiscible in the first fluid, which may now be a gas. If the specific gravity of the fluid at A is S_1 (based on water) and the specific gravity of the manometer liquid is S_2, the equation for pressure at A may be written thus, starting at either A or the upper meniscus and proceeding through the manometer,

$$h_A + h_2 S_1 - h_1 S_2 = 0$$

in which h_A is the unknown pressure, expressed in length units of water, and h_1, h_2 are in length units. If A contains a gas, S_1 is generally so small that $h_2 S_1$ may be neglected.

A general procedure should be followed in working all manometer problems:

1. Start at one end (or any meniscus if the circuit is continuous) and write the pressure there in an appropriate unit (say pounds per square foot) or in an appropriate symbol if it is unknown.
2. Add to this the change in pressure, in the same unit, from one meniscus to the next (plus if the next meniscus is lower, minus if higher). (For pounds per square foot this is the product of the difference in elevation in feet and the specific weight of the fluid in pounds per cubic foot.)

3. Continue until the other end of the gage (or the starting meniscus) is reached and equate the expression to the pressure at that point, known or unknown.

The expression will contain one unknown for a simple manometer or will give a difference in pressures for the differential manometer. In equation form,

$$p_0 - (y_1 - y_0)\gamma_0 - (y_2 - y_1)\gamma_1 - (y_3 - y_2)\gamma_2$$

$$- (y_4 - y_3)\gamma_3 - \cdots - (y_n - y_{n-1})\gamma_{n-1} = p_n$$

in which $y_0, y_1, \ldots, y_n$ are elevations of each meniscus in length units and $\gamma_0, \gamma_1, \gamma_2, \ldots, \gamma_{n-1}$ are specific weights of the fluid columns. The above expression yields the answer in force per unit area and may be converted to other units by use of the conversions in Fig. 2.4.

A differential manometer (Fig. 2.7) determines the difference in pressures at two points A and B, when the actual pressure at any point in the system cannot be determined. Application of the procedure outlined above to Fig. 2.7a produces

$$p_A - h_1\gamma_1 - h_2\gamma_2 + h_3\gamma_3 = p_B \qquad \text{or} \qquad p_A - p_B = h_1\gamma_1 + h_2\gamma_2 - h_3\gamma_3$$

Similarly, for Fig. 2.7b,

$$p_A + h_1\gamma_1 - h_2\gamma_2 - h_3\gamma_3 = p_B \qquad \text{or} \qquad p_A - p_B = -h_1\gamma_1 + h_2\gamma_2 + h_3\gamma_3$$

No formulas for particular manometers should be memorized. It is much more satisfactory to work them out from the general procedure for each case as needed.

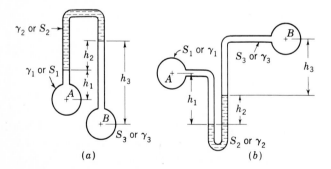

Fig. 2.7 Differential manometers.

If the pressures at A and B are expressed in length of the water column, the above results can be written, for Fig. 2.7a,

$$h_A - h_B = h_1 S_1 + h_2 S_2 - h_3 S_3 \quad \text{units of length } H_2O$$

Similarly, for Fig. 2.7b,

$$h_A - h_B = -h_1 S_1 + h_2 S_2 + h_3 S_3$$

in which S_1, S_2, and S_3 are the applicable specific gravities of the liquids in the system.

EXAMPLE 2.4   In Fig. 2.7a the liquids at A and B are water, and the manometer liquid is oil, sp gr 0.80; $h_1 = 30$ cm, $h_2 = 20$ cm, and $h_3 = 60$ cm. (a) Determine $p_A - p_B$ in pascals. (b) If $p_B = 5$ kg$_f$/cm^2 abs and the barometer reading is 730 mm Hg, find the gage pressure at A in meters of water.

(a) $\quad h_A$ (m H_2O) $- h_1 S_{H_2O} - h_2 S_{oil} + h_3 S_{H_2O} = h_B$ (m H_2O)

$$h_A - (\tfrac{30}{100} \text{ m})(1) - (\tfrac{20}{100} \text{ m})(0.8) + (\tfrac{60}{100} \text{ m})(1) = h_B$$

$$h_A - h_B = -0.14 \text{ m } H_2O$$

$$p_A - p_B = \gamma(h_A - h_B) = (9802 \text{ N/m}^3)(-0.14 \text{ m}) = -1372 \text{ Pa}$$

(b) $\quad h_B = \dfrac{p_B}{\gamma} = 5 \dfrac{kg_f}{cm^2} \dfrac{9.806 \ N}{kg_f} \left(\dfrac{100 \ cm}{m}\right)^2 \dfrac{1}{9802 \ N/m^3} = 50.02 \text{ m } H_2O$

$$h_B \text{ (m } H_2O \text{ abs)} = h_B \text{ (m } H_2O \text{ gage)} + \left(\frac{730}{1000} \text{ m}\right)(13.6)$$

$$= 50.02 + 9.928 = 59.95 \text{ m } H_2O \text{ abs}$$

From (a)

$$h_{A_{abs}} = h_{B_{abs}} - 0.14 \text{ m} = 59.81 \text{ m } H_2O \text{ abs}$$

Micromanometers

Several types of manometers are on the market for determining very small differences in pressure or determining large pressure differences precisely. One type very accurately measures the differences in elevation of two menisci of a manometer. By means of small telescopes with horizontal cross hairs mounted along the tubes on a rack which is raised and lowered by a pinion

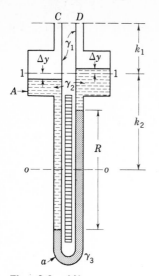

Fig. 2.8 Micromanometer using two gage liquids.

and slow-motion screw so that the cross hairs can be set accurately, the difference in elevation of menisci (the gage difference) can be read with verniers.

With two gage liquids, immiscible in each other and in the fluid to be measured, a large gage difference R (Fig. 2.8) can be produced for a small pressure difference. The heavier gage liquid fills the lower U tube up to 0-0; then the lighter gage liquid is added to both sides, filling the larger reservoirs up to 1-1. The gas or liquid in the system fills the space above 1-1. When the pressure at C is slightly greater than at D, the menisci move as indicated in Fig. 2.8. The volume of liquid displaced in each reservoir equals the displacement in the U tube; thus

$$\Delta y\, A = \frac{R}{2}\, a$$

in which A and a are the cross-sectional areas of reservoir and U tube, respectively. The manometer equation may be written, starting at C, in force per unit area,

$$p_C + (k_1 + \Delta y)\gamma_1 + \left(k_2 - \Delta y + \frac{R}{2}\right)\gamma_2 - R\gamma_3$$

$$- \left(k_2 - \frac{R}{2} + \Delta y\right)\gamma_2 - (k_1 - \Delta y)\gamma_1 = p_D$$

in which γ_1, γ_2, and γ_3 are the specific weights as indicated in Fig. 2.8. Simplifying and substituting for Δy gives

$$p_C - p_D = R\left[\gamma_3 - \gamma_2\left(1 - \frac{a}{A}\right) - \gamma_1\frac{a}{A}\right] \qquad (2.4.1)$$

The quantity in brackets is a constant for specified gage and fluids; hence, the pressure difference is directly proportional to R.

EXAMPLE 2.5 In the micromanometer of Fig. 2.8, the pressure difference is wanted, in pascals, when air is in the system, $S_2 = 1.0$, $S_3 = 1.10$, $a/A = 0.01$, $R = 5$ mm, $t = 20°C$, and the barometer reads 760 mm Hg.

$$\rho_{air} = \frac{p}{RT} = \frac{(0.76 \text{ m})(13.6 \times 9802 \text{ N/m}^3)}{(287 \text{ N·m/kg·K})(273 + 20 \text{ K})} = 1.205 \text{ kg/m}^3$$

$$\gamma_1\frac{a}{A} = (1.205 \text{ kg/m}^3)(9.806 \text{ m/s}^2)(0.01) = 0.118 \text{ N/m}^3$$

$$\gamma_3 - \gamma_2\left(1 - \frac{a}{A}\right) = (9802 \text{ N/m}^3)(1.10 - 0.99) = 1078 \text{ N/m}^3$$

The term $\gamma_1(a/A)$ may be neglected. Substituting into Eq. (2.4.1) gives

$$p_C - p_D = (0.005 \text{ m})(1078 \text{ N/m}^3) = 5.39 \text{ Pa}$$

The inclined manometer (Fig. 2.9) is frequently used for measuring small differences in gas pressures. It is adjusted to read zero, by moving the inclined scale, when A and B are open. Since the inclined tube requires a greater displacement of the meniscus for given pressure difference than a vertical tube, it affords greater accuracy in reading the scale.

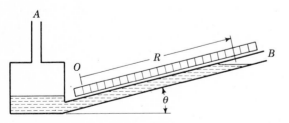

Fig. 2.9 Inclined manometer.

Surface tension causes a capillary rise in small tubes. If a U tube is used with a meniscus in each leg, the surface-tension effects cancel. The capillary rise is negligible in tubes with a diameter of 0.5 in or greater.

2.5 FORCES ON PLANE AREAS

In the preceding sections variations of pressure throughout a fluid have been considered. The distributed forces resulting from the action of fluid on a finite area may be conveniently replaced by a resultant force, insofar as external reactions to the force system are concerned. In this section the magnitude of resultant force and its line of action (pressure center) are determined by integration, by formula, and by use of the concept of the pressure prism.

Horizontal surfaces

A plane surface in a horizontal position in a fluid at rest is subjected to a constant pressure. The magnitude of the force acting on one side of the surface is

$$\int p \, dA = p \int dA = pA$$

The elemental forces $p \, dA$ acting on A are all parallel and in the same sense; therefore, a scalar summation of all such elements yields the magnitude of the resultant force. Its direction is normal to the surface, and *toward* the surface if p is *positive*. To find the line of action of the resultant, i.e., the point in the area where the moment of the distributed force about any axis through the point is zero, arbitrary xy axes may be selected, as in Fig. 2.10. Then, since

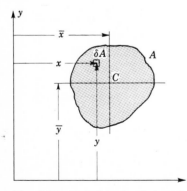

Fig. 2.10 Notation for determining the line of action of a force.

the moment of the resultant must equal the moment of the distributed force system about any axis, say the y axis,

$$pAx' = \int_A xp \, dA$$

in which x' is the distance from the y axis to the resultant. Since p is constant,

$$x' = \frac{1}{A} \int_A x \, dA = \bar{x}$$

in which $\bar{x}$ is the distance to the centroid of the area (see Appendix A). Hence, for a horizontal area subjected to static fluid pressure, the resultant passes through the centroid of the area.

Inclined surfaces

In Fig. 2.11 a plane surface is indicated by its trace $A'B'$. It is inclined $\theta°$ from the horizontal. The intersection of the plane of the area and the free surface is taken as the x axis. The y axis is taken in the plane of the area, with origin O, as shown, in the free surface. The xy plane portrays the arbitrary inclined area. The magnitude, direction, and line of action of the resultant force *due to the liquid*, acting on one side of the area, are sought.

For an element with area δA as a strip with thickness δy with long edges horizontal, the magnitude of force δF acting on it is

$$\delta F = p \, \delta A = \gamma h \, \delta A = \gamma y \sin \theta \, \delta A \tag{2.5.1}$$

Since all such elemental forces are parallel, the integral over the area yields the magnitude of force F, acting on one side of the area,

$$F = \int p \, dA = \gamma \sin \theta \int y \, dA = \gamma \sin \theta \bar{y} A = \gamma \bar{h} A = p_G A \tag{2.5.2}$$

with the relations from Fig. 2.11, $\bar{y} \sin \theta = \bar{h}$, and $p_G = \gamma \bar{h}$, the pressure at the centroid of the area. In words, the magnitude of force exerted on one side of a plane area submerged in a liquid is the product of the area and the pressure at its centroid. In this form, it should be noted, the presence of a free surface is unnecessary. Any means for determining the pressure at the centroid may be used. The sense of the force is to push against the area if p_G is positive. As all force elements are normal to the surface, the line of action of the resultant is also normal to the surface. Any surface may be rotated about

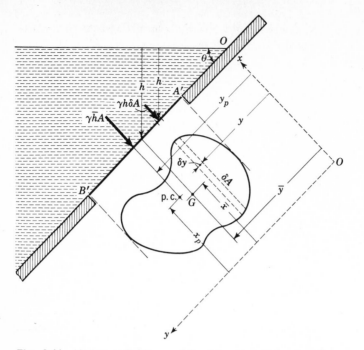

Fig. 2.11 Notation for force of liquid on one side of a plane-inclined area.

any axis through its centroid without changing the magnitude of the resultant if the total area remains submerged in the static liquid.

Center of pressure

The line of action of the resultant force has its piercing point in the surface at a point called the *pressure center*, with coordinates (x_p, y_p) (Fig. 2.11). Unlike that for the horizontal surface, the center of pressure of an inclined surface is not at the centroid. To find the pressure center, the moments of the resultant $x_p F$, $y_p F$ are equated to the moment of the distributed forces about the y axis and x axis, respectively; thus

$$x_p F = \int_A xp \, dA \qquad\qquad (2.5.3)$$

$$y_p F = \int_A yp \, dA \qquad\qquad (2.5.4)$$

The area element in Eq. (2.5.3) should be $\delta x\,\delta y$, and not the strip shown in Fig. 2.11.

Solving for the coordinates of the pressure center results in

$$x_p = \frac{1}{F}\int_A xp\,dA \tag{2.5.5}$$

$$y_p = \frac{1}{F}\int_A yp\,dA \tag{2.5.6}$$

In many applications Eqs. (2.5.5) and (2.5.6) may be evaluated most conveniently through graphical integration; for simple areas they may be transformed into general formulas as follows (see Appendix A):

$$x_p = \frac{1}{\gamma\bar{y}A\,\sin\theta}\int_A x\gamma y\,\sin\theta\,dA = \frac{1}{\bar{y}A}\int_A xy\,dA = \frac{I_{xy}}{\bar{y}A} \tag{2.5.7}$$

In Eqs. (A.10), of Appendix A, and (2.5.7),

$$x_p = \frac{\bar{I}_{xy}}{\bar{y}A} + \bar{x} \tag{2.5.8}$$

When either of the centroidal axes, $x = \bar{x}$ or $y = \bar{y}$, is an axis of symmetry for the surface, $\bar{I}_{xy}$ vanishes and the pressure center lies on $x = \bar{x}$. Since $\bar{I}_{xy}$ may be either positive or negative, the pressure center may lie on either side of the line $x = \bar{x}$. To determine y_p by formula, with Eqs. (2.5.2) and (2.5.6),

$$y_p = \frac{1}{\gamma\bar{y}A\,\sin\theta}\int_A y\gamma y\,\sin\theta\,dA = \frac{1}{\bar{y}A}\int_A y^2\,dA = \frac{I_x}{\bar{y}A} \tag{2.5.9}$$

In the parallel-axis theorem for moments of inertia

$$I_x = I_G + \bar{y}^2 A$$

in which I_G is the second moment of the area about its horizontal centroidal axis. If I_x is eliminated from Eq. (2.5.9),

$$y_p = \frac{I_G}{\bar{y}A} + \bar{y} \tag{2.5.10}$$

or

$$y_p - \bar{y} = \frac{I_G}{\bar{y}A} \tag{2.5.11}$$

I_G is always positive; hence, $y_p - \bar{y}$ is always positive, and the pressure center is always below the centroid of the surface. It should be emphasized that $\bar{y}$ and $y_p - \bar{y}$ are distances in the plane of the surface.

EXAMPLE 2.6 The triangular gate CDE (Fig. 2.12) is hinged along CD and is opened by a normal force P applied at E. It holds oil, sp gr 0.80, above it and is open to the atmosphere on its lower side. Neglecting the weight of the gate, find (*a*) the magnitude of force exerted on the gate by integration and by Eq. (2.5.2); (*b*) the location of pressure center; (*c*) the force P needed to open the gate.

(*a*)   By integration with reference to Fig. 2.12,

$$F = \int_A p \, dA = \gamma \sin \theta \int yx \, dy = \gamma \sin \theta \int_8^{13} xy \, dy + \gamma \sin \theta \int_{13}^{18} xy \, dy$$

When $y = 8$, $x = 0$, and when $y = 13$, $x = 6$, with x varying linearly with y; thus

$$x = ay + b \qquad 0 = 8a + b \qquad 6 = 13a + b$$

in which the coordinates have been substituted to find x in terms of y. Solving for a and b gives

$$a = \tfrac{6}{5} \qquad b = -\tfrac{48}{5} \qquad x = \tfrac{6}{5}(y - 8)$$

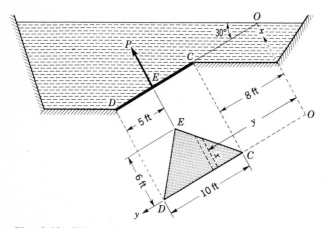

Fig. 2.12 Triangular gate.

Similarly $y = 13$, $x = 6$; $y = 18$, $x = 0$; and $x = \frac{6}{5}(18 - y)$. Hence

$$F = \gamma \sin \theta \, \tfrac{6}{5} \left[\int_8^{13} (y - 8)y \, dy + \int_{13}^{18} (18 - y)y \, dy \right]$$

Integrating and substituting for $\gamma \sin \theta$ leads to

$$F = 62.4 \times 0.8 \times 0.50 \times \tfrac{6}{5} \left[\left(\frac{y^3}{3} - 4y^2 \right)_8^{13} + \left(9y^2 - \frac{y^3}{3} \right)_{13}^{18} \right] = 9734.4 \text{ lb}$$

By Eq. (2.5.2),

$$F = p_G A = \gamma \bar{y} \sin \theta \, A = 62.4 \times 0.80 \times 13 \times 0.50 \times 30 = 9734.4 \text{ lb}$$

(b)   With the axes as shown, $\bar{x} = 2.0$, $\bar{y} = 13$. In Eq. (2.5.8),

$$x_p = \frac{\bar{I}_{xy}}{\bar{y}A} + \bar{x}$$

$\bar{I}_{xy}$ is zero owing to symmetry about the centroidal axis parallel to the x axis; hence $\bar{x} = x_p = 2.0$ ft. In Eq. (2.5.11),

$$y_p - \bar{y} = \frac{I_G}{\bar{y}A} = 2 \times \frac{1 \times 6 \times 5^3}{12 \times 13 \times 30} = 0.32 \text{ ft}$$

i.e., the pressure center is 0.32 ft below the centroid, measured in the plane of the area.

(c)   When moments about CD are taken and the action of the oil is replaced by the resultant,

$$P \times 6 = 9734.4 \times 2 \qquad P = 3244.8 \text{ lb}$$

The pressure prism

The concept of the pressure prism provides another means for determining the magnitude and location of the resultant force on an inclined plane surface. The volume of the pressure prism is the magnitude of the force, and the resultant force passes through the centroid of the prism. The surface is taken as the base of the prism, and its altitude at each point is determined by the pressure γh laid off to an appropriate scale (Fig. 2.13). Since the pressure increases linearly with distance from the free surface, the upper surface of the

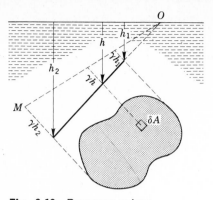

Fig. 2.13 Pressure prism.

prism is in a plane with its trace OM shown in Fig. 2.13. The force acting on an elemental area δA is

$$\delta F = \gamma h\, \delta A = \delta \mho \qquad (2.5.12)$$

which is an element of volume of the pressure prism. After integrating, $F = \mho$, the volume of the pressure prism equals the magnitude of the resultant force acting on one side of the surface.

From Eqs. (2.5.5) and (2.5.6),

$$x_p = \frac{1}{\mho} \int_\mho x\, d\mho \qquad y_p = \frac{1}{\mho} \int_\mho y\, d\mho \qquad (2.5.13)$$

which show that x_p, y_p are distances to the *centroid* of the pressure prism [Appendix A, Eq. (A.5)]. Hence, the line of action of the resultant passes through the centroid of the pressure prism. For some simple areas the pressure prism is more convenient than either integration or formula. For example, a rectangular area with one edge in the free surface has a wedge-shaped prism. Its centroid is one-third the altitude from the base; hence, the pressure center is one-third the altitude from its lower edge.

Effects of atmospheric pressure on forces on plane areas

In the discussion of pressure forces the pressure datum was not mentioned. The pressures were computed by $p = \gamma h$, in which h is the vertical distance below the free surface. Therefore, the datum taken was gage pressure zero, or the local atmospheric pressure. When the opposite side of the surface is open to the atmosphere, a force is exerted on it by the atmosphere equal to the product of the atmospheric pressure p_0 and the area, or p_0A, based on absolute zero as datum. On the liquid side the force is

$$\int (p_0 + \gamma h) \, dA = p_0 A + \gamma \int h \, dA$$

The effect p_0A of the atmosphere acts equally on both sides and in no way contributes to the resultant force or its location.

So long as the same pressure datum is selected for all sides of a free body, the resultant and moment can be determined by constructing a free surface at pressure zero on this datum and using the above methods.

EXAMPLE 2.7 An application of pressure forces on plane areas is given in the design of a gravity dam. The maximum and minimum compressive stresses in the base of the dam are computed from the forces which act on the dam. Figure 2.14 shows a cross section through a concrete dam where the specific weight of concrete has been taken as 2.5γ and γ is the specific weight of water.

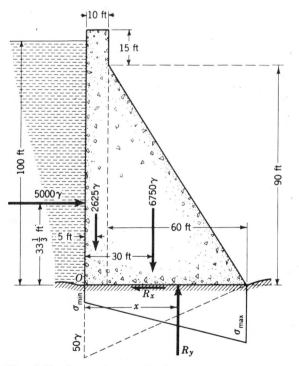

Fig. 2.14 Concrete gravity dam.

A 1-ft section of dam is considered as a free body; the forces are due to the concrete, the water, the foundation pressure, and the hydrostatic uplift. Determining amount of hydrostatic uplift is beyond the scope of this treatment, but it will be assumed to be one-half the hydrostatic head at the upstream edge, decreasing linearly to zero at the downstream edge of the dam. Enough friction or shear stress must be developed at the base of the dam to balance the thrust due to the water; that is, $R_x = 5000\gamma$. The resultant upward force on the base equals the weight of the dam less the hydrostatic uplift, $R_y = 6750\gamma + 2625\gamma - 1750\gamma = 7625\gamma$ lb. The position of R_y is such that the free body is in equilibrium. For moments around O,

$$\Sigma M_0 = 0 = R_y x - 5000\gamma(33.33) - 2625\gamma(5) - 6750\gamma(30) + 1750\gamma(23.33)$$

and

$$x = 44.8 \text{ ft}$$

It is customary to assume that the foundation pressure varies linearly over the base of the dam, i.e., that the pressure prism is a trapezoid with a volume equal to R_y; thus

$$\frac{\sigma_{max} + \sigma_{min}}{2} 70 = 7625\gamma$$

in which σ_{max}, σ_{min} are the maximum and minimum compressive stresses in pounds per square foot. The centroid of the pressure prism is at the point where $x = 44.8$ ft. By taking moments about O to express the position of the centroid in terms of σ_{max} and σ_{min},

$$44.8 = \frac{\sigma_{min} 70 \times \frac{70}{2} + (\sigma_{max} - \sigma_{min})\frac{70}{2} \times \frac{2}{3}(70)}{(\sigma_{max} + \sigma_{min})\frac{70}{2}}$$

Simplifying gives

$$\sigma_{max} = 11.75\sigma_{min}$$

Then

$$\sigma_{max} = 210\gamma = 12,500 \text{ lb/ft}^2 \qquad \sigma_{min} = 17.1\gamma = 1067 \text{ lb/ft}^2$$

When the resultant falls within the middle third of the base of the dam, σ_{min} will always be a compressive stress. Owing to the poor tensile properties of concrete, good design requires the resultant to fall within the middle third of the base.

2.6 FORCE COMPONENTS ON CURVED SURFACES

When the elemental forces $p\,\delta A$ vary in direction, as in the case of a curved surface, they must be added as vector quantities; i.e., their components in three mutually perpendicular directions are added as scalars, and then the three components are added vectorially. With two horizontal components at right angles and with the vertical component—all easily computed for a curved surface—the resultant can be determined. The lines of action of the components are readily determined, and so the resultant and its line of action can be completely determined.

Horizontal component of force on a curved surface

The horizontal component of pressure force on a curved surface is equal to the pressure force exerted on a projection of the curved surface. The vertical plane of projection is normal to the direction of the component. The surface of Fig. 2.15 represents any three-dimensional surface, and δA an element of its area, its normal making the angle θ with the negative x direction. Then

$$\delta F_x = p\,\delta A\,\cos\theta$$

is the x component of force exerted on one side of δA. Summing up the x components of force over the surface gives

$$F_x = \int_A p\,\cos\theta\,dA \tag{2.6.1}$$

$\cos\theta\,\delta A$ is the projection of δA onto a plane perpendicular to x. The element

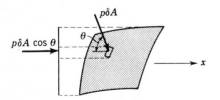

Fig. 2.15 Horizontal component of force on a curved surface.

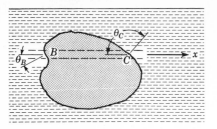

Fig. 2.16 Projections of area elements on opposite sides of a body.

of force on the projected area is $p \cos \theta \, \delta A$, which is also in the x direction. Projecting each element on a plane perpendicular to x is equivalent to projecting the curved surface as a whole onto the vertical plane. Hence, the force acting on this projection of the curved surface is the horizontal component of force exerted on the curved surface, in the direction normal to the plane of projection. To find the horizontal component at right angles to the x direction, the curved surface is projected onto a vertical plane parallel to x and the force on the projection is determined.

When the horizontal component of pressure force on a closed body is to be found, the projection of the curved surface on a vertical plane is always zero, since on opposite sides of the body the area-element projections have opposite signs, as indicated in Fig. 2.16. Let a small cylinder of cross section δA with axis parallel to x intersect the closed body at B and C. If the element of area of the body cut by the prism at B is δA_B and at C is δA_C, then

$$\delta A_B \cos \theta_B = -\delta A_C \cos \theta_C = \delta A$$

as $\cos \theta_C$ is negative. Hence, with the pressure the same at each end of the cylinder,

$$p \, \delta A_B \cos \theta_B + p \, \delta A_C \cos \theta_C = 0$$

and similarly for all other area elements.

To find the line of action of a horizontal component of force on a curved surface, the resultant of the parallel force system composed of the force components from each area element is required. This is exactly the resultant of the force on the projected area, since the two force systems have an identical distribution of elemental horizontal force components. Hence, the pressure center is located on the projected area by the methods of Sec. 2.5.

EXAMPLE 2.8 The equation of an ellipsoid of revolution submerged in water is $x^2/4 + y^2/4 + z^2/9 = 1$. The center of the body is located 2 m below the free surface. Find the horizontal force components acting on the curved surface that is located in the first octant. Consider the xz plane to be horizontal and y to be positive upward.

The projection of the surface on the yz plane has an area of $\pi/4 \times 2 \times 3$ m². Its centroid is located $2 - 4/3\pi \times 2$ m below the free surface. Hence

$$F_x = -\left(\frac{\pi}{4} \times 6\right)\left(2 - \frac{8}{3\pi}\right)\gamma = -(5.425)\,\mathrm{m^3} \times 9802 \, \mathrm{N/m^3} = -53{,}200 \text{ N}$$

Similarly,

$$F_z = -\left(\frac{\pi}{4} \times 4\right)\left(2 - \frac{8}{3\pi}\right)\gamma = -(3.617)\,\mathrm{m^3} \times 9802 \, \mathrm{N/m^3} = -35{,}400 \text{ N}$$

Vertical component of force on a curved surface

The vertical component of pressure force on a curved surface is equal to the weight of liquid vertically above the curved surface and extending up to the free surface. The vertical component of force on a curved surface can be determined by summing up the vertical components of pressure force on elemental areas δA of the surface. In Fig. 2.17 an area element is shown with the force $p\,\delta A$ acting normal to it. (Let θ be the angle the normal to the area element makes with the vertical.) Then the vertical component of force acting on the area element is $p \cos\theta\,\delta A$, and the vertical component of force on the curved surface is

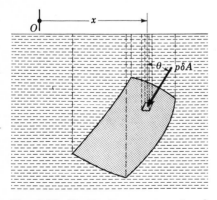

Fig. 2.17 Vertical component of force on a curved surface.

given by

$$F_v = \int_A p \cos \theta \, dA \qquad (2.6.2)$$

When p is replaced by its equivalent γh, in which h is the distance from the area element to the free surface, and it is noted that $\cos \theta \, \delta A$ is the projection of δA on a horizontal plane, Eq. (2.6.2) becomes

$$F_v = \gamma \int_A h \cos \theta \, dA = \gamma \int_\upsilon d\upsilon \qquad (2.6.3)$$

in which $\delta \upsilon$ is the volume of the prism of height h and base $\cos \theta \, \delta A$, or the volume of liquid vertically above the area element.   Integrating gives

$$F_v = \gamma \upsilon \qquad (2.6.4)$$

When the liquid is below the curved surface (Fig. 2.18) and the pressure magnitude is known at some point, for example, O, an *imaginary* or equivalent free surface s-s can be constructed p/γ above O, so that the product of specific weight and vertical distance to any point in the tank is the pressure at the point.   The weight of the imaginary volume of liquid vertically above the curved surface is then the vertical component of pressure force on the curved surface.   In constructing an imaginary free surface, the imaginary liquid must be of the same specific weight as the liquid in contact with the curved surface; otherwise, the pressure distribution over the surface will not be correctly represented.   With an imaginary liquid above a surface, the pressure at a point on the curved surface is equal on both sides, but the elemental force

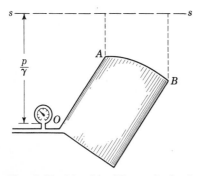

Fig. 2.18 Liquid with equivalent free surface.

components in the vertical direction are opposite in sign. Hence, the direction of the vertical force component is reversed when an imaginary fluid is above the surface. In some cases a confined liquid may be above the curved surface, and an imaginary liquid must be added (or subtracted) to determine the free surface.

The line of action of the vertical component is determined by equating moments of the elemental vertical components about a convenient axis with the moment of the resultant force. With the axis at O (Fig. 2.17),

$$F_v \bar{x} = \gamma \int_{\mho} x \, d\mho$$

in which $\bar{x}$ is the distance from O to the line of action. Then, since $F_v = \gamma \mho$,

$$\bar{x} = \frac{1}{\mho} \int_{\mho} x \, d\mho$$

the distance to the centroid of the volume. Therefore, the line of action of the vertical force passes through the centroid of the volume, real or imaginary, that extends above the curved surface up to the real or imaginary free surface.

EXAMPLE 2.9 A cylindrical barrier (Fig. 2.19) holds water as shown. The contact between cylinder and wall is smooth. Considering a 1-ft length of cylinder, determine (a) its weight and (b) the force exerted against the wall.

(a)   For equilibrium the weight of the cylinder must equal the vertical component of force exerted on it by the water. (The imaginary free surface for CD is at elevation A.) The vertical force on BCD is

$$F_{v_{BCD}} = \left(\frac{\pi r^2}{2} + 2r^2 \right) \gamma = (2\pi + 8)\gamma$$

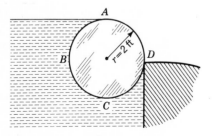

Fig. 2.19 Semifloating body.

The vertical force on AB is

$$F_{v_{AB}} = -\left(r^2 - \frac{\pi r^2}{4}\right)\gamma = -(4 - \pi)\gamma$$

Hence, the weight per foot of length is

$$F_{v_{BCD}} + F_{v_{AB}} = (3\pi + 4)\gamma = 838 \text{ lb}$$

(b)   The force exerted against the wall is the horizontal force on ABC minus the horizontal force on CD. The horizontal components of force on BC and CD cancel; the projection of BCD on a vertical plane is zero. Hence,

$$F_H = F_{H_{AB}} = 2\gamma = 124.8 \text{ lb}$$

since the projected area is 2 ft² and the pressure at the centroid of the projected area is 62.4 lb/ft².

To find external reactions due to pressure forces, the action of the fluid may be replaced by the two horizontal components and one vertical component acting along their lines of action.

Tensile stress in a pipe

A circular pipe under the action of an internal pressure is in tension around its periphery. Assuming that no longitudinal stress occurs, the walls are in tension, as shown in Fig. 2.20. A section of pipe of unit length is considered, i.e., the ring between two planes normal to the axis and unit length apart. If one-half of this ring is taken as a free body, the tensions per unit length at top and bottom are, respectively, T_1 and T_2, as shown in the figure. The horizontal component of force acts through the pressure center of the projected

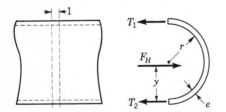

Fig. 2.20 Tensile stress in pipe.

area and is $2pr$ in which p is the pressure at the centerline and r is the internal pipe radius.

For high pressures the pressure center may be taken at the pipe center; then $T_1 = T_2$, and

$$T = pr \qquad (2.6.5)$$

in which T is the tensile force per unit length. For wall thickness e, the *tensile stress* in the pipe wall is

$$\sigma = \frac{T}{e} = \frac{pr}{e} \qquad (2.6.6)$$

For larger variations in pressure between top and bottom of pipe the location of pressure center y is computed, and two equations are needed,

$$T_1 + T_2 = 2pr \qquad 2rT_1 - 2pry = 0$$

in which the second equation is the moment equation about the lower end of the free body, neglecting the vertical component of force. Solving gives

$$T_1 = py \qquad T_2 = p(2r - y)$$

EXAMPLE 2.10 A 4.0-in-ID steel pipe has a $\frac{1}{4}$-in wall thickness. For an allowable tensile stress of 10,000 lb/in² what is the maximum pressure?

From Eq. (2.6.6)

$$p = \frac{\sigma e}{r} = \frac{(10,000 \text{ lb/in}^2)(0.25 \text{ in})}{2.0 \text{ in}} = 1250 \text{ lb/in}^2$$

Tensile stress in a thin spherical shell

If a thin spherical shell is subjected to an internal pressure, neglecting the weight of the fluid within the sphere, the stress in its walls can be found by considering the forces on a free body consisting of a hemisphere cut from the sphere by a vertical plane. The fluid component of force normal to the plane acting on the inside of the hemisphere is $p\pi r^2$, with r the radius. The stress σ times the cut wall area $2\pi re$, with e the thickness, must balance the fluid force; hence

$$\sigma = \frac{pr}{2e}$$

2.7 BUOYANT FORCE

The resultant force exerted on a body *by a static fluid* in which it is submerged or floating is called the *buoyant force*. The buoyant force always acts vertically upward. There can be no horizontal component of the resultant because the vertical projection of the submerged body or submerged portion of the floating body is always zero.

   The buoyant force on a submerged body is the difference between the vertical component of pressure force on its underside and the vertical component of pressure force on its upper side. In Fig. 2.21 the upward force on the bottom is equal to the weight of liquid, real or imaginary, which is vertically above the surface ABC, indicated by the weight of liquid within $ABCEFA$. The downward force on the upper surface equals the weight of liquid $ADCEFA$. The difference between the two forces is a force, vertically upward, due to the weight of fluid $ABCD$ that is *displaced* by the solid. In equation form

$$F_B = \mathcal{U}\gamma \tag{2.7.1}$$

in which F_B is the buoyant force, $\mathcal{U}$ is the volume of fluid displaced, and γ is the specific weight of fluid. The same formula holds for floating bodies when $\mathcal{U}$ is taken as the volume of liquid displaced. This is evident from inspection of the floating body in Fig. 2.21.

   In Fig. 2.22a, the vertical force exerted on an element of the body in the form of a vertical prism of cross section δA is

$$\delta F_B = (p_2 - p_1)\, \delta A = \gamma h\, \delta A = \gamma\, \delta \mathcal{U}$$

in which $\delta \mathcal{U}$ is the volume of the prism. Integrating over the complete body

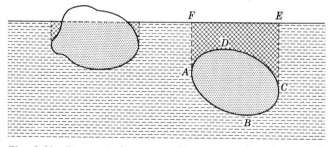

Fig. 2.21 Buoyant force on floating and submerged bodies.

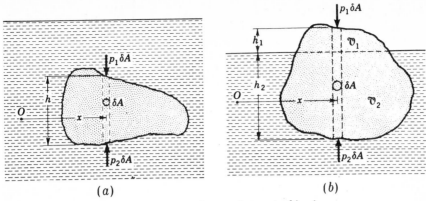

Fig. 2.22 Vertical force components on element of body.

gives

$$F_B = \gamma \int_{\mathcal{V}} d\mathcal{V} = \gamma \mathcal{V}$$

when γ is considered constant throughout the volume.

To find the line of action of the buoyant force, moments are taken about a convenient axis O and are equated to the moment of the resultant; thus,

$$\gamma \int x \, d\mathcal{V} = \gamma \mathcal{V} \bar{x} \qquad \text{or} \qquad \bar{x} = \frac{1}{\mathcal{V}} \int x \, d\mathcal{V}$$

in which $\bar{x}$ is the distance from the axis to the line of action. This equation yields the distance to the centroid of the volume; hence *the buoyant force acts through the centroid of the displaced volume of fluid.* This holds for both submerged and floating bodies. The centroid of the displaced volume of fluid is called the *center of buoyancy.*

When the body floats at the interface of a static two-fluid system (Fig. 2.22b) the buoyant force on a vertical prism of cross section δA is

$$\delta F_B = (p_2 - p_1) \, \delta A = (\gamma_2 h_2 + \gamma_1 h_1) \, \delta A$$

in which γ_1, γ_2 are the specific weights of the lighter and heavier fluids, respectively. Integrating over the area yields

$$F_B = \gamma_2 \int h_2 \, dA + \gamma_1 \int h_1 \, dA = \gamma_2 \mathcal{V}_2 + \gamma_1 \mathcal{V}_1$$

where $\mathcal{V}_1$ is the volume of lighter fluid displaced, and $\mathcal{V}_2$ is the volume of heavier fluid displaced. To locate the line of action of the buoyant force, moments are taken

$$F_B \bar{x} = \gamma_1 \int x \, d\mathcal{V}_1 + \gamma_2 \int x \, d\mathcal{V}_2$$

or

$$\bar{x} = \frac{\gamma_1 \int x \, d\mathcal{V}_1 + \gamma_2 \int x \, d\mathcal{V}_2}{\gamma_1 \mathcal{V}_1 + \gamma_2 \mathcal{V}_2} = \frac{\gamma_1 \bar{x}_1 \mathcal{V}_1 + \gamma_2 \bar{x}_2 \mathcal{V}_2}{\gamma_1 \mathcal{V}_1 + \gamma_2 \mathcal{V}_2}$$

in which $\bar{x}_1$, $\bar{x}_2$ are distances to centroids of volumes $\mathcal{V}_1$, $\mathcal{V}_2$, respectively. The resultant does not, in general, pass through the centroid of the whole volume.

In solving a statics problem involving submerged or floating objects, the object is generally taken as a free body, and a free-body diagram is drawn. The action of the fluid is replaced by the buoyant force. The weight of the object must be shown (acting through its center of gravity) as well as all other contact forces.

Weighing an odd-shaped object suspended in two different fluids yields sufficient data to determine its weight, volume, specific weight, and specific gravity. Figure 2.23 shows two free-body diagrams for the same object suspended and weighted in two fluids. F_1, F_2 are the weights submerged; γ_1, γ_2 are the specific weights of the fluids. W and $\mathcal{V}$, the weight and volume of the object, are to be found.

The equations of equilibrium are written

$$F_1 + \mathcal{V}\gamma_1 = W \qquad F_2 + \mathcal{V}\gamma_2 = W$$

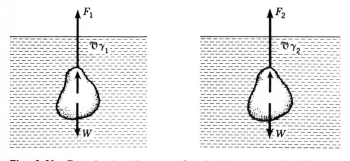

Fig. 2.23 Free-body diagram for body suspended in a fluid.

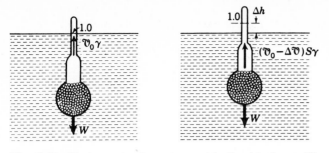

Fig. 2.24 Hydrometer in water and in liquid of specific gravity S.

and solved

$$\mathcal{V} = \frac{F_1 - F_2}{\gamma_2 - \gamma_1} \qquad W = \frac{F_1\gamma_2 - F_2\gamma_1}{\gamma_2 - \gamma_1}$$

A *hydrometer* uses the principle of buoyant force to determine specific gravities of liquids. Figure 2.24 shows a hydrometer in two liquids. It has a stem of prismatic cross section a. Considering the liquid on the left to be distilled water, $S = 1.00$, the hydrometer floats in equilibrium when

$$\mathcal{V}_0\gamma = W \tag{2.7.2}$$

in which $\mathcal{V}_0$ is the volume submerged, γ is the specific weight of water, and W is the weight of hydrometer. The position of the liquid surface is marked 1.00 on the stem to indicate unit specific gravity S. When the hydrometer is floated in another liquid, the equation of equilibrium becomes

$$(\mathcal{V}_0 - \Delta\mathcal{V})\,S\gamma = W \tag{2.7.3}$$

in which $\Delta\mathcal{V} = a\,\Delta h$. Solving for Δh with Eqs. (2.7.2) and (2.7.3) gives

$$\Delta h = \frac{\mathcal{V}_0}{a}\frac{S - 1}{S} \tag{2.7.4}$$

from which the stem can be marked off to read specific gravities.

EXAMPLE 2.11 A piece of ore weighing 0.15 kg$_f$ in air is found to weigh 0.11 kg$_f$ when submerged in water. What is its volume in cubic centimeters and its specific gravity?

The buoyant force due to air may be neglected. From Fig. 2.23

$$0.15 \text{ kg}_f \frac{9.806 \text{ N}}{1 \text{ kg}_f} = 0.11 \text{ kg}_f \frac{9.806 \text{ N}}{1 \text{ kg}_f} + (9802 \text{ N/m}^3) \, \mathcal{V}$$

$$\mathcal{V} = 0.00004 \text{ m}^3 = 40 \text{ cm}^3$$

$$S = \frac{W}{\gamma \mathcal{V}} = \frac{0.15 \text{ kg}_f \, 9.806 \text{ N/kg}_f}{9802 \text{ N/m}^3 \, 0.00004 \text{ m}^3} = 3.75$$

2.8 STABILITY OF FLOATING AND SUBMERGED BODIES

A body floating in a static liquid has vertical stability. A small upward displacement decreases the volume of liquid displaced, resulting in an unbalanced downward force which tends to return the body to its original position. Similarly, a small downward displacement results in a greater buoyant force, which causes an unbalanced upward force.

A body has linear stability when a small linear displacement in any direction sets up restoring forces tending to return the body to its original position. It has rotational stability when a restoring couple is set up by any small angular displacement.

Methods for determining rotational stability are developed in the following discussion. A body may float in stable, unstable, or neutral equilibrium. When a body is in unstable equilibrium, any small angular displacement sets up a couple that tends to increase the angular displacement. With the body in neutral equilibrium, any small angular displacement sets up no couple whatever. Figure 2.25 illustrates the three cases of equilibrium: (a) a light piece of wood with a metal weight at its bottom is stable; (b) when the metal weight is at the top, the body is in equilibrium but any slight angular displacement causes the body to assume the position in a; (c) a homogeneous sphere or right-circular cylinder is in equilibrium for any angular rotation; i.e., no couple results from an angular displacement.

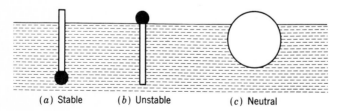

(a) Stable   (b) Unstable   (c) Neutral

Fig. 2.25 Examples of stable, unstable, and neutral equilibrium.

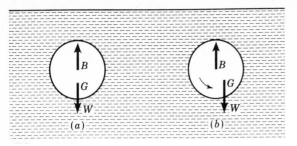

Fig. 2.26 Rotationally stable submerged body.

A completely submerged object is rotationally stable only when its center of gravity is below the center of buoyancy, as in Fig. 2.26a. When the object is rotated counterclockwise, as in Fig. 2.26b, the buoyant force and weight produce a couple in the clockwise direction.

Normally, when a body is too heavy to float, it submerges and goes down until it rests on the bottom. Although the specific weight of a liquid increases slightly with depth, the higher pressure tends to cause the liquid to compress the body or to penetrate into pores of solid substances, thus decreasing the buoyancy of the body. A ship, for example, is sure to go to the bottom once it is completely submerged, owing to compression of air trapped in its various parts.

Determination of rotational stability of floating objects

Any floating object with center of gravity below its center of buoyancy (centroid of displaced volume) floats in stable equilibrium, as in Fig. 2.25a. Certain floating objects, however, are in stable equilibrium when their center of gravity is above the center of buoyancy. The stability of prismatic bodies is first considered, followed by an analysis of general floating bodies for small angles of tip.

Figure 2.27a is a cross section of a body with all other parallel cross sections identical. The center of buoyancy is always at the centroid of the displaced volume, which is at the centroid of the cross-sectional area below liquid surface in this case. Hence, when the body is tipped, as in Fig. 2.27b, the center of buoyancy is at the centroid B′ of the trapezoid ABCD; the buoyant force acts upward through B′, and the weight acts downward through G, the center of gravity of the body. When the vertical through B′ intersects the original centerline above G, as at M, a restoring couple is produced and the body is in stable equilibrium. The intersection of the buoyant force and the centerline is called the *metacenter*, designated M. When M is above G, the body is stable;

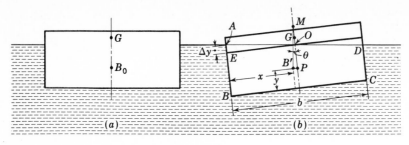

Fig. 2.27 Stability of a prismatic body.

when below G, it is unstable; and when at G, it is in neutral equilibrium. The distance $\overline{MG}$ is called the *metacentric height* and is a direct measure of the stability of the body. The restoring couple is

$$W\overline{MG} \sin \theta$$

in which θ is the angular displacement and W the weight of the body.

EXAMPLE 2.12 In Fig. 2.27 a scow 20 ft wide and 60 ft long has a gross weight of 225 short tons (2000 lb). Its center of gravity is 1.0 ft above the water surface. Find the metacentric height and restoring couple when $\Delta y = 1.0$ ft.
    The depth of submergence h in the water is

$$h = \frac{225 \times 2000}{20 \times 60 \times 62.4} = 6.0 \text{ ft}$$

The centroid in the tipped position is located with moments about AB and BC,

$$x = \frac{5 \times 20 \times 10 + 2 \times 20 \times \frac{1}{2} \times \frac{20}{3}}{6 \times 20} = 9.46 \text{ ft}$$

$$y = \frac{5 \times 20 \times \frac{5}{2} + 2 \times 20 \times \frac{1}{2} \times 5\frac{2}{3}}{6 \times 20} = 3.03 \text{ ft}$$

By similar triangles AEO and $B'PM$,

$$\frac{\Delta y}{b/2} = \frac{\overline{B'P}}{\overline{MP}}$$

$\Delta y = 1, b/2 = 10, \overline{B'P} = 10 - 9.46 = 0.54$ ft; then

$$\overline{MP} = \frac{0.54 \times 10}{1} = 5.40 \text{ ft}$$

G is 7.0 ft from the bottom; hence

$$\overline{GP} = 7.00 - 3.03 = 3.97 \text{ ft}$$

and

$$\overline{MG} = \overline{MP} - \overline{GP} = 5.40 - 3.97 = 1.43 \text{ ft}$$

The scow is stable since $\overline{MG}$ is positive; the righting moment is

$$W\overline{MG} \sin \theta = 225 \times 2000 \times 1.43 \times \frac{1}{\sqrt{101}} = 64,000 \text{ lb} \cdot \text{ft}$$

Nonprismatic cross sections

For a floating object of variable cross section, such as a ship (Fig. 2.28a), a convenient formula can be developed for determination of metacentric height for very small angles of rotation θ. The horizontal shift in center of buoyancy r (Fig. 2.28b) is determined by the change in buoyant forces due to the wedge being submerged, which causes an upward force on the left, and by the other wedge decreasing the buoyant force by an equal amount ΔF_B on the right. The force system, consisting of the original buoyant force at B and the couple $\Delta F_B \times s$ due to the wedges, must have as resultant the equal buoyant force at B'. With moments about B to determine the shift r,

$$\Delta F_B \times s = Wr \tag{2.8.1}$$

The amount of the couple can be determined with moments about O, the centerline of the body at the liquid surface. For an element of area δA on the horizontal section through the body at the liquid surface, an element of volume of the wedge is $x\theta \, \delta A$; the buoyant force due to this element is $\gamma x\theta \, \delta A$; and its moment about O is $\gamma\theta x^2 \, \delta A$, in which θ is the small angle of tip in radians. By integrating over the complete original horizontal area at the liquid surface, the couple is determined to be

$$\Delta F_B \times s = \gamma\theta \int_A x^2 \, dA = \gamma\theta I \tag{2.8.2}$$

in which I is the moment of inertia of the area about the axis yy (Fig. 2.28a).

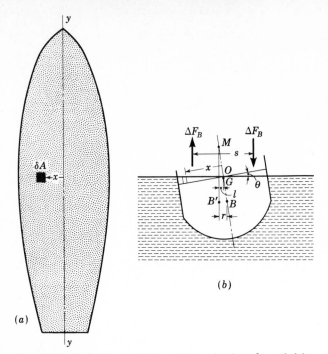

Fig. 2.28 Stability relations in a body of variable cross section.

Substitution into Eq. (2.8.1) produces

$$\gamma\theta I = Wr = \mathcal{V}\gamma r$$

in which $\mathcal{V}$ is the total volume of liquid displaced.

Since θ is very small,

$$\overline{MB} \sin\theta = \overline{MB}\theta = r \qquad \text{or} \qquad \overline{MB} = \frac{r}{\theta} = \frac{I}{\mathcal{V}}$$

The metacentric height is then

$$\overline{MG} = \overline{MB} \mp \overline{GB}$$

or

$$\overline{MG} = \frac{I}{\mathcal{V}} \mp \overline{GB} \qquad\qquad (2.8.3)$$

The minus sign is used if G is above B, the plus sign if G is below B.

EXAMPLE 2.13 A barge displacing 1000 metric tons has the horizontal cross section at the waterline shown in Fig. 2.29. Its center of buoyancy is 2.0 m below the water surface, and its center of gravity is 0.5 m below the water surface. Determine its metacentric height for rolling (about yy axis) and for pitching (about xx axis).

$$\overline{GB} = 2 - 0.5 = 1.5 \text{ m}$$

$$\mathcal{V} = \frac{1000(1000 \text{ kg}_f)(9.806 \text{ N/kg}_f)}{9802 \text{ N/m}^3} = 1000 \text{ m}^3$$

$$I_{yy} = \tfrac{1}{12}(24 \text{ m})(10 \text{ m})^3 + 4(\tfrac{1}{12})(6 \text{ m})(5 \text{ m})^3 = 2250 \text{ m}^4$$

$$I_{xx} = \tfrac{1}{12}(10 \text{ m})(24 \text{ m})^3 + 2(\tfrac{1}{36})(10 \text{ m})(6 \text{ m})^3 + (60 \text{ m}^2)(14 \text{ m})^2 = 23{,}400 \text{ m}^4$$

For rolling

$$\overline{MG} = \frac{I}{\mathcal{V}} - \overline{GB} = \frac{2250}{1000} - 1.5 = 0.75 \text{ m}$$

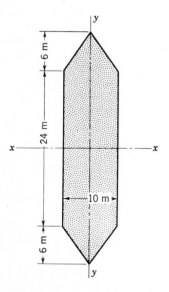

Fig. 2.29 Horizontal cross section of a ship at the waterline.

For pitching

$$\overline{MG} = \frac{I}{\mathcal{U}} - GB = \frac{23,400}{1000} - 1.5 = 21.9 \text{ m}$$

EXAMPLE 2.14 A homogeneous cube of specific gravity S_c floats in a liquid of specific gravity S. Find the range of specific-gravity ratios S_c/S for it to float with sides vertical.

In Fig. 2.30, b is the length of one edge of the cube. The depth of submergence z is determined by application of the buoyant-force equation

$$b^3 \gamma S_c = b^2 z \gamma S$$

in which γ is the specific weight of water. Solving for depth of submergence gives

$$z = b \frac{S_c}{S}$$

The center of buoyancy is $z/2$ from the bottom, and the center of gravity is $b/2$ from the bottom. Hence

$$\overline{GB} = \frac{b - z}{2} = \frac{b}{2} \left(1 - \frac{S_c}{S} \right)$$

Applying Eq. (2.8.3) gives

$$\overline{MG} = \frac{I}{\mathcal{U}} - \overline{GB} = \frac{1}{12} \frac{b \times b^3}{zb^2} - \frac{b - z}{2}$$

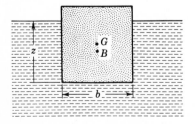

Fig. 2.30 Cube floating in liquid.

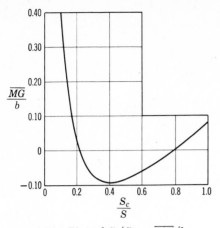

Fig. 2.31 Plot of S_c/S vs. $\overline{MG}/b$.

or

$$\overline{MG} = \frac{b}{12}\frac{S}{S_c} - \frac{b}{2}\left(1 - \frac{S_c}{S}\right)$$

When $\overline{MG}$ equals zero, $S_c/S = 0.212, 0.788$. Substitution shows that $\overline{MG}$ is positive for

$$0 < \frac{S_c}{S} < 0.212 \qquad 0.788 < \frac{S_c}{S} < 1.0$$

Figure 2.31 is a graph of $\overline{MG}/b$ vs. S_c/S.

2.9 RELATIVE EQUILIBRIUM

In fluid statics the variation of pressure is simple to compute, thanks to the absence of shear stresses. For fluid motion such that no layer moves relative to an adjacent layer, the shear stress is also zero throughout the fluid. A fluid with a translation at uniform velocity still follows the laws of static variation of pressure. When a fluid is being accelerated so that no layer moves relative to an adjacent one, i.e., when the fluid moves as if it were a solid, no shear stresses occur and variation in pressure can be determined by writing the equation of motion for an appropriate free body. Two cases are of interest, a uniform linear acceleration and a uniform rotation about a vertical axis. When moving thus, the fluid is said to be in *relative equilibrium*.

Although relative equilibrium is not a fluid-statics phenomenon, it is discussed here because of the similarity of the relationships.

Uniform linear acceleration

A liquid in an open vessel is given a uniform linear acceleration **a** as in Fig. 2.32. After some time the liquid adjusts to the acceleration so that it moves as a solid; i.e., the distance between any two fluid particles remains fixed, and hence no shear stresses occur.

By selecting a cartesian coordinate system with y vertically upward and x such that the acceleration vector **a** is in the xy plane, the z axis is normal to **a** and there is no acceleration component in the z direction. An element of fluid in the form of a small rectangular parallelepiped having edges δx, δy, δz parallel to the coordinate axes is taken as a free body, Fig. 2.33. The center of the element is at (x,y,z), and the pressure there is p. The rate of change of p with respect to x, y, and z is to be found and is first obtained for the element. Integration then yields the pressure variation throughout the fluid. The equation of motion for the x direction is written first:

$$\Sigma f_x = ma_x$$

$$\left(p - \frac{\partial p}{\partial x}\frac{\delta x}{2}\right)\delta y\,\delta z - \left(p + \frac{\partial p}{\partial x}\frac{\delta x}{2}\right)\delta y\,\delta z = \frac{\gamma}{g}\,\delta x\,\delta y\,\delta z\,a_x$$

which reduces to

$$\frac{\partial p}{\partial x} = -\gamma\frac{a_x}{g}$$

when δx, δy, δz are allowed to approach zero. Following the same procedure

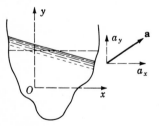

Fig. 2.32 Acceleration with free surface.

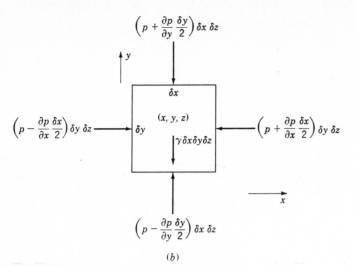

(b)

Fig. 2.33 Uniform linear acceleration of a fluid.

for the z direction (remembering that $a_z = 0$ because of the choice of axes) gives

$$\frac{\partial p}{\partial z} = 0$$

In the vertical direction the weight of the element $\gamma\ \delta x\ \delta y\ \delta z$ is taken into account; thus

$$\left(p - \frac{\partial p}{\partial y}\frac{\delta y}{2}\right)\delta x\ \delta z - \left(p + \frac{\partial p}{\partial y}\frac{\delta y}{2}\right)\delta x\ \delta z - \gamma\ \delta x\ \delta y\ \delta z = \frac{\gamma}{g}\ \delta x\ \delta y\ \delta z\ a_y$$

Simplifying leads to

$$\frac{\partial p}{\partial y} = -\gamma\left(1 + \frac{a_y}{g}\right)$$

Since p is a function of position (x,y,z), its total differential is (Appendix B)

$$dp = \frac{\partial p}{\partial x}\ dx + \frac{\partial p}{\partial y}\ dy + \frac{\partial p}{\partial z}\ dz$$

Substituting for the partial differentials gives

$$dp = -\gamma \frac{a_x}{g} dx - \gamma \left(1 + \frac{a_y}{g}\right) dy \qquad (2.9.1)$$

which can be integrated for an incompressible fluid,

$$p = -\gamma \frac{a_x}{g} x - \gamma \left(1 + \frac{a_y}{g}\right) y + c$$

To evaluate the constant of integration c, let $x = 0$, $y = 0$, $p = p_0$; then $c = p_0$ and

$$p = p_0 - \gamma \frac{a_x}{g} x - \gamma \left(1 + \frac{a_y}{g}\right) y \qquad (2.9.2)$$

When the accelerated incompressible fluid has a free surface, its equation is given by setting $p = 0$ in Eq. (2.9.2). Solving Eq. (2.9.2) for y gives

$$y = -\frac{a_x}{a_y + g} x + \frac{p_0 - p}{\gamma(1 + a_y/g)} \qquad (2.9.3)$$

The lines of constant pressure, $p = $ const, have the slope

$$-\frac{a_x}{a_y + g}$$

and are parallel to the free surface. The y intercept of the free surface is

$$\frac{p_0}{\gamma(1 + a_y/g)}$$

For an isothermal gas $p/\gamma = p_0/\gamma_0$ with p and p_0 in absolute units. Substitution into Eq. (2.9.1) yields, for $x = y = 0$, $p = p_0$, $\gamma = \gamma_0$,

$$\frac{p_0}{\gamma_0} \ln \frac{p}{p_0} = -\frac{a_x}{g} x - \left(1 + \frac{a_y}{g}\right) y$$

or

$$p = p_0 \exp\left[-\frac{xa_x/g + (1 + a_y/g)y}{p_0/\gamma_0}\right] \qquad (2.9.4)$$

The compressible equations apply only to closed containers.

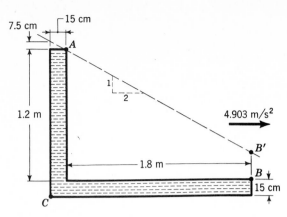

Fig. 2.34 Tank completely filled with liquid.

EXAMPLE 2.15 The tank in Fig. 2.34 is filled with oil, sp gr 0.8, and accelerated as shown. There is a small opening in the tank at A. Determine the pressure at B and C and the acceleration a_x required to make the pressure at B zero.

The planes of constant pressure have the slope

$$\tan\theta = \frac{a_x}{g} = \frac{4.903}{9.806} = 0.5$$

and at A the pressure is zero. The plane through A passes 0.3 m vertically above B; hence

$$p_B = (0.3\text{ m})\,(9802\text{ N/m}^3)\,(0.8) = 2.352\text{ kN/m}^2$$

Similarly, C is vertically below the zero pressure plane a distance 1.425 m, and

$$p_c = (1.425\text{ m})\,(9802\text{ N/m}^3)\,(0.8) = 11.174\text{ kN/m}^2$$

For zero pressure at B,

$$\tan\theta = \frac{1.2}{1.8} = \frac{a_x}{9.806}$$

and

$$a_x = \tfrac{2}{3}(9.806) = 6.537\text{ m/s}^2$$

EXAMPLE 2.16 A cubical box, 2 ft on a side, half filled with oil, sp gr 0.90, is accelerated along an inclined plane at an angle of 30° with the horizontal, as shown in Fig. 2.35. Find the slope of free surface and the pressure along the bottom.

In the coordinate system as indicated in the figure,

$$a_x = 8.05 \cos 30° = 6.98 \text{ ft/s}^2 \qquad a_y = 8.05 \sin 30° = 4.02 \text{ ft/s}^2$$

The slope of the free surface, from Eq. (2.9.3), is

$$-\frac{a_x}{a_y + g} = -\frac{6.98}{4.02 + 32.2} = -0.192$$

Since $\tan^{-1} 0.192 = 10°52'$ the free surface is inclined 40°52' to the bottom of the box. The depth parallel to a side is less on the right-hand side by 2 tan 40°52', or 1.73 ft. If s is the distance of A from the inclined plane, then from the known liquid volume,

$$4 \left(\frac{1.73}{2} + s \right) = 4 \text{ ft}^3$$

or $s = 0.135$. The xy coordinates of A are

$$x = 2 \cos 30° - 0.135 \sin 30° = 1.665 \text{ ft}$$

$$y = 2 \sin 30° + 0.135 \cos 30° = 1.117 \text{ ft}$$

By substitution in Eq. (2.9.2), the pressure p_0 at the origin is obtained,

$$0 = p_0 - 0.9 \times 62.4 \times 6.98 \times \frac{1.665}{32.2} - 0.9 \times 62.4 \left(1 + \frac{4.02}{32.2} \right) 1.117$$

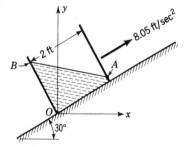

Fig. 2.35 Uniform acceleration along an inclined plane.

or

$$p_0 = 90.73 \text{ lb/ft}^2$$

Let t be the distance along the bottom from the origin; then $x = 0.866t$, $y = 0.5t$ for the bottom. By using Eq. (2.9.2) again

$$p = 90.73 - 42.07t$$

which is the pressure t ft from O along the bottom.

Uniform rotation about a vertical axis

Rotation of a fluid, moving as a solid, about an axis is called *forced-vortex* motion. Every particle of fluid has the same angular velocity. This motion is to be distinguished from *free-vortex* motion, where each particle moves in a circular path with a speed varying inversely as the distance from the center. Free-vortex motion is discussed in Chaps. 7 and 9. A liquid in a container, when rotated about a vertical axis at constant angular velocity, moves like a solid after some time interval. No shear stresses exist in the liquid, and the only acceleration that occurs is directed radially inward toward the axis of rotation. The equation of motion in the vertical direction on a free body shows that hydrostatic conditions prevail along any vertical line; hence, the pressure at any point in the liquid is given by the product of specific weight and vertical distance from the free surface. In equation form, Fig. 2.36,

$$\frac{\partial p}{\partial y} = -\gamma$$

In the equation of motion tangent to the circular path of a particle, the acceleration is zero, and the pressure does not change along the path.

In the equation of motion in the radial (horizontal) direction (Fig. 2.36), with a free body of length δr and cross-sectional area δA, if the pressure at r is p, then at the opposite face the pressure is $p + (\partial p/\partial r)\,\delta r$. The acceleration is $-\omega^2 r$; hence

$$p\,\delta A - \left(p + \frac{\partial p}{\partial r}\,\delta r\right)\delta A = \frac{\delta A\,\delta r\,\gamma}{g}\,(-\omega^2 r)$$

Simplifying and dividing through by the volume of the element $\delta A\,\delta r$ yields

$$\frac{\partial p}{\partial r} = \frac{\gamma}{g}\,\omega^2 r$$

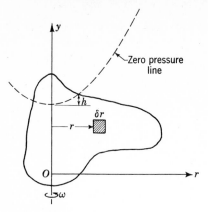

Fig. 2.36 Rotation of fluid about a vertical axis.

Since p is a function of y and r only, the total differential dp is

$$dp = \frac{\partial p}{\partial y} dy + \frac{\partial p}{\partial r} dr$$

Substituting for $\partial p/\partial y$ and $\partial p/\partial r$ results in

$$dp = -\gamma\, dy + \frac{\gamma}{g}\omega^2 r\, dr \tag{2.9.5}$$

For a liquid ($\gamma \approx$ const) integration yields

$$p = \frac{\gamma}{g}\omega^2 \frac{r^2}{2} - \gamma y + c$$

in which c is the constant of integration. If the value of pressure at the origin $(r = 0, y = 0)$ is p_0, then $c = p_0$ and

$$p = p_0 + \gamma \frac{\omega^2 r^2}{2g} - \gamma y \tag{2.9.6}$$

When the particular horizontal plane ($y = 0$) for which $p_0 = 0$ is selected and Eq. (2.9.6) is divided by γ,

$$h = \frac{p}{\gamma} = \frac{\omega^2 r^2}{2g} \tag{2.9.7}$$

which shows that the head, or vertical depth, varies as the square of the radius. The surfaces of equal pressure are paraboloids of revolution.

When a free surface occurs in a container that is being rotated, the fluid volume underneath the paraboloid of revolution is the original fluid volume. The shape of the paraboloid depends only upon the angular velocity ω.

For a circular cylinder rotating about its axis (Fig. 2.37) the rise of liquid from its vertex to the wall of the cylinder is, from Eq. (2.9.7), $\omega^2 r_0^2/2g$. Since a paraboloid of revolution has a volume equal to one-half its circumscribing cylinder, the volume of the liquid above the horizontal plane through the vertex is

$$\pi r_0^2 \times \frac{1}{2} \frac{\omega^2 r_0^2}{2g}$$

When the liquid is at rest, this liquid is also above the plane through the vertex to a uniform depth of

$$\frac{1}{2} \frac{\omega^2 r_0^2}{2g}$$

Hence, the liquid rises along the walls the same amount as the center drops, thereby permitting the vertex to be located when ω, r_0, and depth before rotation are given.

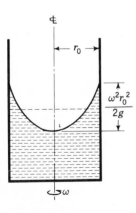

Fig. 2.37 Rotation of circular cylinder about its axis.

EXAMPLE 2.17 A liquid, sp gr 1.2, is rotated at 200 rpm about a vertical axis. At one point A in the fluid 1 m from the axis, the pressure is 70 kPa. What is the pressure at a point B 2 m higher than A and 1.5 m from the axis?
When Eq. (2.9.6) is written for the two points,

$$p_A = p_0 + \gamma \frac{\omega^2 r_A{}^2}{2g} - \gamma y \qquad p_B = p_0 + \gamma \frac{\omega^2 r_B{}^2}{2g} - \gamma(y+2)$$

Then $\omega = 200 \times 2\pi/60 = 20.95$ rad/s, $\gamma = 1.2 \times 9802 = 11{,}762$ N/m³, and

$$r_A = 1 \text{ m} \qquad r_B = 1.5 \text{ m}$$

When the second equation is subtracted from the first and the values are substituted,

$$70{,}000 - p_B = (2 \text{ m})(11{,}762 \text{ N/m}^3) + \frac{11{,}762 \text{ N/m}^3}{2 \times 9.806 \text{ m/s}^2} (20.95/\text{s})^2$$

$$\times [1 \text{ m}^2 - (1.5 \text{ m})^2]$$

Hence

$$p_B = 375.5 \text{ kPa}$$

If a closed container with no free surface or with a partially exposed free surface is rotated uniformly about some vertical axis, an *imaginary* free surface can be constructed, consisting of a paraboloid of revolution of shape given by Eq. (2.9.7). The vertical distance from any point in the fluid to this free surface is the pressure head at the point.
For an isothermal gas $\gamma = p\gamma_0/p_0$. Equation (2.9.5) becomes

$$\frac{p_0}{\gamma_0} \frac{dp}{p} = -dy + \frac{\omega^2 r}{g} dr$$

After integration, for $p = p_0$, $y = 0$, $r = 0$,

$$\frac{p_0}{\gamma_0} \ln \frac{p}{p_0} = -y + \frac{\omega^2 r^2}{2g}$$

p and p_0 must be in absolute pressure units for the compressible case.

EXAMPLE 2.18 A straight tube 4 ft long, closed at the bottom and filled with water, is inclined 30° with the vertical and rotated about a vertical axis through

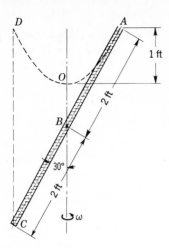

Fig. 2.38 Rotation of inclined tube of liquid about a vertical axis.

its midpoint 8.02 rad/s. Draw the paraboloid of zero pressure, and determine the pressure at the bottom and midpoint of the tube.

In Fig. 2.38, the zero-pressure paraboloid passes through point A. If the origin is taken at the vertex, that is, $p_0 = 0$, Eq. (2.9.7) becomes

$$h = \frac{\omega^2 r^2}{2g} = \frac{8.02^2}{64.4} (2 \sin 30°)^2 = 1.0 \text{ ft}$$

which locates the vertex at O, 1.0 ft below A. The pressure at the bottom of the tube is $\gamma \times \overline{CD}$, or

$$4 \cos 30° \times 62.4 = 216 \text{ lb/ft}^2$$

At the midpoint, $\overline{OB} = 0.732$ ft, and

$$p_B = 0.732 \times 62.4 = 45.6 \text{ lb/ft}^2$$

Fluid pressure forces in relative equilibrium

The magnitude of the force acting on a plane area in contact with a fluid accelerating as a rigid body can be obtained by integration over the surface

$$F = \int p \, dA$$

The nature of the acceleration and orientation of the surface govern the particular variation of p over the surface. When the pressure varies linearly over the plane surface (linear acceleration), the magnitude of force is given by the product of pressure at the centroid and area since the volume of the pressure prism is given by $p_G A$. For nonlinear distributions the magnitude and line of action can be found by integration.

PROBLEMS

2.1 Prove that the pressure is the same in all directions at a point in a static fluid for the three-dimensional case.

2.2 The container of Fig. 2.39 holds water and air as shown. What is the pressure at A, B, C, and D in pounds per square foot and in pascals?

2.3 The tube in Fig. 2.40 is filled with oil. Determine the pressure at A and B in meters of water.

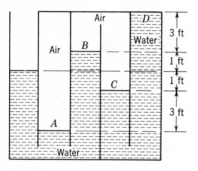

Fig. 2.39

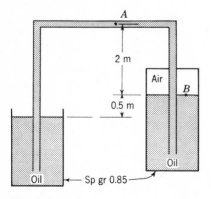

Fig. 2.40

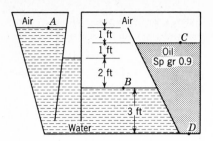

Fig. 2.41

2.4 Calculate the pressure at A, B, C, and D of Fig. 2.41 in pounds per square inch.

2.5 Derive the equations that give the pressure and density at any elevation in a static gas when conditions are known at one elevation and the temperature gradient β is known.

2.6 By a limiting process as $\beta \rightarrow 0$, derive the isothermal case from the results of Prob. 2.5.

2.7 By use of the results of Prob. 2.5, determine the pressure and density at 5000 ft elevation when $p = 14.5$ psia, $t = 68°F$, and $\beta = -0.003°F/ft$ at elevation 1000 ft for air.

2.8 For isothermal air at 0°C, determine the pressure and density at 10,000 ft when the pressure is 1 kg_f/cm^2 abs at sea level.

2.9 In isothermal air at 80°F what is the vertical distance for reduction of density by 10 percent?

2.10 Express a pressure of 8 psi in (a) inches of mercury, (b) feet of water, (c) feet of acetylene tetrabromide, sp gr 2.94, (d) pascals.

2.11 A bourdon gage reads 2 psi suction, and the barometer is 29.5 in Hg. Express the pressure in six other customary ways.

2.12 Express 3 atm in meters of water gage, barometer reading 750 mm.

2.13 Bourdon gage A inside a pressure tank (Fig. 2.42) reads 12 psi. Another

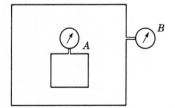

Fig. 2.42

bourdon gage B outside the pressure tank and connected with it reads 20 psi, and an aneroid barometer reads 30 in Hg. What is the absolute pressure measured by A in inches of mercury?

2.14 Determine the heights of columns of water; kerosene, sp gr 0.83; and acetylene tetrabromide, sp gr 2.94, equivalent to 18 cm Hg.

2.15 In Fig. 2.6a for a reading $h = 20$ in determine the pressure at A in pounds per square inch. The liquid has a specific gravity of 1.90.

2.16 Determine the reading h in Fig. 2.6b for $p_A = 20$ kPa suction if the liquid is kerosene, sp gr 0.83.

2.17 In Fig. 2.6b for $h = 8$ in and barometer reading 29 in, with water the liquid, find p_A in feet of water absolute.

2.18 In Fig. 2.6c $S_1 = 0.86$, $S_2 = 1.0$, $h_2 = 8.3$ cm, $h_1 = 17$ cm. Find p_A in millimeters of mercury gage. If the barometer reading is 29.5 in, what is p_A in meters of water absolute?

2.19 Gas is contained in vessel A of Fig. 2.6c. With water the manometer fluid and $h_1 = 5$ in, determine the pressure at A in inches of mercury.

2.20 In Fig. 2.7a $S_1 = 1.0$, $S_2 = 0.95$, $S_3 = 1.0$, $h_1 = h_2 = 30$ cm, and $h_3 = 1$ m. Compute $p_A - p_B$ in centimeters of water.

2.21 In Prob. 2.20 find the gage difference h_2 for $p_A - p_B = -39$ cm H$_2$O.

2.22 In Fig. 2.7b $S_1 = S_3 = 0.83$, $S_2 = 13.6$, $h_1 = 16$ cm, $h_2 = 8$ cm, and $h_3 = 12$ cm. (a) Find p_A if $p_B = 10$ psi. (b) For $p_A = 20$ psia and a barometer reading of 720 mm find p_B in meters of water gage.

2.23 Find the gage difference h_2 in Prob. 2.22 for $p_A = p_B$.

2.24 In Fig. 2.43, A contains water, and the manometer fluid has a specific gravity

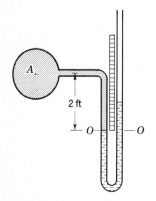

Fig. 2.43

of 2.94. When the left meniscus is at zero on the scale, p_A = 10 cm H₂O. Find the reading of the right meniscus for p_A = 8 kPa with no adjustment of the U tube or scale.

2.25 The Empire State Building is 1250 ft high. What is the pressure difference in pounds per square inch of a water column of the same height?

2.26 What is the pressure at a point 10 m below the free surface in a fluid that has a variable density in kilograms per cubic meter given by $\rho = 450 + ah$, in which $a = 12$ kg/m⁴ and h is the distance in meters measured from the free surface?

2.27 A vertical gas pipe in a building contains gas, ρ = 0.002 slug/ft³ and p = 3.0 in H₂O gage in the basement. At the top of the building 800 ft higher, determine the gas pressure in inches water gage for two cases: (*a*) gas assumed incompressible and (*b*) gas assumed isothermal. Barometric pressure 34 ft H₂O; t = 70°F.

2.28 In Fig. 2.8 determine R, the gage difference, for a difference in gas pressure of 1 cm H₂O. γ_2 = 62.4 lb/ft³; γ_3 = 65.5 lb/ft³; a/A = 0.01.

2.29 The inclined manometer of Fig. 2.9 reads zero when A and B are at the same pressure. The diameter of reservoir is 2.0 in and that of the inclined tube $\frac{1}{4}$ in. For θ = 30°, gage fluid sp gr 0.832, find $p_A - p_B$ in pounds per square inch as a function of gage reading R in feet.

2.30 Determine the weight W that can be sustained by the 100 kg$_f$ force acting on the piston of Fig. 2.44.

2.31 Neglecting the weight of the container (Fig. 2.45), find (*a*) the force tending to lift the circular top CD and (*b*) the compressive load on the pipe wall at A-A.

2.32 Find the force of oil on the top surface CD of Fig. 2.45 if the liquid level in the open pipe is reduced by 1 m.

2.33 The container shown in Fig. 2.46 has a circular cross section. Determine the upward force on the surface of the cone frustum $ABCD$. What is the downward force on the plane EF? Is this force equal to the weight of the fluid? Explain.

2.34 The cylindrical container of Fig. 2.47 weighs 400 N when empty. It is filled with water and supported on the piston. (*a*) What force is exerted on the upper end of the cylinder? (*b*) If an additional 600-N weight is placed on the cylinder, how much will the water force against the top of the cylinder be increased?

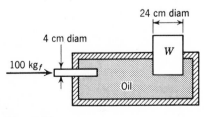

Fig. 2.44

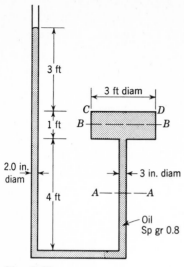

Fig. 2.45

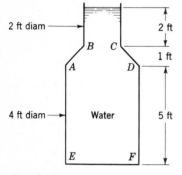

Fig. 2.46

2.35 A barrel 2 ft in diameter filled with water has a vertical pipe of 0.50 in diameter attached to the top. Neglecting compressibility, how many pounds of water must be added to the pipe to exert a force of 1000 lb on the top of the barrel?

2.36 A vertical right-angled triangular surface has a vertex in the free surface of a liquid (Fig. 2.48). Find the force on one side (*a*) by integration and (*b*) by formula.

2.37 Determine the magnitude of the force acting on vertical triangle *ABC* of Fig. 2.49 (*a*) by integration and (*b*) by formula.

2.38 Find the moment about *AB* of the force acting on one side of the vertical surface *ABC* of Fig. 2.48. $\gamma = 9000 \text{ N/m}^3$.

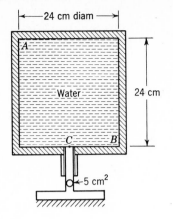

Fig. 2.47

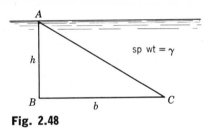

Fig. 2.48

2.39 Find the moment about AB of the force acting on one side of the vertical surface ABC of Fig. 2.49.

2.40 Locate a horizontal line below AB of Fig. 2.49 such that the magnitude of pressure force on the surface ABC is equal above and below the line.

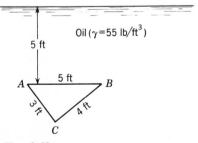

Fig. 2.49

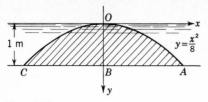

Fig. 2.50

2.41   Determine the force acting on one side of the vertical surface $OABCO$ of Fig. 2.50. $\gamma = 9500 \text{ N/m}^3$.

2.42   Calculate the force exerted by water on one side of the vertical annular area shown in Fig. 2.51.

2.43   Determine the moment at A required to hold the gate as shown in Fig. 2.52.

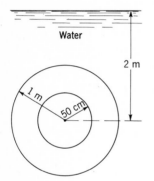

Fig. 2.51

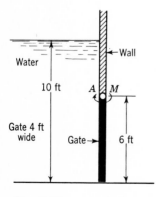

Fig. 2.52

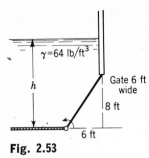

Fig. 2.53

2.44 If there is water on the other side of the gate (Fig. 2.52) up to A, determine the resultant force due to water on both sides of the gate, including its line of action.

2.45 The shaft of the gate in Fig. 2.53 will fail at a moment of $135 \text{ kN} \cdot \text{m}$. Determine the maximum value of liquid depth h.

2.46 The dam of Fig. 2.54 has a strut AB every 5 m. Determine the compressive force in the strut, neglecting the weight of the dam.

2.47 Locate the distance of the pressure center below the liquid surface in the triangular area ABC of Fig. 2.49 by integration and by formula.

2.48 By integration locate the pressure center horizontally in Fig. 2.49.

2.49 By using the pressure prism, determine the resultant force and location for the triangle of Fig. 2.48.

2.50 By integration, determine the pressure center for Fig. 2.48.

2.51 Locate the pressure center for the annular area of Fig. 2.51.

2.52 Locate the pressure center for the gate of Fig. 2.52.

2.53 A vertical square area 5 by 5 ft is submerged in water with upper edge 2 ft below the surface. Locate a horizontal line on the surface of the square such that (a) the force on the upper portion equals the force on the lower portion and (b) the moment of force about the line due to the upper portion equals the moment due to the lower portion.

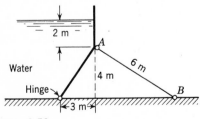

Fig. 2.54

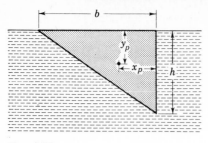

Fig. 2.55

2.54 An equilateral triangle with one edge in a water surface extends downward at a 45° angle. Locate the pressure center in terms of the length of a side b.

2.55 In Fig. 2.53 develop the expression for y_p in terms of h.

2.56 Locate the pressure center of the vertical area of Fig. 2.50.

2.57 Locate the pressure center for the vertical area of Fig. 2.55.

2.58 Demonstrate the fact that the magnitude of the resultant force on a totally submerged plane area is unchanged if the area is rotated about an axis through its centroid.

2.59 The gate of Fig. 2.56 weighs 300 lb/ft normal to the paper. Its center of gravity

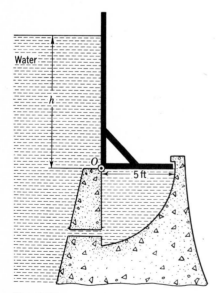

Fig. 2.56

is 1.5 ft from the left face and 2.0 ft above the lower face. It is hinged at O. Determine the water-surface position for the gate just to start to come up. (Water surface is below the hinge.)

2.60 Find h of Prob. 2.59 for the gate just to come up to the vertical position shown.

2.61 Determine the value of h and the force against the stop when this force is a maximum for the gate of Prob. 2.59.

2.62 Determine y of Fig. 2.57 so that the flashboards will tumble when water reaches their top.

2.63 Determine the hinge location y of the rectangular gate of Fig. 2.58 so that it will open when the liquid surface is as shown.

2.64 By use of the pressure prism, show that the pressure center approaches the centroid of an area as its depth of submergence is increased.

2.65 (*a*) Find the magnitude and line of action of force on each side of the gate of Fig. 2.59. (*b*) Find the resultant force due to the liquid on both sides of the gate. (*c*) Determine F to open the gate if it is uniform and weighs 6000 lb.

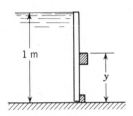

Fig. 2.57

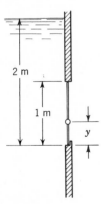

Fig. 2.58

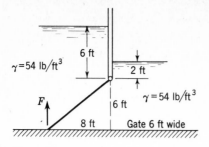

Fig. 2.59

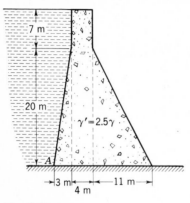

Fig. 2.60

2.66 For linear stress variation over the base of the dam of Fig. 2.60, (*a*) locate where the resultant crosses the base and (*b*) compute the maximum and minimum compressive stresses at the base. Neglect hydrostatic uplift.

2.67 Work Prob. 2.66 with the addition that the hydrostatic uplift varies linearly from 20 m at *A* to zero at the toe of the dam.

2.68 Find the moment *M* at *O* (Fig. 2.61) to hold the gate closed.

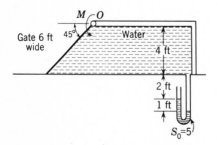

Fig. 2.61

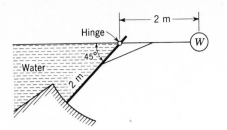

Fig. 2.62

2.69 The gate shown in Fig. 2.62 is in equilibrium. Compute W, the weight of counter-weight per meter of width, neglecting the weight of the gate. Is the gate in stable equilibrium?

2.70 The gate of Fig. 2.63 weighs 150 lb/ft normal to the page. It is in equilibrium as shown. Neglecting the weight of the arm and brace supporting the counterweight, (*a*) find W and (*b*) determine whether the gate is in stable equilibrium. The weight is made of concrete, sp gr 2.50.

2.71 The plane gate (Fig. 2.64) weighs 2000 N/m normal to the paper, with its center of gravity 2 m from the hinge at O. (*a*) Find h as a function of θ for equilibrium of the gate. (*b*) Is the gate in stable equilibrium for any values of θ?

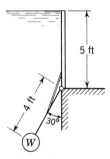

Fig. 2.63

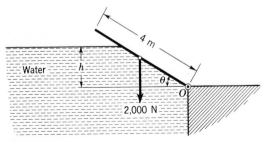

Fig. 2.64

Fig. 2.65

2.72   A 15-ft-diameter pressure pipe carries liquid at 150 psi. What pipe-wall thickness is required for maximum stress of 10,000 psi?

2.73   To obtain the same flow area, which pipe system requires the least steel, a single pipe or four pipes having half the diameter? The maximum allowable pipe-wall stress is the same in each case.

2.74   A thin-walled hollow sphere 3 m in diameter holds gas at 15 kg_f/cm^2. For allowable stress of 60 MPa determine the minimum wall thickness.

2.75   A cylindrical container 8 ft high and 4 ft in diameter provides for pipe tension with two hoops a foot from each end. When it is filled with water, what is the tension in each hoop due to the water?

2.76   A 2-cm-diameter steel ball covers a 1-cm-diameter hole in a pressure chamber where the pressure is 300 kg_f/cm^2. What force is required to lift the ball from the opening?

2.77   If the horizontal component of force on a curved surface did *not* equal the force on a projection of the surface onto a vertical plane, what conclusions could you draw regarding the propulsion of a boat (Fig. 2.65)?

2.78   (*a*) Determine the horizontal component of force acting on the radial gate (Fig. 2.66) and its line of action. (*b*) Determine the vertical component of force and its line of action. (*c*) What force F is required to open the gate, neglecting its weight? (*d*) What is the moment about an axis normal to the paper and through point O?

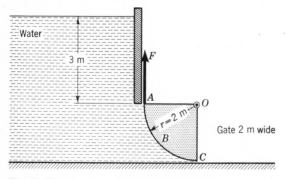

Fig. 2.66

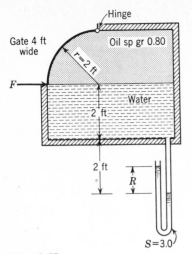

Fig. 2.67

2.79 Calculate the force F required to hold the gate of Fig. 2.67 in a closed position, $R = 2$ ft.

2.80 Calculate the force F required to open or hold closed the gate of Fig. 2.67 when $R = 1.5$ ft.

2.81 What is R of Fig. 2.67 for no force F required to hold the gate closed or to open it?

2.82 Find the vertical component of force on the curved gate of Fig. 2.68, including its line of action.

2.83 What is the force on the surface whose trace is OA of Fig. 2.50? The length normal to the paper is 3 m.

2.84 A right-circular cylinder with a diameter of 2 ft is illustrated in Fig. 2.69. The

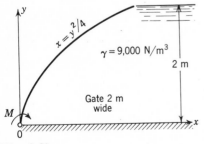

Fig. 2.68

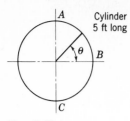

Fig. 2.69

pressure, in pounds per square foot, due to flow around the cylinder varies over the segment ABC as $p = 2\rho(1 - 4\sin^2\theta) + 10$. Calculate the force on ABC.

2.85 If the pressure variation on the cylinder in Fig. 2.69 is $p = 2\rho \times [1 - 4(1 + \sin\theta)^2] + 10$, determine the force on the cylinder.

2.86 Determine the moment M to hold the gate of Fig. 2.68, neglecting its weight.

2.87 Find the resultant force, including its line of action, acting on the outer surface of the first quadrant of a spherical shell of radius 60 cm with center at the origin. Its center is 1 m below the water surface.

2.88 The volume of the ellipsoid given by $x^2/a^2 + y^2/b^2 + z^2/c^2 = 1$ is $4\pi abc/3$, and the area of the ellipse $x^2/a^2 + z^2/c^2 = 1$ is πac. Determine the vertical force on the surface given in Example 2.8.

2.89 A log holds the water as shown in Fig. 2.70. Determine (*a*) the force per foot pushing it against the dam, (*b*) the weight of the log per foot of length, and (*c*) its specific gravity.

2.90 The cylinder of Fig. 2.71 is filled with liquid as shown. Find (*a*) the horizontal component of force on AB per unit of length, including its line of action, and (*b*) the vertical component of force on AB per unit of length, including its line of action.

2.91 The cylinder gate of Fig. 2.72 is made up from a circular cylinder and a plate, hinged at the dam. The gate position is controlled by pumping water into or out of the cylinder. The center of gravity of the empty gate is on the line of symmetry 4 ft from

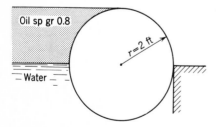

Fig. 2.70

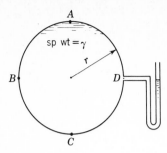

Fig. 2.71

the hinge. It is in equilibrium when empty in the position shown. How many cubic feet of water must be added per foot of cylinder to hold the gate in its position when the water surface is raised 3 ft?

2.92 A hydrometer weighs 0.035 N and has a stem 5 mm in diameter. Compute the distance between specific gravity markings 1.0 and 1.1.

2.93 Design a hydrometer to read specific gravities in the range from 0.80 to 1.10 when the scale is to be 3 in long.

2.94 A sphere 1 ft in diameter, sp gr 1.4, is submerged in a liquid having a density varying with the depth y below the surface given by $\rho = 2 + 0.1y$. Determine the equilibrium position of the sphere in the liquid.

2.95 Repeat the calculations for Prob. 2.94 for a horizontal circular cylinder with a specific gravity of 1.4 and a diameter of 1 ft.

2.96 A cube, 2 ft on an edge, has its lower half of sp gr 1.4 and upper half of sp gr 0.6. It is submerged into a two-layered fluid, the lower sp gr 1.2 and the upper sp gr 0.9. Determine the height of the top of the cube above the interface.

2.97 Determine the density, specific volume, and volume of an object that weighs 3 N in water and 4 N in oil, sp gr 0.83.

2.98 Two cubes of the same size, 1 m³, one of sp gr 0.80, the other of sp gr 1.1,

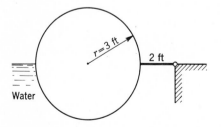

Fig. 2.72

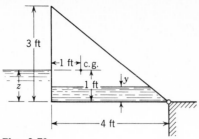

Fig. 2.73

are connected by a short wire and placed in water. What portion of the lighter cube is above the water surface, and what is the tension in the wire?

2.99 In Fig. 2.73 the hollow triangular prism is in equilibrium as shown when $z = 1$ ft and $y = 0$. Find the weight of prism per foot of length and z in terms of y for equilibrium. Both liquids are water. Determine the value of y for $z = 1.5$ ft.

2.100 How many pounds of concrete, $\gamma = 25$ kN/m^3, must be attached to a beam having a volume of 0.1 m^3 and sp gr 0.65 to cause both to sink in water?

2.101 Two beams, each 6 ft by 12 by 4 in, are attached at their ends and float as shown in Fig. 2.74. Determine the specific gravity of each beam.

2.102 A wooden cylinder 60 cm in diameter, sp gr 0.50, has a concrete cylinder 60 cm long of the same diameter, sp gr 2.50, attached to one end. Determine the length of wooden cylinder for the system to float in stable equilibrium with axis vertical.

2.103 What are the proportions r_0/h of a right-circular cylinder of specific gravity S so that it will float in water with end faces horizontal in stable equilibrium?

2.104 Will a beam 10 ft long with square cross section, sp gr 0.75, float in stable equilibrium in water with two sides horizontal?

2.105 Determine the metacentric height of the torus shown in Fig. 2.75.

2.106 Determine whether the thick-walled cylinder of Fig. 2.76 is stable in the position shown.

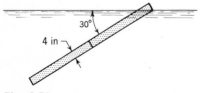

Fig. 2.74

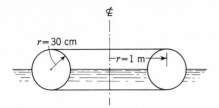

Fig. 2.75

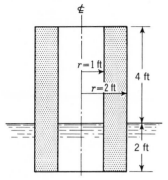

Fig. 2.76

2.107 A spherical balloon 15 m in diameter is open at the bottom and filled with hydrogen. For barometer reading of 28 in Hg and 20°C, what is the total weight of the balloon and the load to hold it stationary?

2.108 A tank of liquid $S = 0.86$ is accelerated uniformly in a horizontal direction so that the pressure decreases within the liquid 1 psi/ft in the direction of motion. Determine the acceleration.

2.109 The free surface of a liquid makes an angle of 20° with the horizontal when accelerated uniformly in a horizontal direction. What is the acceleration?

2.110 In Fig. 2.77, $a_x = 8.05$ ft/s², $a_y = 0$. Find the imaginary free liquid surface and the pressure at B, C, D, and E.

2.111 In Fig. 2.77, $a_x = 0$, $a_y = -16.1$ ft/s². Find the pressure at B, C, D, and E.

2.112 In Fig. 2.77, $a_x = 8.05$ ft/s², $a_y = 16.1$ ft/s². Find the imaginary free surface and the pressure at B, C, D, and E.

2.113 In Fig. 2.78, $a_x = 9.806$ m/s², $a_y = 0$. Find the pressure at A, B, and C.

2.114 In Fig. 2.78, $a_x = 4.903$ m/s², $a_y = 4.903$ m/s². Find the pressure at A, B, and C.

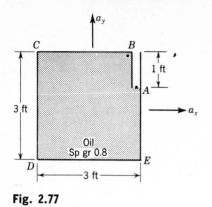

Fig. 2.77

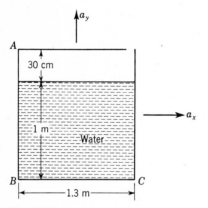

Fig. 2.78

2.115 A circular cross-sectional tank of 6-ft depth and 4-ft diameter is filled with liquid and accelerated uniformly in a horizontal direction. If one-third of the liquid spills out, determine the acceleration.

2.116 Derive an expression for pressure variation in a constant-temperature gas undergoing an acceleration a_x in the x direction.

2.117 The tube of Fig. 2.79 is filled with liquid, sp gr 2.40. When it is accelerated to the right 8.05 ft/s², draw the imaginary free surface and determine the pressure at A. For $p_A = 8$ psi vacuum determine a_x.

2.118 A cubical box 1 m on an edge, open at the top and half filled with water, is placed on an inclined plane making a 30° angle with the horizontal. The box alone weighs 500 N and has a coefficient of friction with the plane of 0.30. Determine the acceleration of the box and the angle the free water surface makes with the horizontal.

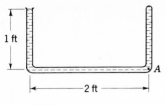

Fig. 2.79

2.119 Show that the pressure is the same in all directions at a point in a liquid moving as a solid.

2.120 A closed box contains two immiscible liquids. Prove that when it is accelerated uniformly in the x direction, the interface and zero-pressure surface are parallel.

2.121 Verify the statement made in Sec. 2.9 on uniform rotation about a vertical axis that when a fluid rotates in the manner of a solid body, no shear stresses exist in the fluid.

2.122 A vessel containing liquid, sp gr 1.2, is rotated about a vertical axis. The pressure at one point 2 ft radially from the axis is the same as at another point 4 ft from the axis and with elevation 2 ft higher. Calculate the rotational speed.

2.123 The U tube of Fig. 2.79 is rotated about a vertical axis 6 in to the right of A at such a speed that the pressure at A is zero gage. What is the rotational speed?

2.124 Locate the vertical axis of rotation and the speed of rotation of the U tube of Fig. 2.79 so that the pressure of liquid at the midpoint of the U tube and at A are both zero.

2.125 An incompressible fluid of density ρ moving as a solid rotates at speed ω about an axis inclined at $\theta°$ with the vertical. Knowing the pressure at one point in the fluid, how do you find the pressure at any other point?

2.126 A right-circular cylinder of radius r_0 and height h_0 with axis vertical is open at the top and filled with liquid. At what speed must it rotate so that half the area of the bottom is exposed?

2.127 A liquid rotating about a *horizontal* axis as a solid has a pressure of 10 psi at the axis. Determine the pressure variation along a vertical line through the axis for density ρ and speed ω.

2.128 Determine the equation for the surfaces of constant pressure for the situation described in Prob. 2.127.

2.129 Prove by integration that a paraboloid of revolution has a volume equal to half its circumscribing cylinder.

2.130 A tank containing two immiscible liquids is rotated about a vertical axis. Prove that the interface has the same shape as the zero pressure surface.

2.131   A hollow sphere of radius r_0 is filled with liquid and rotated about its vertical axis at speed ω. Locate the circular line of maximum pressure.

2.132   A gas following the law $pp^{-n} = \text{const}$ is rotated about a vertical axis as a solid. Derive an expression for pressure in a radial direction for speed ω, pressure p_0, and density ρ_0 at a point on the axis.

2.133   A vessel containing water is rotated about a vertical axis with an angular velocity of 50 rad/s. At the same time the container has a downward acceleration of 16.1 ft/s². What is the equation for a surface of constant pressure?

2.134   The U tube of Fig. 2.79 is rotated about a vertical axis through A at such a speed that the water in the tube begins to vaporize at the closed end above A, which is at 70°F. What is the angular velocity? What would happen if the angular velocity were increased?

2.135   A cubical box 4 ft on an edge is open at the top and filled with water. When it is accelerated upward 8.05 ft/s², find the magnitude of water force on one side of the box.

2.136   A cube 1 ft on an edge is filled with liquid, sp gr 0.65, and is accelerated downward 8.05 ft/s². Find the resultant force on one side of the cube due to liquid pressure.

2.137   Find the force on side OB of Fig. 2.35 for the situation described in Example 2.16.

2.138   A cylinder 2 ft in diameter and 6 ft long is accelerated uniformly along its axis in a horizontal direction 16.1 ft/s². It is filled with liquid, $\gamma = 50$ lb/ft³, and has a pressure along its axis of 10 psi before acceleration starts. Find the horizontal net force exerted against the liquid in the cylinder.

2.139   A closed cube, 30 cm on an edge, has a small opening at the center of its top. When it is filled with water and rotated uniformly about a vertical axis through its center at ω rad/s, find the force on a side due to the water in terms of ω.

2.140   The normal stress is the same in all directions at a point in a fluid

(*a*) *only* when the fluid is frictionless
(*b*) *only* when the fluid is frictionless and incompressible
(*c*) *only* when the fluid has zero viscosity and is at rest
(*d*) when there is no motion of one fluid layer relative to an adjacent layer
(*e*) regardless of the motion of one fluid layer relative to an adjacent layer

2.141   The pressure in the air space above an oil (sp gr 0.75) surface in a tank is 2 psi. The pressure 5.0 ft below the surface of the oil, in feet of water, is

(*a*) 7.0   (*b*) 8.37   (*c*) 9.62   (*d*) 11.16   (*e*) none of these answers

2.142   The pressure, in centimeters of mercury gage, equivalent to 8 cm H₂O plus 6 cm manometer fluid, sp gr 2.94, is

(*a*) 1.03   (*b*) 1.88   (*c*) 2.04   (*d*) 3.06   (*e*) none of these answers

2.143   The differential equation for pressure variation in a static fluid may be written (y measured vertically upward)

(*a*) $dp = -\gamma\,dy$ (*b*) $d\rho = -\gamma\,dy$ (*c*) $dy = -\rho\,dp$ (*d*) $dp = -\rho\,dy$
(*e*) $dp = -y\,d\rho$

2.144   In an isothermal atmosphere, the pressure

(*a*) remains constant   (*b*) decreases linearly with elevation
(*c*) increases exponentially with elevation   (*d*) varies in the same way as the density
(*e*) and density remain constant

2.145   Select the correct statement.

(*a*) Local atmospheric pressure is always below standard atmospheric pressure.
(*b*) Local atmospheric pressure depends upon elevation of locality only.
(*c*) Standard atmospheric pressure is the mean local atmospheric pressure at sea level.
(*d*) A barometer reads the difference between local and standard atmospheric pressure.
(*e*) Standard atmospheric pressure is 34 in Hg abs.

2.146   Select the three pressures that are equivalent.

(*a*) 10.0 psi, 23.1 ft H_2O, 4.91 in Hg   (*b*) 10.0 psi, 4.33 ft H_2O, 20.3 in Hg
(*c*) 10.0 psi, 20.3 ft H_2O, 23.1 in Hg   (*d*) 4.33 psi, 10.0 ft H_2O, 20.3 in Hg
(*e*) 4.33 psi, 10.0 ft H_2O, 8.83 in Hg

2.147   When the barometer reads 730 mm Hg, 10 kPa suction is the same as

(*a*) -10.2 m H_2O (*b*) 0.075 m Hg   (*c*) 8.91 m H_2O abs   (*d*) 107 kPa abs
(*e*) none of these answers

2.148   With the barometer reading 29 in Hg, 7.0 psia is equivalent to

(*a*) 0.476 atm   (*b*) 0.493 atm   (*c*) 7.9 psi suction   (*d*) 7.7 psi
(*e*) 13.8 in Hg abs

2.149   In Fig. 2.6*b* the liquid is oil, sp gr 0.80. When $h = 2$ ft, the pressure at A may be expressed as

(*a*) -1.6 ft H_2O abs   (*b*) 1.6 ft H_2O (*c*) 1.6 ft H_2O suction
(*d*) 2.5 ft H_2O vacuum   (*e*) none of these answers

2.150   In Fig. 2.6*c* air is contained in the pipe, water is the manometer liquid, and $h_1 = 50$ cm, $h_2 = 20$ cm. The pressure at A is

(*a*)   10.14 m H_2O abs   (*b*) 0.2 m H_2O vacuum   (*c*) 0.2 m H_2O (*d*) 4901 Pa
(*e*) none of these answers

2.151   In Fig. 2.7*a*, $h_1 = 2.0$ ft, $h_2 = 1.0$ ft, $h_3 = 4.0$ ft, $S_1 = 0.80$, $S_2 = 0.65$, $S_3 = 1.0$. Then $h_B - h_A$ in feet of water is

(*a*) -3.05 (*b*) -1.75 (*c*) 3.05   (*d*) 6.25   (*e*) none of these answers

2.152   In Fig. 2.7*b*, $h_1 = 1.5$ ft, $h_2 = 1.0$ ft, $h_3 = 2.0$ ft, $S_1 = 1.0$, $S_2 = 3.0$, $S_3 = 1.0$.

Then $p_A - p_B$ in pounds per square inch is

(*a*) -1.08 (*b*) 1.52 (*c*) 8.08 (*d*) 218 (*e*) none of these answers

2.153   A mercury-water manometer has a gage difference of 50 cm (difference in elevation of menisci). The difference in pressure, measured in meters of water, is

(*a*) 0.5 (*b*) 6.3 (*c*) 6.8 (*d*) 7.3 (*e*) none of these answers

2.154   In the inclined manometer of Fig. 2.9 the reservoir is so large that its surface may be assumed to remain at a fixed elevation. $\theta = 30°$. Used as a simple manometer for measuring air pressure, it contains water, and $R = 1.2$ ft. The pressure at A, in inches of water, is

(*a*) 7.2 (*b*) 7.2 vacuum   (*c*) 12.5 (*d*) 14.4 (*e*) none of these answers

2.155   A closed cubical box, 1 m on each edge, is half filled with water, the other half being filled with oil, sp gr 0.75. When it is accelerated vertically upward 4.903 m/s², the pressure difference between bottom and top, in kilopascals, is

(*a*) 4.9 (*b*) 11 (*c*) 12.9 (*d*) 14.7 (*e*) none of these answers

2.156   When the box of Prob. 2.155 is accelerated uniformly in a horizontal direction parallel to one side, 16.1 ft/s², the slope of the interface is

(*a*) 0 (*b*) $-\frac{1}{4}$ (*c*) $-\frac{1}{2}$ (*d*) -1 (*e*) none of these answers

2.157   When the minimum pressure in the box of Prob. 2.156 is zero gage, the maximum pressure in meters of water is

(*a*) 0.94 (*b*) 1.125 (*c*) 1.31 (*d*) 1.5 (*e*) none of these answers

2.158   The magnitude of force on one side of a circular surface of unit area, with centroid 10 ft below a free water surface, is

(*a*) less than 10γ
(*b*) dependent upon orientation of the area
(*c*) greater than 10γ
(*d*) the product of γ and the vertical distance from free surface to pressure center
(*e*) none of the above

2.159   A rectangular surface 3 by 4 ft has the lower 3-ft edge horizontal and 6 ft below a free oil surface, sp gr 0.80. The surface is inclined 30° with the horizontal. The force on one side of the surface is

(*a*) 38.4γ (*b*) 48γ (*c*) 51.2γ (*d*) 60γ (*e*) none of these answers

2.160   The pressure center of the surface of Prob. 2.159 is vertically below the liquid surface

(*a*) 10.133 ft   (*b*) 5.133 ft   (*c*) 5.067 ft   (*d*) 5.00 ft   (*e*) none of these answers

2.161 The pressure center is

(*a*) at the centroid of the submerged area
(*b*) the centroid of the pressure prism
(*c*) independent of the orientation of the area
(*d*) a point on the line of action of the resultant force
(*e*) always above the centroid of the area

2.162 What is the force exerted on the vertical annular area enclosed by concentric circles of radii 1.0 and 2.0 m? The center is 3.0 m below a free water surface. γ = sp wt.

(*a*) $3\pi\gamma$ (*b*) $9\pi\gamma$ (*c*) $10.25\pi\gamma$ (*d*) $12\pi\gamma$ (*e*) none of these answers

2.163 The pressure center for the annular area of Prob. 2.162 is below the centroid of the area

(*a*) 0 m   (*b*) 0.42 m   (*c*) 0.44 m   (*d*) 0.47 m   (*e*) none of these answers

2.164 A vertical triangular area has one side in a free surface, with vertex downward. Its altitude is *h*. The pressure center is below the free surface

(*a*) $h/4$ (*b*) $h/3$ (*c*) $h/2$ (*d*) $2h/3$ (*e*) $3h/4$

2.165 A vertical gate 4 by 4 m holds water with free surface at its top. The moment about the bottom of the gate is

(*a*) 42.7γ (*b*) 57γ (*c*) 64γ (*d*) 85.3γ (*e*) none of these answers

2.166 The magnitude of the resultant force acting on both sides of the gate (Fig. 2.80) is

(*a*) 768γ (*b*) 1593γ (*c*) 1810γ (*d*) 3820γ (*e*) none of these answers

2.167 The line of action of the resultant force on both sides of the gate in Fig. 2.80 is above the bottom of the gate

(*a*) 2.67 ft   (*b*) 3.33 ft   (*c*) 3.68 ft   (*d*) 4.00 ft   (*e*) none of these answers

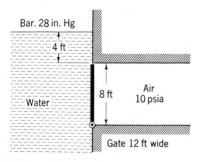

Fig. 2.80

2.168   The horizontal component of force on a curved surface is equal to the

(a) weight of liquid vertically above the curved surface
(b) weight of liquid retained by the curved surface
(c) product of pressure at its centroid and area
(d) force on a vertical projection of the curved surface
(e) scalar sum of all elemental horizontal components

2.169   A pipe 16 ft in diameter is to carry water at 200 psi. For an allowable tensile stress of 8000 psi, the thickness of pipe wall is

(a) 1.2 in   (b) 1.6 in   (c) 2.4 in   (d) 3.2 in   (e) none of these answers

2.170   The vertical component of pressure force on a submerged curved surface is equal to

(a) its horizontal component
(b) the force on a vertical projection of the curved surface
(c) the product of pressure at centroid and surface area
(d) the weight of liquid vertically above the curved surface
(e) none of the above answers

2.171   The vertical component of force on the quadrant of the cylinder AB (Fig. 2.81) is

(a) 224γ (b) 96.5γ (c) 81γ (d) 42.5γ (e) none of these answers

2.172   The vertical component of force on the upper half of a horizontal right-circular cylinder, 3 ft in diameter and 10 ft long, filled with water, and with a pressure of 0.433 psi at the axis, is

(a) -458 lb   (b) -331 lb   (c) 124.8 lb   (d) 1872 lb   (e) none of these answers

2.173   A cylindrical wooden barrel is held together by hoops at top and bottom. When the barrel is filled with liquid, the ratio of tension in the top hoop to tension in the bottom hoop, due to the liquid, is

(a) $\frac{1}{2}$ (b) 1   (c) 2   (d) 3   (e) none of these answers

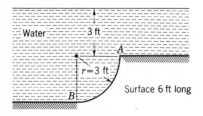

Fig. 2.81

2.174   A 5-cm-ID pipe with 5-mm wall thickness carries water at 20 kg$_f$/cm². The tensile stress in the pipe wall, in megapascals, is

(a) 4.9   (b) 9.8   (c) 19.6   (d) 39.2   (e) none of these answers

2.175   A slab of wood 4 by 4 by 1 ft, sp gr 0.50, floats in water with a 400-lb load on it. The volume of slab submerged, in cubic feet, is

(a) 1.6   (b) 6.4   (c) 8.0   (d) 14.4   (e) none of these answers

2.176   The line of action of the buoyant force acts through the

(a) center of gravity of any submerged body
(b) centroid of the volume of any floating body
(c) centroid of the displaced volume of fluid
(d) centroid of the volume of fluid vertically above the body
(e) centroid of the horizontal projection of the body

2.177   Buoyant force is

(a) the resultant force on a body due to the fluid surrounding it
(b) the resultant force acting on a floating body
(c) the force necessary to maintain equilibrium of a submerged body
(d) a nonvertical force for nonsymmetrical bodies
(e) equal to the volume of liquid displaced

2.178   A body floats in stable equilibrium

(a) when its metacentric height is zero
(b) *only* when its center of gravity is below its center of buoyancy
(c) when $\overline{GB} - I/\mho$ is positive and G is above B
(d) when $I/\mho$ is positive
(e) when the metacenter is above the center of gravity

2.179   A closed cubical metal box 3 ft on an edge is made of uniform sheet and weighs 1200 lb. Its metacentric height when placed in oil, sp gr 0.90, with sides vertical, is

(a) 0 ft   (b) −0.08 ft   (c) 0.62 ft   (d) 0.78 ft   (e) none of these answers

2.180   Liquid in a cylinder 10 m long is accelerated horizontally $20g$ m/s² along the axis of the cylinder. The difference in pressure intensities at the ends of the cylinder, in pascals, if γ = sp wt of liquid is

(a) 20γ (b) 200γ (c) $20g\gamma$ (d) $200\gamma/g$ (e) none of these answers

2.181   When a liquid rotates at constant angular velocity about a vertical axis as a rigid body, the pressure

(a) decreases as the square of the radial distance
(b) increases linearly as the radial distance
(c) decreases as the square of increase in elevation along any vertical line

(d) varies inversely as the elevation along any vertical line

(e) varies as the square of the radial distance

2.182 When a liquid rotates about a vertical axis as a rigid body so that points on the axis have the same pressure as points 2 ft higher and 2 ft from the axis, the angular velocity in radians per second is

(a) 8.02   (b) 11.34   (c) 64.4   (d) not determinable from data given

(e) none of these answers

2.183 A right-circular cylinder, open at the top, is filled with liquid, sp gr 1.2, and rotated about its vertical axis at such speed that half the liquid spills out. The pressure at the center of the bottom is

(a) zero   (b) one-fourth its value when cylinder was full

(c) indeterminable; insufficient data

(d) greater than a similar case with water as liquid

(e) none of these answers

2.184 A forced vortex

(a) turns in an opposite direction to a free vortex

(b) always occurs in conjunction with a free vortex

(c) has the velocity decreasing with the radius

(d) occurs when fluid rotates as a solid

(e) has the velocity decreasing inversely with the radius

3

FLUID-FLOW CONCEPTS AND
BASIC EQUATIONS

The statics of fluids, treated in the preceding chapter, is almost an exact science, specific weight (or density) being the only quantity that must be determined experimentally. On the other hand, the nature of *flow* of a real fluid is very complex. Since the basic laws describing the complete motion of a fluid are not easily formulated and handled mathematically, recourse to experimentation is required. By an analysis based on mechanics, thermodynamics, and orderly experimentation, large hydraulic structures and efficient fluid machines have been produced.

This chapter introduces the concepts needed for analysis of fluid motion. The basic equations that enable us to predict fluid behavior are stated or derived: these are equations of motion, continuity, and momentum, and the first and second laws of thermodynamics as applied to steady flow of a perfect gas. In this chapter the control-volume approach is utilized in the derivation of the continuity, energy, and momentum equations. Viscous effects, the experimental determination of losses, and the dimensionless presentation of loss data are presented in Chap. 5 after dimensional analysis has been introduced in Chap. 4. In general, one-dimensional-flow theory is developed in this chapter, with applications limited to incompressible cases where viscous effects do not predominate. Chapter 6 deals with compressible flow, and Chap. 7 with two- and three-dimensional flow.

3.1 THE CONCEPTS OF SYSTEM AND CONTROL VOLUME

The free-body diagram was used in Chap. 2 as a convenient way to show forces exerted on some arbitrary fixed mass. This is a special case of a *system*.

109

A system refers to a definite mass of material and distinguishes it from all other matter, called its *surroundings*. The boundaries of a system form a closed surface, and this surface may vary with time, so that it contains the same mass during changes in its condition; e.g., a slug of gas may be confined in a cylinder and be compressed by motion of a piston; the system boundary coinciding with the end of the piston then moves with the piston. The system may contain an infinitesimal mass or a large finite mass of fluids and solids at the will of the investigator.

The law of conservation of mass states that the mass within a system remains constant with time (disregarding relativity effects). In equation form

$$\frac{dm}{dt} = 0 \tag{3.1.1}$$

with m the total mass.

Newton's second law of motion is usually expressed for a system as

$$\Sigma \mathbf{F} = \frac{d}{dt}(m\mathbf{v}) \tag{3.1.2}$$

in which it must be remembered that m is the constant mass of the system. $\Sigma\mathbf{F}$ refers to the resultant of all external forces acting on the system, including body forces, such as gravity, and $\mathbf{v}$ is the velocity of the center of mass of the system.

A *control volume* refers to a region in space and is useful in the analysis of situations where flow occurs into and out of the space. The boundary of a control volume is its *control surface*. The size and shape of the control volume are entirely arbitrary, but frequently they are made to coincide with solid boundaries in parts, and in other parts they are drawn normal to the flow directions as a matter of simplification. By superposition of a uniform velocity on a system and its surroundings a convenient situation for application of the control volume may sometimes be found, e.g., determination of sound-wave velocity in a medium. The control-volume concept is used in the derivation of continuity, momentum, and energy equations, as well as in the solution of many types of problems. The control volume is also referred to as an *open system*.

Regardless of the nature of the flow, all flow situations are subject to the following relationships, which may be expressed in analytic form:

1. Newton's laws of motion, which must hold for every particle at every instant.

2. The continuity relationship, i.e., the law of conservation of mass.
3. The first and second laws of thermodynamics.
4. Boundary conditions, analytical statements that a real fluid has zero velocity relative to a boundary at a boundary or that frictionless fluids cannot penetrate a boundary.

Other relations and equations may enter, such as an equation of state or Newton's law of viscosity.

In the derivation that follows the control-volume concept is related to the system in terms of a general property of the system. It is then applied specifically to obtain continuity, energy, and linear-momentum relationships.

To formulate the relationship between equations applied to a system and those applied to a control volume, consider some general flow situation, Fig. 3.1, in which the velocity of a fluid is given relative to an xyz coordinate system. At time t consider a certain mass of fluid that is contained within a system, having the dotted-line boundaries indicated. Also consider a control

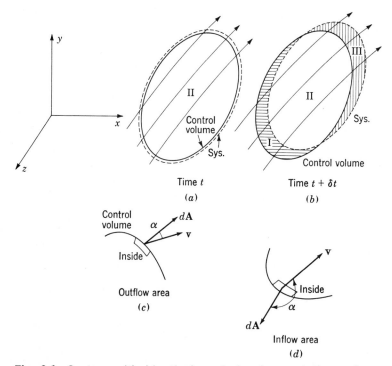

Fig. 3.1 System with identical control volume at time t in a velocity field.

volume, fixed relative to the xyz axes, that exactly coincides with the system at time t. At time $t + \delta t$ the system has moved somewhat since each mass particle moves at the velocity associated with its location.

Let N be the total amount of some property (mass, energy, momentum) within the system at time t, and let η be the amount of this property, per unit mass, throughout the fluid. The time rate of increase of N for the system is now formulated in terms of the control volume.

At $t + \delta t$, Fig. 3.1b, the system comprises volumes II and III, while at time t it occupies volume II, Fig. 3.1a. The increase in property N in the system in time δt is given by

$$N_{\mathrm{sys}_{t+\delta t}} - N_{\mathrm{sys}_t} = \left(\int_{\mathrm{II}} \eta\rho\, d\mathcal{U} + \int_{\mathrm{III}} \eta\rho\, d\mathcal{U} \right)_{t+\delta t} - \left(\int_{\mathrm{II}} \eta\rho\, d\mathcal{U} \right)_t$$

in which $d\mathcal{U}$ is the element of volume. Rearrangement, after adding and subtracting

$$\left(\int_{\mathrm{I}} \eta\rho\, d\mathcal{U} \right)_{t+\delta t}$$

to the right, then dividing through by δt leads to

$$\frac{N_{\mathrm{sys}_{t+\delta t}} - N_{\mathrm{sys}_t}}{\delta t} = \frac{\left(\int_{\mathrm{II}} \eta\rho\, d\mathcal{U} + \int_{\mathrm{I}} \eta\rho\, d\mathcal{U} \right)_{t+\delta t} - \left(\int_{\mathrm{II}} \eta\rho\, d\mathcal{U} \right)_t}{\delta t}$$

$$+ \frac{\left(\int_{\mathrm{III}} \eta\rho\, d\mathcal{U} \right)_{t+\delta t}}{\delta t} - \frac{\left(\int_{\mathrm{I}} \eta\rho\, d\mathcal{U} \right)_{t+\delta t}}{\delta t} \qquad (3.1.3)$$

The term on the left is the average time rate of increase of N within the system during time δt. In the limit as δt approaches zero, it becomes dN/dt. If the limit is taken as δt approaches zero for the first term on the right-hand side of the equation, the first two integrals are the amount of N in the control volume at $t + \delta t$ and the third integral is the amount of N in the control volume at time t. The limit is

$$\frac{\partial}{\partial t} \int_{cv} \eta\rho\, d\mathcal{U}$$

the partial being needed as the volume is held constant (the control volume) as $\delta t \to 0$.

The next term, which is the time rate of flow of N out of the control volume, in the limit, may be written

$$\lim_{\delta t \to 0} \frac{\left(\int_{\mathrm{III}} \eta\rho\, d\mathcal{U} \right)_{t+\delta t}}{\delta t} = \int_{\substack{\mathrm{Outflow\ area}}} \eta\rho\mathbf{V} \cdot d\mathbf{A} = \int \eta\rho v \cos \alpha\, dA \qquad (3.1.4)$$

in which $d\mathbf{A}$, Fig. 3.1c, is the vector representing an area element of the outflow area. It has a direction normal to the surface-area element of the control volume, positive outward; α is the angle between the velocity vector and the elemental area vector.

Similarly, the last term of Eq. (3.1.3), which is the rate of flow of N into the control volume, is, in the limit,

$$\lim_{\delta t \to 0} \frac{(\int_I \eta\rho \, d\mathcal{U})_{t+\delta t}}{\delta t} = -\int_{\text{Inflow area}} \eta\rho\mathbf{V} \cdot d\mathbf{A} = -\int \eta\rho \cos \alpha \, dA \qquad (3.1.5)$$

The minus sign is needed as $\mathbf{V} \cdot d\mathbf{A}$ (or $\cos \alpha$) is negative for inflow, Fig. 3.1d. The last two terms of Eq. (3.1.3), given by Eqs. (3.1.4) and (3.1.5), may be combined into the single term which is an integral over the complete control-volume surface (cs)

$$\lim_{\delta t \to 0} \left(\frac{(\int_{III} \eta\rho \, d\mathcal{U})_{t+\delta t}}{\delta t} - \frac{(\int_I \eta\rho \, d\mathcal{U})_{t+\delta t}}{\delta t} \right) = \int_{cs} \eta\rho\mathbf{V} \cdot d\mathbf{A} = \int_{cs} \eta\rho v \cos \alpha \, dA$$

Where there is no inflow or outflow, $\mathbf{V} \cdot d\mathbf{A} = 0$; hence the equation can be evaluated over the whole control surface.[1]

Collecting the reorganized terms of Eq. (3.1.3) gives

$$\frac{dN}{dt} = \frac{\partial}{\partial t} \int_{cv} \eta\rho \, d\mathcal{U} + \int_{cs} \eta\rho\mathbf{V} \cdot d\mathbf{A} \qquad (3.1.6)$$

In words, this equation states that the time rate of increase of N within a system is just equal to the time rate of increase of the property N within the control volume (fixed relative to xyz) plus the net rate of efflux of N across the control-volume boundary.

Equation (3.1.6) is used throughout this chapter in converting laws and principles from the system form to the control-volume form. The system form, which in effect follows the motion of the particles, is referred to as the Lagrangian method of analysis; the control-volume approach is called the Eulerian method of analysis, as it observes flow from a reference system fixed relative to the control volume.

Since the xyz frame of reference may be given an arbitrary constant velocity without affecting the dynamics of the system and its surroundings, Eq. (3.1.6) is valid if the control volume, fixed in size and shape, has a uniform velocity of translation.

---

[1] This derivation was developed by Professor William Mirsky of the Department of Mechanical Engineering, The University of Michigan.

3.2 APPLICATION OF THE CONTROL VOLUME TO CONTINUITY, ENERGY, AND MOMENTUM

In this section the general relation of system and control volume to a property, developed in Sec. 3.1, is applied first to continuity, then to energy, and finally to linear momentum. In the following sections the uses of equations are brought out and illustrated.

Continuity

The continuity equations are developed from the general principle of conservation of mass, Eq. (3.1.1), which states that the mass within a system remains constant with time; i.e.,

$$\frac{dm}{dt} = 0$$

In Eq. (3.1.6) let N be the mass of the system m. Then η is the mass per unit mass, or $\eta = 1$

$$0 = \frac{\partial}{\partial t} \int_{cv} \rho \, d\upsilon + \int_{cs} \rho \mathbf{v} \cdot d\mathbf{A} \qquad (3.2.1)$$

In words the continuity equation for a control volume states that the time rate of increase of mass within a control volume is just equal to the net rate of mass inflow to the control volume. This equation is examined further in Sec. 3.4.

Energy equation

The first law of thermodynamics for a system states that the heat Q_H added to a system minus the work W done by the system depends only upon the initial and final states of the system. The difference in states of the system, being independent of the path from initial to final state, must be a property of the system. It is called the internal energy E. The first law in equation form is

$$Q_H - W = E_2 - E_1 \qquad (3.2.2)$$

The internal energy per unit mass is called e; hence, applying Eq. (3.1.6),

$N = E$ and $\eta = \rho e / \rho$,

$$\frac{dE}{dt} = \frac{\partial}{\partial t} \int_{cv} \rho e \, d\mathcal{V} + \int_{cs} \rho e \mathbf{v} \cdot d\mathbf{A} \tag{3.2.3}$$

or by use of Eq. (3.2.2)

$$\frac{\delta Q_H}{\delta t} - \frac{\delta W}{\delta t} = \frac{dE}{dt} = \frac{\partial}{\partial t} \int_{cv} \rho e \, d\mathcal{V} + \int_{cs} \rho e \mathbf{v} \cdot d\mathbf{A} \tag{3.2.4}$$

The work done by the system on its surroundings may be broken into two parts: the work W_{pr} done by pressure forces on the moving boundaries, and the work W_S done by shear forces such as the torque exerted on a rotating shaft. The work done by pressure forces in time δt is

$$\delta W_{pr} = \delta t \int p \mathbf{v} \cdot d\mathbf{A} \tag{3.2.5}$$

By use of the definitions of the work terms, Eq. (3.2.4) becomes

$$\frac{\delta Q_H}{\delta t} - \frac{\delta W_S}{\delta t} = \frac{\partial}{\partial t} \int_{cv} \rho e \, d\mathcal{V} + \int_{cs} \left(\frac{p}{\rho} + e \right) \rho \mathbf{v} \cdot d\mathbf{A} \tag{3.2.6}$$

In the absence of nuclear, electrical, magnetic, and surface-tension effects, the internal energy e of a pure substance is the sum of potential, kinetic, and "intrinsic" energies. The intrinsic energy u per unit mass is due to molecular spacing and forces (dependent upon p, ρ, or T):

$$e = gz + \frac{v^2}{2} + u \tag{3.2.7}$$

Linear momentum equation

Newton's second law for a system, Eq. (3.1.2), is used as the basis for finding the linear-momentum equation for a control volume by use of Eq. (3.1.6). Let N be the linear momentum $m\mathbf{v}$ of the system and let η be the linear momentum per unit mass $\rho\mathbf{v}/\rho$. Then by use of Eqs. (3.1.2) and (3.1.6)

$$\Sigma \mathbf{F} = \frac{d(m\mathbf{v})}{dt} = \frac{\partial}{\partial t} \int_{cv} \rho \mathbf{v} \, d\mathcal{V} + \int_{cs} \rho \mathbf{v} \mathbf{v} \cdot d\mathbf{A} \tag{3.2.8}$$

In words, the resultant force acting on a control volume is equal to the time rate of increase of linear momentum within the control volume plus the net efflux of linear momentum from the control volume.

Equations (3.2.1), (3.2.6), and (3.2.8) provide the relationships for analysis of many of the problems of fluid mechanics. In effect, they provide a bridge from the solid-dynamics relations of the system to the convenient control-volume relations of fluid flow.

Flow characteristics and definitions are next discussed, before the basic control-volume equations are examined and applied.

3.3 FLOW CHARACTERISTICS; DEFINITIONS

Flow may be classified in many ways, such as turbulent, laminar; real, ideal; reversible, irreversible; steady, unsteady; uniform, nonuniform; rotational, irrotational. In this and the following section various types of flow are distinguished.

Turbulent-flow situations are most prevalent in engineering practice. In turbulent flow the fluid particles (small molar masses) move in very irregular paths, causing an exchange of momentum from one portion of the fluid to another in a manner somewhat similar to the molecular momentum transfer described in Sec. 1.3 but on a much larger scale. The fluid particles can range in size from very small (say a few thousand molecules) to very large (thousands of cubic feet in a large swirl in a river or in an atmospheric gust). In a situation in which the flow could be either turbulent or nonturbulent (laminar), the turbulence sets up greater shear stresses throughout the fluid and causes more irreversibilities or losses. Also, in turbulent flow, the losses vary as the square of the velocity, while in laminar flow, they vary as the first power of the velocity.

In *laminar* flow, fluid particles move along smooth paths in laminas, or layers, with one layer gliding smoothly over an adjacent layer. Laminar flow is governed by Newton's law of viscosity [Eq. (1.1.1) or extensions of it to three-dimensional flow], which relates shear stress to rate of angular deformation. In laminar flow, the action of viscosity damps out turbulent tendencies (see Sec. 5.3 for criteria for laminar flow). Laminar flow is not stable in situations involving combinations of low viscosity, high velocity, or large flow passages and breaks down into turbulent flow. An equation similar in form to Newton's law of viscosity may be written for turbulent flow:

$$\tau = \eta \frac{du}{dy} \tag{3.3.1}$$

The factor η, however, is not a fluid property alone but depends upon the fluid motion and the density. It is called the *eddy viscosity*.

In many practical flow situations, both viscosity and turbulence contribute to the shear stress:

$$\tau = (\mu + \eta) \frac{du}{dy} \tag{3.3.2}$$

Experimentation is required to determine this type of flow.

An *ideal fluid* is frictionless and incompressible and should not be confused with a perfect gas (Sec. 1.6). The assumption of an ideal fluid is helpful in analyzing flow situations involving large expanses of fluids, as in the motion of an airplane or a submarine. A frictionless fluid is nonviscous, and its flow processes are reversible.

The layer of fluid in the immediate neighborhood of an actual flow boundary that has had its velocity relative to the boundary affected by viscous shear is called the *boundary layer*. Boundary layers may be laminar or turbulent, depending generally upon their length, the viscosity, the velocity of the flow near them, and the boundary roughness.

Adiabatic flow is that flow of a fluid in which no heat is transferred to or from the fluid. *Reversible adiabatic* (frictionless adiabatic) flow is called *isentropic* flow.[1] To proceed in an orderly manner into the analysis of fluid flow requires a clear understanding of the terminology involved. Several of the more important technical terms are defined and illustrated in this section.

Steady flow occurs when conditions at any point in the fluid do not change with the time. For example, if the velocity at a certain point is 10 ft/s in the $+x$ direction in steady flow, it remains exactly that amount and in that direction indefinitely. This can be expressed as $\partial \mathbf{v}/\partial t = 0$, in which space ($x$, y, z coordinates of the point) is held constant. Likewise, in steady flow there is no change in density ρ, pressure p, or temperature T, with time at any point; thus

$$\frac{\partial \rho}{\partial t} = 0 \qquad \frac{\partial p}{\partial t} = 0 \qquad \frac{\partial T}{\partial t} = 0$$

In turbulent flow, owing to the erratic motion of the fluid particles, there are always small fluctuations occurring at any point. The definition for steady flow must be generalized somewhat to provide for these fluctuations. To illustrate this, a plot of velocity against time, at some point in turbulent

---

[1] An isentropic process, however, can occur in irreversible flow with the proper amount of heat transfer (isentropic = constant entropy).

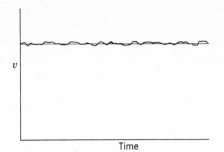

Fig. 3.2 Velocity at a point in steady turbulent flow.

flow, is given in Fig. 3.2. When the temporal mean velocity

$$v_t = \frac{1}{t} \int_0^t v \, dt$$

indicated in the figure by the horizontal line, does not change with the time, the flow is said to be steady. The same generalization applies to density, pressure, temperature, etc., when they are substituted for v in the above formula.

The flow is *unsteady* when conditions at any point change with the time, $\partial \mathbf{v}/\partial t \neq 0$. Water being pumped through a fixed system at a constant rate is an example of steady flow. Water being pumped through a fixed system at an increasing rate is an example of unsteady flow.

Uniform flow occurs when at every point the velocity vector is identical (in magnitude and direction) for any given instant, or, in equation form, $\partial \mathbf{v}/\partial s = 0$, in which time is held constant and δs is a displacement in any direction. The equation states that there is no change in the velocity vector in any direction throughout the fluid at any one instant. It says nothing about the change in velocity at a point with time.

In flow of a real fluid in an open or closed conduit, the definition of uniform flow may also be extended in most cases even though the velocity vector at the boundary is always zero. When all parallel cross sections through the conduit are identical (i.e., when the conduit is prismatic) and the average velocity at each cross section is the same at any given instant, the flow is said to be *uniform*.

Flow such that the velocity vector varies from place to place at any instant ($\partial \mathbf{v}/\partial s \neq 0$) is *nonuniform* flow. A liquid being pumped through a long straight pipe has uniform flow. A liquid flowing through a reducing section or through a curved pipe has nonuniform flow.

Examples of steady and unsteady flow and of uniform and nonuniform flow are: liquid flow through a long pipe at a constant rate is *steady uniform* flow; liquid flow through a long pipe at a decreasing rate is *unsteady uniform* flow; flow through an expanding tube at a constant rate is *steady nonuniform* flow; and flow through an expanding tube at an increasing rate is *unsteady nonuniform* flow.

Rotation of a fluid particle about a given axis, say the z axis, is defined as the average angular velocity of two infinitesimal line elements in the particle that are at right angles to each other and to the given axis. If the fluid particles within a region have rotation about any axis, the flow is called *rotational flow*, or *vortex flow*. If the fluid within a region has no rotation, the flow is called *irrotational flow*. It is shown in texts on hydrodynamics that if a fluid is at rest and is frictionless, any later motion of this fluid will be irrotational.

One-dimensional flow neglects variations or changes in velocity, pressure, etc., transverse to the main flow direction. Conditions at a cross section are expressed in terms of average values of velocity, density, and other properties. Flow through a pipe, for example, may usually be characterized as one-dimensional. Many practical problems can be handled by this method of analysis, which is much simpler than two- and three-dimensional methods of analysis. In *two-dimensional* flow all particles are assumed to flow in parallel planes along identical paths in each of these planes; hence, there are no changes in flow normal to these planes. The flow net, developed in Chap. 7, is the most useful method for analysis of two-dimensional-flow situations. *Three-dimensional* flow is the most general flow in which the velocity components u, v, w in mutually perpendicular directions are functions of space coordinates and time x, y, z, and t. Methods of analysis are generally complex mathematically, and only simple geometrical flow boundaries can be handled.

A *streamline* is a continuous line drawn through the fluid so that it has the direction of the velocity vector at every point. There can be no flow across a streamline. Since a particle moves in the direction of the streamline at any instant, its displacement $\delta \mathbf{s}$, having components δx, δy, δz, has the direction of the velocity vector $\mathbf{q}$ with components u, v, w in the x, y, z directions, respectively. Then

$$\frac{\delta x}{u} = \frac{\delta y}{v} = \frac{\delta z}{w}$$

states that the corresponding components are proportional and hence that $\delta \mathbf{s}$ and $\mathbf{q}$ have the same direction. Expressing the displacements in differential form

$$\frac{dx}{u} = \frac{dy}{v} = \frac{dz}{w} \tag{3.3.3}$$

produces the differential equations of a streamline. Equations (3.3.3) are two independent equations. Any continuous line that satisfies them is a streamline.

In steady flow, since there is no change in direction of the velocity vector at any point, the streamline has a fixed inclination at every point and is, therefore, *fixed in space*. A particle always moves tangent to the streamline; hence, in steady flow the *path of a particle* is a streamline. In unsteady flow, since the direction of the velocity vector at any point may change with time, a streamline may shift in space from instant to instant. A particle then follows one streamline one instant, another one the next instant, and so on, so that the path of the particle may have no resemblance to any given instantaneous streamline.

A dye or smoke is frequently injected into a fluid in order to trace its subsequent motion. The resulting dye or smoke trails are called *streak* lines. In steady flow a streak line is a streamline and the path of a particle.

Streamlines in two-dimensional flow can be obtained by inserting fine, bright particles (aluminum dust) into the fluid, brilliantly lighting one plane, and taking a photograph of the streaks made in a short time interval. Tracing on the picture continuous lines that have the direction of the streaks at every point portrays the streamlines for either steady or unsteady flow.

In illustration of an incompressible two-dimensional flow, as in Fig. 3.3, the streamlines are drawn so that per unit time the volume flowing between adjacent streamlines is the same if unit depth is considered normal to the plane of the figure. Hence, when the streamlines are closer together, the velocity must be greater, and vice versa. If v is the average velocity between two adjacent streamlines at some position where they are h apart, the flow rate Δq is

$$\Delta q = vh \tag{3.3.4}$$

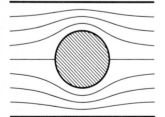

Fig. 3.3 Streamlines for steady flow around a cylinder between parallel walls.

At any other position on the chart where the distance between streamlines is h_1, the average velocity is $v_1 = \Delta q/h_1$. By increasing the number of streamlines drawn, i.e., by decreasing Δq, in the limiting case the velocity at a point is obtained.

A *stream tube* is the tube made by all the streamlines passing through a small, closed curve. In steady flow it is fixed in space and can have no flow through its walls because the velocity vector has no component normal to the tube surface.

EXAMPLE 3.1 In two-dimensional, incompressible steady flow around an airfoil the streamlines are drawn so that they are 1 cm apart at a great distance from the airfoil, where the velocity is 40 m/s. What is the velocity near the airfoil, where the streamlines are 0.75 cm apart?

$$(40 \text{ m/s})\,(0.01 \text{ m})\,(1 \text{ m}) = 0.40 \text{ m}^3/\text{s} = v\,0.0075 \text{ m}^2$$

and

$$v = \frac{0.40 \text{ m}^3/\text{s}}{0.0075 \text{ m}^2} = 53.3 \text{ m/s}$$

3.4 CONTINUITY EQUATION

The use of Eq. (3.2.1) is developed in this section. First, consider steady flow through a portion of the stream tube of Fig. 3.4. The control volume comprises the walls of the stream tube between sections 1 and 2, plus the end

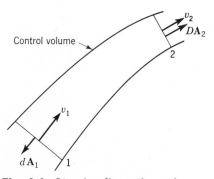

Fig. 3.4 Steady flow through a stream tube.

areas of sections 1 and 2. Because the flow is steady, the first term of Eq. (3.2.1) is zero; hence

$$\int_{cs} \rho \mathbf{v} \cdot d\mathbf{A} = 0 \tag{3.4.1}$$

which states that the net mass outflow from the control volume must be zero. At section 1 the net mass outflow is $\rho_1 \mathbf{v}_1 \cdot d\mathbf{A}_1 = -\rho_1 v_1\, dA_1$, and at section 2 it is $\rho_2 \mathbf{v}_2 \cdot d\mathbf{A}_2 = \rho_2 v_2\, dA_2$. Since there is no flow through the wall of the stream tube,

$$\rho_1 v_1\, dA_1 = \rho_2 v_2\, dA_2 \tag{3.4.2}$$

is the continuity equation applied to two sections along a stream tube in steady flow.

For a collection of stream tubes, as in Fig. 3.5, if ρ_1 is the average density at section 1 and ρ_2 the average density at section 2, then

$$\dot{m} = \rho_1 V_1 A_1 = \rho_2 V_2 A_2 \tag{3.4.3}$$

in which V_1, V_2 represent average velocities over the cross sections and $\dot{m}$ is the rate of mass flow. The average velocity over a cross section is given by

$$V = \frac{1}{A} \int v\, dA$$

If the discharge Q is defined as

$$Q = AV \tag{3.4.4}$$

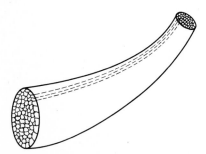

Fig. 3.5 Collection of stream tubes between fixed boundaries.

the continuity equation may take the form

$$\dot{m} = \rho_1 Q_1 = \rho_2 Q_2 \tag{3.4.5}$$

For incompressible, steady flow

$$Q = A_1 V_1 = A_2 V_2 \tag{3.4.6}$$

is a useful form of the equation.

For constant-density flow, steady or unsteady, Eq. (3.2.1) becomes

$$\int_{cs} \mathbf{v} \cdot d\mathbf{A} = 0 \tag{3.4.7}$$

which states that the net volume efflux is zero (this implies that the control volume is filled with liquid at all times).

EXAMPLE 3.2 At section 1 of a pipe system carrying water (Fig. 3.6) the velocity is 3.0 ft/s, and the diameter is 2.0 ft. At section 2 the diameter is 3.0 ft. Find the discharge and the velocity at section 2.

From Eq. (3.4.6)

$$Q = V_1 A_1 = 3.0\pi = 9.42 \text{ ft}^3/\text{s}$$

and

$$V_2 = \frac{Q}{A_2} = \frac{9.42}{2.25\pi} = 1.33 \text{ ft/s}$$

For two- and three-dimensional-flow studies, differential expressions of

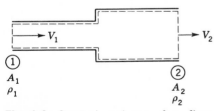

Fig. 3.6 Control volume for flow through series pipes.

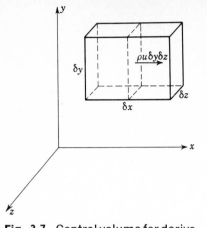

Fig. 3.7 Control volume for derivation of three-dimensional continuity equation in cartesian coordinates.

the continuity equation must be used. For three-dimensional cartesian coordinates, Eq. (3.2.1) is applied to the control-volume element $\delta x \, \delta y \, \delta z$ of Fig. 3.7 with center at (x,y,z), where the velocity components in the x, y, z directions are u, v, w, respectively, and ρ is the density. Consider first the flux through the pair of faces normal to the x direction. On the right-hand face the flux outward is

$$\left[\rho u + \frac{\partial}{\partial x} (\rho u) \frac{\delta x}{2} \right] \delta y \, \delta z$$

since both ρ and u are assumed to vary continuously throughout the fluid. In the expression, $\rho u \, \delta y \, \delta z$ is the mass flux through the center face normal to the x axis. The second term is the rate of increase of mass flux with respect to x, multiplied by the distance $\delta x/2$ to the right-hand face. Similarly on the left-hand face the flux into the volume is

$$\left[\rho u - \frac{\partial}{\partial x} (\rho u) \frac{\delta x}{2} \right] \delta y \, \delta z$$

since the step is $-\delta x/2$. The net flux out through these two faces is

$$\frac{\partial}{\partial x} (\rho u) \, \delta x \, \delta y \, \delta z$$

The other two directions yield similar expressions; hence the net mass outflow is

$$\left[\frac{\partial}{\partial x}(\rho u) + \frac{\partial}{\partial y}(\rho v) + \frac{\partial}{\partial z}(\rho w)\right] \delta x \, \delta y \, \delta z$$

which takes the place of the right-hand part of Eq. (3.2.1). The left-hand part of Eq. (3.2.1) becomes, for an element,

$$-\frac{\partial \rho}{\partial t} \delta x \, \delta y \, \delta z$$

When these two expressions are used in Eq. (3.2.1), after dividing through by the volume element and taking the limit as $\delta x \, \delta y \, \delta z$ approaches zero, the continuity equation at a point becomes

$$\frac{\partial}{\partial x}(\rho u) + \frac{\partial}{\partial y}(\rho v) + \frac{\partial}{\partial z}(\rho w) = -\frac{\partial \rho}{\partial t} \tag{3.4.8}$$

which must hold for every point in the flow, steady or unsteady, compressible or incompressible.[1] For incompressible flow, however, it simplifies to

$$\frac{\partial u}{\partial x} + \frac{\partial v}{\partial y} + \frac{\partial w}{\partial z} = 0 \tag{3.4.9}$$

Equations (3.4.8) and (3.4.9) may be compactly written in vector notation. By using fixed unit vectors in x, y, z directions, **i**, **j**, **k**, respectively, the operator $\boldsymbol{\nabla}$ is defined as

$$\boldsymbol{\nabla} = \mathbf{i}\frac{\partial}{\partial x} + \mathbf{j}\frac{\partial}{\partial y} + \mathbf{k}\frac{\partial}{\partial z} \tag{3.4.10}$$

and the velocity vector **q** is given by

$$\mathbf{q} = \mathbf{i}u + \mathbf{j}v + \mathbf{k}w \tag{3.4.11}$$

[1] Equation (3.4.8) can be derived from Eq. (3.2.1) by application of Gauss' theorem. See L. Page, "Introduction to Theoretical Physics," 2d ed., pp. 32–36, Van Nostrand, Princeton, N.J., 1935.

Then

$$\nabla \cdot (\rho\mathbf{q}) = \left(\mathbf{i}\frac{\partial}{\partial x} + \mathbf{j}\frac{\partial}{\partial y} + \mathbf{k}\frac{\partial}{\partial z}\right) \cdot (\mathbf{i}\rho u + \mathbf{j}\rho v + \mathbf{k}\rho w)$$

$$= \frac{\partial}{\partial x}(\rho u) + \frac{\partial}{\partial y}(\rho v) + \frac{\partial}{\partial z}(\rho w)$$

because $\mathbf{i} \cdot \mathbf{i} = 1$, $\mathbf{i} \cdot \mathbf{j} = 0$, etc. Equation (3.4.8) becomes

$$\nabla \cdot \rho\mathbf{q} = -\frac{\partial\rho}{\partial t} \qquad (3.4.12)$$

and Eq. (3.4.9) becomes

$$\nabla \cdot \mathbf{q} = 0 \qquad (3.4.13)$$

The dot product $\nabla \cdot \mathbf{q}$ is called the *divergence* of the velocity vector $\mathbf{q}$. In words it is the net volume efflux per unit volume at a point and must be zero for incompressible flow. See Sec. 7.2 for further discussion of the operator ∇.

In two-dimensional flow, generally assumed to be in planes parallel to the xy plane, $w = 0$, and there is no change with respect to z, so $\partial/\partial z = 0$, which reduces the three-dimensional equations given for continuity.

EXAMPLE 3.3 The velocity distribution for a two-dimensional incompressible flow is given by

$$u = -\frac{x}{x^2 + y^2} \qquad v = -\frac{y}{x^2 + y^2}$$

Show that it satisfies continuity.

In two dimensions the continuity equation is, from Eq. (3.4.9),

$$\frac{\partial u}{\partial x} + \frac{\partial v}{\partial y} = 0$$

Then

$$\frac{\partial u}{\partial x} = -\frac{1}{x^2 + y^2} + \frac{2x^2}{(x^2 + y^2)^2} \qquad \frac{\partial v}{\partial y} = -\frac{1}{x^2 + y^2} + \frac{2y^2}{(x^2 + y^2)^2}$$

and their sum does equal zero, satisfying continuity.

3.5 EULER'S EQUATION OF MOTION ALONG A STREAMLINE

In addition to the continuity equation, other general controlling equations are Euler's equation, Bernoulli's equation, the energy equation, the momentum equations, and the first and second laws of thermodynamics. In this section Euler's equation is derived in differential form. In Sec. 3.9 it is integrated to obtain Bernoulli's equation. The first law of thermodynamics is then developed for steady flow, and some of the interrelations of the equations are explored, including an introduction to the second law of thermodynamics. In Chap. 7 Euler's equation is derived for general three-dimensional flow. Here it is restricted to flow along a streamline.

In Fig. 3.8 a prismatic-shaped fluid particle of mass $\rho \, \delta A \, \delta s$ is moving along a streamline in the $+s$ direction. To simplify the development of the equation of motion for this particle it is assumed that the viscosity is zero or that the fluid is frictionless. This eliminates all shear forces from consideration, leaving as forces to take into consideration the body force due to the pull of gravity and surface forces on the end areas of the particle. The gravity force is $\rho g \, \delta A \, \delta s$. On the upstream face the pressure force is $p \, \delta A$ in the $+s$ direction; on the downstream face it is $[p + (\partial p/\partial s) \, \delta s] \, \delta A$ and acts in the $-s$ direction. Any forces on the sides of the element are normal to s and do not enter the equation. The body-force component in the s direction is $-\rho g \, \delta A \, \delta s \cos \theta$. Substituting into Newton's second law of motion, $\Sigma f_s = \delta m \, a_s$, gives

$$p \, \delta A - \left(p + \frac{\partial p}{\partial s} \, \delta s\right) \delta A - \rho g \, \delta A \, \delta s \cos \theta = \rho \, \delta A \, \delta s \, a_s$$

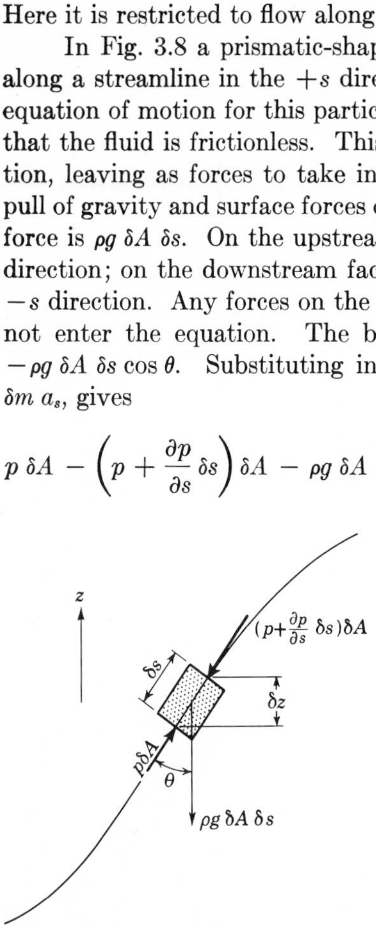

Fig. 3.8 Force components on a fluid particle in the direction of the streamline.

a_s is the acceleration of the fluid particle along the streamline. Dividing through by the mass of the particle, $\rho \, \delta A \, \delta s$, and simplifying leads to

$$\frac{1}{\rho} \frac{\partial p}{\partial s} + g \cos \theta + a_s = 0 \qquad (3.5.1)$$

δz is the increase in elevation of the particle for a displacement δs. From Fig. 3.8,

$$\frac{\delta z}{\delta s} = \cos \theta = \frac{\partial z}{\partial s}$$

The acceleration a_s is dv/dt. In general, if v depends upon s and time t, $v = v(s,t)$,

$$dv = \frac{\partial v}{\partial s} \, ds + \frac{\partial v}{\partial t} \, dt$$

s becomes a function of t in describing the motion of a particle, so one may divide by dt, obtaining

$$a_s = \frac{dv}{dt} = \frac{\partial v}{\partial s} \frac{ds}{dt} + \frac{\partial v}{\partial t} \qquad (3.5.2)$$

By substituting for $\cos \theta$ and a_s in Eq. (3.5.1)

$$\frac{1}{\rho} \frac{\partial p}{\partial s} + g \frac{\partial z}{\partial s} + v \frac{\partial v}{\partial s} + \frac{\partial v}{\partial t} = 0 \qquad (3.5.3)$$

If the flow is steady, $\partial v/\partial t = 0$, yielding

$$\frac{1}{\rho} \frac{\partial p}{\partial s} + g \frac{\partial z}{\partial s} + v \frac{\partial v}{\partial s} = 0 \qquad (3.5.4)$$

Since p, z, and v now are functions of s only, the partials may be replaced by total derivatives:

$$\frac{dp}{\rho} + g \, dz + v \, dv = 0 \qquad (3.5.5)$$

This is one form of Euler's equation of motion and requires three important assumptions: (1) motion along a streamline, (2) frictionless fluid, and (3) steady flow. It can be integrated if ρ is known as a function of p or is constant.

3.6 REVERSIBILITY, IRREVERSIBILITY, AND LOSSES

A process may be defined as the path of the succession of states through which the system passes, such as the changes in velocity, elevation, pressure, density, temperature, etc. The expansion of air in a cylinder as the piston moves out and heat is transferred through the walls is an example of a process. Normally, the process causes some change in the surroundings, e.g., displacing it or transferring heat to or from its boundaries. When a process can be made to take place in such a manner that it can be *reversed*, i.e., made to return to its original state without a final change in either the system or its surroundings, it is said to be *reversible*. In any actual flow of a real fluid or change in a mechanical system, the effects of viscous friction, Coulomb friction, unrestrained expansion, hysteresis, etc., prohibit the process from being reversible. It is, however, an ideal to be strived for in design processes, and their efficiency is usually defined in terms of their nearness to reversibility.

When a certain process has a sole effect upon its surroundings that is equivalent to the raising of a weight, it is said to have done *work* on its surroundings. Any actual process is *irreversible*. The difference between the amount of work a substance can do by changing from one state to another state along a path reversibly and the actual work it produces for the same path is the *irreversibility* of the process. It may be defined in terms of work per unit mass or weight or work per unit time. Under certain conditions the irreversibility of a process is referred to as its *lost work*,[1] i.e., the loss of ability to do work because of friction and other causes. In this book when *losses* are referred to, they mean irreversibility or lost work and do not mean an actual loss of energy.

EXAMPLE 3.4 A hydroelectric plant has a head (difference in elevation of headwater and tail water) of 50 m and a flow of 5 m³/s of water through the turbine, which rotates at 180 rpm. The torque in the turbine shaft is measured to be 1.16×10^5 N·m, and the output of the generator is 2100 kW. Determine the irreversibility, or losses, and the reversible power for the system. $g = 9.8$ m/s².

The potential energy of the water is 50 m·N/N. Hence for perfect conversion the reversible power is $\gamma QH = 9802$ N/m³ $\times$ 5 m³/s $\times$ 50 m·N/N = 2,450,500 N·m/s = 2450.5 kW. The rate of work by the turbine is

$$T\omega = (1.16 \times 10^5 \text{ N·m}) \left(\tfrac{180}{60}\right) (2\pi \text{ rad/s}) = 2186.5 \text{ kW}$$

The irreversibility through the turbine is then 2450.5 kW − 2186.5 kW =

---

[1] Reference to a text on thermodynamics is advised for a full discussion of these concepts.

264 kW or

$$\frac{264 \text{ kW}}{5 \text{ m}^3/\text{s}} \frac{1000 \text{ N} \cdot \text{m/s}}{1 \text{ kW}} \frac{1}{9802 \text{ N/m}^3} = 5.39 \text{ m} \cdot \text{N/N}$$

The irreversibility through the generator is

$$2186.5 - 2100 = 86.5 \text{ kW}$$

or

$$\frac{86.5 \text{ kW}}{5 \text{ m}^3/\text{s}} \frac{1000 \text{ N} \cdot \text{m/s}}{1 \text{ kW}} \frac{1}{9802 \text{ N/m}^3} = 1.76 \text{ m} \cdot \text{N/N}$$

Efficiency of the turbine η_t is

$$\eta_t = 100 \times \frac{50 \text{ m} \cdot \text{N/N} - 5.39 \text{ m} \cdot \text{N/N}}{50 \text{ m} \cdot \text{N/N}} = 89.22\%$$

and efficiency of the generator η_g is

$$\eta_g = 100 \times \frac{50 - 1.76}{50} = 96.48\%$$

3.7 THE STEADY-STATE ENERGY EQUATION

When Eq. (3.2.6) is applied to steady flow through a control volume similar to Fig. 3.9, the volume integral drops out and it becomes

$$\frac{\delta Q_H}{\delta t} + \left(\frac{p_1}{\rho_1} + gz_1 + \frac{v_1^2}{2} + u_1\right)\rho_1 v_1 A_1 = \frac{\delta W_s}{\delta t} + \left(\frac{p_2}{\rho_2} + gz_2 + \frac{v_2^2}{2} + u_2\right)\rho_2 v_2 A_2$$

Since the flow is steady in this equation, it is convenient to divide through by the mass per second flowing through the system $\rho_1 A_1 v_1 = \rho_2 A_2 v_2$, getting

$$q_H + \frac{p_1}{\rho_1} + gz_1 + \frac{v_1^2}{2} + u_1 = w_s + \frac{p_2}{\rho_2} + gz_2 + \frac{v_2^2}{2} + u_2 \qquad (3.7.1)$$

q_H is the heat added per unit mass of fluid flowing, and w_s is the shaft work per unit mass of fluid flowing. This is the *energy* equation for steady flow through a control volume.

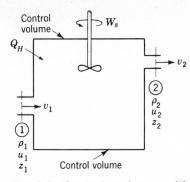

Fig. 3.9 Control volume with flow across control surface normal to surface.

The energy equation (3.7.1) in differential form, for flow through a stream tube (Fig. 3.10) with no shaft work, is

$$d\,\frac{p}{\rho} + g\,dz + v\,dv + du - dq_H = 0 \tag{3.7.2}$$

Rearranging gives

$$\frac{dp}{\rho} + g\,dz + v\,dv + du + p\,d\,\frac{1}{\rho} - dq_H = 0 \tag{3.7.3}$$

For frictionless flow the sum of the first three terms equals zero from the Euler

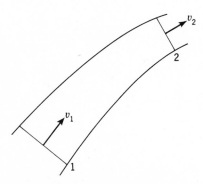

Fig. 3.10 Steady-stream tube as control volume.

equation (3.5.5); the last three terms are one form of the first law of thermo-dynamics for a system,

$$dq_H = p \, d\frac{1}{\rho} + du \tag{3.7.4}$$

Now, for reversible flow, *entropy s* per unit mass is defined by

$$ds = \left(\frac{dq_H}{T}\right)_{\text{rev}} \tag{3.7.5}$$

in which T is the absolute temperature. Entropy is shown to be a fluid prop-erty in texts on thermodynamics. In this equation it may have the units Btu per slug per degree Rankine or foot-pounds per slug per degree Rankine, as heat may be expressed in foot-pounds (1 Btu = 778 ft·lb). In SI units s is in kilocalories per kilogram per kelvin or joules per kilogram per kelvin (1 kcal = 4187 J). Since Eq. (3.7.4) is for a frictionless fluid (reversible), dq_H can be eliminated from Eqs. (3.7.4) and (3.7.5),

$$T \, ds = du + p \, d\frac{1}{\rho} \tag{3.7.6}$$

which is a very important thermodynamic relation. Although it was derived for a reversible process, since all terms are thermodynamic properties, it must also hold for irreversible-flow cases as well. By use of Eq. (3.7.6) together with the *Clausius inequality*[1] and various combinations of Euler's equation and the first law, a clearer understanding of entropy and losses is gained.

3.8   INTERRELATIONSHIPS BETWEEN EULER'S EQUATIONS AND THE THERMODYNAMIC RELATIONS

The first law in differential form, from Eq. (3.7.3), with shaft work included, is

$$dw_s + \frac{dp}{\rho} + v \, dv + g \, dz + du + p \, d\frac{1}{\rho} - dq_H = 0 \tag{3.8.1}$$

Substituting for $du + p \, d(1/\rho)$ in Eq. (3.7.6) gives

$$dw_s + \frac{dp}{\rho} + v \, dv + g \, dz + T \, ds - dq_H = 0 \tag{3.8.2}$$

[1] See any text on thermodynamics.

The Clausius inequality states that

$$ds \geq \frac{dq_H}{T}$$

or

$$T\, ds \geq dq_H \tag{3.8.3}$$

Thus $T\, ds - dq_H \geq 0$. The equals sign applies to a reversible process [or a frictionless fluid, yielding Eq. (3.5.4) with a work term]. If the quantity called losses or irreversibilities is identified as

$$d\,(\text{losses}) \equiv T\, ds - dq_H \tag{3.8.4}$$

it is seen that d (losses) is positive in irreversible flow, is zero in reversible flow, and can never be negative. Substituting Eq. (3.8.4) into Eq. (3.8.2) yields

$$dw_s + \frac{dp}{\rho} + v\, dv + g\, dz + d\,(\text{losses}) = 0 \tag{3.8.5}$$

This is a most important form of the energy equation. In general, the losses must be determined by experimentation. It implies that some of the available energy is converted into intrinsic energy during an irreversible process. This equation, in the absence of the shaft work, differs from Euler's equation by the loss term only. In integrated form,

$$\frac{v_1^2}{2} + gz_1 = \int_1^2 \frac{dp}{\rho} + \frac{v_2^2}{2} + gz_2 + w_s + \text{losses}_{1-2} \tag{3.8.6}$$

If work is done on the fluid in the control volume, as with a pump, then w_s is negative. Section 1 is upstream, and section 2 is downstream.

EXAMPLE 3.5 The cooling-water plant for a large building is located on a small lake fed by a stream, as shown in Fig. 3.11a. The design low-stream flow is 5 cfs, and at this condition the only outflow from the lake is 5 cfs via a gated structure near the discharge channel for the cooling-water system. The temperature of the incoming stream is 80°F. The flow rate of the cooling system is 4490 gpm, and the building's heat exchanger raises the cooling-water temperature by 10°F. What is the temperature of the cooling water

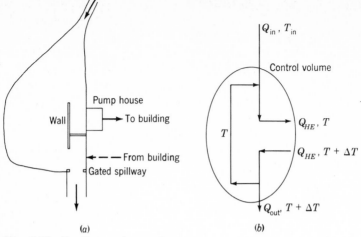

(a)   (b)

Fig. 3.11 Cooling-water system.

recirculated through the lake, neglecting heat losses to the atmosphere and lake bottom, if these conditions exist for a prolonged period?

A heat balance may be written for the lake, Fig. 3.11b, with energy in = energy out. Let T be the average temperature of the lake and ΔT be the temperature rise through the heat exchanger

$$Q_{in}T_{in} + Q_{HE}(T + \Delta T) = Q_{HE}T + Q_{in}(T + \Delta T)$$

or

$$5 \times 80 + \frac{4490 \text{ gpm}}{449 \text{ gpm}/1 \text{ cfs}}(T + 10) = \tfrac{4490}{449}T + 5(T + 10)$$

or

$$T = 90°F$$

The temperature leaving the lake is 100°F.

3.9 THE BERNOULLI EQUATION

Integration of Eq. (3.5.5) for constant density yields the Bernoulli equation

$$gz + \frac{v^2}{2} + \frac{p}{\rho} = \text{const} \tag{3.9.1}$$

The constant of integration (called the *Bernoulli constant*) in general varies from one streamline to another but remains constant along a *streamline* in *steady, frictionless, incompressible* flow. These four assumptions are needed and must be kept in mind when applying this equation. Each term has the dimensions $(L/T)^2$ or the units square feet (or square meters) per second per second, which is equivalent to foot-pounds per slug (meter-newtons per kilogram):

$$\text{ft}\cdot\text{lb/slug} = \frac{\text{ft-lb}}{\text{lb-s}^2/\text{ft}} = \text{ft}^2/\text{s}^2$$

as 1 slug $\equiv$ 1 lb·s²/ft. Therefore Eq. (3.9.1) is energy per unit mass. When it is divided through by g,

$$z + \frac{v^2}{2g} + \frac{p}{\gamma} = \text{const} \tag{3.9.2}$$

since $\gamma = \rho g$, or

$$z_1 + \frac{v_1^2}{2g} + \frac{p_1}{\gamma} = z_2 + \frac{v_2^2}{2g} + \frac{p_2}{\gamma} \tag{3.9.3}$$

it can be interpreted as energy per unit weight, or foot-pounds per pound (or meter-newtons per newton). This form is particularly convenient for dealing with liquid problems with a free surface. Multiplying Eq. (3.9.1) by ρ gives

$$\gamma z + \frac{\rho v^2}{2} + p = \text{const} \tag{3.9.4}$$

which is convenient for gas flow, since elevation changes are frequently unimportant and γz may be dropped out. In this form each term is foot-pounds per cubic foot (meter-newtons per cubic meter), or energy per unit volume.

Each of the terms of Bernoulli's equation may be interpreted as a form of energy. In Eq. (3.9.1) the first term is potential energy per unit mass. With reference to Fig. 3.12 the work needed to lift W newtons a distance z meters is Wz. The mass of WN weight is W/g kg; hence the potential energy per kilogram is

$$\frac{Wz}{W/g} = gz$$

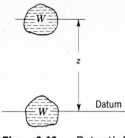

Fig. 3.12   Potential energy.

The next term, $v^2/2$, is interpreted as follows. Kinetic energy of a particle of mass is $\delta m \, v^2/2$. To place this on a unit mass basis, divide by δm; thus $v^2/2$ is meter-newtons per kilogram kinetic energy.

The last term, p/ρ, is the *flow work* or *flow energy* per unit mass. Flow work is net work done by the fluid element on its surroundings while it is flowing. For example, in Fig. 3.13, imagine a turbine consisting of a vaned unit that rotates as fluid passes through it, exerting a torque on its shaft. For a small rotation the pressure drop across a vane times the exposed area of vane is a force on the rotor. When multiplied by the distance from center of force to axis of the rotor, a torque is obtained. Elemental work done is $p \, \delta A \, ds$ by $\rho \, \delta A \, ds$ units of mass of flowing fluid; hence, the work per unit mass is p/ρ. The three energy terms in Eq. (3.9.1) are referred to as *available energy*.

By applying Eq. (3.9.3) to two points on a streamline,

$$z_1 + \frac{p_1}{\gamma} + \frac{v_1^2}{2g} = z_2 + \frac{p_2}{\gamma} + \frac{v_2^2}{2g} \qquad\qquad (3.9.5)$$

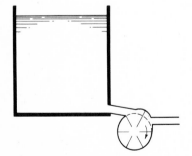

Fig. 3.13 Work done by sustained pressure.

or

$$z_1 - z_2 + \frac{p_1 - p_2}{\gamma} + \frac{v_1^2 - v_2^2}{2g} = 0$$

This equation shows that it is the difference in potential energy, flow energy, and kinetic energy that actually has significance in the equation. Thus $z_1 - z_2$ is independent of the particular elevation datum, as it is the difference in elevation of the two points. Similarly $p_1/\gamma - p_2/\gamma$ is the difference in pressure heads expressed in units of length of the fluid flowing and is not altered by the particular pressure datum selected. Since the velocity terms are not linear, their datum is fixed.

EXAMPLE 3.6 Water is flowing in an open channel at a depth of 4 ft and a velocity of 8.02 ft/s. It then flows down a chute into another open channel, where the depth is 2 ft and the velocity is 40.1 ft/s. Assuming frictionless flow, determine the difference in elevation of the channel floors.

If the difference in elevation of floors is y, Bernoulli's equation from the upper water surface to the lower water surface may be written

$$\frac{V_1^2}{2g} + \frac{p_1}{\gamma} + z_1 = \frac{V_2^2}{2g} + \frac{p_2}{\gamma} + z_2$$

V_1 and V_2 are average velocities. With gage pressure zero as datum and the floor of the lower channel as elevation datum, then $z_1 = y + 4$, $z_2 = 2$, $V_1 = 8.02$, $V_2 = 40.1$, $p_1 = p_2 = 0$,

$$\frac{8.02^2}{64.4} + 0 + y + 4 = \frac{40.1^2}{64.4} + 0 + 2$$

and $y = 22$ ft.

Modification of assumptions underlying Bernoulli's equation

Under special conditions each of the four assumptions underlying Bernoulli's equation may be waived.

1.   When all streamlines originate from a reservoir, where the energy content is everywhere the same, the constant of integration does not change from one streamline to another and points 1 and 2 for application of Bernoulli's

equation may be selected arbitrarily, i.e., not necessarily on the same streamline.

2. In the flow of a gas, as in a ventilation system, where the change in pressure is only a small fraction (a few percent) of the absolute pressure, the gas may be considered incompressible. Equation (3.9.3) may be applied, with an average specific weight γ.

3. For unsteady flow with gradually changing conditions, e.g., emptying a reservoir, Bernoulli's equation may be applied without appreciable error.

4. Bernoulli's equation is of use in analyzing real fluid cases by first neglecting viscous shear to obtain theoretical results. The resulting equation may then be modified by a coefficient, determined by experiment, which corrects the theoretical equation so that it conforms to the actual physical case. In general, losses are handled by use of the energy equation, Eq. (3.8.6).

EXAMPLE 3.7 (*a*) Determine the velocity of efflux from the nozzle in the wall of the reservoir of Fig. 3.14. (*b*) Find the discharge through the nozzle.

(*a*) The jet issues as a cylinder with atmospheric pressure around its periphery. The pressure along its centerline is at atmospheric pressure for all practical purposes. Bernoulli's equation is applied between a point on the water surface and a point downstream from the nozzle,

$$\frac{V_1^2}{2g} + \frac{p_1}{\gamma} + z_1 = \frac{V_2^2}{2g} + \frac{p_2}{\gamma} + z_2$$

With the pressure datum as local atmospheric pressure, $p_1 = p_2 = 0$; with the elevation datum through point 2, $z_2 = 0$, $z_1 = H$. The velocity on the

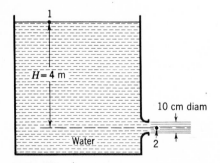

Fig. 3.14 Flow through nozzle from reservoir.

surface of the reservoir is zero (practically); hence

$$0 + 0 + H = \frac{V_2^2}{2g} + 0 + 0$$

and

$$V_2 = \sqrt{2gH} = \sqrt{2 \times 9.806 \times 4} = 8.86 \text{ m/s}$$

which states that the velocity of efflux is equal to the velocity of free fall from the surface of the reservoir. This is known as *Torricelli's theorem*.

(b) The discharge Q is the product of velocity of efflux and area of stream,

$$Q = A_2 V_2 = \pi (0.05 \text{ m})^2 \, (8.86 \text{ m/s}) = 0.07 \text{ m}^3/\text{s}$$

EXAMPLE 3.8 A venturi meter, consisting of a converging portion followed by a throat portion of constant diameter and then a gradually diverging portion, is used to determine rate of flow in a pipe (Fig. 3.15). The diameter at section 1 is 6.0 in and at section 2 is 4.0 in. Find the discharge through the pipe when $p_1 - p_2 = 3$ psi and oil, sp gr 0.90, is flowing.

From the continuity equation, Eq. (3.4.6),

$$Q = A_1 V_1 = A_2 V_2 = \frac{\pi}{16} V_1 = \frac{\pi}{36} V_2$$

in which Q is the discharge (volume per unit time flowing). By applying Eq. (3.9.3) for $z_1 = z_2$,

$$p_1 - p_2 = 3 \times 144 = 432 \text{ lb/ft}^2 \qquad \gamma = 0.90 \times 62.4 = 56.16 \text{ lb/ft}^3$$

$$\frac{p_1 - p_2}{\gamma} = \frac{V_2^2}{2g} - \frac{V_1^2}{2g} \qquad \text{or} \qquad \frac{432}{56.16} = \frac{Q^2}{\pi^2} \frac{1}{2g} (36^2 - 16^2)$$

Solving for discharge gives $Q = 2.20$ cfs.

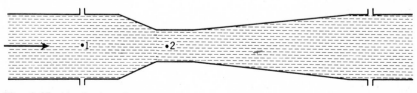

Fig. 3.15 Venturi meter.

3.10 APPLICATION OF THE BERNOULLI AND ENERGY EQUATIONS TO STEADY FLUID-FLOW SITUATIONS

For an incompressible fluid Eq. (3.8.6) may be simplified to

$$\frac{p_1}{\gamma} + \frac{v_1^2}{2g} + z_1 = \frac{p_2}{\gamma} + \frac{v_2^2}{2g} + z_2 + \text{losses}_{1-2} \tag{3.10.1}$$

in which each term now is energy in foot-pounds per pound or meter-newtons per newton, including the loss term. The work term has been omitted but may be inserted if needed.

Kinetic energy correction factor

In dealing with flow situations in open- or closed-channel flow, the so-called *one-dimensional* form of analysis is frequently used. The whole flow is considered to be one large stream tube with average velocity V at each cross section. The kinetic energy per unit weight given by $V^2/2g$, however, is not the average of $v^2/2g$ taken over the cross section. It is necessary to compute a correction factor α for $V^2/2g$, so that $\alpha V^2/2g$ is the average kinetic energy per unit weight passing the section. Referring to Fig. 3.16, the kinetic energy passing the cross section per unit time is

$$\gamma \int_A \frac{v^2}{2g} v \, dA$$

in which $\gamma v \, \delta A$ is the weight per unit time passing δA and $v^2/2g$ is the kinetic energy per unit weight. By equating this to the kinetic energy per unit time

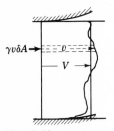

Fig. 3.16 Velocity distribution and average velocity.

passing the section, in terms of $\alpha V^2/2g$,

$$\alpha \frac{V^2}{2g} \gamma V A = \gamma \int_A \frac{v^3}{2g} dA$$

By solving for α, the *kinetic-energy correction factor*,

$$\alpha = \frac{1}{A} \int_A \left(\frac{v}{V}\right)^3 dA \qquad (3.10.2)$$

Bernoulli's equation becomes

$$z_1 + \frac{p_1}{\gamma} + \alpha_1 \frac{V_1^2}{2g} = z_2 + \frac{p_2}{\gamma} + \alpha_2 \frac{V_2^2}{2g} \qquad (3.10.3)$$

For laminar flow in a pipe, $\alpha = 2$, as shown in Sec. 5.2. For turbulent flow[1] in a pipe, α varies from about 1.01 to 1.10 and is usually neglected except for precise work.

EXAMPLE 3.9 The velocity distribution in turbulent flow in a pipe is given approximately by Prandtl's one-seventh power law,

$$\frac{v}{v_{\max}} = \left(\frac{y}{r_0}\right)^{1/7}$$

with y the distance from the pipe wall and r_0 the pipe radius. Find the kinetic-energy correction factor.

The average velocity V is expressed by

$$\pi r_0^2 V = 2\pi \int_0^{r_0} rv \, dr$$

in which $r = r_0 - y$. By substituting for r and v,

$$\pi r_0^2 V = 2\pi v_{\max} \int_0^{r_0} (r_0 - y) \left(\frac{y}{r_0}\right)^{1/7} dy = \pi r_0^2 v_{\max} \frac{98}{120}$$

or

$$V = \frac{98}{120} v_{\max} \qquad \frac{v}{V} = \frac{120}{98} \left(\frac{y}{r_0}\right)^{1/7}$$

[1] V. L. Streeter, The Kinetic Energy and Momentum Correction Factors for Pipes and Open Channels of Great Width, *Civ. Eng. N.Y.*, vol. 12, no. 4, pp. 212–213, 1942.

By substituting into Eq. (3.10.2)

$$\alpha = \frac{1}{\pi r_0^2} \int_0^{r_0} 2\pi r \left(\frac{120}{98}\right)^3 \left(\frac{y}{r_0}\right)^{3/7} dr$$

$$= 2 \left(\frac{120}{98}\right)^3 \frac{1}{r_0^2} \int_0^{r_0} (r_0 - y) \left(\frac{y}{r_0}\right)^{3/7} dy = 1.06$$

All the terms in the energy equation (3.10.1) except the term losses are *available energy*. For real fluids flowing through a system the available energy decreases in the downstream direction; it is available to do work, as in passing through a water turbine. A plot showing the available energy along a stream tube portrays the *energy grade line* (see Sec. 10.1). A plot of the two terms $z + p/\gamma$ along a stream tube portrays the *hydraulic grade line*. The energy grade line always slopes downward in real fluid flow, except at a pump or other source of energy. Reductions in energy grade line are also referred to as *head losses*.

EXAMPLE 3.10 The siphon of Fig. 3.17 is filled with water and discharging at 2.80 cfs. Find the losses from point 1 to point 3 in terms of the velocity head $V^2/2g$. Find the pressure at point 2 if two-thirds of the losses occur between points 1 and 2.

The energy equation is applied to the control volume between points 1 and 3, with elevation datum at point 3 and gage pressure zero for pressure datum:

$$\frac{V_1^2}{2g} + \frac{p_1}{\gamma} + z_1 = \frac{V_3^2}{2g} + \frac{p_3}{\gamma} + z_3 + \text{losses}$$

or

$$0 + 0 + 4 = \frac{V_3^2}{2g} + 0 + 0 + \frac{KV_3^2}{2g}$$

in which the losses from 1 to 3 have been expressed as $KV_3^2/2g$. From the discharge

$$V_3 = \frac{Q}{A} = \frac{2.80}{\pi/9} = 8.02 \text{ ft/s}$$

and $V_3^2/2g = 1.0$ ft. Hence $K = 3$, and the losses are $3V_3^2/2g$, or 3 ft-lb/lb.

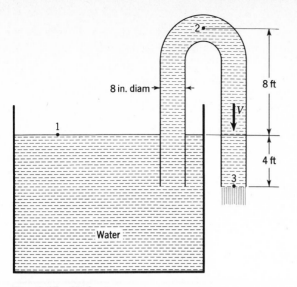

Fig. 3.17 Siphon.

The energy equation applied to the control volume between points 1 and 2, with losses $2V_3^2/2g = 2.0$ ft, is

$$0 + 0 + 0 = 1 + \frac{p_2}{\gamma} + 8 + 2$$

The pressure at 2 is -11 ft H_2O, or 4.76 psi vacuum.

EXAMPLE 3.11 The device shown in Fig. 3.18 is used to determine the velocity of liquid at point 1. It is a tube with its lower end directed upstream and its other leg vertical and open to the atmosphere. The impact of liquid against the opening 2 forces liquid to rise in the vertical leg to the height Δz above the free surface. Determine the velocity at 1.

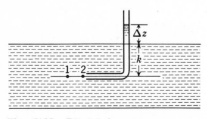

Fig. 3.18 Pitot tube.

Point 2 is a stagnation point, where the velocity of the flow is reduced to zero. This creates an impact pressure, called the dynamic pressure, which forces the fluid into the vertical leg. By writing Bernoulli's equation between points 1 and 2, neglecting losses, which are very small,

$$\frac{V_1^2}{2g} + \frac{p_1}{\gamma} + 0 = 0 + \frac{p_2}{\gamma} + 0$$

p_1/γ is given by the height of fluid above point 1 and equals k ft of fluid flowing. p_2/γ is given by the manometer as $k + \Delta z$, neglecting capillary rise. After substituting these values into the equation, $V_1^2/2g = \Delta z$ and

$$V_1 = \sqrt{2g\,\Delta z}$$

This is the pitot tube in a simple form.

Examples of compressible flow are given in Chap. 6.

3.11 APPLICATIONS OF THE LINEAR-MOMENTUM EQUATION

Newton's second law, the equation of motion, was developed into the linear-momentum equation in Sec. 3.2,

$$\Sigma \mathbf{F} = \frac{\partial}{\partial t} \int_{cv} \rho \mathbf{v}\, d\mathcal{U} + \int_{cs} \rho \mathbf{v}\mathbf{v} \cdot d\mathbf{A} \tag{3.11.1}$$

This vector relation may be applied for any component, say the x direction, reducing to

$$\Sigma F_x = \frac{\partial}{\partial t} \int_{cv} \rho v_x\, d\mathcal{U} + \int_{cs} \rho v_x \mathbf{v} \cdot d\mathbf{A} \tag{3.11.2}$$

In selecting the arbitrary control volume, it is generally advantageous to take the surface normal to the velocity wherever it cuts across the flow. In addition, if the velocity is constant over the surface, the surface integral can be dispensed with. In Fig. 3.19, with the control surface as shown, and with steady flow, the force F_x acting on the control volume is given by Eq. (3.11.2) as

$$F_x = \rho_2 A_2 V_2 V_{x2} - \rho_1 A_1 V_1 V_{x1}$$

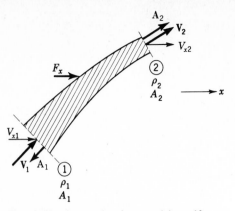

Fig. 3.19 Control volume with uniform inflow and outflow normal to control surface.

or

$$F_x = \rho Q (V_{x2} - V_{x1})$$

as the mass per second entering and leaving is $\rho Q = \rho_1 Q_1 = \rho_2 Q_2$.

   When the velocity varies over a plane cross section of the control surface, by introduction of a momentum correction factor β, the average velocity may be utilized, Fig. 3.20,

$$\int_A \rho v^2 \, dA = \beta \rho V^2 A \tag{3.11.3}$$

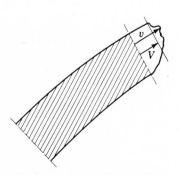

Fig. 3.20 Nonuniform flow through a control surface.

in which β is dimensionless. Solving for β yields

$$\beta = \frac{1}{A} \int_A \left(\frac{v}{V}\right)^2 dA \tag{3.11.4}$$

which is analogous to α, the kinetic-energy correction factor, Eq. (3.10.2). For laminar flow in a straight round tube, β is shown to equal $\frac{4}{3}$ in Chap. 5. It equals 1 for uniform flow and cannot have a value less than 1.

In applying Eq. (3.11.1) or a component equation such as Eq. (3.11.2) care should be taken to define the control volume and the forces acting on it clearly. Also the sign of the inflow or outflow term must be carefully evaluated. The first example is an unsteady one using Eq. (3.11.2) and the general continuity equation (3.2.1).

EXAMPLE 3.12   The horizontal pipe of Fig. 3.21 is filled with water for the distance x. A jet of constant velocity V_1 impinges against the filled portion. Fluid frictional force on the pipe wall is given by $\tau_0 \pi D x$, with $\tau_0 = \rho f V_2^2 / 8$ [see Eq. (5.10.2)]. Determine the equations to analyze this flow condition when initial conditions are known; that is, $t = 0$, $x = x_0$, $V_2 = V_{2_0}$. Specifically for $V_1 = 20$ m/s, $D_1 = 6$ cm, $V_{2_0} = 50$ cm/s, $D_2 = 25$ cm, $x_0 = 100$ m, $\rho = 997.3$ kg/m^3, and $f = 0.02$, find the rate of change of V_2 and x with time.

The continuity and momentum equations are used to analyze this unsteady-flow problem. Take as control volume the inside surface of the pipe, with the two end sections l ft apart, as shown. The continuity equation

$$0 = \frac{\partial}{\partial t} \int_{cv} \rho \, d\mathcal{V} + \int_{cs} \rho \mathbf{v} \cdot d\mathbf{A}$$

becomes, using $A_1 = \pi D_1^2 / 4$, $A_2 = \pi D_2^2 / 4$,

$$\frac{\partial}{\partial t} [\rho A_2 x + \rho A_1 (l - x)] + \rho (V_2 A_2 - V_1 A_1) = 0$$

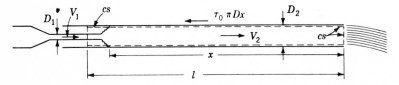

Fig. 3.21   Jet impact on pipe flowing full over partial length.

After simplifying,

$$\frac{\partial x}{\partial t}(A_2 - A_1) + V_2 A_2 - V_1 A_1 = 0$$

The momentum equation for the horizontal direction x,

$$\Sigma F_x = \frac{\partial}{\partial t}\int_{cv} \rho v_x \, d\mathcal{V} + \int_{cs} \rho v_x \mathbf{v} \cdot d\mathbf{A}$$

becomes

$$-\rho\frac{fV_2{}^2 \pi D_2 x}{8} = \frac{\partial}{\partial t}\big[\rho A_2 x V_2 + \rho A_1 (l - x) V_1\big] + \rho A_2 V_2{}^2 - \rho A_1 V_1{}^2$$

which simplifies to

$$\frac{fV_2{}^2 \pi D_2 x}{8} + A_2 \frac{\partial}{\partial t}(xV_2) - A_1 V_1 \frac{\partial x}{\partial t} + A_2 V_2{}^2 - A_1 V_1{}^2 = 0$$

As t is the only independent variable, the partials may be replaced by totals. The continuity equation is

$$\frac{dx}{dt} = -\frac{V_2 A_2 - V_1 A_1}{A_2 - A_1}$$

By expanding the momentum equation, and substituting for dx/dt, it becomes

$$\frac{dV_2}{dt} = \frac{1}{xA_2}\left[A_1 V_1{}^2 - A_2 V_2{}^2 - \frac{fV_2{}^2 \pi D_2 x}{8} + \frac{(A_2 V_2 - A_1 V_1)^2}{A_2 - A_1}\right]$$

These two equations, being nonlinear, can be solved simultaneously by numerical methods, such as Runge-Kutta methods described in Appendix E, when initial conditions are known. The rate of change of x and V_2 can be found directly from the equations for the specific problem

$$\frac{dx}{dt} = 0.692 \text{ m/s} \qquad \frac{dV_2}{dt} = 0.0496 \text{ m/s}^2$$

EXAMPLE 3.13 In Fig. 3.22 a fluid jet impinges on a body as shown; the momentum per second of each of the jets is given by **M** and is the vector located

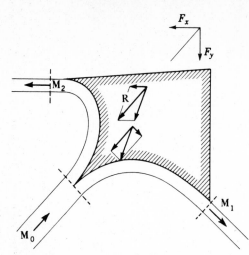

Fig. 3.22 Solution of linear-momentum problem by addition of vectors.

at the center of the jets. By vector addition find the resultant force needed to hold the body at rest.

The vector form of the linear momentum equation (3.11.1) is to be applied to a control volume comprising the fluid bounded by the body and the three dotted cross sections. As the problem is steady, Eq. (3.11.1) reduces to

$$\Sigma \mathbf{F} = \int \rho \mathbf{v} \mathbf{v} \cdot d\mathbf{A} = \sum_{\text{out}} \mathbf{M}_i$$

By taking $\mathbf{M}_1$ and $\mathbf{M}_0$ first, the vector $\mathbf{M}_1 - \mathbf{M}_0$ is the net momentum efflux for these two vectors, shown graphically on their lines of action. The resultant of these two vectors is then added to momentum efflux $\mathbf{M}_2$, along its line of action, to obtain $\mathbf{R}$. $\mathbf{R}$ is the momentum efflux over the control surface and is just equal to the force that must be exerted on the control surface. The same force must then be exerted on the body, to resist the control-volume force on it.

EXAMPLE 3.14 The reducing bend of Fig. 3.23 is in a vertical plane. Water is flowing, $D_1 = 6$ ft, $D_2 = 4$ ft, $Q = 300$ cfs, $W = 18,000$ lb, $z = 10$ ft, $\theta = 120°$, $p_1 = 40$ psi, $x = 6$ ft, and losses through the bend are $0.5 \ V_2^2/2g$ ft·lb/lb. Determine F_x, F_y, and the line of action of the resultant force. $\beta_1 = \beta_2 = 1$.

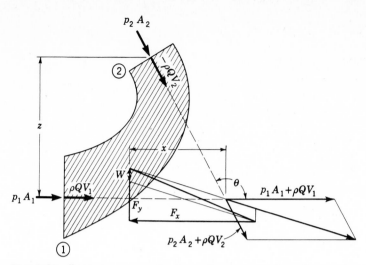

Fig. 3.23 Forces on a reducing elbow, including the vector solution.

The inside surface of the reducing bend comprises the control-volume surface for the portion of the surface with no flow across it. The normal sections 1 and 2 complete the control surface.

$$V_1 = \frac{Q}{A_1} = \frac{300}{\pi(6^2)/4} = 10.61 \text{ ft/s} \qquad V_2 = \frac{Q}{A_2} = \frac{300}{\pi(4^2)/4} = 23.88 \text{ ft/s}$$

By application of the energy equation, Eq. (3.10.1),

$$\frac{p_1}{\gamma} + \frac{V_1^2}{2g} + z_1 = \frac{p_2}{\gamma} + \frac{V_2^2}{2g} + z_2 + \text{losses}_{1-2}$$

$$\frac{40 \times 144}{62.4} + \frac{10.61^2}{64.4} + 0 = \frac{p_2}{62.4} + \frac{23.88^2}{64.4} + 10 + 0.5 \times \frac{23.88^2}{64.4}$$

from which $p_2 = 4420 \text{ lb/ft}^2 = 30.7 \text{ psi}$.
To determine F_x, Eq. (3.11.2) yields

$$p_1 A_1 - p_2 A_2 \cos\theta - F_x = \rho Q(V_2 \cos\theta - V_1)$$

$$40 \times 144\pi(3^2) - 4420\pi(2^2) \cos 120° - F_x$$

$$= 1.935 \times 300(23.88 \cos 120° - 10.61)$$

Since $\cos 120° = -0.5$,

$$162,900 + 27,750 - F_x = 580.5(-11.94 - 10.61)$$

$$F_x = 203,740 \text{ lb}$$

For the y direction

$$\Sigma F_y = \rho Q(V_{y2} - V_{y1})$$

$$F_y - W - p_2 A_2 \sin \theta = \rho Q V_2 \sin \theta$$

$$F_y - 18,000 - 4420\pi(2^2) \sin 120° = 1.935 \times 300 \times 23.88 \sin 120°$$

$$F_y = 78,100 \text{ lb}$$

To find the line of action of the resultant force, using the momentum flux vectors (Fig. 3.23), $\rho Q V_1 = 6160$ lb, $\rho Q V_2 = 13,860$ lb, $p_1 A_1 = 162,900$ lb, $p_2 A_2 = 55,560$ lb. Combining these vectors and the weight W in Fig. 3.23 yields the final force, 218,000 lb, which must be opposed by F_x and F_y.

As demonstrated in Example 3.14, a change in direction of a pipeline causes forces to be exerted on the line unless the bend or elbow is anchored in place. These forces are due to both static pressure in the line and dynamic reactions in the turning fluid stream. Expansion joints are placed in large pipelines to avoid stress in the pipe in an axial direction, whether caused by fluid or by temperature change. These expansion joints permit relatively free movement of the line in an axial direction, and hence the static and dynamic forces must be provided for at the bends.

EXAMPLE 3.15 A jet of water 8 cm in diameter with a velocity of 40 m/s is discharged in a horizontal direction from a nozzle mounted on a boat. What force is required to hold the boat stationary?

When the control volume is selected as shown in Fig. 3.24, the net efflux of momentum is [Eq. (3.11.2)]

$$F_x = \rho Q(V_{x2} - V_{x1}) = \rho Q V = 997.3 \text{ kg/m}^3 \times \frac{\pi}{4}(0.08 \text{ m})^2(40 \text{ m/s})^2$$

$$= 8021 \text{ N}$$

The force exerted against the boat is 8021 N in the x direction.

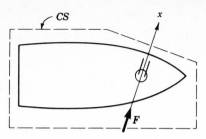

Fig. 3.24 Nozzle mounted on boat.

EXAMPLE 3.16 Find the force exerted by the nozzle on the pipe of Fig. 3.25a. Neglect losses. The fluid is oil, sp gr 0.85, and $p_1 = 100$ psi.

To determine the discharge, Bernoulli's equation is written for the stream from section 1 to the downstream end of the nozzle, where the pressure is zero.

$$z_1 + \frac{V_1^2}{2g} + \frac{(100 \text{ lb/in}^2)(144 \text{ in}^2/\text{ft}^2)}{0.85 \times 62.4 \text{ lb/ft}^3} = z_2 + \frac{V_2^2}{2g} + 0$$

Since $z_1 = z_2$, and $V_2 = (D_1/D_2)^2 V_1 = 9V_1$, after substituting,

$$\frac{V_1^2}{2g}(1 - 81) + \frac{(100 \text{ lb/in}^2)(144 \text{ in}^2/\text{ft}^2)}{0.85 \times 62.4 \text{ lb/ft}^3} = 0$$

and

$$V_1 = 14.78 \text{ ft/s} \qquad V_2 = 133 \text{ ft/s} \qquad Q = 14.78 \frac{\pi}{4}\left(\frac{1}{4}\right)^2 = 0.725 \text{ ft}^3/\text{s}$$

Let P_x (Fig. 3.25b) be the force exerted on the liquid control volume by the nozzle; then, with Eq. (3.11.2),

$$(100 \text{ lb/in}^2)\frac{\pi}{4}(3 \text{ in})^2 - P_x = (1.935 \text{ slugs/ft}^3)(0.85)(0.725 \text{ ft}^3/\text{s})$$

$$\times (133 \text{ ft/s} - 14.78 \text{ ft/s})$$

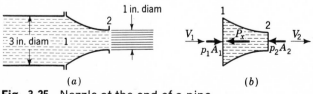

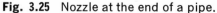

(a) (b)

Fig. 3.25 Nozzle at the end of a pipe.

or $P_x = 565$ lb. The oil exerts a force on the nozzle of 565 lb to the right, and a tension force of 565 lb is exerted by the nozzle on the pipe.

In many situations an unsteady-flow problem can be converted to a steady-flow problem by superposing a constant velocity upon the system and its surroundings, i.e., by changing the reference velocity. The dynamics of a system and its surroundings are unchanged by the superposition of a constant velocity; hence, pressures and forces are unchanged. In the next flow situation studied, advantage is taken of this principle.

The momentum theory for propellers

The action of a propeller is to change the momentum of the fluid within which it is submerged and thus to develop a thrust that is used for propulsion. Propellers cannot be designed according to the momentum theory, although some of the relations governing them are made evident by its application. A propeller, with its slipstream and velocity distributions at two sections a fixed distance from it, is shown in Fig. 3.26. The propeller may be either (1) stationary in a flow as indicated or (2) moving to the left with a velocity V_1 through a stationary fluid since the relative picture is the same. The fluid is assumed to be frictionless and incompressible.

The flow is undisturbed at section 1 upstream from the propeller and is

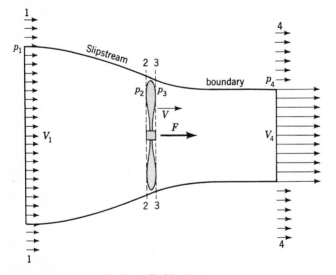

Fig. 3.26 Propeller in a fluid stream.

accelerated as it approaches the propeller, owing to the reduced pressure on its upstream side. In passing through the propeller, the fluid has its pressure increased, which further accelerates the flow and reduces the cross section at 4. The velocity V does not change across the propeller, from 2 to 3. The pressure at 1 and 4 is that of the undisturbed fluid, which is also the pressure along the slipstream boundary.

When the momentum equation (3.11.2) is applied to the control volume within sections 1 and 4 and the slipstream boundary of Fig. 3.26, the force F exerted by the propeller is the only external force acting in the axial direction since the pressure is everywhere the same on the control surface. Therefore

$$F = \rho Q (V_4 - V_1) = (p_3 - p_2) A \qquad (3.11.5)$$

in which A is the area swept over by the propeller blades. The force on the propeller must be equal and opposite to the force on the fluid. After substituting $Q = AV$ and simplifying,

$$\rho V (V_4 - V_1) = p_3 - p_2 \qquad (3.11.6)$$

When Bernoulli's equation is written for the stream between sections 1 and 2 and between sections 3 and 4,

$$p_1 + \tfrac{1}{2}\rho V_1{}^2 = p_2 + \tfrac{1}{2}\rho V^2 \qquad p_3 + \tfrac{1}{2}\rho V^2 = p_4 + \tfrac{1}{2}\rho V_4{}^2$$

since $z_1 = z_2 = z_3 = z_4$. In solving for $p_3 - p_2$, with $p_1 = p_4$,

$$p_3 - p_2 = \tfrac{1}{2}\rho (V_4{}^2 - V_1{}^2) \qquad (3.11.7)$$

Eliminating $p_3 - p_2$ in Eqs. (3.11.6) and (3.11.7) gives

$$V = \frac{V_1 + V_4}{2} \qquad (3.11.8)$$

which shows that the velocity through the propeller area is the average of the velocities upstream and downstream from it.

The useful work per unit time done by a propeller moving through still fluid is the product of propeller thrust and velocity, i.e.,

$$\text{Power} = FV_1 = \rho Q (V_4 - V_1) V_1 \qquad (3.11.9)$$

The power input is that required to increase the velocity of fluid from V_1 to

V_4, or the useful work plus the kinetic energy per unit time remaining in the slipstream.

$$\text{Power input} = \rho \frac{Q}{2} (V_4^2 - V_1^2)$$

$$= \rho Q (V_4 - V_1)V_1 + \rho \frac{Q}{2} (V_4 - V_1)^2 \qquad (3.11.10)$$

With the ratio of Eqs. (3.11.9) and (3.11.10) used to obtain the theoretical efficiency e_t,

$$e_t = \frac{2V_1}{V_4 + V_1} = \frac{V_1}{V} \qquad (3.11.11)$$

If $\Delta V = V_4 - V_1$ is the increase in slipstream velocity, substituting into Eq. (3.11.11) produces

$$e_t = \frac{V_1}{V_1 + \Delta V/2} \qquad (3.11.12)$$

which shows that maximum efficiency is obtained with a propeller that increases the velocity of slipstream as little as possible, or for which $\Delta V/V_1$ is a minimum.

Owing to compressibility effects, the efficiency of an airplane propeller drops rapidly with speeds above 400 mph. Airplane propellers under optimum conditions have actual efficiencies close to the theoretical efficiencies, in the neighborhood of 85 percent. Ship propeller efficiencies are less, around 60 percent, owing to restrictions in diameter.

The windmill can be analyzed by application of the momentum relations. The jet has its speed reduced, and the diameter of slipstream is increased.

EXAMPLE 3.17 An airplane traveling 400 km/h through still air, $\gamma = 12$ N/m³, discharges 1000 m³/s through its two 2.25-m-diameter propellers. Determine (a) the theoretical efficiency, (b) the thrust, (c) the pressure difference across the propellers, and (d) the theoretical power required.

$$(a) \quad V_1 = \frac{400 \text{ km}}{1 \text{ h}} \frac{1000 \text{ m}}{1 \text{ km}} \frac{1 \text{ h}}{3600 \text{ s}} = 111.11 \text{ m/s}$$

$$V = \frac{500 \text{ m}^3/\text{s}}{(\pi/4)(2.25^2)} = 126 \text{ m/s}$$

From Eq. (3.11.11)

$$e_t = \frac{V_1}{V} = \frac{111.11}{126} = 88.2\%$$

(b)   From Eq. (3.11.8)

$$V_4 = 2V - V_1 = 2 \times 126 - 111.11 = 140.9 \text{ m/s}$$

The thrust from the propellers is, from Eq. (3.11.5),

$$F = \frac{12 \text{ N/m}^3}{9.806 \text{ m/s}^2} (1000 \text{ m}^3/\text{s})(140.9 - 111.11 \text{ m/s}) = 36,500 \text{ N}$$

(c)   The pressure difference, from Eq. (3.11.6), is

$$p_3 - p_2 = \frac{12 \text{ N/m}^3}{9.806 \text{ m/s}^2} (126 \text{ m/s})(140.9 - 111.11 \text{ m/s}) = 4600 \text{ N/m}^2$$

(d)   The theoretical power is

$$FV_1 = (36,500 \text{ N})(111.11 \text{ m/s}) \frac{1 \text{ kW}}{1000 \text{ N} \cdot \text{m/s}} = 4050 \text{ kW}$$

Jet propulsion

The propeller is one form of jet propulsion in that it creates a jet and by so doing has a thrust exerted upon it that is the propelling force. In jet engines, air (initially at rest) is taken into the engine and burned with a small amount of fuel; the gases are then ejected with a much higher velocity than in a propeller slipstream. The jet diameter is necessarily smaller than the propeller slipstream. For the mechanical energy only, the theoretical efficiency is given by the ratio of useful work to work input or by useful work divided by the sum of useful work and kinetic energy per unit time remaining in the jet. If the mass of fuel burned is neglected, the propelling force F [Eq. (3.11.5)] is

$$F = \rho Q(V_2 - V_1) = \rho Q V_{\text{abs}} \tag{3.11.13}$$

in which V_{abs} (Fig. 3.27) is the absolute velocity of fluid in the jet and ρQ is the mass per unit time being discharged. The useful work is FV_1, in which V_1 is the speed of the body. The kinetic energy per unit time being discharged

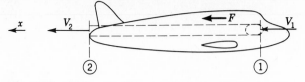

Fig. 3.27 Walls of flow passages through jet engines taken as impenetrable part of control surface for plane when viewed as a steady-state problem.

in the jet is $\gamma Q V_{abs}^2/2g = \rho Q V_{abs}^2/2$, since γQ is the weight per unit time being discharged and $V_{abs}^2/2g$ is the kinetic energy per unit weight. Hence, the theoretical mechanical efficiency is

$$e_t = \frac{\text{output}}{\text{output} + \text{loss}} = \frac{F V_1}{F V_1 + \rho Q V_{abs}^2/2}$$

$$= \frac{\rho Q V_{abs} V_1}{\rho Q V_{abs} V_1 + \rho Q V_{abs}^2/2} = \frac{1}{1 + V_{abs}/2V_1} \tag{3.11.14}$$

which is the same expression as that for efficiency of the propeller. It is obvious that, other things being equal, V_{abs}/V_1 should be as small as possible. For a given speed V_1, the resistance force F is determined by the body and fluid in which it moves; hence, for V_{abs} in Eq. (3.11.13) to be very small, ρQ must be very large.

An example is the type of propulsion system to be used on a boat. If the boat requires a force of 400 lb to move it through water at 15 mph, first a method of jet propulsion can be considered in which water is taken in at the bow and discharged out the stern by a 100 percent efficient pumping system. To analyze the propulsion system the problem is converted to steady state by superposition of the boat speed $- V_1$ on boat and surroundings (Fig. 3.28).

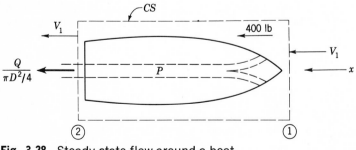

Fig. 3.28 Steady-state flow around a boat.

If a 6-in-diameter jet pipe is used, $V_2 = 16Q/\pi$. By use of Eq. (3.11.2), for $V_1 = 15$ mi/h $= 22$ ft/s,

$$400 = 1.935Q \left(\frac{16Q}{\pi} - 22 \right)$$

Hence $Q = 8.89$ ft³/s, $V_{abs} = 23.2$, and the efficiency is

$$e_t = \frac{1}{1 + V_{abs}/2V_1} = \frac{1}{1 + 23.2/44} = 65.5\%$$

The horsepower required is

$$\frac{FV_1}{550e_t} = \frac{400 \times 22}{550 \times 0.655} = 24.4$$

With an 8-in-diameter jet pipe, $V_2 = 9Q/\pi$, and

$$400 = 1.935Q \left(\frac{9Q}{\pi} - 22 \right)$$

so that $Q = 13.14$ ft³/s, $V_2 = 15.72$, $e_t = 73.7$ percent, and the horsepower required is 21.7.

With additional enlarging of the jet pipe and the pumping of more water with less velocity head, the efficiency can be further increased. The type of pump best suited for large flows at small head is the axial-flow propeller pump. Increasing the size of pump and jet pipe would increase weight greatly and take up useful space in the boat; the logical limit is to drop the propeller down below or behind the boat and thus eliminate the jet pipe, which is the usual propeller for boats. Jet propulsion of a boat by a jet pipe is practical, however, in very shallow water where a propeller would be damaged by striking bottom or other obstructions.

To take the weight of fuel into account in jet propulsion of aircraft, let $\dot{m}_{air}$ be the mass of air per unit time and r the ratio of mass of fuel burned to mass of air. Then (Fig. 3.27), the propulsive force F is

$$F = \dot{m}_{air}(V_2 - V_1) + r\dot{m}_{air}V_2$$

The second term on the right is the mass of fuel per unit time multiplied by its change in velocity. Rearranging gives

$$F = \dot{m}_{air}[V_2(1 + r) - V_1] \tag{3.11.15}$$

Defining the mechanical efficiency again as the useful work divided by the sum of useful work and kinetic energy remaining gives

$$e_t = \frac{FV_1}{FV_1 + \dot{m}_{air}(1 + r)(V_2 - V_1)^2/2}$$

and by Eq. (3.11.15)

$$e_t = \frac{1}{1 + \dfrac{(1 + r)[(V_2/V_1) - 1]^2}{2[(1 + r)(V_2/V_1) - 1]}} \tag{3.11.16}$$

The efficiency becomes unity for $V_1 = V_2$, as the combustion products are then brought to rest and no kinetic energy remains in the jet.

EXAMPLE 3.18 An airplane consumes 1 lb_m fuel for each 20 lb_m air and discharges hot gases from the tail pipe at $V_2 = 6000$ ft/s. Determine the mechanical efficiency for airplane speeds of 1000 and 500 ft/s.

For 1000 ft/s, $V_2/V_1 = \frac{6000}{1000} = 6$, $r = 0.05$. From Eq. (3.11.16),

$$e_t = \frac{1}{1 + \dfrac{(1 + 0.05)(6 - 1)^2}{2[6(1 + 0.05) - 1]}} = 0.287$$

For 500 ft/s, $V_2/V_1 = \frac{6000}{500} = 12$, and

$$e_t = \frac{1}{1 + \dfrac{(1 + 0.05)(12 - 1)^2}{2[12(1 + 0.05) - 1]}} = 0.154$$

Jet propulsion of aircraft or missiles

Propulsion through air or water in each case is caused by reaction to the formation of a jet behind the body. The various means include the propeller, turbojet, turboprop, ram jet, and rocket motor, which are briefly described in the following paragraphs.

The momentum relations for a propeller determine that its theoretical efficiency increases as the speed of the aircraft increases and the absolute velocity of the slipstream decreases. As the speed of the blade tips approaches the speed of sound, however, compressibility effects greatly increase the drag on the blades and thus decrease the overall efficiency of the propulsion system.

A *turbojet* is an engine consisting of a compressor, a combustion chamber, a turbine, and a jet pipe. Air is scooped through the front of the engine and is compressed, and fuel is added and burned with a great excess of air. The air and combustion gases then pass through a gas turbine that drives the compressor. Only a portion of the energy of the hot gases is removed by the turbine, since the only means of propulsion is the issuance of the hot gas through the jet pipe. The overall efficiency of a jet engine increases with speed of the aircraft. Although there is very little information available on propeller systems near the speed of sound, it appears that the overall efficiencies of the turbojet and propeller systems are about the same at the speed of sound.

The *turboprop* is a system combining thrust from a propeller with thrust from the ejection of hot gases. The gas turbine must drive both compressor and propeller. The proportion of thrust between the propeller and the jet may be selected arbitrarily by the designer.

The *ram jet* is a high-speed engine that has neither compressor nor turbine. The ram pressure of the air forces air into the front of the engine, where some of the kinetic energy is converted into pressure energy by enlarging the flow cross section. It then enters a combustion chamber, where fuel is burned, and the air and gases of combustion are ejected through a jet pipe. It is a supersonic device requiring very high speed for compression of the air. An intermittent ram jet was used by the Germans in the V-1 *buzz bomb*. Air is admitted through spring-closed flap valves in the nose. Fuel is ignited to build up pressure that closes the flap valves and ejects the hot gases as a jet. The ram pressure then opens the valves in the nose to repeat the cycle. The cyclic rate is around 40 per second. Such a device must be launched at high speed to initiate operation of the ram jet.

Rocket mechanics

The rocket motor carries with it an oxidizing agent to mix with its fuel so that it develops a thrust that is independent of the medium through which it travels. In contrast, a gas turbine can eject a mass many times the mass of fuel it carries because it takes in air to mix with the fuel.

To determine the acceleration of a rocket during flight, Fig. 3.29, it is convenient to take the control volume as the outer surface of the rocket, with a plane area normal to the jet across the nozzle exit. The control volume has a velocity equal to the velocity of the rocket at the instant the analysis is made. Let R be the air resistance, m_R the mass of the rocket body, m_f the mass of fuel, $\dot{m}$ the rate at which fuel is being burned, and v_r the exit-gas ve-

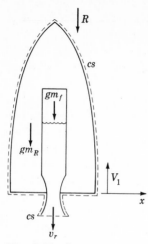

Fig. 3.29 Control surface for analysis of rocket acceleration. Frame of reference has the velocity V_1 of the rocket.

locity relative to the rocket. V_1 is the actual velocity of the rocket (and of the frame of reference), and V is the velocity of the rocket relative to the frame of reference. V is zero, but $dV/dt = dV_1/dt$ is the rocket acceleration. The basic linear momentum equation for the y direction (vertical motion)

$$\Sigma F_y = \frac{\partial}{\partial t} \int_{cv} \rho v_y \, d\mathcal{V} + \int_{cs} \rho v_y \mathbf{V} \cdot d\mathbf{A} \tag{3.11.17}$$

becomes

$$-R - (m_R + m_f)g = \frac{\partial}{\partial t} \left[(m_R + m_f) V \right] - \dot{m} v_r \tag{3.11.18}$$

Since V is a function of t only, the equation can be written as a total differential equation

$$\frac{dV}{dt} = \frac{dV_1}{dt} = \frac{\dot{m} v_r - g(m_f + m_R) - R}{m_R + m_f} \tag{3.11.19}$$

The mass of propellant reduces with time; for constant burning rate $\dot{m}$, $m_f = m_{f_0} - \dot{m} t$, with m_{f_0} the initial mass of fuel and oxidizer. Gravity is a function of y, and the air resistance depends on the Reynolds number and Mach number (Chap. 4), as well as on the shape and size of the rocket.

By considering the mass of rocket and fuel together [Eq. (3.11.19)], the thrust $\dot{m}v_r$ minus the weight and the air resistance just equals the combined mass times its acceleration.

The theoretical efficiency of a rocket motor (based on available energy) is shown to increase with rocket speed. E represents the available energy in the propellant per unit mass. When the propellant is ignited, its available energy is converted into kinetic energy; $E = v_r^2/2$, in which v_r is the jet velocity relative to the rocket, or $v_r = \sqrt{2E}$. For rocket speed V_1 referred to axes fixed in the earth, the useful power is $\dot{m}v_rV_1$. The kinetic energy being used up per unit time is due to mass loss $\dot{m}V_1^2/2$ of the unburned propellant and to the burning $\dot{m}E$, or

$$\text{Available energy input per unit time} = \dot{m}\left(E + \frac{V_1^2}{2}\right) \qquad (3.11.20)$$

The mechanical efficiency e is

$$e = \frac{\dot{m}V_1\sqrt{2E}}{\dot{m}(E + V_1^2/2)} = \frac{2v_r/V_1}{1 + (v_r/V_1)^2} \qquad (3.11.21)$$

When $v_r/V_1 = 1$, the maximum efficiency $e = 1$ is obtained. In this case the absolute velocity of ejected gas is zero.

When the thrust on a vertical rocket is greater than the total weight plus resistance, the rocket accelerates. Its mass is continuously reduced. To lift a rocket off its pad, its thrust $\dot{m}v_r$ must exceed its total weight.

EXAMPLE 3.19 (*a*) Determine the burning time for a rocket that initially weighs 500,000 kg$_f$, of which 70 percent is propellant. It consumes fuel at a constant rate, and its initial thrust is 10 percent greater than its weight. $v_r = 3300$ m/s. (*b*) Considering g constant at 9.8 m/s^2 and the flight to be vertical without air resistance, find the speed of the rocket at burnout time, its height above ground, and the maximum height it will attain.

(*a*) From the thrust relation
$$\dot{m}v_r = 1.1W_0 = (550,000 \text{ kg}_f)(9.806 \text{ N/kg}_f) = 5,393,300 \text{ N} = \dot{m}3300$$
and $\dot{m} = 1634.3$ kg/s. The available mass of propellant is 350,000 kg; hence the burning time is

$$\frac{350,000 \text{ kg}}{1634.3 \text{ kg/s}} = 214.2 \text{ s}$$

(b)   From Eq. (3.11.19)

$$\frac{dV_1}{dt} = \frac{(1634.3 \text{ kg/s})(3300 \text{ m/s}) - (9.8 \text{ m/s}^2)[350{,}000 \text{ kg} - (1634.3 \text{ kg/s})t + 150{,}000 \text{ kg}]}{150{,}000 \text{ kg} + 350{,}000 \text{ kg} - (1634.3 \text{ kg/s})t}$$

Simplifying gives

$$\frac{dV_1}{dt} = \frac{299.95 + 9.8t}{305.94 - t} = -9.8 + \frac{3298.16}{305.94 - t}$$

$$V_1 = -9.8t - 3298.16 \ln (305.94 - t) + \text{const}$$

when $t = 0$, $V_1 = 0$; hence

$$V_1 = -9.8t - 3298.16 \ln \left(1 - \frac{t}{305.94} \right)$$

when $t = 214.2$, $V_1 = 1873.24$ m/s. The height at $t = 214.2$ s is

$$y = \int_0^{214.2} V \, dt = -9.8 \frac{t^2}{2} \Big]_0^{214.2} - 3298.16 \int_0^{214.2} \ln \left(1 - \frac{t}{305.94} \right) dt$$

$$= 117.22 \text{ km}$$

The rocket will glide $V_1^2/2g$ ft higher after burnout, or

$$117{,}220 \text{ m} + \frac{1873.24^2}{2 \times 9.8} \text{ m} = 296.25 \text{ km}$$

Fixed and moving vanes

The theory of turbomachines is based on the relationships between jets and vanes. The mechanics of transfer of work and energy from fluid jets to moving vanes is studied as an application of the momentum principles. When a free jet impinges onto a smooth vane that is curved, as in Fig. 3.30, the jet is deflected, its momentum is changed, and a force is exerted on the vane. The jet is assumed to flow onto the vane in a tangential direction, without shock; and furthermore the frictional resistance between jet and vane is neglected. The velocity is assumed to be uniform throughout the jet upstream and downstream from the vane. Since the jet is open to the air, it has the same pressure

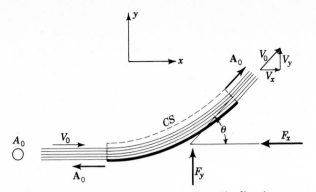

Fig. 3.30   Free jet impinging on smooth, fixed vane.

at each end of the vane. When the small change in elevation between ends, if any, is neglected, application of Bernoulli's equation shows that the magnitude of the velocity is unchanged for *fixed* vanes.

EXAMPLE 3.20   Find the force exerted on a fixed vane when a jet discharging 2 ft³/s water at 150 ft/s is deflected through 135°.

By referring to Fig. 3.30 and by applying Eq. (3.11.2) in the x and y directions, it is found that

$$-F_x = \rho V_0 \cos \theta\, V_0 A_0 + \rho V_0 (-V_0 A_0) \qquad F_y = \rho V_0 \sin \theta\, V_0 A_0$$

Hence,

$$F_x = -(1.935 \text{ slugs/ft}^3)\,(2 \text{ ft}^3/\text{s})\,(150 \cos 135° - 150 \text{ ft/s}) = 990 \text{ lb}$$

$$F_y = (1.935 \text{ slugs/ft}^3)\,(2 \text{ ft}^3/\text{s})\,(150 \sin 135° \text{ ft/s}) = 410 \text{ lb}$$

The force components on the fixed vane are then equal and opposite to F_x and F_y.

EXAMPLE 3.21   Fluid issues from a long slot and strikes against a smooth inclined flat plate (Fig. 3.31). Determine the division of flow and the force exerted on the plate, neglecting losses due to impact.

As there are no changes in elevation or pressure before and after impact, the magnitude of the velocity leaving is the same as the initial speed of the jet. The division of flow Q_1, Q_2 can be computed by applying the momentum

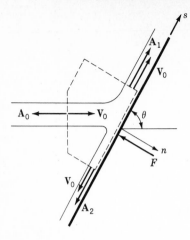

Fig. 3.31 Two-dimensional jet impinging on an inclined fixed plane surface.

equation in the s direction, parallel to the plate. No force is exerted on the fluid by the plate in this direction; hence, the final momentum component must equal the initial momentum component in the s direction. The steady-state momentum equation for the s direction, from Eq. (3.11.2), yields

$$\Sigma F_s = \int_{cs} \rho v_s \mathbf{V} \cdot d\mathbf{A} = 0 = \rho V_0 V_0 A_1 + \rho V_0 \cos \theta \, (-V_0 A_0) + \rho(-V_0) V_0 A_2$$

By substituting $Q_1 = V_0 A_1$, $Q_2 = V_0 A_2$, and $Q_0 = V_0 A_0$, it reduces to

$$Q_1 - Q_2 = Q_0 \cos \theta$$

and with the continuity equation,

$$Q_1 + Q_2 = Q_0$$

The two equations can be solved for Q_1 and Q_2,

$$Q_1 = \frac{Q_0}{2} (1 + \cos \theta) \qquad Q_2 = \frac{Q_0}{2} (1 - \cos \theta)$$

The force F exerted on the plate must be normal to it. For the momentum equation normal to the plate, Fig. 3.31,

$$\Sigma F_n = \int_{cs} \rho v_n \mathbf{V} \cdot d\mathbf{A} = -F = \rho V_0 \sin \theta \, (-V_0 A_0)$$

$$F = \rho Q_0 V_0 \sin \theta$$

Moving vanes

Turbomachinery utilizes the forces resulting from the motion over moving vanes. No work can be done on or by a fluid that flows over a fixed vane. When vanes can be displaced, work can be done either on the vane or on the fluid. In Fig. 3.32a a moving vane is shown with fluid flowing onto it tangentially. Forces exerted on the fluid by the vane are indicated by F_x and F_y. To analyze the flow the problem is reduced to steady state by superposition of vane velocity u to the left (Fig. 3.32b) on both vane and fluid. The control volume then encloses the fluid in contact with the vane, with its control surface normal to the flow at sections 1 and 2. Figure 3.32c shows the *polar vector diagram* for flow through the vane. The absolute-velocity vectors originate at the origin O, and the relative-velocity vector $\mathbf{V}_0 - \mathbf{u}$ is turned

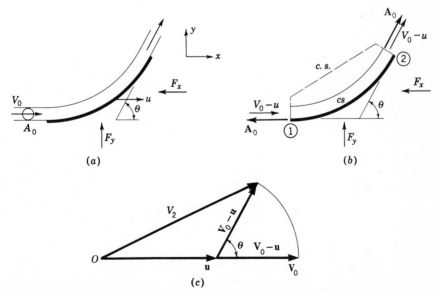

Fig. 3.32 (a) Moving vane. (b) Vane flow viewed as steady-state problem by superposition of velocity u to the left. (c) Polar vector diagram.

through the angle θ of the vane as shown. $\mathbf{V}_2$ is the final absolute velocity leaving the vane. The relative velocity $v_r = V_0 - u$ is unchanged in magnitude as it traverses the vane. The mass per unit time is given by $\rho A_0 v_r$ and is not the mass rate being discharged from the nozzle. If a *series of vanes* is employed, as on the periphery of a wheel, arranged so that one or another of the jets intercept all flow from the nozzle and so that the velocity is substantially u, then the mass per second is the total mass per second being discharged. Application of Eq. (3.11.2) to the control volume of Fig. 3.32b yields

$$\Sigma F_x = \int_{cs} \rho v_x \mathbf{V} \cdot d\mathbf{A} = -F_x = \rho(V_0 - u)\cos\theta\left[(V_0 - u)A_0\right]$$

$$+ \rho(V_0 - u)\left[-(V_0 - u)A_0\right]$$

or

$$F_x = \rho(V_0 - u)^2 A_0(1 - \cos\theta)$$

$$\Sigma F_y = \int_{cs} \rho v_y \mathbf{V} \cdot d\mathbf{A} = F_y = \rho(V_0 - u)\sin\theta\left[(V_0 - u)A_0\right]$$

or

$$F_y = \rho(V_0 - u)^2 A_0 \sin\theta$$

These relations are for the single vane. For a series of vanes they become

$$F_x = \rho Q_0(V_0 - u)(1 - \cos\theta) \qquad F_y = \rho Q_0(V_0 - u)\sin\theta$$

EXAMPLE 3.22 Determine for a single moving vane of Fig. 3.33a the force components due to the water jet and the rate of work done on the vane.

Figure 3.33b is the steady-state reduction with a control volume shown. The polar vector diagram is shown in Fig. 3.33c. By applying Eq. (3.11.2) in the x and y directions to the control volume of Fig. 3.33b

$$-F_x = (997.3 \text{ kg/m}^3)(60 \text{ m/s})(\cos 170°)\left[(60 \text{ m/s})(0.001 \text{ m}^2)\right]$$

$$+ (997.3 \text{ kg/m}^3)(60 \text{ m/s})\left[(-60 \text{ m/s})(0.001 \text{ m}^2)\right]$$

$$F_x = 3590 \text{ N}$$

$$F_y = (997.3 \text{ kg/m}^3)(60 \text{ m/s})(\sin 170°)\left[(60 \text{ m/s})(0.001 \text{ m}^2)\right] = 625 \text{ N}$$

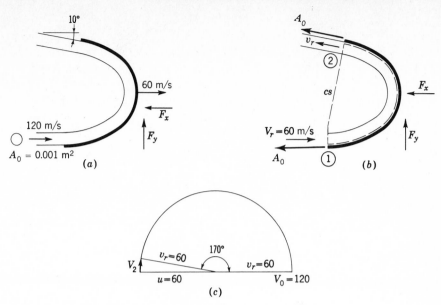

Fig. 3.33 Jet acting on a moving vane.

The power exerted on the vane is

$$uF_x = (60 \text{ m/s})(3590 \text{ N}) = 215.4 \text{ kW}$$

EXAMPLE 3.23 Determine the horsepower that can be obtained from a series of vanes (Fig. 3.34a), curved through 150°, moving 60 ft/s away from a 3.0-cfs water jet having a cross section of 0.03 ft². Draw the polar vector diagram and calculate the energy remaining in the jet.

The jet velocity is $V_0 = 3/0.03 = 100$ ft/s. The steady-state vane control volume is shown in Fig. 3.34b and the polar vector diagram in Fig. 3.34c. The force on the series of vanes in the x direction is

$$F_x = (1.935 \text{ slugs/ft}^3)(3 \text{ ft}^3/\text{s})(40 \text{ ft/s})(1 - \cos 150°) = 433 \text{ lb}$$

The horsepower is

$$\text{hp} = \frac{(433 \text{ lb})(60 \text{ ft/s})}{550 \text{ ft} \cdot \text{lb/s}} = 47.3$$

The components of absolute velocity leaving the vane are, from Fig. 3.34c,

$$V_{2x} = 60 - 40 \cos 30° = 25.4 \text{ ft/s} \qquad V_{2y} = 40 \sin 30° = 20 \text{ ft/s}$$

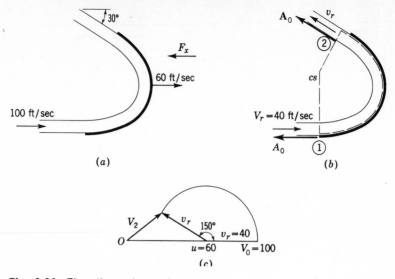

Fig. 3.34 Flow through moving vanes.

and the exit-velocity head is

$$\frac{V_2^2}{2g} = \frac{25.4^2 + 20^2}{64.4} = 16.2 \text{ ft} \cdot \text{lb/lb}$$

The kinetic energy remaining in the jet, in foot-pounds per second, is

$$Q\gamma \frac{V_2^2}{2g} = (3 \text{ ft}^3/\text{s})(62.4 \text{ lb/ft}^3)(16.2 \text{ ft}) = 3030 \text{ ft} \cdot \text{lb/s}$$

The initial kinetic energy available was

$$(3 \text{ ft}^3/\text{s})(62.4 \text{ lb/ft}^3) \frac{100^2}{64.4} \text{ ft} = 29,030 \text{ ft} \cdot \text{lb/s}$$

which is the sum of the work done and the energy remaining per second.

When a vane or series of vanes moves toward a jet, work is done by the vane system on the fluid, thereby increasing the energy of the fluid. Figure 3.35 illustrates this situation; the polar vector diagram shows the exit velocity to be greater than the entering velocity.

In turbulent flow, losses generally must be determined from experimental tests on the system or a geometrically similar model of the system. In the

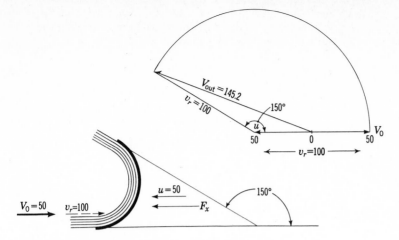

Fig. 3.35 Vector diagram for vane doing work on a jet.

following two cases, application of the continuity, energy, and momentum equations permits the losses to be evaluated analytically.

Losses due to sudden expansion in a pipe

The losses due to sudden enlargement in a pipeline may be calculated with both the energy and momentum equations. For steady, incompressible, turbulent flow through the control volume between sections 1 and 2 of the sudden expansion of Fig. 3.36a, b, the small shear force exerted on the walls between the two sections may be neglected. By assuming uniform velocity over the flow cross sections, which is approached in turbulent flow, application of Eq. (3.11.2) produces

$$p_1 A_1 - p_2 A_2 = \rho V_2 (V_2 A_2) + \rho V_1 (-V_1 A_1)$$

At section 1 the radial acceleration of fluid particles in the eddy along the

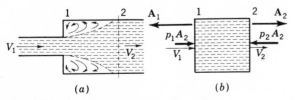

(a)   (b)

Fig. 3.36 Sudden expansion in a pipe.

surface is small, and so generally a hydrostatic pressure variation occurs across the section. The energy equation (3.10.1) applied to sections 1 and 2, with the loss term h_l, is (for $\alpha = 1$)

$$\frac{V_1^2}{2g} + \frac{p_1}{\gamma} = \frac{V_2^2}{2g} + \frac{p_2}{\gamma} + h_l$$

Solving for $(p_1 - p_2)/\gamma$ in each equation and equating the results give

$$\frac{V_2^2 - V_2 V_1}{g} = \frac{V_2^2 - V_1^2}{2g} + h_l$$

As $V_1 A_1 = V_2 A_2$,

$$h_l = \frac{(V_1 - V_2)^2}{2g} = \frac{V_1^2}{2g}\left(1 - \frac{A_1}{A_2}\right)^2 \tag{3.11.22}$$

which indicates that the losses in turbulent flow are proportional to the square of the velocity.

Hydraulic jump

The hydraulic jump is the second application of the basic equations to determine losses due to a turbulent-flow situation. Under proper conditions a rapidly flowing stream of liquid in an open channel suddenly changes to a slowly flowing stream with a larger cross-sectional area and a sudden rise in elevation of liquid surface. This phenomenon is known as the *hydraulic jump* and is an example of steady nonuniform flow. In effect, the rapidly flowing liquid jet expands (Fig. 3.37) and converts kinetic energy into potential energy and losses or irreversibilities. A roller develops on the inclined surface of the expanding liquid jet and draws air into the liquid. The surface of the jump is very rough and turbulent, the losses being greater as the jump height

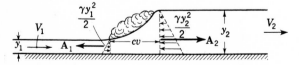

Fig. 3.37 Hydraulic jump in a rectangular channel.

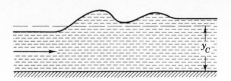

Fig. 3.38 Standing wave.

is greater. For small heights, the form of the jump changes to a standing wave (Fig. 3.38). The jump is discussed further in Sec. 11.4.

The relations between the variables for the hydraulic jump in a horizontal rectangular channel are easily obtained by use of the continuity, momentum, and energy equations. For convenience the width of channel is taken as unity. The continuity equation (Fig. 3.37) is ($A_1 = y_1$, $A_2 = y_2$)

$$V_1 y_1 = V_2 y_2$$

The momentum equation is

$$\frac{\gamma y_1^2}{2} - \frac{\gamma y_2^2}{2} = \rho V_2 (A_2 V_2) + \rho V_1 (-A_1 V_1)$$

and the energy equation (for points on the liquid surface) is

$$\frac{V_1^2}{2g} + y_1 = \frac{V_2^2}{2g} + y_2 + h_j$$

in which h_j represents losses due to the jump. Eliminating V_2 in the first two equations leads to

$$y_2 = -\frac{y_1}{2} + \sqrt{\left(\frac{y_1}{2}\right)^2 + \frac{2V_1^2 y_1}{g}} \qquad\qquad (3.11.23)$$

in which the plus sign has been taken before the radical (a negative y_2 has no physical significance). The depths y_1 and y_2 are referred to as *conjugate* depths. Solving the energy equation for h_j and eliminating V_1 and V_2 give

$$h_j = \frac{(y_2 - y_1)^3}{4 y_1 y_2} \qquad\qquad (3.11.24)$$

The hydraulic jump, which is a very effective device for creating irreversibilities, is commonly used at the ends of chutes or the bottoms of spillways to destroy much of the kinetic energy in the flow. It is also an effective mixing chamber, because of the violent agitation that takes place in the roller. Experimental measurements of hydraulic jumps show that the equations yield the correct value of y_2 to within 1 percent.

EXAMPLE 3.24   If 12 m³/s of water per meter of width flows down a spillway onto a horizontal floor and the velocity is 20 m/s, determine the downstream depth required to cause a hydraulic jump and the losses in power by the jump per meter of width

$$y_1 = \frac{12 \text{ m}^2/\text{s}}{20 \text{ m/s}} = 0.6 \text{ m}$$

Substituting into Eq. (3.11.23) gives

$$y_2 = -0.3 + \sqrt{0.3^2 + \frac{2 \times 20^2 \times 0.6}{9.806}} = 7 \text{ m}$$

With Eq. (3.11.24),

$$\text{Losses} = \frac{(7 - 0.6)^3}{4 \times 0.6 \times 7} = 15.6 \text{ m} \cdot \text{N/N}$$

Power/m $= \gamma Q (\text{losses}) = (9802 \text{ N/m}^3)(12 \text{ m}^3/\text{s})(15.6 \text{ m}) = 1840 \text{ kW}$

EXAMPLE 3.25   Find the head H in the reservoir of Fig. 3.39 needed to accelerate the flow of oil, $S = 0.85$, at the rate of 0.5 ft/s² when the flow is

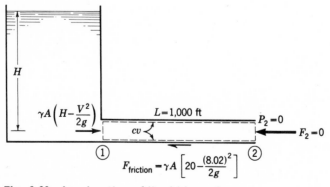

Fig. 3.39 Acceleration of liquid in a pipe.

8.02 ft/s. At 8.02 ft/s the steady-state head on the pipe is 20 ft. Neglect entrance loss.

The oil may be considered to be incompressible and to be moving uniformly in the pipeline. By applying Eq. (3.11.2), the last term is zero, as the net efflux is zero,

$$\gamma A \left(H - \frac{V^2}{2g} \right) - \gamma A \left(20 - \frac{8.02^2}{2g} \right) = \frac{\partial}{\partial t} \left(\frac{\gamma}{g} ALV \right)$$

or

$$H = 20 + \frac{1000}{32.2} \times 0.5 = 35.52 \text{ ft}$$

3.12 THE MOMENT-OF-MOMENTUM EQUATION

The general unsteady linear momentum equation applied to a control volume, Eq. (3.11.1), is

$$\mathbf{F} = \frac{\partial}{\partial t} \int_{cv} \rho \mathbf{v} \, d\mathcal{V} + \int_{cs} \rho \mathbf{v} \mathbf{v} \cdot d\mathbf{A} \qquad (3.12.1)$$

The moment of a force $\mathbf{F}$ about a point O (Fig. 3.40) is given by

$$\mathbf{r} \times \mathbf{F}$$

which is the cross, or vector, product of $\mathbf{F}$ and the position vector $\mathbf{r}$ of a point on the line of action of the vector from O. The cross product of two vectors is a vector at right angles to the plane defined by the first two vectors and with magnitude

$$Fr \sin \theta$$

which is the product of F and the shortest distance from O to the line of action of $\mathbf{F}$. The sense of the final vector follows the right-hand rule. In Fig. 3.40 the force tends to cause a counterclockwise rotation around O. If this were a right-hand screw thread turning in this direction, it would tend to come up, and so the vector is likewise directed up out of the paper. If one curls the fingers of the right hand in the direction the force tends to cause rotation, the thumb yields the direction, or sense, of the vector.

Fig. 3.40 Notation for moment of a vector.

By taking $\mathbf{r} \times \mathbf{F}$, using Eq. (3.12.1),

$$\mathbf{r} \times \mathbf{F} = \frac{\partial}{\partial t} \int_{cv} \rho \mathbf{r} \times \mathbf{v} \, d\mathcal{V} + \int_{cs} (\rho \mathbf{r} \times \mathbf{v})(\mathbf{v} \cdot d\mathbf{A}) \tag{3.12.2}$$

The left-hand side of the equation is the torque exerted by any forces on the control volume, and terms on the right-hand side represent the rate of change of *moment of momentum* within the control volume plus the net efflux of moment of momentum from the control volume. This is the general moment-of-momentum equation for a control volume. It has great value in the analysis of certain flow problems, e.g., in turbomachinery, where torques are more significant than forces.

When Eq. (3.12.2) is applied to a case of flow in the xy plane, with r the shortest distance to the tangential component of the velocity v_t, as in Fig. 3.41, and v_n is the normal component of velocity,

$$F_t r = T_z = \int_{cs} \rho r v_t v_n \, dA + \frac{\partial}{\partial t} \int_{cv} \rho r v_t \, d\mathcal{V} \tag{3.12.3}$$

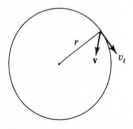

Fig. 3.41 Notation for two-dimensional flow.

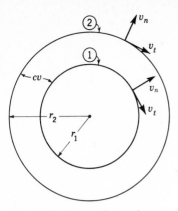

Fig. 3.42 Flow through annular passage such as a centrifugal pump impeller.

in which T_z is the torque. A useful form of Eq. (3.12.3) applied to an annular control volume, in steady flow (Fig. 3.42), is

$$T_z = \int_{A_2} \rho_2 r_2 v_{t_2} v_{n_2} \, dA_2 - \int_{A_1} \rho_1 r_1 v_{t_1} v_{n_1} \, dA_1 \tag{3.12.4}$$

For complete circular symmetry, where r, ρ, v_t, and v_n are constant over the inlet and outlet control surfaces, it takes the simple form

$$T_z = \rho Q[(rv_t)_2 - (rv_t)_1] \tag{3.12.5}$$

since $\int \rho v_n \, dA = \rho Q$, the same at inlet or outlet.

EXAMPLE 3.26 The sprinkler shown in Fig. 3.43 discharges water upward and outward from the horizontal plane so that it makes an angle of $\theta°$ with the t axis when the sprinkler arm is at rest. It has a constant cross-sectional flow area of A_0 and discharges q cfs starting with $\omega = 0$ and $t = 0$. The resisting torque due to bearings and seals is the constant T_0, and the moment of inertia of the rotating empty sprinkler head is I_s. Determine the equation for ω as a function of time.

Equation (3.12.2) may be applied. The control volume is the cylindical area enclosing the rotating sprinkler head. The inflow is along the axis, so that it has no moment of momentum; hence the torque $-T_0$ due to friction is equal to the time rate of change of moment of momentum of sprinkler head and fluid within the sprinkler head plus the net efflux of moment of momentum

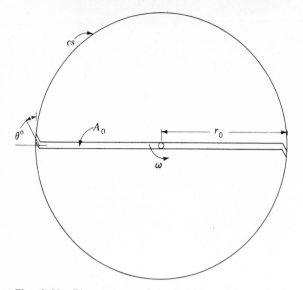

Fig. 3.43 Plan view of sprinkler and control surface.

from the control volume. Let $V_r = q/2A_0$

$$-T_0 = 2\frac{d}{dt}\int_0^{r_0} A_0\rho\omega r^2\,dr + I_s\frac{d\omega}{dt} - \frac{2\rho qr_0}{2}\,(V_r\cos\theta - \omega r_0)$$

The total derivative may be used. Simplifying gives

$$\frac{d\omega}{dt}\,(I_s + \tfrac{2}{3}\rho A_0 r_0^3) = \rho q r_0(V_r\cos\theta - \omega r_0) - T_0$$

For rotation to start, $\rho q r_0 V_r \cos\theta$ must be greater than T_0. The equation is easily integrated to find ω as a function of t. The final value of ω is obtained by setting $d\omega/dt = 0$ in the equation.

EXAMPLE 3.27 A turbine discharging 10 m³/s is to be designed so that a torque of 10,000 N·m is to be exerted on an impeller turning at 200 rpm that takes all the moment of momentum out of the fluid. At the outer periphery of the impeller, $r = 1$ m. What must the tangential component of velocity be at this location?

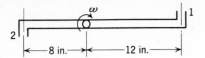

Fig. 3.44 Rotating jet system.

Equation (3.12.5) is

$$T = \rho Q(rv_t)_{in}$$

in this case, since the outflow has $v_t = 0$. By solving for $v_{t_{in}}$

$$v_{t_{in}} = \frac{T}{\rho Q r} = \frac{10{,}000 \text{ N} \cdot \text{m}}{(997.3 \text{ kg/m}^3)(10 \text{ m}^3/\text{s})(1 \text{ m})} = 1.003 \text{ m/s}$$

EXAMPLE 3.28 The sprinkler of Fig. 3.44 discharges 0.01 cfs through each nozzle. Neglecting friction, find its speed of rotation. The area of each nozzle opening is 0.001 ft².

The fluid entering the sprinkler has no moment of momentum, and no torque is exerted on the system externally; hence the moment of momentum of fluid leaving must be zero. Let ω be the speed of rotation; then the moment of momentum leaving is

$$\rho Q_1 r_1 v_{t1} + \rho Q_2 r_2 v_{t2}$$

in which v_{t1} and v_{t2} are absolute velocities. Then

$$v_{t1} = v_{r1} - \omega r_1 = \frac{Q_1}{0.001} - \omega r_1 = 10 - \omega$$

and

$$v_{t2} = v_{r2} - \omega r_2 = 10 - \tfrac{2}{3}\omega$$

For moment of momentum to be zero,

$$\rho Q(r_1 v_{t1} + r_2 v_{t2}) = 0 \quad \text{or} \quad 10 - \omega + \tfrac{2}{3}(10 - \tfrac{2}{3}\omega) = 0$$

and $\omega = 11.54$ rad/s or 110.2 rpm.

PROBLEMS

3.1 A pipeline leads from one water reservoir to another which has its water surface 10 m lower. For a discharge of 0.5 m³/s determine the losses in meter-newtons per kilogram and in kilowatts.

3.2 A pump which is located 10 ft above the surface of a lake expels a jet of water vertically upward a distance of 50 ft. If 0.5 cfs is being pumped by a 5-hp electric motor running at rated capacity, what is the efficiency of the motor-pump combination? What is the irreversibility of the pump system when comparing the zenith of the jet and the lake surface? What is the irreversibility after the water falls to the lake surface?

3.3 A blower delivers 2 m³/s air, $\rho = 1.3$ kg/m³, at an increase in pressure of 10 cm water. It is 72 percent efficient. Determine the irreversibility of the blower in meter-newtons per kilogram and in kilowatts, and determine the torque in the shaft if the blower turns at 1800 rpm.

3.4 A three-dimensional velocity distribution is given by $u = -x, v = 2y, w = 5 - z$. Find the equation of the streamline through $(2,1,1)$.

3.5 A two-dimensional flow can be described by $u = -y/b^2, v = x/a^2$. Verify that this is the flow of an incompressible fluid and that the ellipse $x^2/a^2 + y^2/b^2 = 1$ is a streamline.

3.6 Oil is flowing in a laminar way in a pipeline at the rate of 300 gpm (gallons per minute). The irreversibilities are measured at 20 ft-lb/lb$_m$. (*a*) What losses should be expected if the flow is reduced to 200 gpm? (*b*) What losses should be encountered for a flow of 450 gpm?

3.7 In a flow of liquid through a pipeline the losses are 3 kW for average velocity of 2 m/s and 6 kW for 3 m/s. What is the nature of the flow?

3.8 When tripling the flow in a line causes the losses to increase by 7.64 times, how do the losses vary with velocity and what is the nature of the flow?

3.9 In two-dimensional flow around a circular cylinder (Fig. 3.3), the discharge between streamlines is 0.01 cfs/ft. At a great distance the streamlines are 0.20 in apart, and at a point near the cylinder they are 0.12 in apart. Calculate the magnitude of the velocity at these two points.

3.10 A pipeline carries oil, sp gr 0.86, at $V = 2$ m/s through 20-cm-ID pipe. At another section the diameter is 5 cm. Find the velocity at this section and the mass rate of flow in kilograms per second.

3.11 Hydrogen is flowing in a 2.0-in-diameter pipe at the mass rate of 0.03 lb$_m$/s. At section 1 the pressure is 40 psia and $t = 40°F$. What is the average velocity?

3.12 A nozzle with a base diameter of 8 cm and with 2-cm-diameter tip discharges 10 l/s. Derive an expression for the fluid velocity along the axis of the nozzle. Measure the distance x along the axis from the plane of the larger diameter.

3.13 An 18-ft-diameter pressure pipe has a velocity of 10 ft/s. After passing through a reducing bend the flow is in a 16-ft-diameter pipe. If the losses vary as the square of the velocity, how much greater are they through the 16-ft pipe than through the 18-ft pipe per 1000 ft of pipe?

3.14 Does the velocity distribution of Prob. 3.4 for incompressible flow satisfy the continuity equation?

3.15 Does the velocity distribution

$$\mathbf{q} = \mathbf{i}(5x) + \mathbf{j}(5y) + \mathbf{k}(-10z)$$

satisfy the law of mass conservation for incompressible flow?

3.16 Consider a cube with 1-m edges parallel to the coordinate axes located in the first quadrant with one corner at the origin. By using the velocity distribution of Prob. 3.15, find the flow through each face and show that no mass is being accumulated within the cube if the fluid is of constant density.

3.17 Find the flow (per foot in the z direction) through each side of the square with corners at $(0,0)$, $(0,1)$, $(1,1)$, $(1,0)$ due to

$$\mathbf{q} = \mathbf{i}(16y - 12x) + \mathbf{j}(12y - 9x)$$

and show that continuity is satisfied.

3.18 Show that the velocity

$$\mathbf{q} = \mathbf{i}\frac{4x}{x^2 + y^2} + \mathbf{j}\frac{4y}{x^2 + y^2}$$

satisfies continuity at every point except the origin.

3.19 Problem 3.18 is a velocity distribution that is everywhere radial from the origin with magnitude $v_r = 4/r$. Show that the flow through each circle concentric with the origin (per unit length in the z direction) is the same.

3.20 Perform the operation $\nabla \cdot \mathbf{q}$ on the velocity vectors of Probs. 3.15, 3.17, and 3.18.

3.21 By introducing the following relationships between cartesian coordinates and plane polar coordinates, obtain a form of the continuity equation in plane polar coordinates:

$$x^2 + y^2 = r^2 \qquad \frac{y}{x} = \tan\theta \qquad \frac{\partial}{\partial x} = \frac{\partial}{\partial r}\frac{\partial r}{\partial x} + \frac{\partial}{\partial \theta}\frac{\partial \theta}{\partial x}$$

$$u = v_r \cos\theta - v_\theta \sin\theta \qquad v = v_r \sin\theta + v_\theta \cos\theta$$

Does the velocity given in Prob. 3.19 satisfy the equation that has been derived?

3.22 A standpipe 20 ft in diameter and 50 ft high is filled with water. How much potential energy is in this water if the elevation datum is taken 10 ft below the base of the standpipe?

3.23 How much work could be obtained from the water of Prob. 3.22 if run through a 100 percent efficient turbine that discharged into a reservoir with elevation 30 ft below the base of the standpipe?

3.24 What is the kinetic-energy flux in meter-newtons per second of 0.01 m^3/s of oil, sp gr 0.80, discharging through a 5-cm diameter nozzle?

3.25 By neglecting air resistance, determine the height a vertical jet of water will rise, with velocity 40 ft/s.

3.26 If the water jet of Prob. 3.25 is directed upward 45° with the horizontal and air resistance is neglected, how high will it rise and what is the velocity at its high point?

3.27 Show that the work a liquid can do by virtue of its pressure is $\int p \, d\mathcal{V}$, in which $\mathcal{V}$ is the volume of liquid displaced.

3.28 What angle α of jet is required to reach the roof of the building of Fig. 3.45 with minimum jet velocity V_0 at the nozzle? What is the value of V_0?

3.29 The velocity distribution in laminar flow in a pipe is given by

$$v = V_{max}\left[1 - (r/r_0)^2\right]$$

Determine the average velocity and the kinetic-energy correction factor.

3.30 For highly turbulent flow the velocity distribution in a pipe is given by

$$\frac{v}{v_{max}} = \left(\frac{y}{r_0}\right)^{1/9}$$

with y the wall distance and r_0 the pipe radius. Determine the kinetic-energy correction factor for this flow.

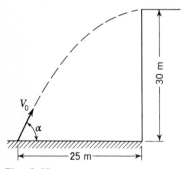

Fig. 3.45

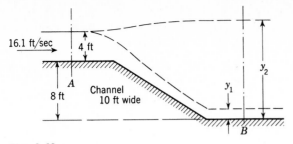

Fig. 3.46

3.31 The velocity distribution between two parallel plates separated by a distance a is

$$u = -10\frac{y}{a} + 20\frac{y}{a}\left(1 - \frac{y}{a}\right)$$

in which u is the velocity component parallel to the plate and y is measured from, and normal to, the lower plate. Determine the volume rate of flow and the average velocity. What is the time rate of flow of kinetic energy between the plates? In what direction is the kinetic energy flowing?

3.32 What is the efflux of kinetic energy out of the cube given by Prob. 3.16 for the velocity prescribed in Prob. 3.15?

3.33 Water is flowing in a channel, as shown in Fig. 3.46. Neglecting all losses, determine the two possible depths of flow y_1 and y_2.

3.34 High-velocity water flows up an inclined plane as shown in Fig. 3.47. Neglecting all losses, calculate the two possible depths of flow at section B.

3.35 If the losses from section A to section B of Fig. 3.46 are 1.5 ft-lb/lb, determine the two possible depths at section B.

3.36 In Fig. 3.47 the situation exists under which each kilogram of water increases in temperature 0.0006°C due to losses incurred in flowing between A and B. Determine the lower depth of flow at section B.

3.37 Neglecting all losses, in Fig. 3.46 the channel narrows in the drop to 6 ft wide at

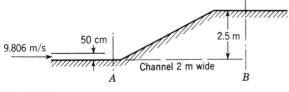

Fig. 3.47

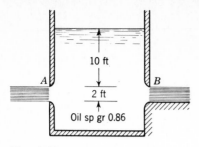

Fig. 3.48

section B. For uniform flow across section B, determine the two possible depths of flow.

3.38   In Fig. 3.47 the channel changes in width from 2 m at section A to 3 m at section B. For losses of 0.3 m·N/N between sections A and B, find the two possible depths at section B.

3.39   Some steam locomotives had scoops installed that took water from a tank between the tracks and lifted it into the water reservoir in the tender. To lift the water 12 ft with a scoop, neglecting all losses, what speed is required? *Note:* Consider the locomotive stationary and the water moving toward it, to reduce to a steady-flow situation.

3.40   In Fig. 3.48 oil discharges from a "two-dimensional" slot as indicated at A into the air. At B oil discharges from under a gate onto a floor. Neglecting all losses, determine the discharges of A and at B per foot of width. Why do they differ?

3.41   At point A in a pipeline carrying water the diameter is 1 m, the pressure 1 kg$_f$/cm^2, and the velocity 1 m/s. At point B, 2 m higher than A, the diameter is 0.5 m, and the pressure 0.2 kg$_f$/cm^2. Determine the direction of flow.

3.42   Neglecting losses, determine the discharge in Fig. 3.49.

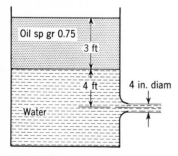

Fig. 3.49

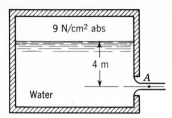

9 N/cm² abs

4 m

A

Water

Fig. 3.50

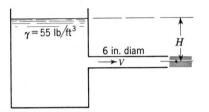

$\gamma = 55$ lb/ft³

6 in. diam

H

$\rightarrow V$

Fig. 3.51

3.43 For losses of 0.1 m ·N/N, find the velocity at A in Fig. 3.50. Barometer reading is 750 mm Hg.

3.44 The losses in Fig. 3.51 for $H = 20$ ft are $3V^2/2g$ ft·lb/lb. What is the discharge?

3.45 For flow of 750 gpm in Fig. 3.51, determine H for losses of $15V^2/2g$ ft·lb/lb.

3.46 For 1500-gpm flow and $H = 32$ ft in Fig. 3.51, calculate the losses through the system in velocity heads, $KV^2/2g$.

3.47 In Fig. 3.52 the losses up to section A are $4V_1^2/2g$, and the nozzle losses are $0.05V_2^2/2g$. Determine the discharge and the pressure at A. $H = 8$ m.

3.48 For pressure at A of 25,000 Pa in Fig. 3.52 with the losses in Prob. 3.47, determine the discharge and the head H.

3.49 The pumping system shown in Fig. 3.53 must have pressure of 5 psi in the

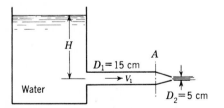

H

$D_1 = 15$ cm

A

$\rightarrow V_1$

Water

$D_2 = 5$ cm

Fig. 3.52

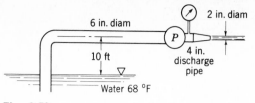

Fig. 3.53

discharge line when cavitation is incipient at the pump inlet. Calculate the length of pipe from the reservoir to the pump for this operating condition if the loss in this pipe can be expressed as $(V_1^2/2g)(0.03L/D)$. What horsepower is being supplied to the fluid by the pump? What percentage of this power is being used to overcome losses? Barometer reads 30 in Hg.

3.50 Neglecting losses and surface-tension effects, derive an equation for the water surface r of the jet of Fig. 3.54, in terms of y/H.

3.51 In the siphon of Fig. 3.55, $h_1 = 1$ m, $h_2 = 3$ m, $D_1 = 3$ m, $D_2 = 5$ m, and the losses to section 2 are $2.6V_2^2/2g$, with 10 percent of the losses occurring before section 1. Find the discharge and the pressure at section 1.

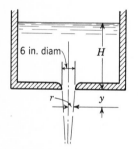

Fig. 3.54

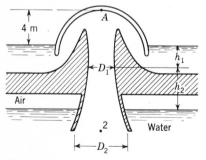

Fig. 3.55

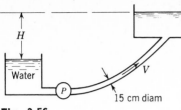

Fig. 3.56

3.52 Find the pressure at A of Prob. 3.51 if it is a stagnation point (velocity zero).

3.53 The siphon of Fig. 3.17 has a nozzle 6 in long attached at section 3, reducing the diameter to 6 in. For no losses, compute the discharge and the pressure at sections 2 and 3.

3.54 With exit velocity V_E in Prob. 3.53 and losses from 1 to 2 of $1.7V_2^2/2g$, from 2 to 3 of $0.9V_2^2/2g$, and through the nozzle $0.06V_E^2/2g$, calculate the discharge and the pressure at sections 2 and 3.

3.55 Determine the shaft horsepower for an 80 percent efficient pump to discharge 30 l/s through the system of Fig. 3.56. The system losses, exclusive of pump losses, are $10V^2/2g$, and $H = 16$ m.

3.56 The fluid horsepower $(Q\gamma H_p/550)$ produced by the pump of Fig. 3.56 is 10. For $H = 60$ ft and system losses of $8\ V^2/2g$, determine the discharge and the pump head H_p. Draw the hydraulic and energy grade lines.

3.57 If the overall efficiency of the system and turbine in Fig. 3.57 is 80 percent, what horsepower is produced for $H = 200$ ft and $Q = 1000$ cfs?

3.58 Losses through the system of Fig. 3.57 are $4V^2/2g$, exclusive of the turbine. The turbine is 90 percent efficient and runs at 240 rpm. To produce 1000 hp for $H = 300$ ft, determine the discharge and torque in the turbine shaft. Draw the energy and hydraulic grade lines.

3.59 Neglecting losses, find the discharge through the venturi meter of Fig. 3.58.

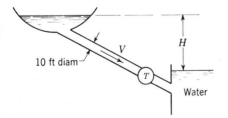

Fig. 3.57

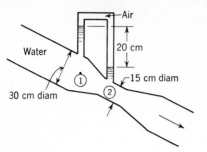

Fig. 3.58

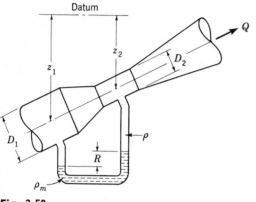

Fig. 3.59

3.60   For the venturi meter and manometer installation shown in Fig. 3.59 derive an expression relating the volume rate of flow with the manometer reading.

3.61   With losses of $0.2V_1^2/2g$ between sections 1 and 2 of Fig. 3.58, calculate the flow in gallons per minute.

3.62   In Fig. 3.60 determine V for $R = 12$ in.

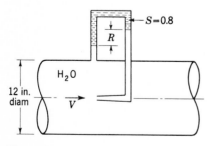

Fig. 3.60

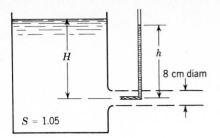

Fig. 3.61

3.63 In Fig. 3.61 $H = 6$ m, and $h = 5.75$ m. Calculate the discharge and the losses in meter-newtons per newton and in watts.

3.64 Neglecting losses, calculate H in terms of R for Fig. 3.62.

3.65 For losses of $0.1H$ through the nozzle of Fig. 3.62, what is the gage difference R in terms of H?

3.66 A liquid flows through a long pipeline with losses of 4 ft ·lb/lb per 100 ft of pipe. What is the slope of the hydraulic and energy grade lines?

3.67 In Fig. 3.63, 100 l/s water flows from section 1 to section 2 with losses of $0.4(V_1 - V_2)^2/2g$; $p_1 = 75{,}000$ Pa. Compute p_2 and plot the energy and hydraulic grade lines through the diffuser.

3.68 In an isothermal, reversible flow at 200°F, 2 Btu/s heat is added to 14 slugs/s flowing through a control volume. Calculate the entropy increase in foot-pounds per slug per degree Rankine.

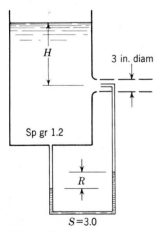

Fig. 3.62

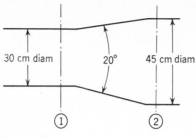

30 cm diam   20°   45 cm diam

①   ②

Fig. 3.63

3.69 In isothermal flow of a real fluid through a pipe system the losses are 20 m $\cdot$N/kg per 100 m, and 0.02 kcal/s per 100 m heat transfer from the fluid is required to hold the temperature at 10°C. What is the entropy change Δs in meter-newtons per kilogram per Kelvin of pipe system for 4 kg/s flowing?

3.70 Determine the momentum correction factor for the velocity distribution of Prob. 3.29.

3.71 Calculate the average velocity and momentum correction factor for the velocity distribution in a pipe,

$$\frac{v}{v_{\max}} = \left(\frac{y}{r_0}\right)^{1/n}$$

with y the wall distance and r_0 the pipe radius.

3.72 By introducing $V + v'$ for v into Eq. (3.11.4) show that $\beta \geq 1$. The term v' is the variation of v from the average velocity V and can be positive or negative.

3.73 Determine the time rate of x momentum passing out of the cube of Prob. 3.16. (*Hint:* Consider all six faces of the cube.)

3.74 Calculate the y-momentum efflux from the figure described in Prob. 3.17 for the velocity given there.

3.75 If gravity acts in the negative z direction, determine the z component of the force acting on the fluid within the cube described in Prob. 3.16 for the velocity specified there.

3.76 Find the y component of the force acting on the control volume given in Prob. 3.17 for the velocity given there. Consider gravity to be acting in the negative y direction.

3.77 What force components F_x, F_y are required to hold the black box of Fig. 3.64 stationary? All pressures are zero gage.

3.78 What force F (Fig. 3.65) is required to hold the plate for oil flow, sp gr 0.83, for $V_0 = 20$ m/s?

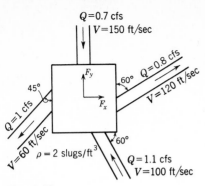

$Q=0.7$ cfs
$V=150$ ft/sec

F_y

$60°$ $Q=0.8$ cfs
$V=120$ ft/sec

$45°$

F_x

$Q=1$ cfs
$V=60$ ft/sec $\rho=2$ slugs/ft^3

$60°$

$Q=1.1$ cfs
$V=100$ ft/sec

Fig. 3.64

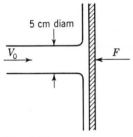

5 cm diam

V_0

F

Fig. 3.65

3.79 How much is the apparent weight of the tank full of water (Fig. 3.66) increased by the steady jet flow into the tank?

3.80 Does a nozzle on a fire hose place the hose in tension or in compression?

3.81 When a jet from a nozzle is used to aid in maneuvering a fireboat, can more force be obtained by directing the jet against a solid surface such as a wharf than by allowing it to discharge into air?

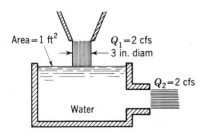

Area $=1$ ft^2

$Q_1=2$ cfs
3 in. diam

$Q_2=2$ cfs

Water

Fig. 3.66

3.82 Work Example 3.16 with the flow direction reversed, and compare results.

3.83 In the reducing bend of Fig. 3.23, $D_1 = 4$ m, $D_2 = 3$ m, $\theta = 135°$, $Q = 50$ m^3/s, $W = 40{,}000$ kg$_f$, $z = 2$ m, $p_2 = 1.4$ MPa, $x = 2.2$ m, and losses may be neglected. Find the force components and the line of action of the force which must be resisted by an anchor block.

3.84 20 ft^3/s of water flows through an 18-in-diameter pipeline that contains a horizontal 90° bend. The pressure at the entrance to the bend is 10 psi. Determine the force components, parallel and normal to the approach velocity, required to hold the bend in place. Neglect losses.

3.85 Oil, sp gr 0.83, flows through a 90° expanding pipe bend from 40- to 60-cm-diameter pipe. The pressure at the bend entrance is 1.3 kg$_f$/cm^2, and losses are to be neglected. For 0.6 m^3/s, determine the force components (parallel and normal to the approach velocity) necessary to support the bend.

3.86 Work Prob. 3.85 with elbow losses of $0.6V_1^2/2g$, with V_1 the approach velocity, and compare results.

3.87 A 4-in-diameter steam line carries saturated steam at 1400 ft/s velocity. Water is entrained by the steam at the rate of 0.3 lb/s. What force is required to hold a 90° bend in place owing to the entrained water?

3.88 Neglecting losses, determine the x and y components of force needed to hold the tee (Fig. 3.67) in place. The plane of the tee is horizontal.

3.89 Determine the net force on the sluice gate shown in Fig. 3.68. Neglect losses. By noting that the pressure at A and B is atmospheric, sketch the pressure distribution on the surface AB. Is it a hydrostatic distribution? How is it related to the force just calculated?

3.90 The vertical reducing section shown in Fig. 3.69 contains oil, sp gr 0.86, flowing

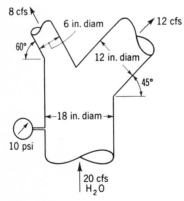

Fig. 3.67

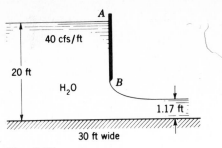

Fig. 3.68

upward at the rate of 0.5 m³/s. The pressure at the larger section is 1.5 kg$_f$/cm². Neglecting losses but including gravity, determine the force on the contraction.

3.91 Apply the momentum and energy equations to a windmill as if it were a propeller, noting that the slipstream is slowed down and expands as it passes through the blades. Show that the velocity through the plane of the blades is the average of the velocities in the slipstream at the downstream and upstream sections. By defining the theoretical efficiency (neglecting all losses) as the power output divided by the power available in an undisturbed jet having the area at the plane of the blades, determine the maximum theoretical efficiency of a windmill.

3.92 An airplane with propeller diameter of 8.0 ft travels through still air ($\rho =$ 0.0022 slug/ft³) at 210 mph. The speed of air through the plane of the propeller is 280 mph relative to the airplane. Calculate (*a*) the thrust on the plane, (*b*) the kinetic energy per second remaining in the slipstream, (*c*) the theoretical horsepower required to drive the propeller, (*d*) the propeller efficiency, and (*e*) the pressure difference across the blades.

3.93 A boat traveling at 40 km/h has a 60-cm-diameter propeller that discharges 4.5 m³/s through its blades. Determine the thrust on the boat, the theoretical efficiency of the propulsion system, and the power input to the propeller.

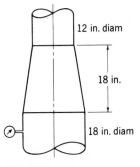

Fig. 3.69

3.94 A ship propeller has a theoretical efficiency of 60 percent. If it is 4 ft in diameter and the ship travels 20 mph, what is the thrust developed and what is the theoretical horsepower required?

3.95 A jet-propelled airplane traveling 1000 km/h takes in 40 kg/s air and discharges it at 550 m/s relative to the airplane. Neglecting the weight of fuel, find the thrust produced.

3.96 A jet-propelled airplane travels 700 mph. It takes in 165 lb_m/s of air, burns 3 lb_m/s of fuel, and develops 8000 lb of thrust. What is the velocity of the exhaust gas?

3.97 What is the theoretical mechanical efficiency of the jet engine of Prob. 3.96?

3.98 A boat requires a 1800-kg_f thrust to keep it moving at 25 km/h. How many cubic meters of water per second must be taken in and ejected through a 45-cm pipe to maintain this motion? What is the overall efficiency if the pumping system is 60 percent efficient?

3.99 In Prob. 3.98 what would be the required discharge if water were taken from a tank inside the boat and ejected from the stern through 45-cm pipe?

3.100 Determine the size of jet pipe and the theoretical power necessary to produce a thrust of 1000 kg_f on a boat moving 12 m/s when the propulsive efficiency is 68 percent.

3.101 In Fig. 3.70, a jet, $\rho = 2$ slugs/ft³, is deflected by a vane through 180°. Assume that the cart is frictionless and free to move in a horizontal direction. The cart weighs 200 lb. Determine the velocity and the distance traveled by the cart 10 s after the jet is directed against the vane. $A_0 = 0.02$ ft²; $V_0 = 100$ ft/s.

3.102 A rocket burns 260 lb_m/s fuel, ejecting hot gases at 6560 ft/s relative to the rocket. How much thrust is produced at 1500 and 3000 mph?

3.103 What is the mechanical efficiency of a rocket moving at 1200 m/s that ejects gas at 1800 m/s relative to the rocket?

3.104 Can a rocket travel faster than the velocity of ejected gas? What is the mechanical efficiency when it travels 12,000 ft/s and the gas is ejected at 8000 ft/s relative to the rocket? Is a positive thrust developed?

3.105 In Example 3.19 what is the thrust just before the completion of burning?

3.106 Neglecting air resistance, what velocity would a vertically directed V-2 rocket

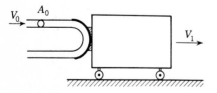

Fig. 3.70

attain in 68 s if it starts from rest, initially weighs 13,000 kg$_f$, burns 124 kg/s fuel, and ejects gas at $v_r = 1950$ m/s? Consider $g = 9.8$ m/s².

3.107 What altitude has the rocket of Prob. 3.106 reached after 68 s?

3.108 If the fuel supply is exhausted after 68 s (burnout), what is the maximum height of the rocket of Prob. 3.106?

3.109 What is the thrust of the rocket of Prob. 3.106 after 68 s?

3.110 Draw the polar vector diagram for a vane, angle θ, doing work on a jet. Label all vectors.

3.111 Determine the resultant force exerted on the vane of Fig. 3.30. $A_0 = 0.1$ ft²; $V_0 = 100$ ft/s; $\theta = 60°$, $\gamma = 55$ lb/ft³. How can the line of action be determined?

3.112 In Fig. 3.31, 45 percent of the flow is deflected in one direction. What is the plate angle θ?

3.113 A flat plate is moving with velocity u into a jet, as shown in Fig. 3.71. Derive the expression for power required to move the plate.

3.114 At what speed u should the cart of Fig. 3.71 move away from the jet in order to produce maximum power from the jet?

3.115 Calculate the force components F_x, F_y needed to hold the stationary vane of Fig. 3.72. $Q_0 = 80$ l/s; $\rho = 1000$ kg/m³; $V_0 = 100$ m/s.

3.116 If the vane of Fig. 3.72 moves in the x direction at $u = 40$ ft/s, for $Q = 2$ ft³/s, $\rho = 1.935$ slugs/ft³, $V_0 = 150$ ft/s, what are the force components F_x, F_y?

3.117 For the flow divider of Fig. 3.73 find the force components for the following conditions: $Q_0 = 10$ l/s, $Q_1 = 3$ l/s, $\theta_0 = 45°$, $\theta_1 = 30°$, $\theta_2 = 120°$, $V_0 = 10$ m/s, $\rho = 830$ kg/m³.

3.118 Solve the preceding problem by graphical vector addition.

3.119 At what speed u should the vane of Fig. 3.32 travel for maximum power from the jet? What should be the angle θ for maximum power?

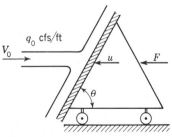

q_0 cfs/ft

V_0

u

F

θ

Fig. 3.71

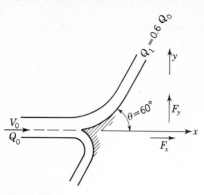

Fig. 3.72

3.120 Draw the polar vector diagram for the moving vane of Fig. 3.32 for $V_0 =$ 100 ft/s, $u = 60$ ft/s, and $\theta = 160°$.

3.121 Draw the polar vector diagram for the moving vane of Fig. 3.32 for $V_0 =$ 40 m/s, $u = -20$ m/s, and $\theta = 150°$.

3.122 What horsepower can be developed from (a) a single vane and (b) a series of vanes (Fig. 3.32) when $A_0 = 10$ in², $V_0 = 240$ ft/s, $u = 80$ ft/s, and $\theta = 173°$, for water flowing?

3.123 Determine the blade angles θ_1 and θ_2 of Fig. 3.74 so that the flow enters the vane tangent to its leading edge and leaves with no x component of absolute velocity.

3.124 Determine the vane angle required to deflect the absolute velocity of a jet 120° (Fig. 3.75).

3.125 In Prob. 3.39 for pickup of 30 l/s water at locomotive speed of 60 km/h, what force is exerted parallel to the tracks?

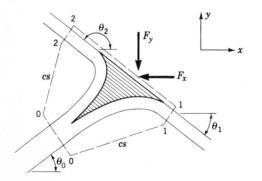

Fig. 3.73

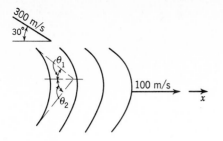

Fig. 3.74

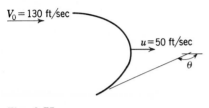

Fig. 3.75

3.126 Figure 3.76 shows an orifice called a Borda mouthpiece. The tube is long enough for the fluid velocity near the bottom of the tank to be nearly zero. Calculate the ratio of the jet area to the tube area.

3.127 Determine the irreversibility in foot-pounds per pound mass for 6 ft^3/s flow of liquid, $\rho = 1.6$ slugs/ft^3, through a sudden expansion from a 12- to 24-in-diameter pipe. $g = 30$ ft/s^2.

3.128 Air flows through a 60-cm-diameter duct at $p = 70$ kPa, $t = 10°$C, $V = 60$ m/s. The duct suddenly expands to 75 cm diameter. Considering the gas as incompressible, calculate the losses in meter-newtons per newton of air and the pressure difference in centimeters of water.

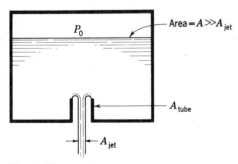

Fig. 3.76

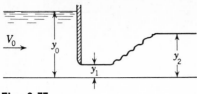

Fig. 3.77

3.129   What are the losses when 180 cfs water discharges from a submerged 48-in-diameter pipe into a reservoir?

3.130   Show that in the limiting case, as $y_1 = y_2$ in Eq. (3.11.23), the relation $V = \sqrt{gy}$ is obtained.

3.131   A jump occurs in a 6-m-wide channel carrying 15 m³/s water at a depth of 30 cm. Determine y_2, V_2, and the losses in meter-newtons per newton, in kilowatts, and in kilocalories per kilogram.

3.132   Derive an expression for a hydraulic jump in a channel having an equilateral triangle as its cross section (symmetric with the vertical).

3.133   Derive Eq. (3.11.24).

3.134   Assuming no losses through the gate of Fig. 3.77 and neglecting $V_0^2/2g$, for $y_0 = 20$ ft and $y_1 = 2$ ft, find y_2 and losses through the jump. What is the basis for neglecting $V_0^2/2g$?

3.135   Under the same assumptions as in Prob. 3.134, for $y_1 = 45$ cm and $y_2 = 2$ m, determine y_0.

3.136   Under the same assumptions as in Prob. 3.134, $y_0 = 22$ ft and $y_2 = 8$ ft. Find the discharge per foot.

3.137   For losses down the spillway of Fig. 3.78 of 2 m·N/N and discharge per meter of 10 m³/s, determine the floor elevation for the jump to occur.

3.138   Water is flowing through the pipe of Fig. 3.79 with velocity $V = 8.02$ ft/s and

El 50 m

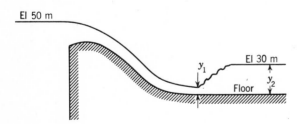

El 30 m

Fig. 3.78

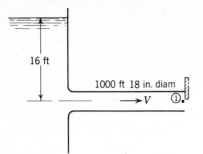

Fig. 3.79

losses of 10 ft ·lb/lb up to section 1. When the obstruction at the end of the pipe is removed, calculate the acceleration of water in the pipe.

3.139   Water fills the piping system of Fig. 3.80. At one instant $p_1 = 10$ psi, $p_2 = 0$, $V_1 = 10$ ft/s, and the flow rate is increasing by 3000 gpm/min. Find the force F_x required to hold the piping system stationary.

3.140   If in Fig. 3.66 Q_2 is 1.0 cfs, what is the vertical force to support the tank? Assume that overflow has not occurred. The tank weighs 20 lb, and water depth is 1 ft.

3.141   In Fig. 3.42, $r_1 = 10$ cm, $r_2 = 16$ cm, $v_{t1} = 0$, and $v_{t2} = 3$ m/s for a centrifugal pump impeller discharging 0.2 m³/s of water. What torque must be exerted on the impeller?

3.142   In a centrifugal pump 400 gpm water leaves an 8-in-diameter impeller with a tangential velocity component of 30 ft/s. It enters the impeller in a radial direction. For pump speed of 1200 rpm and neglecting all losses, determine the torque in the pump shaft, the horsepower input, and the energy added to the flow in foot-pounds per pound.

3.143   A water turbine at 240 rpm discharges 40 m³/s. To produce 40,000 kW, what must be the tangential component of velocity at the entrance to the impeller at $r = 1.6$ m? All whirl is taken from the water when it leaves the turbine. Neglect all losses. What head is required for the turbine?

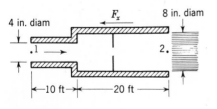

Fig. 3.80

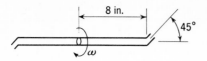

Fig. 3.81

3.144   The symmetrical sprinkler of Fig. 3.81 has a total discharge of 14 gpm and is frictionless. Determine its rpm if the nozzle tips are $\frac{1}{4}$ in diameter.

3.145   What torque would be required to hold the sprinkler of Prob. 3.144 stationary? Total flow 2 l/s water.

3.146   If there is a torque resistance of 0.50 lb·ft in the shaft of Prob. 3.144, what is its speed of rotation?

3.147   For torque resistance of $0.01\omega^2$ in the shaft, determine the speed of rotation of the sprinkler of Prob. 3.144.

3.148   A *reversible* process requires that

(*a*) there be no heat transfer
(*b*) Newton's law of viscosity be satisfied
(*c*) temperature of system and surroundings be equal
(*d*) there be no viscous or Coloumb friction in the system
(*e*) heat transfer occurs from surroundings to system only

3.149   An *open system* implies

(*a*) the presence of a free surface
(*b*) that a specified mass is considered
(*c*) the use of a control volume
(*d*) no interchange between system and surroundings
(*e*) none of the above answers

3.150   A *control volume* refers to

(*a*) a fixed region in space   (*b*) a specified mass   (*c*) an isolated system
(*d*) a reversible process only   (*e*) a closed system

3.151   Which three of the following are synonymous?

1. losses
2. irreversibilities
3. energy losses
4. available energy losses
5. drop in hydraulic grade line

(*a*) 1, 2, 3,   (*b*) 1, 2, 5   (*c*) 1, 2, 4   (*d*) 2, 3, 4   (*e*) 3, 4, 5

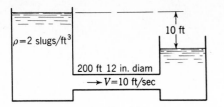

Fig. 3.82

3.152 Irreversibility of the system of Fig. 3.82 is

(*a*) 9.2 hp   (*b*) 36.8 hp   (*c*) 8.45 ft   (*d*) 11.55 ft
(*e*) none of these answers

3.153 Isentropic flow is

(*a*) irreversible adiabatic flow   (*b*) perfect-gas flow   (*c*) ideal-fluid flow
(*d*) reversible adiabatic flow   (*e*) frictionless reversible flow

3.154 One-dimensional flow is

(*a*) steady uniform flow
(*b*) uniform flow
(*c*) flow which neglects changes in a transverse direction
(*d*) restricted to flow in a straight line
(*e*) none of these answers

3.155 The continuity equation may take the form

(*a*) $Q = pAv$ (*b*) $\rho_1 A_1 = \rho_2 A_2$ (*c*) $p_1 A_1 v_1 = p_2 A_2 v_2$
(*d*) $\nabla \cdot \mathbf{p} = 0$ (*e*) $A_1 v_1 = A_2 v_2$

3.156 The first law of thermodynamics, for steady flow,

(*a*) accounts for all energy entering and leaving a control volume
(*b*) is an energy balance for a specified mass of fluid
(*c*) is an expression of the conservation of linear momentum
(*d*) is primarily concerned with heat transfer
(*e*) is restricted in its application to perfect gases

3.157 Entropy, for reversible flow, is defined by the expression

(*a*) $ds = du + p\, d(1/\rho)$ (*b*) $ds = T\, dq_H$ (*c*) $s = u + pv_s$
(*d*) $ds = dq_H/T$ (*e*) none of these answers

3.158 The equation $d\,(\text{losses}) = T\, ds$ is restricted to

(*a*) isentropic flow   (*b*) reversible flow   (*c*) adiabatic flow
(*d*) perfect-gas flow   (*e*) none of these answers

3.159 In turbulent flow

(a) the fluid particles move in an orderly manner
(b) cohesion is more effective than momentum transfer in causing shear stress
(c) momentum transfer is on a molecular scale only
(d) one lamina of fluid glides smoothly over another
(e) the shear stresses are generally larger than in a similar laminar flow

3.160 The ratio $\eta = \tau/(du/dy)$ for turbulent flow is

(a) a physical property of the fluid
(b) dependent upon the flow and the density
(c) the viscosity divided by the density
(d) a function of temperature and pressure of fluid
(e) independent of the nature of the flow

3.161 Turbulent flow generally occurs for cases involving

(a) very viscous fluids   (b) very narrow passages or capillary tubes
(c) very slow motions   (d) combinations of (a), (b), and (c)
(e) none of these answers

3.162 In laminar flow
(a) experimentation is required for the simplest flow cases
(b) Newton's law of viscosity applies
(c) the fluid particles move in irregular and haphazard paths
(d) the viscosity is unimportant
(e) the ratio $\tau/(du/dy)$ depends upon the flow

3.163 An ideal fluid is

(a) very viscous   (b) one which obeys Newton's law of viscosity
(c) a useful assumption in problems in conduit flow
(d) frictionless and incompressible   (e) none of these answers

3.164 Which of the following must be fulfilled by the flow of any fluid, real or ideal?

1. Newton's law of viscosity
2. Newton's second law of motion
3. The continuity equation
4. $\tau = (\mu + \eta)\, du/dy$
5. Velocity at boundary must be zero relative to boundary
6. Fluid cannot penetrate a boundary

(a) 1, 2, 3   (b) 1, 3, 6   (c) 2, 3, 5   (d) 2, 3, 6   (e) 2, 4, 5

3.165 Steady flow occurs when

(a) conditions do not change with time at any point
(b) conditions are the same at adjacent points at any instant
(c) conditions change steadily with the time
(d) $\partial v/\partial t$ is constant   (e) $\partial v/\partial s$ is constant

3.166 Uniform flow occurs

(a) whenever the flow is steady   (b) when $\partial\mathbf{v}/\partial t$ is everywhere zero
(c) only when the velocity vector at any point remains constant
(d) when $\partial\mathbf{v}/\partial s = 0$
(e) when the discharge through a curved pipe of constant cross-sectional area is constant

3.167 Select the correct practical example of steady nonuniform flow:

(a) motion of water around a ship in a lake
(b) motion of a river around bridge piers
(c) steadily increasing flow through a pipe
(d) steadily decreasing flow through a reducing section
(e) constant discharge through a long, straight pipe

3.168 A streamline

(a) is the line connecting the midpoints of flow cross sections
(b) is defined for uniform flow only
(c) is drawn normal to the velocity vector at every point
(d) is always the path of a particle   (e) is fixed in space in steady flow

3.169 In two-dimensional flow around a cylinder the streamlines are 2 in apart at a great distance from the cylinder, where the velocity is 100 ft/s. At one point near the cylinder the streamlines are 1.5 in apart. The average velocity there is

(a) 75 ft/s   (b) 133 ft/s   (c) 150 ft/s   (d) 200 ft/s   (e) 300 ft/s

3.170 An oil has a specific gravity of 0.80. Its density in kilograms per cubic meter is

(a) 400   (b) 414   (c) 800   (d) 1000   (e) 25,800

3.171 The continuity equation

(a) requires that Newton's second law of motion be satisfied at every point in the fluid
(b) expresses the relation between energy and work
(c) states that the velocity at a boundary must be zero relative to the boundary for a real fluid
(d) relates the momentum per unit volume for two points on a streamline
(e) relates mass rate of flow along a stream tube

3.172 Water has an average velocity of 10 ft/s through a 24-in pipe. The discharge through the pipe, in cubic feet per second, is

(a) 7.85   (b) 31.42   (c) 40   (d) 125.68   (e) none of these answers

3.173 The assumptions about flow required in deriving the equation $gz + v^2/2 + \int dp/\rho = \text{const}$ are that it is

(a) steady, frictionless, incompressible, along a streamline
(b) uniform, frictionless, along a streamline, ρ a function of p

(c) steady, uniform, incompressible, along a streamline
(d) steady, frictionless, ρ a function of p, along a streamline
(e) none of these answers

3.174 The equation $z + p/\gamma + v^2/2g = C$ has the units of

(a) m·N/s   (b) N   (c) m·N/kg   (d) m·N/m³   (e) m·N/N

3.175 The work that a liquid is capable of doing by virtue of its sustained pressure is, in foot-pounds per pound,

(a) z (b) p (c) p/γ (d) $v^2/2g$ (e) $\sqrt{2gh}$

3.176 The velocity head is

(a) $v^2/2g$ (b) z (c) v (d) $\sqrt{2gh}$ (e) none of these answers

3.177 The kinetic-energy correction factor

(a) applies to the continuity equation   (b) has the units of velocity head

(c) is expressed by $\dfrac{1}{A}\displaystyle\int_A \dfrac{v}{V}\,dA$

(d) is expressed by $\dfrac{1}{A}\displaystyle\int_A \left(\dfrac{v}{V}\right)^2 dA$

(e) is expressed by $\dfrac{1}{A}\displaystyle\int_A \left(\dfrac{v}{V}\right)^3 dA$

3.178 The kinetic-energy correction factor for the velocity distribution given by Fig. 1.1 is

(a) 0   (b) 1   (c) $\frac{4}{3}$ (d) 2   (e) none of these answers

3.179 The equation $\Sigma F_x = \rho Q(V_{x_{out}} - V_{x_{in}})$ requires the following assumptions for its derivation:

1. Velocity constant over the end cross sections
2. Steady flow
3. Uniform flow
4. Compressible fluid
5. Frictionless fluid

(a) 1, 2   (b) 1, 5   (c) 1, 3   (d) 3, 5   (e) 2, 4

3.180 The momentum correction factor is expressed by

(a) $\dfrac{1}{A}\displaystyle\int_A \dfrac{v}{V}\,dA$ (b) $\dfrac{1}{A}\displaystyle\int_A \left(\dfrac{v}{V}\right)^2 dA$ (c) $\dfrac{1}{A}\displaystyle\int_A \left(\dfrac{v}{V}\right)^3 dA$ (d) $\dfrac{1}{A}\displaystyle\int_A \left(\dfrac{v}{V}\right)^4 dA$

(e) none of these answers

3.181   The momentum correction factor for the velocity distribution given by Fig. 1.1 is

(*a*) 0   (*b*) 1   (*c*) $\frac{4}{3}$ (*d*) 2   (*e*) none of these answers

3.182   The velocity over one-third of a cross section is zero and is uniform over the remaining two-thirds of the area. The momentum correction factor is

(*a*) 1   (*b*) $\frac{4}{3}$ (*c*) $\frac{3}{2}$ (*d*) $\frac{9}{4}$ (*e*) none of these answers

3.183   The magnitude of the resultant force necessary to hold a 20-cm-diameter 90° elbow under no-flow conditions when the pressure is 10 kg$_f$/cm^2 is, in kilonewtons,

(*a*) 61.5   (*b*) 43.5   (*c*) 30.8   (*d*) 0   (*e*) none of these answers

3.184   A 12-in-diameter 90° elbow carries water with average velocity of 15 ft/s and pressure of -5 psi. The force component in the direction of the approach velocity necessary to hold the elbow in place is, in pounds,

(*a*) -342 (*b*) 223   (*c*) 565   (*d*) 907   (*e*) none of these answers

3.185   A 5-cm-diameter 180° bend carries a liquid, $\rho = 1000$ kg/m^3 at 6 m/s at a pressure of zero gage. The force tending to push the bend off the pipe is, in newtons,

(*a*) 0   (*b*) 70.5   (*c*) 141   (*d*) 515   (*e*) none of these answers

3.186   The thickness of wall for a large high-pressure pipeline is determined by consideration of

(*a*) axial tensile stresses in the pipe
(*b*) forces exerted by dynamic action at bends
(*c*) forces exerted by static and dynamic action at bends
(*d*) circumferential pipe wall tension
(*e*) temperature stresses

3.187   Select from the following list the correct assumptions for analyzing flow of a jet that is deflected by a fixed or moving vane:

1. The momentum of the jet is unchanged.
2. The absolute speed does not change along the vane.
3. The fluid flows onto the vane without shock.
4. The flow from the nozzle is steady.
5. Friction between jet and vane is neglected.
6. The jet leaves without velocity.
7. The velocity is uniform over the cross section of the jet before and after contacting the vane.

(*a*) 1, 3, 4, 5   (*b*) 2, 3, 5, 6   (*c*) 3, 4, 5, 6   (*d*) 3, 4, 5, 7   (*e*) 3, 5, 6, 7

3.188   When a steady jet impinges on a fixed inclined plane surface

(*a*) the momentum in the direction of the approach velocity is unchanged
(*b*) no force is exerted on the jet by the vane

(c) the flow is divided into parts directly proportional to the angle of inclination of the surface

(d) the speed is reduced for that portion of the jet turned through more than 90° and increased for the other portion

(e) the momentum component is unchanged parallel to the surface

3.189 A jet with initial velocity of 100 ft/s in the $+x$ direction is deflected by a fixed vane with a blade angle of 120°. The velocity components leaving the vane parallel to and normal to the approach velocity are

(a) $v_x = -50, v_y = 86.6$ (b) $v_x = 100, v_y = 0$ (c) $v_x = 50, v_y = 50$
(d) $v_x = 50, v_y = 86.6$ (e) $v_x = -86.6, v_y = 50$

3.190 An oil jet, sp gr 0.80, discharges 10 kg/s onto a fixed vane that turns the flow through 90°. The speed of the jet is 30 m/s as it leaves the vane. The force component on the vane in the direction of the approach velocity is, in newtons,

(a) 424   (b) 300   (c) 240   (d) 212   (e) none of these answers

3.191 A water jet having a velocity of 120 ft/s and cross-sectional area 0.05 ft² flows onto a vane moving 40 ft/s in the same direction as the jet. The mass having its momentum changed per unit time, in slugs per second, is

(a) 4   (b) 7.74   (c) 11.61   (d) 15.48   (e) none of these answers

3.192 A jet having a velocity of 30 m/s flows onto a vane, angle $\theta = 150°$, having a velocity of 15 m/s in the same direction as the jet. The final absolute velocity components parallel and normal to the approach velocity are

(a) $v_x = 2.01, v_y = 7.5$ (b) $v_x = 7.2, v_y = 13$ (c) $v_x = -11, v_y = 15$
(d) $v_x = 4.39, v_y = 10.6$ (e) none of these answers

3.193 A vane moves toward a nozzle 30 ft/s, and the jet issuing from the nozzle has a velocity of 40 ft/s. The vane angle is $\theta = 90°$. The absolute velocity components of the jet as it leaves the vane, parallel and normal to the undisturbed jet, are

(a) $v_x = 10, v_y = 10$ (b) $v_x = -30, v_y = 10$ (c) $v_x = -30, v_y = 40$
(d) $v_x = -30, v_y = 70$ (e) none of these answers

3.194 A force of 250 N is exerted upon a moving blade in the direction of its motion, $u = 20$ m/s. The power obtained in kilowatts is

(a) 0.5   (b) 30   (c) 50   (d) 100   (e) none of these answers

3.195 A series of moving vanes, $u = 50$ ft/s, $\theta = 90°$, intercepts a jet, $Q = 1$ ft³/s, $\rho = 1.5$ slugs/ft³, $V_0 = 100$ ft/s. The work done on the vanes, in foot-pounds per second, is

(a) 1875   (b) 2500   (c) 3750   (d) 7500   (e) none of these answers

3.196 The kilowatts available in a water jet of cross-sectional area 0.004 m² and velocity 20 m/s is

(a) 0.495   (b) 16.0   (c) 17.2   (d) 32   (e) none of these answers

3.197 A ship moves through water at 30 ft/s. The velocity of water in the slipstream behind the boat is 20 ft/s, and the propeller diameter is 3.0 ft. The theoretical efficiency of the propeller is, in percent,

(*a*) 0   (*b*) 60   (*c*) 75   (*d*) 86   (*e*) none of these answers

3.198 The thrust on the ship of Prob. 3.197, in kilograms force, is

(*a*) 822   (*b*) 2480   (*c*) 3300   (*d*) 4963   (*e*) none of these answers

3.199 A rocket exerts a constant horizontal thrust of 40 lb on a missile for 3 s. If the missile weighs 8 lb and starts from rest, its speed at the end of the period, neglecting the downward acceleration of gravity and reduction in weight of the rocket, is, in feet per second,

(*a*) 386   (*b*) 483   (*c*) 580   (*d*) 600   (*e*) none of these answers

3.200 What is the reduction in weight of the rocket of Prob. 3.199 if the jet leaves at 6000 ft/s relative to the rocket?

(*a*) 0.02 lb   (*b*) 0.04 lb   (*c*) 0.32 lb   (*d*) 0.64 lb   (*e*) none of these answers

3.201 A glass tube with a 90° bend is open at both ends. It is inserted into a flowing stream of oil, sp gr 0.90, so that one opening is directed upstream and the other is directed upward. Oil inside the tube is 5 cm higher than the surface of flowing oil. The velocity measured by the tube is, in meters per second,

(*a*) 0.89   (*b*) 0.99   (*c*) 1.10   (*d*) 1.40   (*e*) none of these answers

3.202 In Fig. 8.4 the gage difference R' for $v_1 = 5$ ft/s, $S = 0.08$, $S_0 = 1.2$ is, in feet,

(*a*) 0.39   (*b*) 0.62   (*c*) 0.78   (*d*) 1.17   (*e*) none of these answers

3.203 The theoretical velocity of oil, sp gr 0.75, flowing from an orifice in a reservoir under a head of 4 m, is in meters per second,

(*a*) 6.7   (*b*) 8.86   (*c*) 11.8   (*d*) not determinable from data given
(*e*) none of these answers

3.204 In which of the following cases is it possible for flow to occur from low pressure to high pressure?

(*a*) flow through a converging section   (*b*) adiabatic flow in a horizontal pipe
(*c*) flow of a liquid upward in a vertical pipe   (*d*) flow of air downward in a pipe
(*e*) impossible in a constant-cross-section conduit

3.205 The head loss in turbulent flow in a pipe

(*a*) varies directly as the velocity
(*b*) varies inversely as the square of the velocity
(*c*) varies inversely as the square of the diameter

(d) depends upon the orientation of the pipe

(e) varies approximately as the square of the velocity

3.206   The losses due to a sudden expansion are expressed by

(a) $\dfrac{V_1^2 - V_2^2}{2g}$ (b) $\dfrac{V_1 - V_2}{2g}$ (c) $\dfrac{V_2^2 - V_1^2}{g}$ (d) $\dfrac{(V_1 - V_2)^2}{g}$

(e) $\dfrac{(V_1 - V_2)^2}{2g}$

3.207   If all losses are neglected, the pressure at the summit of a siphon

(a) is a minimum for the siphon

(b) depends upon height of summit above upstream reservoir only

(c) is independent of the length of the downstream leg

(d) is independent of the discharge through the siphon

(e) is independent of the liquid density

3.208   The depth conjugate to $y = 1$ ft and $V = 20$ ft/s is

(a) 2.32 ft   (b) 4.5 ft   (c) 5.0 ft   (d) 5.5 ft   (e) none of these answers

3.209   The depth conjugate to $y = 3$ m and $V = 8$ m/s is

(a) 4.55 m   (b) 4.9 m   (c) 7.04 m   (d) 9.16 m   (e) none of these answers

3.210   The depth conjugate to $y = 10$ ft and $V = 1$ ft/s is

(a) 0.06 ft   (b) 1.46 ft   (c) 5.06 ft   (d) 10.06 ft   (e) none of these answers

3.211   The continuity equation in ideal-fluid flow

(a) states that the net rate of inflow into any small volume must be zero

(b) states that the energy is constant along a streamline

(c) states that the energy is constant everywhere in the fluid

(d) applies to irrotational flow only

(e) implies the existence of a velocity potential

4

DIMENSIONAL ANALYSIS AND DYNAMIC SIMILITUDE

Dimensionless parameters significantly deepen our understanding of fluid-flow phenomena in a way which is analogous to the case of a hydraulic jack, where the ratio of piston diameters determines the mechanical advantage, a dimensionless number which is independent of the overall size of the jack. They permit limited experimental results to be applied to situations involving different physical dimensions and often different fluid properties. The concepts of dimensional analysis introduced in this chapter plus an understanding of the mechanics of the type of flow under study make possible this generalization of experimental data. The consequence of such generalization is manifold, since one is now able to describe the phenomenon in its entirety and is not restricted to discussing the specialized experiment that was performed. Thus, it is possible to conduct fewer, although highly selective, experiments to uncover the hidden facets of the problem and thereby achieve important savings in time and money. The results of an investigation can also be presented to other engineers and scientists in a more compact and meaningful way to facilitate their use. Equally important is the fact that through such incisive and uncluttered presentations of information researchers are able to discover new features and missing areas of knowledge of the problem at hand. This directed advancement of our understanding of a phenomenon would be impaired if the tools of dimensional analysis were not available. In the following chapter, dealing primarily with viscous effects, one parameter is highly significant, viz., the Reynolds number. In Chap. 6, dealing with compressible flow, the Mach number is the most important dimensionless parameter. In Chap. 11, dealing with open channels, the Froude number has the greatest significance.

Many of the dimensionless parameters may be viewed as a ratio of a pair of fluid forces, the relative magnitude indicating the relative importance of one of the forces with respect to the other. If some forces in a particular flow situation are very much larger than a few others, it is often possible to neglect the effect of the smaller forces and treat the phenomenon as though it were completely determined by the major forces. This means that simpler, although not necessarily easy, mathematical and experimental procedures can be used to solve the problem. For situations with several forces of the same magnitude, such as inertial, viscous, and gravitational forces, special techniques are required. After a discussion of dimensions, dimensional analysis, and dimensionless parameters, dynamic similitude and model studies are presented.

4.1 DIMENSIONAL HOMOGENEITY AND DIMENSIONLESS RATIOS

Solving practical design problems in fluid mechanics usually requires both theoretical developments and experimental results. By grouping significant quantities into dimensionless parameters it is possible to reduce the number of variables appearing and to make this compact result (equations or data plots) applicable to all similar situations.

If one were to write the equation of motion $\Sigma \mathbf{F} = m\mathbf{a}$ for a fluid particle, including all types of force terms that could act, such as gravity, pressure, viscous, elastic, and surface-tension forces, an equation of the sum of these forces equated to $m\mathbf{a}$, the inertial force, would result. As with all physical equations, each term must have the same dimensions, in this case, force. The division of each term of the equation by any one of the terms would make the equation dimensionless. For example, dividing through by the inertial-force term would yield a sum of dimensionless parameters equated to unity. The relative size of any one parameter, compared with unity, would indicate its importance. If one were to divide the force equation through by a different term, say the viscous-force term, another set of dimensionless parameters would result. Without experience in the flow case it is difficult to determine which parameters will be most useful.

An example of the use of dimensional analysis and its advantages is given by considering the hydraulic jump, treated in Sec. 3.11. The momentum equation for this case

$$\frac{\gamma y_1^2}{2} - \frac{\gamma y_2^2}{2} = \frac{V_1 y_1 \gamma}{g} (V_2 - V_1) \tag{4.1.1}$$

can be rewritten as

$$\frac{\gamma}{2} y_1^2 \left[1 - \left(\frac{y_2}{y_1}\right)^2 \right] = V_1^2 \frac{\gamma}{g} y_1 \left(1 - \frac{y_2}{y_1} \right) \frac{y_1}{y_2}$$

Clearly the right-hand side represents the inertial forces and the left-hand side, the pressure forces that exist due to gravity. These two forces are of equal magnitude since one determines the other in this equation. Furthermore, the term $\gamma y_1{}^2/2$ has the dimensions of force per unit width, and it multiplies a dimensionless number which is specified by the geometry of the hydraulic jump.

If one divides this equation by the geometric term $1 - y_2/y_1$ and a number representative of the gravity forces, one has

$$\frac{V_1{}^2}{gy_1} = \frac{1}{2}\frac{y_2}{y_1}\left(1 + \frac{y_2}{y_1}\right) \qquad (4.1.2)$$

It is now clear that the left-hand side is the ratio of the inertia and gravity forces, even though the explicit representation of the forces has been obscured through the cancellation of terms that are common in both the numerator and denominator. This ratio is equivalent to a dimensionless parameter, called the *Froude number*, which will be discussed in further detail later in this chapter. It is also interesting to note that this ratio of forces is known once the ratio y_2/y_1 is given, regardless of what the values y_2 and y_1 are. From this observation one can obtain an appreciation of the increased scope that Eq. (4.1.2) affords over Eq. (4.1.1) even though one is only a rearrangement of the other.

In writing the momentum equation which led to Eq. (4.1.2) only inertia and gravity forces were included in the original problem statement. But other forces are present, such as surface tension and viscosity. These were neglected as being small in comparison with gravity and inertia forces; however, only experience with the phenomenon, or ones similar to it, would justify such an initial simplification. For example, if viscosity had been included because one was not sure of the magnitude of its effect, the momentum equation would become

$$\frac{\gamma y_1{}^2}{2} - \frac{\gamma y_2{}^2}{2} - F_{\text{viscous}} = V_1 \frac{y_1\gamma}{g}(V_2 - V_1)$$

with the result that

$$\frac{V_1{}^2}{gy_1} + \frac{F_{\text{viscous}}y_2}{\gamma y_1{}^2(y_1 - y_2)} = \frac{1}{2}\frac{y_2}{y_1}\left(1 + \frac{y_2}{y_1}\right)$$

This statement is more complete than that given by Eq. (4.1.2). However, experiments would show that the second term on the left-hand side is usually a small fraction of the first term and could be neglected in making initial tests on a hydraulic jump.

In the last equation one can consider the ratio y_2/y_1 to be a dependent variable which is determined for each of the various values of the force ratios, V_1^2/gy_1 and $F_{\text{viscous}}/\gamma y_1^2$, which are the independent variables. From the previous discussion it appears that the latter variable plays only a minor role in determining the values of y_2/y_1. Nevertheless, if one observed that the ratios of the forces, V_1^2/gy_1 and $F_{\text{viscous}}/\gamma y_1^2$, had the same values in two different tests, one would expect, on the basis of the last equation, that the values of y_2/y_1 would be the same in the two situations. If the ratio of V_1^2/gy_1 was the same in the two tests but the ratio $F_{\text{viscous}}/\gamma y_1$, which has only a minor influence for this case, was not equal, one would conclude that the values of y_2/y_1 for the two cases would be almost the same.

This is the key to much of what follows. For if one can create in a model experiment the same geometric and force ratios that occur on the full-scale unit, then the dimensionless solution for the model is valid also for the prototype. Often, as will be seen, it is not possible to have all the ratios equal in the model and prototype. Then one attempts to plan the experimentation in such a way that the dominant force ratios are as nearly equal as possible. The results obtained with such incomplete modeling are often sufficient to describe the phenomenon in the detail that is desired.

Writing a force equation for a complex situation may not be feasible, and another process, *dimensional analysis*, is then used if one knows the pertinent quantities that enter into the problem.

In a given situation several of the forces may be of little significance, leaving perhaps two or three forces of the same order of magnitude. With three forces of the same order of magnitude, two dimensionless parameters are obtained; one set of experimental data on a geometrically similar model provides the relationships between parameters holding for all other similar flow cases.

4.2 DIMENSIONS AND UNITS

The dimensions of mechanics are force, mass, length, and time, related by Newton's second law of motion,

$$\mathbf{F} = m\mathbf{a} \tag{4.2.1}$$

Force and mass units are discussed in Sec. 1.2. For all physical systems, it would probably be necessary to introduce two more dimensions, one dealing with electromagnetics and the other with thermal effects. For the compressible work in this text, it is unnecessary to include a thermal unit, as the equations of state link pressure, density, and temperature.

Table 4.1 *Dimensions of physical quantities used in fluid mechanics*

Quantity	Symbol	Dimensions (M,L,T)
Length	l	L
Time	t	T
Mass	m	M
Force	F	MLT^{-2}
Velocity	V	LT^{-1}
Acceleration	a	LT^{-2}
Area	A	L^2
Discharge	Q	L^3T^{-1}
Pressure	Δp	$ML^{-1}T^{-2}$
Gravity	g	LT^{-2}
Density	ρ	ML^{-3}
Specific weight	γ	$ML^{-2}T^{-2}$
Dynamic viscosity	μ	$ML^{-1}T^{-1}$
Kinematic viscosity	ν	L^2T^{-1}
Surface tension	σ	MT^{-2}
Bulk modulus of elasticity	K	$ML^{-1}T^{-2}$

Newton's second law of motion in dimensional form is

$$F = MLT^{-2} \tag{4.2.2}$$

which shows that only three of the dimensions are independent. F is the force dimension, M the mass dimension, L the length dimension, and T the time dimension. One common system employed in dimensional analysis is the MLT system. Table 4.1 lists some of the quantities used in fluid flow, together with their symbols and dimensions.

4.3   THE Π THEOREM

The Buckingham[1] Π theorem proves that in a physical problem including n quantities in which there are m dimensions, the quantities can be arranged into $n - m$ independent dimensionless parameters. Let A_1, A_2, A_3, ..., A_n be the quantities involved, such as pressure, viscosity, velocity, etc. All the quantities are known to be essential to the solution, and hence some func-

[1] E. Buckingham, Model Experiments and the Form of Empirical Equations, *Trans. ASME*, vol. 37, pp. 263–296, 1915.

tional relation must exist

$$F(A_1, A_2, A_3, \ldots, A_n) = 0 \tag{4.3.1}$$

If Π_1, Π_2, $\ldots$, represent dimensionless groupings of the quantities A_1, A_2, A_3, $\ldots$, then with m dimensions involved, an equation of the form

$$f(\Pi_1, \Pi_2, \Pi_3, \ldots, \Pi_{n-m}) = 0 \tag{4.3.2}$$

exists.

Proof of the Π theorem may be found in Buckingham's paper, as well as in Sedov's book listed in the references at the end of this chapter. The method of determining the Π parameters is to select m of the A quantities, with different dimensions, that contain among them the m dimensions, and to use them as repeating variables[1] together with one of the other A quantities for each Π. For example, let A_1, A_2, A_3 contain M, L, and T, not necessarily in each one, but collectively. Then, the first Π parameter is made up as

$$\Pi_1 = A_1{}^{x_1} A_2{}^{y_1} A_3{}^{z_1} A_4 \tag{4.3.3}$$

the second one as

$$\Pi_2 = A_1{}^{x_2} A_2{}^{y_2} A_3{}^{z_2} A_4$$

and so on, until

$$\Pi_{n-m} = A_1{}^{x_{n-m}} A_2{}^{y_{n-m}} A_3{}^{z_{n-m}} A_n$$

In these equations the exponents are to be determined so that each Π is dimensionless. The dimensions of the A quantities are substituted, and the exponents M, L, and T are set equal to zero respectively. These produce three equations in three unknowns for each Π parameter, so that the x, y, z exponents can be determined, and hence the Π parameter.

If only two dimensions are involved, then two of the A quantities are selected as repeating variables, and two equations in the two unknown exponents are obtained for each Π term.

In many cases the grouping of A terms is such that the dimensionless arrangement is evident by inspection. The simplest case is that when two quantities have the same dimensions, e.g., length, the ratio of these two terms is the Π parameter.

---

[1] It is essential that no one of the m selected quantities used as repeating variables be derivable from the other repeating variables.

The procedure is best illustrated by several examples.

EXAMPLE 4.1 The discharge through a horizontal capillary tube is thought to depend upon the pressure drop per unit length, the diameter, and the viscosity. Find the form of the equation.

The quantities are listed with their dimensions:

Quantity	Symbol	Dimensions
Discharge	Q	L^3T^{-1}
Pressure drop/length	$\Delta p/l$	$ML^{-2}T^{-2}$
Diameter	D	L
Viscosity	μ	$ML^{-1}T^{-1}$

Then

$$F\left(Q, \frac{\Delta p}{l}, D, \mu\right) = 0$$

Three dimensions are used, and with four quantities there will be one Π parameter:

$$\Pi = Q^{x_1}\left(\frac{\Delta p}{l}\right)^{y_1} D^{z_1}\mu$$

Substituting in the dimensions gives

$$\Pi = (L^3T^{-1})^{x_1}(ML^{-2}T^{-2})^{y_1}L^{z_1}ML^{-1}T^{-1} = M^0L^0T^0$$

The exponents of each dimension must be the same on both sides of the equation. With L first,

$$3x_1 - 2y_1 + z_1 - 1 = 0$$

and similarly for M and T

$$y_1 + 1 = 0$$
$$-x_1 - 2y_1 - 1 = 0$$

from which $x_1 = 1$, $y_1 = -1$, $z_1 = -4$, and

$$\Pi = \frac{Q\mu}{D^4 \, \Delta p/l}$$

After solving for Q,

$$Q = C \frac{\Delta p}{l} \frac{D^4}{\mu}$$

from which dimensional analysis yields no information about the numerical value of the dimensionless constant C; experiment (or analysis) shows that it is $\pi/128$ [Eq. (5.2.10a)].

When dimensional analysis is used, the variables in a problem must be known. In the last example if kinematic viscosity had been used in place of dynamic viscosity, an incorrect formula would have resulted.

EXAMPLE 4.2   A V-notch weir is a vertical plate with a notch of angle ϕ cut into the top of it and placed across an open channel. The liquid in the channel is backed up and forced to flow through the notch. The discharge Q is some function of the elevation H of upstream liquid surface above the bottom of the notch. In addition the discharge depends upon gravity and upon the velocity of approach V_0 to the weir. Determine the form of discharge equation.
    A functional relationship

$$F(Q,H,g,V_0,\phi) = 0$$

is to be grouped into dimensionless parameters. ϕ is dimensionless, hence it is one Π parameter. Only two dimensions are used, L and T. If g and H are the repeating variables,

$$\Pi_1 = H^{x_1}g^{y_1}Q = L^{x_1}(LT^{-2})^{y_1}L^3T^{-1}$$

$$\Pi_2 = H^{x_2}g^{y_2}V_0 = L^{x_2}(LT^{-2})^{y_2}LT^{-1}$$

Then

$$x_1 + y_1 + 3 = 0 \qquad x_2 + y_2 + 1 = 0$$

$$-2y_1 - 1 = 0 \qquad -2y_2 - 1 = 0$$

from which $x_1 = -\frac{5}{2}$, $y_1 = -\frac{1}{2}$, $x_2 = -\frac{1}{2}$, $y_2 = -\frac{1}{2}$, and

$$\Pi_1 = \frac{Q}{\sqrt{g}\,H^{5/2}} \qquad \Pi_2 = \frac{V_0}{\sqrt{gH}} \qquad \Pi_3 = \phi$$

or

$$f\left(\frac{Q}{\sqrt{g}\,H^{5/2}}, \frac{V_0}{\sqrt{gH}}, \phi\right) = 0$$

This may be written

$$\frac{Q}{\sqrt{g}\,H^{5/2}} = f_1\left(\frac{V_0}{\sqrt{gH}}, \phi\right)$$

in which both f, f_1 are unknown functions. After solving for Q,

$$Q = \sqrt{g}\,H^{5/2}f_1\left(\frac{V_0}{\sqrt{gH}}, \phi\right)$$

Either experiment or analysis is required to yield additional information about the function f_1.

If H and V_0 were selected as repeating variables in place of g and H,

$$\Pi_1 = H^{x_1}V_0{}^{y_1}Q = L^{x_1}(LT^{-1})^{x_1}L^3T^{-1}$$
$$\Pi_2 = H^{x_2}V_0{}^{y_2}g = L^{x_2}(LT^{-1})^{y_2}LT^{-2}$$

Then

$$x_1 + y_1 + 3 = 0 \qquad x_2 + y_2 + 1 = 0$$
$$-y_1 - 1 = 0 \qquad -y_2 - 2 = 0$$

from which $x_1 = -2$, $y_1 = -1$, $x_2 = 1$, $y_2 = -2$, and

$$\Pi_1 = \frac{Q}{H^2V_0} \qquad \Pi_2 = \frac{gH}{V_0{}^2} \qquad \Pi_3 = \phi$$

or

$$f\left(\frac{Q}{H^2V_0}, \frac{gH}{V_0{}^2}, \phi\right) = 0$$

Since any of the Π parameters may be inverted or raised to any power without affecting their dimensionless status,

$$Q = V_0 H^2 f_2 \left(\frac{V_0}{\sqrt{gH}}, \phi \right)$$

The unknown function f_2 has the same parameters as f_1, but it could not be the same function. The last form is not very useful, in general, because frequently V_0 may be neglected with V-notch weirs. This shows that a term of minor importance should not be selected as a repeating variable.

Another method of determining alternate sets of Π parameters would be the arbitrary recombination of the first set. If four independent Π parameters Π_1, Π_2, Π_3, Π_4 are known, the term

$$\Pi_a = \Pi_1{}^{a_1} \Pi_2{}^{a_2} \Pi_3{}^{a_3} \Pi_4{}^{a_4}$$

with the exponents chosen at will, would yield a new parameter. Then Π_a, Π_2, Π_3, Π_4 would constitute a new set. This procedure may be continued to find all possible sets.

EXAMPLE 4.3 The losses $\Delta h/l$ per unit length of pipe in turbulent flow through a smooth pipe depend upon velocity V, diameter D, gravity g, dynamic viscosity μ, and density ρ. With dimensional analysis, determine the general form of the equation

$$F \left(\frac{\Delta h}{l}, V, D, \rho, \mu, g \right) = 0$$

Clearly, $\Delta h/l$ is a Π parameter. If V, D, and ρ are repeating variables,

$$\Pi_1 = V^{x_1} D^{y_1} \rho^{z_1} \mu = (LT^{-1})^{x_1} L^{y_1} (ML^{-3})^{z_1} M L^{-1} T^{-1}$$

$$x_1 + y_1 - 3z_1 - 1 = 0$$

$$-x_1 \qquad\qquad - 1 = 0$$

$$z_1 + 1 = 0$$

from which $x_1 = -1$, $y_1 = -1$, $z_1 = -1$.

$$\Pi_2 = V^{x_2}D^{y_2}\rho^{z_2}g = (LT^{-1})^{x_2}L^{y_2}(ML^{-3})^{z_2}LT^{-2}$$

$$x_2 + y_2 - 3z_2 + 1 = 0$$

$$-x_2 \qquad\qquad -2 = 0$$

$$z_2 \qquad = 0$$

from which $x_2 = -2$, $y_2 = 1$, $z_2 = 0$.

$$\Pi_1 = \frac{\mu}{VD\rho} \qquad \Pi_2 = \frac{gD}{V^2} \qquad \Pi_3 = \frac{\Delta h}{l}$$

or

$$f\left(\frac{VD\rho}{\mu}, \frac{V^2}{gD}, \frac{\Delta h}{l}\right) = 0$$

since the Π quantities may be inverted if desired. The first parameter, $VD\rho/\mu$, is the *Reynolds number* **R**, one of the most important of the dimensionless parameters in fluid mechanics. The size of the Reynolds number determines the nature of the flow. It is discussed in Sec. 5.3. Solving for $\Delta h/l$ gives

$$\frac{\Delta h}{l} = f_1\left(\mathbf{R}, \frac{V^2}{gD}\right)$$

The usual formula is

$$\frac{\Delta h}{l} = f(\mathbf{R})\, \frac{1}{D}\frac{V^2}{2g}$$

EXAMPLE 4.4 A fluid-flow situation depends upon the velocity V, the density ρ, several linear dimensions l, l_1, l_2, pressure drop Δp, gravity g, viscosity μ, surface tension σ, and bulk modulus of elasticity K. Apply dimensional analysis to these variables to find a set of Π parameters.

$$F(V,\rho,l,l_1,l_2,\Delta p,g,\mu,\sigma,K) = 0$$

As three dimensions are involved, three repeating variables are selected. For complex situations, V, ρ, and l are generally helpful. There are seven Π parameters:

$$\Pi_1 = V^{x_1}\rho^{y_1}l^{z_1}\,\Delta p \qquad \Pi_2 = V^{x_2}\rho^{y_2}l^{z_2}g$$
$$\Pi_3 = V^{x_3}\rho^{y_3}l^{z_3}\mu \qquad \Pi_4 = V^{x_4}\rho^{y_4}l^{z_4}\sigma$$

$$\Pi_5 = V^{x_5}\rho^{y_5}l^{z_5}K \qquad \Pi_6 = \frac{l}{l_1}$$

$$\Pi_7 = \frac{l}{l_2}$$

By expanding the Π quantities into dimensions,

$$\Pi_1 = (LT^{-1})^{x_1}(ML^{-3})^{y_1}L^{z_1}ML^{-1}T^{-2}$$
$$x_1 - 3y_1 + z_1 - 1 = 0$$
$$-x_1 \qquad\qquad -2 = 0$$
$$y_1 \qquad +1 = 0$$

from which $x_1 = -2$, $y_1 = -1$, $z_1 = 0$.

$$\Pi_2 = (LT^{-1})^{x_2}(ML^{-3})^{y_2}L^{z_2}LT^{-2}$$
$$x_2 - 3y_2 + z_2 + 1 = 0$$
$$-x_2 \qquad\qquad -2 = 0$$
$$y_2 \qquad\qquad = 0$$

from which $x_2 = -2$, $y_2 = 0$, $z_2 = 1$.

$$\Pi_3 = (LT^{-1})^{x_3}(ML^{-3})^{y_3}L^{z_3}ML^{-1}T^{-1}$$
$$x_3 - 3y_3 + z_3 - 1 = 0$$
$$-x_3 \qquad\qquad -1 = 0$$
$$y_3 \qquad +1 = 0$$

from which $x_3 = -1$, $y_3 = -1$, $z_3 = -1$.

$$\Pi_4 = (LT^{-1})^{x_4}(ML^{-3})^{y_4}L^{z_4}MT^{-2}$$
$$x_4 - 3y_4 + z_4 \qquad = 0$$
$$-x_4 \qquad\qquad -2 = 0$$
$$y_4 \qquad +1 = 0$$

from which $x_4 = -2$, $y_4 = -1$, $z_4 = -1$.

$$\Pi_5 = (LT^{-1})^{x_5}(ML^{-3})^{y_5}L^{z_5}ML^{-1}T^{-2}$$
$$x_5 - 3y_5 + z_5 - 1 = 0$$
$$-x_5 \qquad\qquad - 2 = 0$$
$$\qquad y_5 \qquad\quad + 1 = 0$$

from which $x_5 = -2$, $y_5 = -1$, $z_5 = 0$.

$$\Pi_1 = \frac{\Delta p}{\rho V^2} \qquad \Pi_2 = \frac{gl}{V^2} \qquad \Pi_3 = \frac{\mu}{Vl\rho} \qquad \Pi_4 = \frac{\sigma}{V^2\rho l}$$

$$\Pi_5 = \frac{K}{\rho V^2} \qquad \Pi_6 = \frac{l}{l_1} \qquad \Pi_7 = \frac{l}{l_2}$$

and

$$f\left(\frac{\Delta p}{\rho V^2}, \frac{gl}{V^2}, \frac{\mu}{Vl\rho}, \frac{\sigma}{V^2\rho l}, \frac{K}{\rho V^2}, \frac{l}{l_1}, \frac{l}{l_2}\right) = 0$$

It is convenient to invert some of the parameters and to take the square root of Π_5,

$$f_1\left(\frac{\Delta p}{\rho V^2}, \frac{V^2}{gl}, \frac{Vl\rho}{\mu}, \frac{V^2l\rho}{\sigma}, \frac{V}{\sqrt{K/\rho}}, \frac{l}{l_1}, \frac{l}{l_2}\right) = 0$$

The first parameter, usually written $\Delta p/(\rho V^2/2)$, is the *pressure coefficient*; the second parameter is the *Froude* number **F**; the third is the *Reynolds* number **R**; the fourth is the *Weber* number **W**; and the fifth is the *Mach* number **M**. Hence

$$f_1\left(\frac{\Delta p}{\rho V^2}, \mathbf{F}, \mathbf{R}, \mathbf{W}, \mathbf{M}, \frac{l}{l_1}, \frac{l}{l_2}\right) = 0$$

After solving for pressure drop

$$\Delta p = \rho V^2 f_2\left(\mathbf{F}, \mathbf{R}, \mathbf{W}, \mathbf{M}, \frac{l}{l_1}, \frac{l}{l_2}\right)$$

in which f_1, f_2 must be determined from analysis or experiment. By selecting other repeating variables, a different set of Π parameters could be obtained.

Figure 5.32 is a representation of a functional relationship of the type just given as it applies to the flow in pipes. Here the parameters **F**, **W**, and **M** are neglected as being unimportant; l is the pipe diameter D, l_1 is the length of the pipe L, and l_2 is a dimension which is representative of the effective height of the surface roughness of the pipe and is given by ϵ. Thus

$$\frac{\Delta p}{\rho V^2} = f_3 \left(\mathbf{R}, \frac{L}{D}, \frac{\epsilon}{D} \right)$$

The fact that the pressure drop in the pipeline varies linearly with the length (i.e., doubling the length of pipe doubles the loss in pressure) appears reasonable, so that one has

$$\frac{\Delta p}{\rho V^2} = \frac{L}{D} f_4 \left(\mathbf{R}, \frac{\epsilon}{D} \right) \qquad \text{or} \qquad \frac{\Delta p}{\rho V^2 (L/D)} = f_4 \left(\mathbf{R}, \frac{\epsilon}{D} \right)$$

The term on the left-hand side is commonly given the notation $f/2$, as in Fig. 5.32. The curves shown in this figure have f and $\mathbf{R}$ as ordinate and abscissa, respectively, with ϵ/D a parameter which assumes a given value for each curve. The nature of these curves was determined through experiment. Such experiments show that when the parameter $\mathbf{R}$ is below the value of 2000, all the curves for the various values of ϵ/D coalesce into one. Hence f is independent of ϵ/D, and the result is

$$f = f_5(\mathbf{R})$$

This relationship will be predicted in Chap. 5 on the basis of theoretical considerations, but it remained for an experimental verification of these predictions to indicate the power of the theoretical methods.

EXAMPLE 4.5 The thrust due to any one of a family of geometrically similar airplane propellers is to be determined experimentally from a wind-tunnel test on a model. By means of dimensional analysis find suitable parameters for plotting test results.

The thrust F_T depends upon speed of rotation ω, speed of advance V_0, diameter D, air viscosity μ, density ρ, and speed of sound c. The function

$$F(F_T, V_0, D, \omega, \mu, \rho, c) = 0$$

is to be arranged into four dimensionless parameters, since there are seven quantities and three dimensions. Starting first by selecting ρ, ω, and D as repeating variables,

$$\Pi_1 = \rho^{x_1}\omega^{y_1}D^{z_1}F_T = (ML^{-3})^{x_1}(T^{-1})^{y_1}L^{z_1}MLT^{-2}$$

$$\Pi_2 = \rho^{x_2}\omega^{y_2}D^{z_2}V_0 = (ML^{-3})^{x_2}(T^{-1})^{y_2}L^{z_2}LT^{-1}$$

$$\Pi_3 = \rho^{x_3}\omega^{y_3}D^{z_3}\mu = (ML^{-3})^{x_3}(T^{-1})^{y_3}L^{z_3}ML^{-1}T^{-1}$$

$$\Pi_4 = \rho^{x_4}\omega^{y_4}D^{z_4}c = (ML^{-3})^{x_4}(T^{-1})^{y_4}L^{z_4}LT^{-1}$$

By writing the simultaneous equations in x_1, y_1, z_1, etc., as before and solving them,

$$\Pi_1 = \frac{F_T}{\rho\omega^2 D^2} \qquad \Pi_2 = \frac{V_0}{\omega D} \qquad \Pi_3 = \frac{\mu}{\rho\omega D^2} \qquad \Pi_4 = \frac{c}{\omega D}$$

Solving for the thrust parameter leads to

$$\frac{F_T}{\rho\omega^2 D^4} = f_1\left(\frac{V_0}{\omega D}, \frac{\rho\omega D^2}{\mu}, \frac{c}{\omega D}\right)$$

Since the parameters may be recombined to obtain other forms, the second term is replaced by the product of the first and second terms, $VD\rho/\mu$, and the third term is replaced by the first term divided by the third term, V_0/c; thus

$$\frac{F_T}{\rho\omega^2 D^4} = f_2\left(\frac{V_0}{\omega D}, \frac{V_0 D\rho}{\mu}, \frac{V_0}{c}\right)$$

Of the dimensionless parameters, the first is probably of the most importance, since it relates speed of advance to speed of rotation. The second parameter is a Reynolds number and accounts for viscous effects. The last parameter, speed of advance divided by speed of sound, is a Mach number, which would be important for speeds near or higher than the speed of sound. Reynolds effects are usually small, so that a plot of $F_T/\rho\omega^2 D^4$ against $V_0/\omega D$ should be most informative.

The steps in a dimensional analysis may be summarized as follows:

1. Select the pertinent variables. This requires some knowledge of the process.
2. Write the functional relationships, e.g.,

$$F(V, D, \rho, \mu, c, H) = 0$$

3. Select the repeating variables. (Do not make the dependent quantity a repeating variable.) These variables should contain all the m dimensions of the problem. Often one variable is chosen because it specifies the scale, another the kinematic conditions, and in the cases of major interest in this chapter one variable is chosen which is related to the forces or mass of the system, for example, D, V, ρ.

4. Write the Π parameters in terms of unknown exponents, e.g.,

$$\Pi_1 = V^{x_1}D^{y_1}\rho^{z_1}\mu = (LT^{-1})^{x_1}L^{y_1}(ML^{-3})^{z_1}ML^{-1}-T^1$$

5. For each of the Π expressions write the equations of the exponents, so that the sum of the exponents of each dimension will be zero.
6. Solve the equations simultaneously.
7. Substitute back into the Π expressions of step 4 the exponents to obtain the dimensionless Π parameters.
8. Establish the functional relation

$$f_1(\Pi_1, \Pi_2, \Pi_3, \ldots, \Pi_{n-m}) = 0$$

or solve for one of Π's explicitly:

$$\Pi_2 = f(\Pi_1, \Pi_3, \ldots, \Pi_{n-m})$$

9. Recombine, if desired, to alter the forms of the Π parameters, keeping the same number of independent parameters.

Alternate formulation of Π parameters

A rapid method for obtaining Π parameters, developed by Hunsaker and Rightmire (referenced at end of chapter), uses the repeating variables as primary quantities and solves for M, L, and T in terms of them. In Example 4.3 the repeating variables are V, D, and ρ; therefore

$$V = LT^{-1} \qquad D = L \qquad \rho = ML^{-3}$$

$$L = D \qquad T = DV^{-1} \qquad M = \rho D^3 \tag{4.3.4}$$

Now, by use of Eqs. (4.3.4),

$$\mu = ML^{-1}T^{-1} = \rho D^3 D^{-1}D^{-1}V = \rho DV$$

hence the Π parameter is

$$\Pi_1 = \frac{\mu}{\rho D V}$$

Equations (4.3.4) may be used directly to find the other Π parameters. For Π_2

$$g = LT^{-2} = DD^{-2}V^2 = V^2 D^{-1}$$

and

$$\Pi_2 = \frac{g}{V^2 D^{-1}} = \frac{gD}{V^2}$$

This method does not require the repeated solution of three equations in three unknowns for each Π parameter determination.

4.4 DISCUSSION OF DIMENSIONLESS PARAMETERS

The five dimensionless parameters—pressure coefficient, Reynolds number, Froude number, Weber number, and Mach number—are of importance in correlating experimental data. They are discussed in this section, with particular emphasis placed on the relation of pressure coefficient to the other parameters.

Pressure coefficient

The pressure coefficient $\Delta p/(\rho V^2/2)$ is the ratio of pressure to dynamic pressure. When multiplied by area, it is the ratio of pressure force to inertial force, as $(\rho V^2/2)A$ would be the force needed to reduce the velocity to zero. It may also be written as $\Delta h/(V^2/2g)$ by division by γ. For pipe flow the Darcy-Weisbach equation relates losses h_l to length of pipe L, diameter D, and velocity V by a dimensionless friction factor[1] f

$$h_l = f \frac{L}{D} \frac{V^2}{2g} \qquad \text{or} \qquad \frac{fL}{D} = \frac{h_l}{V^2/2g} = f_2\left(\mathbf{R, F, W, M}, \frac{l}{l_1}, \frac{l}{l_2}\right)$$

[1] There are several friction factors in general use. This is the Darcy-Weisbach friction factor, which is four times the size of the *Fanning* friction factor, also called f.

as fL/D is shown to be equal to the pressure coefficient (see Example 4.4). In pipe flow, gravity has no influence on losses; therefore **F** may be dropped out. Similarly surface tension has no effect, and **W** drops out. For steady liquid flow, compressibility is not important, and **M** is dropped. l may refer to D, l_1 to roughness height projection ϵ in the pipe wall, and l_2 to their spacing ϵ'; hence

$$\frac{fL}{D} = f_2\left(\mathbf{R}, \frac{\epsilon}{D}, \frac{\epsilon'}{D}\right) \tag{4.4.1}$$

Pipe-flow problems are discussed in Chaps. 5, 6, and 10. If compressibility is important,

$$\frac{fL}{D} = f_2\left(\mathbf{R,M}, \frac{\epsilon}{D}, \frac{\epsilon'}{D}\right) \tag{4.4.2}$$

Compressible-flow problems are studied in Chap. 6. With orifice flow, studied in Chap. 8, $V = C_v \sqrt{2gH}$,

$$\frac{H}{V^2/2g} = \frac{1}{C_v{}^2} = f_2\left(\mathbf{R,W,M}, \frac{l}{l_1}, \frac{l}{l_2}\right) \tag{4.4.3}$$

in which l may refer to orifice diameter and l_1 and l_2 to upstream dimensions. Viscosity and surface tension are unimportant for large orifices and low-viscosity fluids. Mach number effects may be very important for gas flow with large pressure drops, i.e., Mach numbers approaching unity.

In steady, uniform open-channel flow, discussed in Chap. 5, the Chézy formula relates average velocity V, slope of channel S, and hydraulic radius of cross section R (area of section divided by wetted perimeter) by

$$V = C \sqrt{RS} = C \sqrt{R \frac{\Delta h}{L}} \tag{4.4.4}$$

C is a coefficient depending upon size, shape, and roughness of channel. Then

$$\frac{\Delta h}{V^2/2g} = \frac{2gL}{R} \frac{1}{C^2} = f_2\left(\mathbf{F,R}, \frac{l}{l_1}, \frac{l}{l_2}\right) \tag{4.4.5}$$

since surface tension and compressible effects are usually unimportant.

The drag F on a body is expressed by $F = C_D A \rho V^2/2$, in which A is a typical area of the body, usually the projection of the body onto a plane normal to the flow. Then F/A is equivalent to Δp, and

$$\frac{F}{A\rho V^2/2} = C_D = f_2\left(\mathbf{R}, \mathbf{F}, \mathbf{M}, \frac{l}{l_1}, \frac{l}{l_2}\right) \tag{4.4.6}$$

The term $\mathbf{R}$ is related to *skin-friction* drag due to viscous shear as well as to *form*, or *profile*, drag resulting from *separation* of the flow streamlines from the body; $\mathbf{F}$ is related to wave drag if there is a free surface; for large Mach numbers C_D may vary more markedly with $\mathbf{M}$ than with the other parameters; the length ratios may refer to shape or roughness of the surface.

The Reynolds number

The Reynolds number $VD\rho/\mu$ is the ratio of inertial forces to viscous forces. A *critical* Reynolds number distinguishes among flow regimes, such as laminar or turbulent flow in pipes, in the boundary layer, or around immersed objects. The particular value depends upon the situation. In compressible flow, the Mach number is generally more significant than the Reynolds number.

The Froude number

The Froude number V^2/gl, when multiplied and divided by ρA, is a ratio of dynamic (or inertial) force to weight. With free liquid-surface flow the nature of the flow (rapid[1] or tranquil) depends upon whether the Froude number is greater or less than unity. It is useful in calculations of hydraulic jump, in design of hydraulic structures, and in ship design.

The Weber number

The Weber number $V^2l\rho/\sigma$ is the ratio of inertial forces to surface-tension forces (evident when numerator and denominator are multiplied by l). It is important at gas-liquid or liquid-liquid interfaces and also where these interfaces are in contact with a boundary. Surface tension causes small (capillary) waves and droplet formation and has an effect on discharge of orifices and weirs at very small heads. The effect of surface tension on wave propagation is shown in Fig. 4.1. To the left of the curve's minimum the wave

---

[1] Open-channel flow at depth y is *rapid* when the flow velocity is greater than the speed $\sqrt{gy}$ of an elementary wave in quiet liquid. *Tranquil* flow occurs when the flow velocity is less than $\sqrt{gy}$.

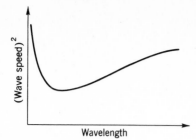

Fig. 4.1 Wave speed vs. wavelength for surface waves.

speed is controlled by surface tension (the waves are called ripples), and to the right of the minimum gravity effects are dominant.

The Mach number

The speed of sound in a liquid is written $\sqrt{K/\rho}$ if K is the bulk modulus of elasticity (Secs. 1.7 and 6.2) or $c = \sqrt{kRT}$ (k is the specific heat ratio and T the absolute temperature, for a perfect gas). V/c or $V/\sqrt{K/\rho}$ is the Mach number. It is a measure of the ratio of inertial forces to elastic forces. By squaring V/c and multiplying by $\rho A/2$ in numerator and denominator, the numerator is the dynamic force and the denominator is the dynamic force at sonic flow. It may also be shown to be a measure of the ratio of kinetic energy of the flow to internal energy of the fluid. It is the most important correlating parameter when velocities are near or above local sonic velocities.

4.5 SIMILITUDE; MODEL STUDIES

Model studies of proposed hydraulic structures and machines are frequently undertaken as an aid to the designer. They permit visual observation of the flow and make it possible to obtain certain numerical data, e.g., calibrations of weirs and gates, depths of flow, velocity distributions, forces on gates, efficiencies and capacities of pumps and turbines, pressure distributions, and losses.

If accurate quantitative data are to be obtained from a model study, there must be dynamic similitude between model and prototype. This similitude requires (1) that there be exact geometric similitude and (2) that the ratio of dynamic pressures at corresponding points be a constant. The second requirement may also be expressed as a kinematic similitude, i.e., the streamlines must be geometrically similar.

Geometric similitude extends to the actual surface roughness of model and prototype. If the model is one-tenth the size of the prototype in every linear dimension, then the height of roughness projections must be in the same ratio. For dynamic pressures to be in the same ratio at corresponding points in model and prototype, the ratios of the various types of forces must be the same at corresponding points. Hence, for strict dynamic similitude, the Mach, Reynolds, Froude, and Weber numbers must be the same in both model and prototype.

Strict fulfillment of these requirements is generally impossible to achieve, except with a 1:1 scale ratio. Fortunately, in many situations only two of the forces are of the same magnitude. Discussion of a few cases will make this clear.

Wind- and water-tunnel tests

This equipment is used to examine the streamlines and the forces that are induced as the fluid flows past a fully submerged body. The type of test that is being conducted and the availability of the equipment determine which kind of tunnel will be used. Because the kinematic viscosity of water is about one-tenth that of air, a water tunnel can be used for model studies at relatively high Reynolds numbers. The drag effect of various parachutes was studied in a water tunnel! At very high air velocities the effects of compressibility, and consequently Mach number, must be taken into consideration, and indeed may be the chief reason for undertaking an investigation. Figure 4.2 shows a model of an aircraft carrier being tested in a low-speed tunnel to study the flow pattern around the ship's superstructure. The model has been inverted and suspended from the ceiling so that the wool tufts can be used to give an indication of the flow direction. Behind the model there is an apparatus for sensing the air speed and direction at various locations along an aircraft's glide path.

Pipe flow

In steady flow in a pipe, viscous and inertial forces are the only ones of consequence; hence, when geometric similitude is observed, the same Reynolds number in model and prototype provides dynamic similitude. The various corresponding pressure coefficients are the same. For testing with fluids having the same kinematic viscosity in model and prototype, the product, VD, must be the same. Frequently this requires very high velocities in small models.

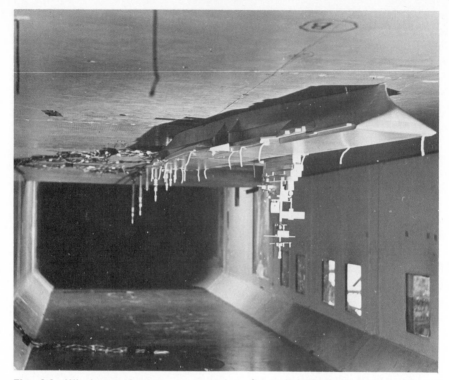

Fig. 4.2 Wind-tunnel tests on an aircraft-carrier superstructure. Model is inverted and suspended from ceiling. (*Photograph taken in Aeronautical and Astronautical Laboratories of The University of Michigan for the Dynasciences Corp.*)

Open hydraulic structures

Structures such as spillways, stilling pools, channel transitions, and weirs generally have forces due to gravity (from changes in elevation of liquid surfaces) and inertial forces that are greater than viscous and turbulent shear forces. In these cases geometric similitude and the same value of Froude's number in model and prototype produce a good approximation to dynamic similitude; thus

$$\frac{V_m{}^2}{g_m l_m} = \frac{V_p{}^2}{g_p l_p}$$

Since gravity is the same, the velocity ratio varies as the square root of the scale ratio $\lambda = l_p/l_m$,

$$V_p = V_m \sqrt{\lambda}$$

The corresponding times for events to take place (as time for passage of a particle through a transition) are related; thus

$$t_m = \frac{l_m}{V_m} \qquad t_p = \frac{l_p}{V_p} \qquad \text{and} \qquad t_p = t_m \frac{l_p}{l_m} \frac{V_m}{V_p} = t_m \sqrt{\lambda}$$

The discharge ratio Q_p/Q_m is

$$\frac{Q_p}{Q_m} = \frac{l_p^3/t_p}{l_m^3/t_m} = \lambda^{5/2}$$

Force ratios, e.g., on gates, F_p/F_m, are

$$\frac{F_p}{F_m} = \frac{\gamma h_p l_p^2}{\gamma h_m l_m^2} = \lambda^3$$

where h is the head. In a similar fashion other pertinent ratios can be derived so that model results can be interpreted as prototype performance.

Figure 4.3 shows a model test conducted to determine the effect of a breakwater on the wave formation in a harbor.

Ship's resistance

The resistance to motion of a ship through water is composed of pressure drag, skin friction, and wave resistance. Model studies are complicated by the three types of forces that are important, inertia, viscosity, and gravity. Skin-friction studies should be based on equal Reynolds numbers in model and prototype, but wave resistance depends upon the Froude number. To satisfy both requirements, model and prototype must be the same size.

The difficulty is surmounted by using a small model and measuring the total drag on it when towed. The skin friction is then computed for the model and subtracted from the total drag. The remainder is stepped up to prototype size by Froude's law, and the prototype skin friction is computed and added to yield total resistance due to the water. Figure 4.4 shows the dramatic change in the wave profile which resulted from a redesigned bow. From such tests it is possible to predict through Froude's law the wave formation and drag that would occur on the prototype.

Fig. 4.3 Model test on a harbor to determine the effect of a breakwater. (*Department of Civil Engineering, The University of Michigan.*)

Hydraulic machinery

The moving parts in a hydraulic machine require an extra parameter to ensure that the streamline patterns are similar in model and prototype. This parameter must relate the throughflow (discharge) to the speed of moving parts. For geometrically similar machines if the vector diagrams of velocity entering or leaving the moving parts are similar, the units are *homologous*; i.e., for practical purposes dynamic similitude exists. The Froude number is unimportant, but the Reynolds number effects (called *scale effects* because it is impossible to maintain the same Reynolds number in homologous units) may cause a discrepancy of 2 or 3 percent in efficiency between model and proto-

Fig. 4.4 Model tests showing the influence of a bulbous bow on bow-wave formation. (*Department of Naval Architecture and Marine Engineering, The University of Michigan.*) $\longrightarrow$

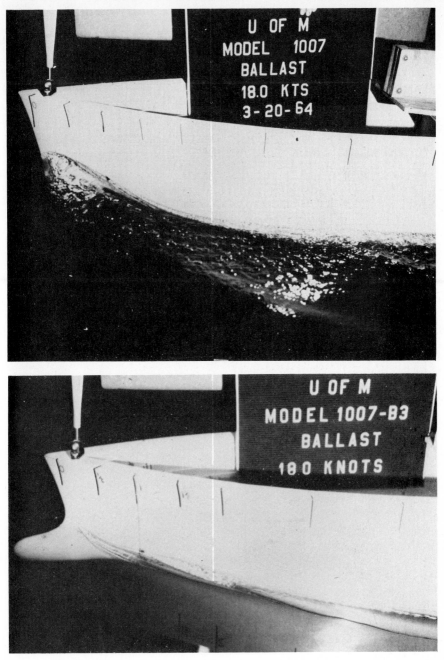

type. The Mach number is also of importance in axial-flow compressors and gas turbines.

EXAMPLE 4.6 The valve coefficients $K = \Delta p/(\rho v^2/2)$ for a 60-cm-diameter valve are to be determined from tests on a geometrically similar 30-cm-diameter valve using atmospheric air at 80°F. The ranges of tests should be for flow of water at 70°F at 1 to 2.5 m/s. What ranges of airflows is needed?

The Reynolds number range for the prototype valve is

$$\left(\frac{VD}{\nu}\right)_{\min} = \frac{(1 \text{ m/s}) (0.6 \text{ m})}{(1.059 \times 10^{-5} \text{ ft}^2/\text{s}) (0.3048 \text{ m/ft})^2} = 610,000$$

$$\left(\frac{VD}{\nu}\right)_{\max} = 610,000 \times 2.5 = 1,525,000$$

For testing with air at 80°F

$$\nu = (1.8 \times 10^{-4} \text{ ft}^2/\text{s}) (0.3048 \text{ m/ft})^2 = 1.672 \times 10^{-5} \text{ m}^2/\text{s}$$

Then the ranges of air velocities are

$$\frac{(V_{\min}) (0.3 \text{ m})}{1.672 \times 10^{-5} \text{ m}^2/\text{s}} = 610,000 \qquad V_{\min} = 30.6 \text{ m/s}$$

$$\frac{(V_{\max}) (0.3 \text{ m})}{1.672 \times 10^{-5} \text{ m}^2/\text{s}} = 1,525,000 \qquad V_{\max} = 85 \text{ m/s}$$

$$Q_{\min} = \frac{\pi}{4} (0.3 \text{ m})^2 (30.6 \text{ m/s}) = 2.16 \text{ m}^3/\text{s}$$

$$Q_{\max} = \frac{\pi}{4} (0.3 \text{ m})^2 (85 \text{ m/s}) = 6.0 \text{ m}^3/\text{s}$$

PROBLEMS

4.1 Show that Eqs. (3.7.6), (3.9.3), and (3.11.13) are dimensionally homogeneous.

4.2 Arrange the following groups into dimensionless parameters:

(a) $\Delta p, \rho, V$ (b) ρ, g, V, F (c) $\mu, F, \Delta p, t$

4.3 By inspection, arrange the following groups into dimensionless parameters:

(a) a, l, t (b) v, l, t (c) A, Q, ω (d) K, σ, A

4.4 Derive the unit of mass consistent with the units inches, minutes, tons.

4.5 In terms of M, L, T, determine the dimensions of radians, angular velocity, power, work, torque, and moment of momentum.

4.6 Find the dimensions of the quantities in Prob. 4.5 in the FLT system.

4.7 Work Example 4.2 using Q and H as repeating variables.

4.8 Using the variables $Q, D, \Delta H/l, \rho, \mu, g$ as pertinent to smooth-pipe flow, arrange them into dimensionless parameters with Q, ρ, μ as repeating variables.

4.9 If the shear stress τ is known to depend upon viscosity and rate of angular deformation du/dy in one-dimensional laminar flow, determine the form of Newton's law of viscosity by dimensional reasoning.

4.10 The variation Δp of pressure in static liquids is known to depend upon specific weight γ and elevation difference Δz. By dimensional reasoning determine the form of the hydrostatic law of variation of pressure.

4.11 When viscous and surface-tension effects are neglected, the velocity V of efflux of liquid from a reservoir is thought to depend upon the pressure drop Δp of the liquid and its density ρ. Determine the form of expression for V.

4.12 The buoyant force F_B on a body is thought to depend upon its volume submerged $\mathcal{V}$ and the gravitational body force acting on the fluid. Determine the form of the buoyant-force equation.

4.13 In a fluid rotated as a solid about a vertical axis with angular velocity ω, the pressure rise p in a radial direction depends upon speed ω, radius r, and fluid density ρ. Obtain the form of equation for p.

4.14 In Example 4.3, work out two other sets of dimensionless parameters by recombination of the dimensionless parameters given.

4.15 Find the dimensionless parameters of Example 4.4 using $\Delta p, \rho$, and l as repeating variables.

4.16 The Mach number **M** for flow of a perfect gas in a pipe depends upon the specific-heat ratio k (dimensionless), the pressure p, the density ρ, and the velocity V. Obtain by dimensional reasoning the form of the Mach number expression.

4.17 Work out the scaling ratio for torque T on a disk of radius r that rotates in fluid of viscosity μ with angular velocity ω and clearance y between disk and fixed plate.

4.18 The velocity at a point in a model of a spillway for a dam is 3.3 ft/s. For a ratio of prototype to model of 10:1, what is the velocity at the corresponding point in the prototype under similar conditions?

4.19   The power input to a pump depends upon the discharge Q, the pressure rise Δp, the fluid density ρ, size D, and efficiency e. Find the expression for power by the use of dimensional analysis.

4.20   The torque delivered by a water turbine depends upon discharge Q, head H, specific weight γ, angular velocity ω, and efficiency e. Determine the form of equation for torque.

4.21   A model of a venturi meter has linear dimensions one-fifth those of the prototype. The prototype operates on water at 20°C, and the model on water at 95°C. For a throat diameter of 60 cm and a velocity at the throat of 6 m/s in the prototype, what discharge is needed through the model for similitude?

4.22   The drag F on a high-velocity projectile depends upon speed V of projectile, density of fluid ρ, acoustic velocity c, diameter of projectile D, and viscosity μ. Develop an expression for the drag.

4.23   The wave drag on a model of a ship is 3.52 lb at a speed of 8 ft/s. For a prototype fifteen times as long what will the corresponding speed and wave drag be if the liquid is the same in each case?

4.24   Determine the specific gravity of spherical particles, $D = \frac{1}{200}$ in, which drop through air at 33°F at a speed U of 0.3 ft/s. The drag force on a small sphere in laminar motion is given by $3\pi\mu DU$.

4.25   A small sphere of radius r_0 and density ρ_0 settles at velocity U in another liquid of density ρ and viscosity μ. The tests are conducted inside vertical tubes of radius r. By dimensional analysis determine a set of dimensionless parameters to be used in determining the influence of the tube wall on the settling velocity.

4.26   The losses in a Y in a 1.2-m-diameter pipe system carrying gas ($\rho = 40 \text{ kg/m}^3$, $\mu = 0.002$ P, $V = 25$ m/s) are to be determined by testing a model with water at 20°C. The laboratory has a water capacity of 75 l/s. What model scale should be used, and how are the results converted into prototype losses?

4.27   Ripples have a velocity of propagation that is dependent upon the surface tension and density of the fluid as well as the wavelength. By dimensional analysis justify the shape of Fig. 4.1 for small wavelengths.

4.28   In very deep water the velocity of propagation of waves depends upon the wavelength, but in shallow water it is independent of this dimension. Upon what variables does the speed of advance depend for shallow-water waves? Is Fig. 4.1 in agreement with this problem?

4.29   If a vertical circular conduit which is not flowing full is rotated at high speed, the fluid will attach itself uniformly to the inside wall as it flows downward (see Sec. 2.9). Under these conditions the radial acceleration of the fluid yields a radial force field which is similar to gravitational attraction, and a hydraulic jump can occur on the inside of the tube, whereby the fluid thickness suddenly changes. Determine a set of dimensionless parameters for studying this rotating hydraulic jump.

4.30 A nearly spherical fluid drop oscillates as it falls. Surface tension plays a dominant role. Determine a meaningful dimensionless parameter for this natural frequency.

4.31 The lift and drag coefficients for a wing are shown in Fig. 5.23. If the wing has a chord of 10 ft, determine the lift and drag per foot of length when the wing is operating at zero angle of attack at a Reynolds number, based on the chord length, of 4.5×10^7 in air at 50°F. What force would be on a 1:20 scale model if the tests were conducted in water at 70°F? What would be the speed of the water? Comment on the desirability of conducting the model tests in water.

4.32 A 1:5 scale model of a water pumping station piping system is to be tested to determine overall head losses. Air at 25°C, 1 kg$_f$/cm^2 abs is available. For a prototype velocity of 50 cm/s in a 4-m-diameter section with water at 15°C, determine the air velocity and quantity needed and how losses determined from the model are converted into prototype losses.

4.33 Full-scale wind-tunnel tests of the lift and drag on hydrofoils for a boat are to be made. The boat will travel at 35 mph through water at 60°F. What velocity of air ($p = 30$ psia, $t = 90$°F) is required to determine the lift and drag? *Note:* The lift coefficient C_L is dimensionless. Lift $= C_L A \rho V^2 / 2$.

4.34 The resistance to ascent of a balloon is to be determined by studying the ascent of a 1:50 scale model in water. How would such a model study be conducted and the results converted to prototype behavior?

4.35 The moment exerted on a submarine by its rudder is to be studied with a 1:20 scale model in a water tunnel. If the torque measured on the model is 5 N·m for a tunnel velocity of 15 m/s, what are the corresponding torque and speed for the prototype?

4.36 For two hydraulic machines to be homologous they must (*a*) be geometrically similar, (*b*) have the same discharge coefficient when viewed as an orifice, $Q_1/(A_1 \sqrt{2gH_1}) = Q_2/(A_2 \sqrt{2gH_2})$, and (*c*) have the same ratio of peripheral speed to fluid velocity, $\omega D/(Q/A)$. Show that the scaling ratios may be expressed as $Q/ND^3 = $ const and $H/(ND)^2 = $ const. N is the rotational speed.

4.37 By use of the scaling ratios of Prob. 4.36, determine the head and discharge of a 1:4 model of a centrifugal pump that produces 20 cfs at 96 ft head when turning 240 rpm. The model operates at 1200 rpm.

4.38 An *incorrect* arbitrary recombination of the Π parameters

$$F\left(\frac{V_0}{\omega D}, \frac{\rho \omega D^2}{\mu}, \frac{c}{\omega D}\right) = 0$$

is

(*a*) $F\left(\dfrac{c}{V_0}, \dfrac{\rho c D}{\mu}, \dfrac{c}{\omega D}\right) = 0$ (*b*) $F\left(\dfrac{V_0}{\omega D}, \dfrac{\rho c D^2}{\mu}, \dfrac{c}{\omega D}\right) = 0$

(c) $F\left(\dfrac{V_0}{\omega D}, \dfrac{V_0 c\rho}{\omega \mu}, \dfrac{\rho c D}{\mu}\right) = 0$ (d) $F\left(\dfrac{V_0 \mu}{\omega^2 D^3 \rho}, \dfrac{V_0 \rho D}{\mu}, \dfrac{c}{\omega D}\right) = 0$

(e) none of these answers

4.39 The repeating variables in a dimensional analysis should

(a) include the dependent variable
(b) have two variables with the same dimensions if possible
(c) exclude one of the dimensions from each variable if possible
(d) include those variables not considered very important factors
(e) satisfy none of these answers

4.40 Select a common dimensionless parameter in fluid mechanics from the following:

(a) angular velocity   (b) kinematic viscosity   (c) specific gravity
(d) specific weight   (e) none of these answers

4.41 Select the quantity in the following that is *not* a dimensionless parameter:

(a) pressure coefficient   (b) Froude number   (c) Darcy-Weisbach friction
factor   (d) kinematic viscosity   (e) Weber number

4.42 Which of the following has the form of a Reynolds number?

(a) $\dfrac{ul}{\nu}$ (b) $\dfrac{VD\mu}{\rho}$ (c) $\dfrac{u_w \nu}{l}$ (d) $\dfrac{V}{gD}$ (e) $\dfrac{\Delta p}{\rho V^2}$

4.43 The Reynolds number may be defined as the ratio of

(a) viscous forces to inertial forces
(b) viscous forces to gravity forces
(c) gravity forces to inertial forces
(d) elastic forces to pressure forces
(e) none of these answers

4.44 The pressure coefficient may take the form

(a) $\dfrac{\Delta p}{\gamma H}$ (b) $\dfrac{\Delta p}{\rho V^2/2}$ (c) $\dfrac{\Delta p}{l\mu V}$ (d) $\Delta p \dfrac{\rho}{\mu^2 l^4}$

(e) none of these answers

4.45 The pressure coefficient is a ratio of pressure forces to

(a) viscous forces
(b) inertial forces
(c) gravity forces
(d) surface-tension forces
(e) elastic-energy forces

4.46 How many Π parameters are needed to express the function
$F(a,V,t,\nu,L) = 0$?

(a) 5   (b) 4   (c) 3   (d) 2   (e) 1

4.47 Which of the following could be a Π parameter of the function $F(Q,H,g,V_0,\phi) = 0$
when Q and g are taken as repeating variables?

(a) Q^2/gH^4 (b) V_0^2/g^2Q (c) $Q/g\phi^2$ (d) $Q/\sqrt{gH}$
(e) none of these answers

4.48 Select the situation in which inertial forces would be *unimportant*:

(a) flow over a spillway crest
(b) flow through an open-channel transition
(c) waves breaking against a sea wall
(d) flow through a long capillary tube
(e) flow through a half-opened valve

4.49 Which two forces are most important in laminar flow between closely spaced
parallel plates?

(a) inertial, viscous   (b) pressure, inertial   (c) gravity, pressure
(d) viscous, pressure   (e) none of these answers

4.50 A dimensionless combination of Δp, ρ, l, Q is

(a) $\sqrt{\dfrac{\Delta p}{\rho}}\dfrac{Q}{l^2}$ (b) $\dfrac{\rho Q}{\Delta p\, l^2}$ (c) $\dfrac{\rho l}{\Delta p\, Q^2}$ (d) $\dfrac{\Delta p\, lQ}{\rho}$ (e) $\sqrt{\dfrac{\rho}{\Delta p}}\dfrac{Q}{l^2}$

4.51 What velocity of oil, $\rho = 1.6$ slugs/ft^3, $\mu = 0.20$ P, must occur in a 1-in-diameter
pipe to be dynamically similar to 10 ft/s water velocity at 68°F in a $\frac{1}{4}$-in-diameter
tube?

(a) 0.60 ft/s   (b) 9.6 ft/s   (c) 4.0 ft/s   (d) 60 ft/s
(e) none of these answers

4.52 The velocity at a point on a model dam crest was measured to be 1 m/s. The
corresponding prototype velocity for $\lambda = 25$ is, in meters per second,

(a) 25   (b) 5   (c) 0.2   (d) 0.04   (e) none of these answers

4.53 The height of a hydraulic jump in a stilling pool was found to be 4.0 in in a model,
$\lambda = 36$. The prototype jump height is

(a) 12 ft   (b) 2 ft   (c) not determinable from data given
(d) less than 4 in   (e) none of these answers

4.54 A ship's model, scale 1:100, had a wave resistance of 10 N at its design speed.
The corresponding prototype wave resistance is, in kilonewtons,

(a) 10   (b) 100   (c) 1000   (d) 10,000   (e) none of these answers

4.55   A 1:5 scale model of a projectile has a drag coefficient of 3.5 at **M** = 2.0. How many times greater would the prototype resistance be when fired at the same Mach number in air of the same temperature and half the density?

(*a*) 0.312   (*b*) 3.12   (*c*) 12.5   (*d*) 25   (*e*) none of these answers

4.56   If the capillary rise Δh of a liquid in a circular tube of diameter D depends upon surface tension σ and specific weight γ, the formula for capillary rise could take the form

(*a*) $\Delta h = \sqrt{\dfrac{\sigma}{\gamma}}\, F\left(\dfrac{\sigma}{\gamma D^2}\right)$ (*b*) $\Delta h = c\left(\dfrac{\sigma}{\gamma D^2}\right)^n$ (*c*) $\Delta h = cD\left(\dfrac{\sigma}{\gamma}\right)^n$

(*d*) $\Delta h = \sqrt{\dfrac{\gamma}{\sigma}}\, F\left(\dfrac{\gamma D^2}{\sigma}\right)$ (*e*) none of these answers

REFERENCES

Bridgman, P. W.: "Dimensional Analysis," Yale University Press, New Haven, Conn., 1931, Paperback Y-82, 1963.

Holt, M.: Dimensional Analysis, sec. 15 in V. L. Streeter (ed.), "Handbook of Fluid Dynamics," McGraw-Hill, New York, 1961.

Hunsaker, J. C., and B. G. Rightmire: "Engineering Applications of Fluid Mechanics," pp. 110, 111, McGraw-Hill, New York, 1947.

Hydraulic Models, *ASCE Man. Eng. Pract.* 25, 1942.

Ipsen, D. C.: "Units, Dimensions, and Dimensionless Numbers," McGraw-Hill, New York, 1960.

Langhaar, H. L.: "Dimensional Analysis and Theory of Models," Wiley, New York, 1951.

Sedov, L. I.: "Similarity and Dimensional Methods in Mechanics," English trans. ed. by M. Holt, Academic, New York, 1959.

5

VISCOUS EFFECTS: FLUID RESISTANCE

In Chap. 3 the basic equations used in the analysis of fluid-flow situations were discussed. The fluid was considered frictionless, or in some cases losses were assumed or computed without probing into their underlying causes. This chapter deals with real fluids, i.e., with situations in which irreversibilities are important. Viscosity is the fluid property that causes shear stresses in a moving fluid; it is also one means by which irreversibilities or losses are developed. Without viscosity in a fluid there is no fluid resistance. Simple cases of steady, laminar, incompressible flow are first developed in this chapter, since in these cases the losses can be computed. The concept of the Reynolds number, introduced in Chap. 4, is then further developed. Turbulent-flow shear relationships are introduced by use of the Prandtl mixing-length theory and are applied to turbulent velocity distributions. This is followed by boundary-layer concepts and by drag on immersed bodies. Resistance to steady, uniform, incompressible, turbulent flow is then examined for open and closed conduits, with a section devoted to open channels and to pipe flow. The chapter closes with a section on lubrication mechanics.

The equations of motion for a real fluid can be developed from consideration of the forces acting on a small element of the fluid, including the shear stresses generated by fluid motion and viscosity. The derivation of these equations, called the *Navier-Stokes equations*, is beyond the scope of this treatment. They are listed, however, for the sake of completeness, and many of the developments of this chapter could be made directly from them. First, Newton's law of viscosity, Eq. (1.1.1), for one-dimensional laminar flow can be generalized to three-dimensional flow (Stokes' law of viscosity)

$$\tau_{xy} = \mu \left(\frac{\partial u}{\partial y} + \frac{\partial v}{\partial x} \right) \qquad \tau_{yz} = \mu \left(\frac{\partial v}{\partial z} + \frac{\partial w}{\partial y} \right) \qquad \tau_{zx} = \mu \left(\frac{\partial w}{\partial x} + \frac{\partial u}{\partial z} \right)$$

The first subscript of the shear stress is the normal direction to the face over which the stress component is acting. The second subscript is the direction of the stress component.

By limiting the Navier-Stokes equations to incompressible flow, with gravity the only body force acting (let h be measured vertically upward), they become

$$-\frac{1}{\rho}\frac{\partial}{\partial x}(p + \gamma h) + \nu \nabla^2 u = \frac{du}{dt}$$

$$-\frac{1}{\rho}\frac{\partial}{\partial y}(p + \gamma h) + \nu \nabla^2 v = \frac{dv}{dt}$$

$$-\frac{1}{\rho}\frac{\partial}{\partial z}(p + \gamma h) + \nu \nabla^2 w = \frac{dw}{dt}$$

in which ν is the kinematic viscosity, assumed to be constant; d/dt is differentiation with respect to the motion

$$\frac{d}{dt} = u\frac{\partial}{\partial x} + v\frac{\partial}{\partial y} + w\frac{\partial}{\partial z} + \frac{\partial}{\partial t}$$

as explained in Sec. 7.3, and the operator ∇^2 is

$$\nabla^2 = \frac{\partial^2}{\partial x^2} + \frac{\partial^2}{\partial y^2} + \frac{\partial^2}{\partial z^2}$$

For a nonviscous fluid, the Navier-Stokes equations reduce to the Euler equations of motion in three dimensions, given by Eqs. (7.3.8), (7.3.9), and (7.3.10). For one-dimensional flow of a real fluid in the x direction, with z vertically upward the equations reduce to (u varies with z and t only)

$$-\frac{1}{\rho}\frac{\partial p}{\partial x} + \frac{\mu}{\rho}\frac{\partial^2 u}{\partial z^2} = \frac{du}{dt} \qquad \frac{\partial p}{\partial y} = 0 \qquad \frac{\partial}{\partial z}(p + \gamma z) = 0$$

and for steady flow the first equation reduces to

$$\frac{\partial p}{\partial x} = \mu \frac{\partial^2 u}{\partial z^2}$$

5.1 LAMINAR, INCOMPRESSIBLE, STEADY FLOW BETWEEN PARALLEL PLATES

The general case of steady flow between parallel inclined plates is first developed for laminar flow, with the upper plate having a constant velocity U (Fig. 5.1). Flow between fixed plates is a special case obtained by setting $U = 0$. In Fig. 5.1 the upper plate moves parallel to the flow direction, and there is a pressure variation in the l direction. The flow is analyzed by taking a thin lamina of unit width as a free body. In steady flow the lamina moves at constant velocity u. The equation of motion yields

$$p \, \delta y - \left(p \, \delta y + \frac{dp}{dl} \, \delta l \, \delta y \right) - \tau \, \delta l + \left(\tau \, \delta l + \frac{d\tau}{dy} \, \delta y \, \delta l \right) + \gamma \, \delta l \, \delta y \sin \theta = 0$$

Dividing through by the volume of the element and simplifying gives

$$\frac{d\tau}{dy} = \frac{d}{dl} (p + \gamma h) \tag{5.1.1}$$

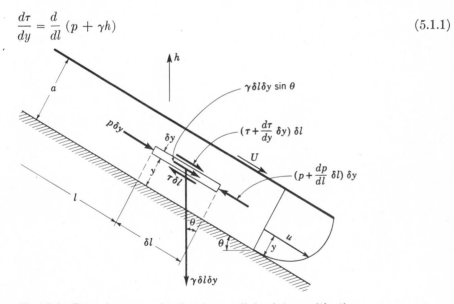

Fig. 5.1 Flow between inclined parallel plates with the upper plate in motion.

in which $\sin\theta = -dh/dl$ has been substituted. Since there is no acceleration in the y direction, the right-hand side of the equation is not a function
of y. Integrating with respect to y yields

$$\tau = y\frac{d}{dl}(p + \gamma h) + A$$

Now, substitution of Newton's law of viscosity [Eq. (1.1.1)],

$$\tau = \mu\frac{du}{dy}$$

for τ gives

$$\frac{du}{dy} = \frac{1}{\mu}\frac{d}{dl}(p + \gamma h)y + \frac{A}{\mu}$$

Integrating again with respect to y leads to

$$u = \frac{1}{2\mu}\frac{d}{dl}(p + \gamma h)y^2 + \frac{A}{\mu}y + B$$

in which A and B are constants of integration. To evaluate them take $y = 0$,
$u = 0$ and $y = a$, $u = U$, and obtain

$$B = 0 \qquad U = \frac{1}{2\mu}\frac{d}{dl}(p + \gamma h)a^2 + \frac{Aa}{\mu} + B$$

Eliminating A and B results in

$$u = \frac{Uy}{a} - \frac{1}{2\mu}\frac{d}{dl}(p + \gamma h)(ay - y^2) \tag{5.1.2}$$

For horizontal plates, $h = C$; for no gradient due to pressure or elevation,
i.e., hydrostatic pressure distribution, $p + \gamma h = C$ and the velocity has a
straight-line distribution. For fixed plates, $U = 0$, and the velocity distribution is parabolic.

The discharge past a fixed cross section is obtained by integration of
Eq. (5.1.2) with respect to y:

$$Q = \int_0^a u\,dy = \frac{Ua}{2} - \frac{1}{12\mu}\frac{d}{dl}(p + \gamma h)a^3 \tag{5.1.3}$$

In general the maximum velocity is not at the midplane.

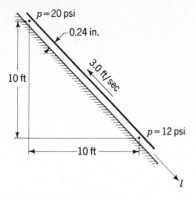

Fig. 5.2 Flow   between
inclined flat plates.

EXAMPLE 5.1 In Fig. 5.2 one plate moves relative to the other as shown.
$\mu = 0.80$ P; $\rho = 1.7$ slugs/ft^3. Determine the velocity distribution, the discharge, and the shear stress exerted on the upper plate.

At the upper point

$$p + \gamma h = (20 \text{ lb/in}^2)\,(144 \text{ in}^2/\text{ft}^2) + (1.7 \text{ slugs/ft}^3)\,(32.2 \text{ ft/s}^2)\,(10 \text{ ft})$$

$$= 3427 \text{ lb/ft}^2$$

and at the lower point

$$p + \gamma h = 12 \times 144 = 1728 \text{ lb/ft}^2$$

to the same datum. Hence

$$\frac{d(p + \gamma h)}{dl} = \frac{1728 - 3427 \text{ lb/ft}^2}{10\sqrt{2} \text{ ft}} = -120 \text{ lb/ft}^3$$

From the figure, $a = 0.24/12 = 0.02$ ft, $U = -3.0$ ft/s, and from Eq. (5.1.2)

$$u = -\frac{(3 \text{ ft/s})\,(y \text{ ft})}{0.02 \text{ ft}} + \frac{(120 \text{ lb/ft}^3)\,(0.02y - y^2 \text{ ft}^2)}{2 \times 0.8 \text{ P} \times \dfrac{1 \text{ lb}\cdot\text{s/ft}^2}{479 \text{ P}}}$$

Simplifying gives

$$u = 566y - 35{,}800y^2 \qquad \text{ft/s}$$

The maximum velocity occurs where $du/dy = 0$, or $y = 0.0079$ ft, and is $u_{\max} = 2.24$ ft/s; the minimum velocity occurs at the upper plate.

The discharge is

$$Q = \int_0^{0.02} u \, dy = 283y^2 - 11{,}933y^3 \Big]_0^{0.02} = 0.0177 \text{ cfs/ft}$$

and is downward.

To find the shear stress on the upper plate,

$$\frac{du}{dy}\bigg]_{y=0.02} = 566 - 71{,}600y \bigg]_{y=0.02} = -866 \text{ lb/ft}^2$$

and

$$\tau = \mu \frac{du}{dy} = \frac{0.80}{479}(-866) = -1.45 \text{ lb/ft}^2$$

This is the fluid shear at the plate; hence the shear force on the plate is 1.45 lb/ft² resisting the motion of the plate.

Losses in laminar flow

An expression for the irreversibilities is developed for one-dimensional, incompressible, steady, laminar flow, in which the equation of motion and the principle of work and energy are utilized. There is no increase in kinetic energy in steady flow in a tube, between parallel plates or in film flow at constant depth. The reduction in $p + \gamma h$, which represents work done on the fluid per unit volume, is converted into irreversibilities by the action of viscous shear. The losses in the length L are $Q \, \Delta(p + \gamma h)$ per unit time.

After examination of the work done on the fluid in one-dimensional flow, an expression for the losses can be developed. First, the equation of motion applied to an element (Fig. 5.3) relates the shear stress and change in $p + \gamma h$. There is no acceleration; hence $\Sigma f_x = 0$, and

$$(p + \gamma h)\, \delta y - \left[p + \gamma h + \frac{d(p + \gamma h)}{dx}\, \delta x\right] \delta y - \tau \, \delta x + \left(\tau + \frac{d\tau}{dy}\, \delta y\right) \delta x = 0$$

Simplifying gives

$$\frac{d(p + \gamma h)}{dx} = \frac{d\tau}{dy} \tag{5.1A}$$

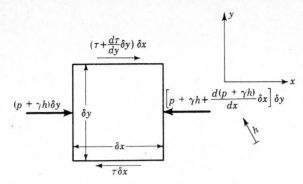

Fig. 5.3 Forces on a fluid element.

which implies that the rate of change of $p + \gamma h$ in the x direction must equal the rate of change of shear in the y direction. Clearly $d(p + \gamma h)/dx$ is independent of y, and $d\tau/dy$ is independent of x.

The work done per unit time, or power input, to a fluid element (Fig. 5.4) for one-dimensional flow consists in the power input to the element by $p + \gamma h$ and by shear stress minus the work per unit time that the element does on the surrounding fluid, or

$$(p + \gamma h)\left(u + \frac{du}{dy}\frac{\delta y}{2}\right)\delta y - \left[p + \gamma h + \frac{d(p + \gamma h)}{dx}\delta x\right]$$

$$\times \left(u + \frac{du}{dy}\frac{\delta y}{2}\right)\delta y + \tau u\,\delta x + \frac{d}{dy}(\tau u)\,\delta y\,\delta x - \tau u\,\delta x$$

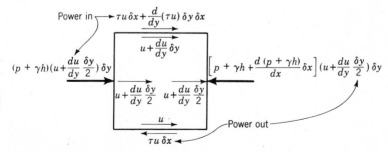

Fig. 5.4 Work done per unit time in a fluid element in one-dimensional motion.

Simplifying leads to

$$\frac{\text{Net power input}}{\text{Unit volume}} = \frac{d}{dy}(\tau u) - u\frac{d(p + \gamma h)}{dx} \tag{5.1.5}$$

Expanding Eq. (5.1.5) and substituting Eq. (5.1.4) gives

$$\frac{\text{Net power input}}{\text{Unit volume}} = \tau\frac{du}{dy} + u\frac{d\tau}{dy} - u\frac{d(p + \gamma h)}{dx} = \tau\frac{du}{dy} \tag{5.1.6}$$

With Newton's law of viscosity,

$$\frac{\text{Net power input}}{\text{Unit volume}} = \tau\frac{du}{dy} = \mu\left(\frac{du}{dy}\right)^2 = \frac{\tau^2}{\mu} \tag{5.1.7}$$

This power is used up by viscous friction and is converted into irreversibilities.

Integrating the expression over a length L between two fixed parallel plates, with Eq. (5.1.2) for $U = 0$ and with Eq. (5.1.7), gives

$$\text{Net power input} = \int_0^a \mu\left(\frac{du}{dy}\right)^2 L\,dy = \mu L \int_0^a \left[\frac{1}{2\mu}\frac{d(p + \gamma h)}{dl}(2y - a)\right]^2 dy$$

$$= \left[\frac{d(p + \gamma h)}{dl}\right]^2 \frac{a^3 L}{12\mu}$$

Substituting for Q from Eq. (5.1.3) for $U = 0$ yields

$$\text{Losses} = \text{net power input} = -Q\frac{d(p + \gamma h)}{dl}L = Q\,\Delta(p + \gamma h)$$

in which $\Delta(p + \gamma h)$ is the drop in $p + \gamma h$ in the length L. The expression for

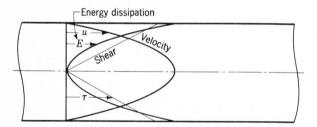

Fig. 5.5 Distribution of velocity, shear, and losses per unit volume for a round tube.

power input per unit volume [Eq. (5.1.7)] is also applicable to laminar flow in a tube. The irreversibilities are greatest when du/dy is greatest. The distribution of shear stress, velocity, and losses per unit volume is shown in Fig. 5.5 for a round tube.

EXAMPLE 5.2 A conveyor-belt device, illustrated in Fig. 5.6, is mounted on a ship and used to pick up undesirable surface contaminants, e.g., oil, from the surface of the sea. Assume the oil film to be thick enough for the supply to be unlimited with respect to the operation of the device. Assume the belt to operate at a steady velocity U and to be long enough for a uniform flow depth to exist. Determine the rate at which oil can be carried up the belt per unit width, in terms of θ, U, and the oil properties μ and γ.

A thin lamina of unit width that moves at velocity u is shown in Fig. 5.6. With the free surface as shown on the belt, and for steady flow at constant depth, the end-pressure effects on the lamina cancel. The equation of motion applied to the element yields

$$-\left(\tau + \frac{d\tau}{dy}\,\delta y\right)\delta l + \tau\,\delta l - \gamma\,\delta y\,\delta l \sin\theta = 0 \qquad \text{or} \qquad \frac{d\tau}{dy} = -\gamma\sin\theta$$

When the shear stress at the surface is recognized as zero, integration yields

$$\tau = \gamma \sin\theta\,(a - y)$$

This equation can be combined with Newton's law of viscosity, $\tau = -\mu\,du/dy$, to give

$$\int_U^u du = -\frac{\gamma\sin\theta}{\mu}\int_0^y (a - y)\,dy$$

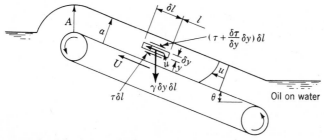

Fig. 5.6 Oil-pickup device.

or

$$u = U - \frac{\gamma \sin \theta}{\mu} \left(ay - \frac{y^2}{2} \right)$$

The flow rate per unit width up the belt can be determined by integration:

$$q = \int_0^a u \, dy = Ua - \frac{\gamma \sin \theta}{\mu} \frac{a^3}{3}$$

This expression shows the flow rate to vary with a. However, a is still a dependent variable that is not uniquely defined by the above equations. The actual depth of flow on the belt is controlled by the end conditions.

The depth for maximum flow rate can be obtained by setting the derivative dq/da to zero and solving for the particular a

$$a = \bar{a} = \left(\frac{U\mu}{\gamma \sin \theta} \right)^{1/2}$$

To attach some physical significance to this particular depth the influence of alternative crest depths may be considered. If the crest depth A, Fig. 5.6, is such that $\bar{a}$ occurs on the belt, then the maximum flow for that belt velocity and slope will be achieved. If A is physically controlled at a depth greater than $\bar{a}$, more flow will temporarily be supplied by the belt than can get away at the crest, causing the belt depth to increase and the flow to decrease correspondingly, until either an equilibrium condition is realized or A is lowered. Alternatively if $A < \bar{a}$, flow off the belt will be less than the maximum flow up the belt at depth $\bar{a}$ and the crest depth will increase to $\bar{a}$. At all times it is assumed that an unlimited supply is available at the bottom. By this reasoning it is seen that $\bar{a}$ is the only physical flow depth that can exist on the belt if the crest depth is free to seek its own level. A similar reasoning at the base leads to the same conclusion.

The discharge, as a function of fluid properties and U and θ, is given by

$$q = U \left(\frac{U\mu}{\gamma \sin \theta} \right)^{1/2} - \frac{\gamma \sin \theta}{3\mu} \left(\frac{U\mu}{\gamma \sin \theta} \right)^{3/2}$$

or

$$q = \left(\frac{\mu}{\gamma \sin \theta} \right)^{1/2} \frac{2}{3} U^{3/2}$$

5.2 LAMINAR FLOW THROUGH CIRCULAR TUBES AND CIRCULAR ANNULI

For steady, incompressible, laminar flow through a circular tube or an annulus, a cylindrical infinitesimal sleeve (Fig. 5.7) is taken as a free body, and the equation of motion is applied in the l direction, with acceleration equal to zero. From the figure,

$$2\pi r\, \delta r\, p - \left(2\pi r\, \delta r\, p + 2\pi r\, \delta r\, \frac{dp}{dl}\, \delta l\right) + 2\pi r\, \delta l\, \tau$$

$$- \left[2\pi r\, \delta l\, \tau + \frac{d}{dr}\, (2\pi r\, \delta l\, \tau)\, \delta r\right] + \gamma 2\pi r\, \delta r\, \delta l \sin \theta = 0$$

Replacing $\sin \theta$ by $-dh/dl$ and dividing by the volume of the free body, $2\pi r\, \delta r\, \delta l$, gives

$$\frac{d}{dl}\, (p + \gamma h) + \frac{1}{r}\frac{d}{dr}\, (\tau r) = 0 \qquad (5.2.1)$$

Since $d(p + \gamma h)/dl$ is not a function of r, the equation may be multiplied

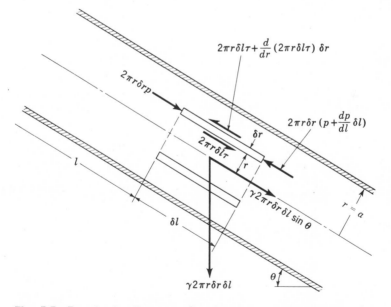

Fig. 5.7 Free-body diagram of cylindrical sleeve element for laminar flow in an inclined circular tube.

by $r \, \delta r$ and integrated with respect to r, yielding

$$\frac{r^2}{2} \frac{d}{dl} (p + \gamma h) + \tau r = A \tag{5.2.2}$$

in which A is the constant of integration. For a circular tube this equation must be satisfied when $r = 0$; hence $A = 0$ for this case. Substituting

$$\tau = -\mu \frac{du}{dr}$$

note that the minus sign is required to obtain the sign of the τ term in Fig. 5.7. (u is considered to decrease with r, hence du/dr is negative.)

$$du = \frac{1}{2\mu} \frac{d}{dl} (p + \gamma h) r \, dr - \frac{A}{\mu} \frac{dr}{r}$$

Another integration gives

$$u = \frac{r^2}{4\mu} \frac{d}{dl} (p + \gamma h) - \frac{A}{\mu} \ln r + B \tag{5.2.3}$$

For the annular case, to evaluate A and B, $u = 0$ when $r = b$, the inner tube radius, and $u = 0$ when $r = a$ (Fig. 5.8). When A and B are eliminated,

$$u = -\frac{1}{4\mu} \frac{d}{dl} (p + \gamma h) \left(a^2 - r^2 + \frac{a^2 - b^2}{\ln b/a} \ln \frac{a}{r} \right) \tag{5.2.4}$$

and for discharge through an annulus (Fig. 5.8),

$$Q = \int_b^a 2\pi r u \, dr = -\frac{\pi}{8\mu} \frac{d}{dl} (p + \gamma h) \left[a^4 - b^4 - \frac{(a^2 - b^2)^2}{\ln a/b} \right] \tag{5.2.5}$$

Circular tube; Hagen-Poiseuille equation

For the circular tube, $A = 0$ in Eq. (5.2.3) and $u = 0$ for $r = a$,

$$u = -\frac{a^2 - r^2}{4\mu} \frac{d}{dl} (p + \gamma h) \tag{5.2.6}$$

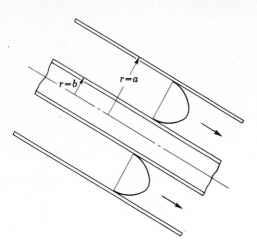

Fig. 5.8 Flow through an annulus.

The maximum velocity $u_{\max}$ is given for $r = 0$ as

$$u_{\max} = -\frac{a^2}{4\mu}\frac{d}{dl}(p + \gamma h) \qquad (5.2.7)$$

Since the velocity distribution is a paraboloid of revolution (Fig. 5.5), its volume is one-half that of its circumscribing cylinder; therefore the average velocity is one-half of the maximum velocity,

$$V = -\frac{a^2}{8\mu}\frac{d}{dl}(p + \gamma h) \qquad (5.2.8)$$

The discharge Q is equal to $V\pi a^2$,

$$Q = -\frac{\pi a^4}{8\mu}\frac{d}{dl}(p + \gamma h) \qquad (5.2.9)$$

The discharge can also be obtained by integration of the velocity u over the area, i.e.,

$$Q = \int_a^0 2\pi r u\, dr$$

For a horizontal tube, $h = $ const; writing the pressure drop Δp in the length

L gives

$$\frac{\Delta p}{L} = -\frac{dp}{dl}$$

and substituting diameter D leads to

$$Q = \frac{\Delta p \pi D^4}{128 \mu L} \qquad (5.2.10a)$$

In terms of average velocity,

$$V = \frac{\Delta p D^2}{32 \mu L} \qquad (5.2.10b)$$

Equation (5.2.10a) can then be solved for pressure drop, which represents losses per unit volume,

$$\Delta p = \frac{128 \mu L Q}{\pi D^4} \qquad (5.2.11)$$

The losses are seen to vary directly as the viscosity, the length, and the discharge and to vary inversely as the fourth power of the diameter. It should be noted that tube roughness does not enter into the equations. Equation (5.2.10a) is known as the *Hagen-Poiseuille equation*; it was determined experimentally by Hagen in 1839 and independently by Poiseuille in 1840. The analytical derivation was made by Wiedemann in 1856.

The results as given by Eqs. (5.2.1) to (5.2.10) are not valid near the entrance of a pipe. If the flow enters the pipe from a reservoir through a well-rounded entrance, the velocity at first is almost uniform over the cross section. The action of wall shear stress (as the velocity must be zero at the wall) is to slow down the fluid near the wall. As a consequence of continuity the velocity must then increase in the central region. The transition length L' for the characteristic parabolic velocity distribution to develop is a function of the Reynolds number. Langhaar[1] developed the theoretical formula

$$\frac{L'}{D} = 0.058\mathbf{R}$$

which agrees well with observation.

---

[1] H. L. Langhaar, Steady Flow in the Transition Length of a Straight Tube, *J. Appl. Mech.*, vol. 9, pp. 55–58, 1942.

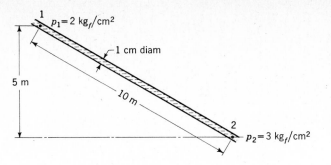

Fig. 5.9 Flow through an inclined tube.

EXAMPLE 5.3 Determine the direction of flow through the tube shown in Fig. 5.9, in which $\gamma = 8000$ N/m³ and $\mu = 0.04$ kg/m·s. Find the quantity flowing in liters per second, and calculate the Reynolds number for the flow.

At section 1

$$p + \gamma h = (2 \text{ kg}_f/\text{cm}^2)(9.806 \text{ N/kg}_f)\left(\frac{100 \text{ cm}}{1 \text{ m}}\right)^2 + (8000 \text{ N/m}^3)(5 \text{ m})$$

$$= 236.12 \text{ kN/m}^2$$

and at section 2

$$p + \gamma h = (3 \text{ kg}_f/\text{cm}^2)(9.806 \text{ N/kg}_f)\left(\frac{100 \text{ cm}}{1 \text{ m}}\right)^2 + 0 = 294.18 \text{ kN/m}^2$$

if the elevation datum is taken through section 2. The flow is from 2 to 1 since the energy is greater at 2 (kinetic energy must be the same at each section) than at 1. To determine the quantity flowing, the expression is written

$$-\frac{d}{dl}(p + \gamma h) = -\frac{236{,}120 - 294{,}180}{10 \text{ m}} \text{ N/m}^2 = 5806 \text{ N/m}^3$$

Substituting into Eq. (5.2.9) gives

$$Q = \frac{(5806 \text{ N/m}^3)(\pi)(0.01 \text{ m})^4}{8 \times 0.04 \text{ kg/m·s}} = 0.00057 \text{ m}^3/\text{s} = 0.57 \text{ l/s}$$

The average velocity is

$$V = \frac{0.00057 \text{ m}^3/\text{s}}{\pi(0.01 \text{ m})^2} = 1.814 \text{ m/s}$$

and the Reynolds number (Sec. 4.4) is

$$\mathbf{R} = \frac{VD\rho}{\mu} = \frac{(1.814 \text{ m/s})\,(0.01 \text{ m})\,(8000 \text{ N/m}^3)}{(0.04 \text{ kg/m}\cdot\text{s})\,(9.806 \text{ m/s}^2)} = 370$$

If the Reynolds number had been above 2000, the Hagen-Poiseuille equation would no longer apply, as discussed in Sec. 5.3.

The kinetic-energy correction factor α [Eq. (3.10.2)] can be determined for laminar flow in a tube by use of Eqs. (5.2.6) and (5.2.7),

$$\frac{u}{V} = 2\frac{u}{u_{\max}} = 2\left[1 - \left(\frac{r}{a}\right)^2\right] \tag{5.2.12}$$

Substituting into the expression for α gives

$$\alpha = \frac{1}{A}\int\left(\frac{u}{V}\right)^3 dA = \frac{1}{\pi a^2}\int_0^a \left\{2\left[1 - \left(\frac{r}{a}\right)^2\right]\right\}^3 2\pi r\,dr = 2 \tag{5.2.13}$$

There is twice as much energy in the flow as in uniform flow at the same average velocity.

5.3   THE REYNOLDS NUMBER

Laminar flow is defined as flow in which the fluid moves in layers, or laminas, one layer gliding smoothly over an adjacent layer with only a molecular interchange of momentum. Any tendencies toward instability and turbulence are damped out by viscous shear forces that resist relative motion of adjacent fluid layers. Turbulent flow, however, has very erratic motion of fluid particles, with a violent transverse interchange of momentum. The nature of the flow, i.e., whether laminar or turbulent, and its relative position along a scale indicating the relative importance of turbulent to laminar tendencies are indicated by the *Reynolds number*. The concept of the Reynolds number and its interpretation are discussed in this section. In Sec. 3.5 an equation of motion was developed with the assumption that the fluid is frictionless, i.e., that the viscosity is zero. More general equations have been developed that include viscosity, by including shear stresses. These equations (see introduction to this chapter) are complicated, nonlinear, partial differential equations for which no general solution has been obtained. In the last century

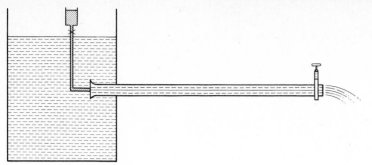

Fig. 5.10 Reynolds apparatus.

Osborne Reynolds[1] studied these equations to try to determine when two different flow situations would be similar.

Two flow cases are said to be *dynamically similar* when

1. They are geometrically similar, i.e., corresponding linear dimensions have a constant ratio and
2. The corresponding streamlines are geometrically similar, or pressures at corresponding points have a constant ratio.

In considering two geometrically similar flow situations, Reynolds deduced that they would be dynamically similar if the general differential equations describing their flow were identical. By changing the units of mass, length, and time in one set of equations and determining the condition that must be satisfied to make them identical to the original equations, Reynolds found that the dimensionless group $ul\rho/\mu$ must be the same for both cases. Of these, u is a characteristic velocity, l a characteristic length, ρ the mass density, and μ the viscosity. This group, or parameter, is now called the Reynolds number **R**,

$$\mathbf{R} = \frac{ul\rho}{\mu} \tag{5.3.1}$$

To determine the significance of the dimensionless group, Reynolds conducted his experiments on flow of water through glass tubes, illustrated in Fig. 5.10. A glass tube was mounted horizontally with one end in a tank and a valve on the opposite end. A smooth bellmouth entrance was attached to the upstream end, with a dye jet arranged so that a fine stream of dye could

---

[1] O. Reynolds, An Experimental Investigation of the Circumstances Which Determine Whether the Motion of Water Shall Be Direct or Sinuous, and of the Laws of Resistance in Parallel Channels, *Trans. R. Soc. Lond.*, vol. 174, 1883.

be ejected at any point in front of the bellmouth. Reynolds took the average velocity V as characteristic velocity and the diameter of tube D as characteristic length, so that $\mathbf{R} = VD\rho/\mu$.

For small flows the dye stream moved as a straight line through the tube, showing that the flow was laminar. As the flow rate increased, the Reynolds number increased, since D, ρ, μ were constant, and V was directly proportional to the rate of flow. With increasing discharge a condition was reached at which the dye stream wavered and then suddenly broke up and was diffused throughout the tube. The flow had changed to turbulent flow with its violent interchange of momentum that had completely disrupted the orderly movement of laminar flow. By careful manipulation Reynolds was able to obtain a value $\mathbf{R} = 12,000$ before turbulence set in. A later investigator, using Reynolds' original equipment, obtained a value of 40,000 by allowing the water to stand in the tank for several days before the experiment and by taking precautions to avoid vibration of the water or equipment. These numbers, referred to as the *Reynolds upper critical numbers*, have no practical significance in that the ordinary pipe installation has irregularities that cause turbulent flow at a much smaller value of the Reynolds number.

Starting with turbulent flow in the glass tube, Reynolds found that it always becomes laminar when the velocity is reduced to make $\mathbf{R}$ less than 2000. This is the *Reynolds lower critical number* for pipe flow and is of practical importance. With the usual piping installation, the flow will change from laminar to turbulent in the range of the Reynolds numbers from 2000 to 4000. For the purpose of this treatment it is assumed that the change occurs at $\mathbf{R} = 2000$. In laminar flow the losses are directly proportional to the average velocity, while in turbulent flow the losses are proportional to the velocity to a power varying from 1.7 to 2.0.

There are many Reynolds numbers in use today in addition to the one for straight round tubes. For example, the motion of a sphere through a fluid may be characterized by $UD\rho/\mu$, in which U is the velocity of sphere, D is the diameter of sphere, and ρ and μ are the fluid density and viscosity.

The Reynolds number may be viewed as a ratio of shear stress τ_t due to turbulence to shear stress τ_v due to viscosity. By applying the momentum equation to the flow through an element of area δA (Fig. 5.11) the apparent shear stress due to turbulence can be determined. If v' is the velocity normal to δA and u' is the difference in velocity, or the velocity fluctuation, on the two sides of the area, then, with Eq. (3.11.1), the shear force δF acting is computed to be

$$\delta F = \rho v' \, \delta A \, u'$$

in which $\rho v' \, \delta A$ is the mass per second having its momentum changed and

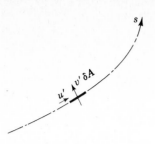

Fig. 5.11 Notation for shear stress due to turbulent flow.

u' is the final velocity minus the initial velocity in the s direction. By dividing through by δA, the shear stress τ_t due to turbulent fluctuations is obtained,

$$\tau_t = \rho u'v' \tag{5.3.2}$$

The shear stress due to viscosity may be written

$$\tau_v = \frac{\mu u'}{l} \tag{5.3.3}$$

in which u' is interpreted as the change in velocity in the distance l, measured normal to the velocity. Then the ratio

$$\frac{\tau_t}{\tau_v} = \frac{v'l\rho}{\mu}$$

has the form of a Reynolds number.

The *nature* of a given flow of an incompressible fluid is characterized by its Reynolds number. For large values of **R** one or all of the terms in the numerator are large compared with the denominator. This implies a large expanse of fluid, high velocity, great density, extremely small viscosity, or combinations of these extremes. The numerator terms are related to *inertial* forces, or to forces set up by acceleration or deceleration of the fluid. The denominator term is the cause of viscous shear forces. Thus, the Reynolds number parameter may also be considered as a ratio of inertial to viscous forces. A large **R** indicates a highly turbulent flow with losses proportional to the square of the velocity. The turbulence may be *fine scale*, composed of a great many small eddies that rapidly convert mechanical energy into irreversibilities through viscous action; or it may be *large scale*, like the huge vortices and swirls in a river or gusts in the atmosphere. The large eddies generate smaller eddies, which in turn create fine-scale turbulence. Turbulent flow may be thought of as a smooth, possibly uniform flow, with a secondary flow superposed on it. A fine-scale turbulent flow has small fluctuations in velocity that occur with high frequency. The root-mean-square value of the fluctuations and the frequency of change of sign of the fluctuations are quantitative measures of turbulence. In general the intensity of turbulence increases as the Reynolds number increases.

For intermediate values of **R** both viscous and inertial effects are important, and changes in viscosity change the velocity distribution and the resistance to flow.

For the same **R**, two geometrically similar closed-conduit systems (one, say, twice the size of the other) will have the same ratio of losses to velocity head. Use of the Reynolds number provides a means for using experimental results with one fluid for predicting results in a similar case with another fluid.

In addition to the applications of laminar flow shown in this and the preceding section, the results may also apply to greatly different situations, because the equations describing the cases are analogous. As an example the two-dimensional laminar flow between closely spaced plates is called Hele-Shaw flow.[1] If some of the space is filled between the plates, by use of dye in the fluid the streamlines for flow around the obstructions are made visible. These streamlines in laminar flow are the same as the streamlines for similar flow of a frictionless (irrotational) fluid around the same obstructions. Likewise the two-dimensional frictionless flow cases (Chap. 7) are analogous and similar to two-dimensional percolation through porous media.

5.4 PRANDTL MIXING LENGTH; VELOCITY DISTRIBUTION IN TURBULENT FLOW

Pressure drop and velocity distribution for several cases of laminar flow were worked out in the preceding sections. In this section the mixing-length theory of turbulence is developed, including its application to several flow situations. The apparent shear stress in turbulent flow is expressed by [Eq. (3.3.2)]

$$\tau = (\mu + \eta) \frac{du}{dy} \tag{5.4.1}$$

[1] H. J. S. Hele-Shaw, Investigation of the Nature of the Surface Resistance of Water and of Streamline Motion under Certain Experimental Conditions, *Trans. Inst. Nav. Archit.*, vol. 40, 1898.

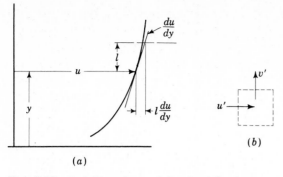

Fig. 5.12 Notation for mixing-length theory.

including direct viscous effects. Prandtl[1] has developed a most useful theory of turbulence called the *mixing-length theory*. In Sec. 5.3 the shear stress, τ, due to turbulence, was shown to be

$$\tau_t = \rho u'v' \tag{5.3.2}$$

in which u', v' are the velocity fluctuations at a point. In Prandtl's[2] theory, expressions for u' and v' are obtained in terms of a mixing-length distance l and the velocity gradient du/dy, in which u is the temporal mean velocity at a point and y is the distance normal to u, usually measured from the boundary. In a gas, one molecule, before striking another, travels an average distance known as the *mean free path* of the gas. Using this as an analogy (Fig. 5.12a), Prandtl assumed that a particle of fluid is displaced a distance l before its momentum is changed by the new environment. The fluctuation u' is then related to l by

$$u' \sim l\frac{du}{dy}$$

which means that the amount of the change in velocity depends upon the changes in temporal mean velocity at two points distant l apart in the y direction. From the continuity equation, he reasoned that there must be a corre-

[1] For an account of the development of turbulence theory the reader is referred to L. Prandtl, "Essentials of Fluid Dynamics," pp. 105–145, Hafner, New York, 1952.

[2] L. Prandtl, Bericht über Untersuchungen zur ausgebildeten Turbulenz, *Z. Angew. Math. Mech.*, vol. 5, no. 2, p. 136, 1925.

lation between u' and v' (Fig. 5.12b), so that v' is proportional to u',

$$v' \sim u' \sim l \frac{du}{dy}$$

By substituting for u' and v' in Eq. (5.3.2) and by letting l absorb the proportionality factor, the defining equation for mixing length is obtained:

$$\tau = \rho l^2 \left(\frac{du}{dy}\right)^2 \tag{5.4.2}$$

τ always acts in the sense that causes the velocity distribution to become more uniform. When Eq. (5.4.2) is compared with Eq. (3.3.1), it is found that

$$\eta = \rho l^2 \frac{du}{dy} \tag{5.4.3}$$

But η is not a fluid property like dynamic viscosity; instead η depends upon the density, the velocity gradient, and the mixing length l. In turbulent flow there is a violent interchange of globules of fluid except at a boundary, or very near to it, where this interchange is reduced to zero; hence, l must approach zero at a fluid boundary. The particular relationship of l to wall distance y is not given by Prandtl's derivation. Von Kármán[1] suggested, after considering similitude relationships in a turbulent fluid, that

$$l = \kappa \frac{du/dy}{d^2u/dy^2} \tag{5.4.4}$$

in which κ is a universal constant in turbulent flow, regardless of the boundary configuration or value of the Reynolds number.

In turbulent flows, η, sometimes referred to as the *eddy viscosity*, is generally much larger than μ. It may be considered as a coefficient of momentum transfer, expressing the transfer of momentum from points where the concentration is high to points where it is lower. It is convenient to utilize a *kinematic eddy viscosity* $\epsilon = \eta/\rho$ which is a property of the flow alone and is analogous to kinematic viscosity.

[1] T. von Kármán, Turbulence and Skin Friction, *J. Aeronaut. Sci.*, vol. 1, no. 1, p. 1, 1934.

Velocity distributions

The mixing-length concept is used to discuss turbulent velocity distributions for the flat plate and the pipe. For turbulent flow over a smooth plane surface (such as the wind blowing over smooth ground) the shear stress in the fluid is constant, say τ_0. Equation (5.4.1) is applicable, but η approaches zero at the surface and μ becomes unimportant away from the surface. If η is negligible for the film thickness $y = \delta$, in which μ predominates, Eq. (5.4.1) becomes

$$\frac{\tau_0}{\rho} = \frac{\mu}{\rho} \frac{u}{y} = \nu \frac{u}{y} \qquad y \leq \delta \tag{5.4.5}$$

The term $\sqrt{\tau_0/\rho}$ has the dimensions of a velocity and is called the shear-stress velocity u_*. Hence

$$\frac{u}{u_*} = \frac{u_* y}{\nu} \qquad y \leq \delta \tag{5.4.6}$$

shows a linear relation between u and y in the laminar film. For $y > \delta$, μ is neglected, and Eq. (5.4.1) produces

$$\tau_0 = \rho l^2 \left(\frac{du}{dy}\right)^2 \tag{5.4.7}$$

Since l has the dimensions of a length and from dimensional consideration would be proportional to y (the only significant linear dimension), assume $l = \kappa y$. Substituting into Eq. (5.4.7) and rearranging gives

$$\frac{du}{u_*} = \frac{1}{\kappa} \frac{dy}{y} \tag{5.4.8}$$

and integration leads to

$$\frac{u}{u_*} = \frac{1}{\kappa} \ln y + \text{const} \tag{5.4.9}$$

It is to be noted that this value of u substituted in Eq. (5.4.4) also determines l proportional to y (d^2u/dy^2 is negative since the velocity gradient decreases as y increases). Equation (5.4.9) agrees well with experiment and, in fact, is also useful when τ is a function of y, because most of the velocity change occurs near the wall where τ is substantially constant. It is quite satisfactory to apply the equation to turbulent flow in pipes.

EXAMPLE 5.4 By integration of Eq. (5.4.9) find the relation between the average velocity V and the maximum velocity u_m in turbulent flow in a pipe.

When $y = r_0$, $u = u_m$, so that

$$\frac{u}{u_*} = \frac{u_m}{u_*} + \frac{1}{\kappa} \ln \frac{y}{r_0}$$

The discharge $V\pi r_0^2$ is obtained by integrating the velocity over the area,

$$V\pi r_0^2 = 2\pi \int_0^{r_0-\delta} ur\, dr = 2\pi \int_\delta^{r_0} \left(u_m + \frac{u_*}{\kappa} \ln \frac{y}{r_0} \right) (r_0 - y)\, dy$$

The integration cannot be carried out to $y = 0$, since the equation holds in the turbulent zone only. The volume per second flowing in the laminar zone is so small that it may be neglected. Then

$$V = 2 \int_{\delta/r_0}^1 \left(u_m + \frac{u_*}{\kappa} \ln \frac{y}{r_0} \right) \left(1 - \frac{y}{r_0} \right) d\frac{y}{r_0}$$

in which the variable of integration is y/r_0. Integrating gives

$$V = 2 \left\{ u_m \left[\frac{y}{r_0} - \frac{1}{2} \left(\frac{y}{r_0} \right)^2 \right] + \frac{u_*}{\kappa} \left[\frac{y}{r_0} \ln \frac{y}{r_0} - \frac{y}{r_0} - \frac{1}{2} \left(\frac{y}{r_0} \right)^2 \ln \frac{y}{r_0} + \frac{1}{4} \left(\frac{y}{r_0} \right)^2 \right] \right\}_{\delta/r_0}^1$$

Since δ/r_0 is very small, such terms as δ/r_0 and $\delta/r_0 \ln (\delta/r_0)$ become negligible $(\lim_{x\to 0} x \ln x = 0)$; thus

$$V = u_m - \frac{3}{2} \frac{u_*}{\kappa} \qquad \text{or} \qquad \frac{u_m - V}{u_*} = \frac{3}{2\kappa}$$

In evaluating the constant in Eq. (5.4.9) following the methods of Bakhmeteff,[1] $u = u_w$, the *wall velocity*, when $y = \delta$. According to Eq. (5.4.6),

$$\frac{u_w}{u_*} = \frac{u_* \delta}{\nu} = N \qquad\qquad (5.4.10)$$

---

[1] B. A. Bakhmeteff, "The Mechanics of Turbulent Flow," Princeton University Press, Princeton, N.J., 1941.

from which it is reasoned that $u_*\delta/\nu$ should have a critical value N at which flow changes from laminar to turbulent, since it is a Reynolds number in form. Substituting $u = u_w$ when $y = \delta$ into Eq. (5.4.9) and using Eq. (5.4.10) yields

$$\frac{u_w}{u_*} = N = \frac{1}{\kappa}\ln\delta + \text{const} = \frac{1}{\kappa}\ln\frac{N\nu}{u_*} + \text{const}$$

Eliminating the constant gives

$$\frac{u}{u_*} = \frac{1}{\kappa}\ln\frac{yu_*}{\nu} + N - \frac{1}{\kappa}\ln N$$

or

$$\frac{u}{u_*} = \frac{1}{\kappa}\ln\frac{yu_*}{\nu} + A \tag{5.4.11}$$

in which $A = N - (1/\kappa)\ln N$ has been found experimentally by plotting u/u_* against $\ln yu_*/\nu$. For flat plates $\kappa = 0.417$, $A = 5.84$, but for smooth-wall pipes Nikuradse's[1] experiments yield $\kappa = 0.40$ and $A = 5.5$.

Prandtl has developed a convenient exponential velocity-distribution formula for turbulent pipe flow,

$$\frac{u}{u_m} = \left(\frac{y}{r_0}\right)^n \tag{5.4.12}$$

in which n varies with the Reynolds number. This empirical equation is valid only at some distance from the wall. For **R** less than 100,000, $n = \frac{1}{7}$, and for greater values of **R**, n decreases. The velocity-distribution equations, Eqs. (5.4.11) and (5.4.12), both have the fault of a nonzero value of du/dy at the center of the pipe.

EXAMPLE 5.5 Find an approximate expression for mixing-length distribution in turbulent flow in a pipe from Prandtl's one-seventh-power law.

Writing a force balance for steady flow in a round tube (Fig. 5.13) gives

$$\tau = -\frac{dp}{dl}\frac{r}{2}$$

[1] J. Nikuradse, Gesetzmassigkeiten der turbulenten Strömung in glatten Rohren, *Ver. Dtsch. Ing. Forschungsh.*, vol. 356, 1932.

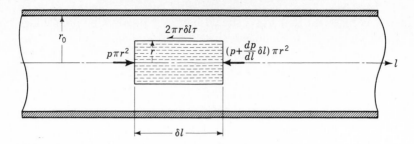

Fig. 5.13 Free-body diagram for steady flow through a round tube.

At the wall

$$\tau_0 = -\frac{dp}{dl}\frac{r_0}{2}$$

hence

$$\tau = \tau_0 \frac{r}{r_0} = \tau_0 \left(1 - \frac{y}{r_0}\right) = \rho l^2 \left(\frac{du}{dy}\right)^2$$

Solving for l gives

$$l = \frac{u_* \sqrt{1 - y/r_0}}{du/dy}$$

From Eq. (5.4.12)

$$\frac{u}{u_m} = \left(\frac{y}{r_0}\right)^{1/7}$$

the approximate velocity gradient is obtained,

$$\frac{du}{dy} = \frac{u_m}{r_0}\frac{1}{7}\left(\frac{y}{r_0}\right)^{-6/7} \qquad \text{and} \qquad \frac{l}{r_0} = \frac{u_*}{u_m} 7 \left(\frac{y}{r_0}\right)^{6/7} \sqrt{1 - \frac{y}{r_0}}$$

The dimensionless *velocity deficiency*, $(u_m - u)/u_*$, is a function of y/r_0 only for large Reynolds numbers (Example 5.4) whether the pipe surface is smooth or rough. From Eq. (5.4.9), evaluating the constant for $u = u_m$ when

$y = r_0$ gives

$$\frac{u_m - u}{u_*} = \frac{1}{\kappa} \ln \frac{r_0}{y} \qquad (5.4.13)$$

For rough pipes, the velocity may be assumed to be u_w at the wall distance $y_w = m\epsilon'$, in which ϵ' is a typical height of the roughness projections and m is a form coefficient depending upon the nature of the roughness. Substituting into Eq. (5.4.13) and eliminating u_m/u_* between the two equations leads to

$$\frac{u}{u_*} = \frac{1}{\kappa} \ln \frac{y}{\epsilon'} + \frac{u_w}{u_*} - \frac{1}{\kappa} \ln m \qquad (5.4.14)$$

in which the last two terms on the right-hand side are constant for a given type of roughness,

$$\frac{u}{u_*} = \frac{1}{\kappa} \ln \frac{y}{\epsilon'} + B \qquad (5.4.15)$$

In Nikuradse's experiments with sand-roughened pipes constant-size sand particles (those passing a given screen and being retained on a slightly finer screen) were glued to the inside pipe walls. If ϵ' represents the diameter of sand grains, experiment shows that $\kappa = 0.40$, $B = 8.48$.

5.5 RATE PROCESSES

The violent interchange of fluid globules in turbulence also tends to transfer any uneven concentration within the fluid, such as salinity, temperature, dye coloring, or sediment concentration. Studies[1] indicate that the transfer coefficient is roughly proportional to, but probably larger than, the eddy viscosity for turbulent diffusions of concentrations other than momentum.

If T is the temperature, H the heat transfer per unit area per unit time, and c_p the specific heat at constant pressure (e.g., Btu per unit of temperature per unit of mass), then

$$H = -c_p \eta \frac{\partial T}{\partial y} = -c_p \rho l^2 \frac{\partial u}{\partial y} \frac{\partial T}{\partial y} \qquad (5.5.1)$$

in which $c_p \eta$ is the eddy conductivity. For transfer of material substances, such as salinity, dye, or sediment, if C is the concentration per unit volume

[1] See footnote 1, p. 259.

(e.g., pounds of salt per cubic foot, number of particles per cubic foot) and c the rate of transfer per unit area per unit time (e.g., pounds of salt per square foot per second, number of sediment particles per square foot per second), then

$$c = -\epsilon_c \frac{\partial C}{\partial y} \tag{5.5.2}$$

and ϵ_c is proportional to ϵ.

EXAMPLE 5.6   A tank of liquid containing fine solid particles of uniform size is agitated so that the kinematic eddy viscosity may be considered constant. If the fall velocity of the particles in still liquid is v_f and the concentration of particles is C_0 at $y = y_0$ (y measured from the bottom), find the distribution of solid particles vertically throughout the liquid.

By using Eq. (5.5.2) to determine the rate per second carried upward by turbulence per unit of area at the level y, the amount per second falling across this surface by settling is equated to it for steady conditions. Those particles in the height v_f above the unit area will fall out in a second, that is, $C v_f$ particles cross the level downward per second per unit area. From Eq. (5.5.2) $-\epsilon_c \, dC/dy$ particles are carried upward due to the turbulence and the higher concentration below; hence

$$C v_f = -\epsilon_c \frac{dC}{dy} \qquad \text{or} \qquad \frac{dC}{C} = -\frac{v_f}{\epsilon_c} dy$$

Integrating gives

$$\ln C = -\frac{v_f}{\epsilon_c} y + \text{const}$$

For $C = C_0$, $y = y_0$,

$$C = C_0 \exp\left[-\frac{v_f}{\epsilon_c}(y - y_0)\right]$$

5.6   BOUNDARY-LAYER CONCEPTS

In 1904 Prandtl[1] developed the concept of the boundary layer. It provides an important link between ideal-fluid flow and real-fluid flow. *For fluids*

---

[1] L. Prandtl, Über Flussigkeitsbewegung bei sehr kleiner Reibung, *Verh. III Int. Math.-Kongr., Heidelb*, 1904.

having relatively small viscosity, the effect of internal friction in a fluid is appreciable only in a narrow region surrounding the fluid boundaries. From this hypothesis, the flow outside of the narrow region near the solid boundaries may be considered as ideal flow or potential flow. Relations within the boundary-layer region can be computed from the general equations for viscous fluids, but use of the momentum equation permits the developing of approximate equations for boundary-layer growth and drag. In this section the boundary layer is described and the momentum equation applied to it. Two-dimensional flow along a flat plate is studied by means of the momentum relationships for both the laminar and the turbulent boundary layer. The phenomenon of separation of the boundary layer and formation of the wake is described.

Description of the boundary layer

When motion is started in a fluid having very small viscosity, the flow is essentially irrotational (Sec. 3.3) in the first instants. Since the fluid at the boundaries has zero velocity relative to the boundaries, there is a steep velocity gradient from the boundary into the flow. This velocity gradient in a real fluid sets up near the boundary shear forces that reduce the flow relative to the boundary. That fluid layer which has had its velocity affected by the boundary shear is called the *boundary layer*. The velocity in the boundary layer approaches the velocity in the main flow asymptotically. The boundary layer is very thin at the upstream end of a streamlined body at rest in an otherwise uniform flow. As this layer moves along the body, the continual action of shear stress tends to slow down additional fluid particles, causing the thickness of the boundary layer to increase with distance from the upstream point. The fluid in the layer is also subjected to a pressure gradient, determined from the potential flow, that increases the momentum of the layer if the pressure decreases downstream and decreases its momentum if the pressure increases downstream (*adverse* pressure gradient). The flow outside the boundary layer may also bring momentum into the layer.

For smooth upstream boundaries the boundary layer starts out as a *laminar boundary layer* in which the fluid particles move in smooth layers. As the thickness of the laminar boundary layer increases, it becomes unstable and finally transforms into a *turbulent boundary layer* in which the fluid particles move in haphazard paths, although their velocity has been reduced by the action of viscosity at the boundary. When the boundary layer has become turbulent, there is still a very thin layer next to the boundary that has laminar motion. It is called the *laminar sublayer*.

Various definitions of boundary-layer thickness δ have been suggested. The most basic definition refers to the displacement of the main flow due to slowing down of fluid particles in the boundary zone. This thickness δ_1, called

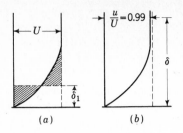

Fig. 5.14 Definitions of boundary-layer thickness.

the *displacement thickness*, is expressed by

$$U\delta_1 = \int_0^\delta (U - u)\, dy \tag{5.6.1}$$

in which δ is that value of y at which $u = U$ in the undisturbed flow. In Fig. 5.14a, the line $y = \delta_1$ is drawn so that the shaded areas are equal. This distance is, in itself, not the distance that is strongly affected by the boundary but is the amount the main flow must be shifted away from the boundary. In fact, that region is frequently taken as $3\delta_1$. Another definition, expressed by Fig. 5.14b, is the distance to the point where $u/U = 0.99$.

Momentum equation applied to the boundary layer

By following von Kármán's method[1] the principle of momentum can be applied directly to the boundary layer in steady flow along a flat plate. In Fig. 5.15 a control volume is taken enclosing the fluid above the plate, as shown, extending the distance x along the plate. In the y direction it extends a distance h so great that the velocity is undisturbed in the x direction, although some flow occurs along the upper surface, leaving the control volume.

The momentum equation for the x direction is

$$\Sigma F_x = \frac{\partial}{\partial t} \int_{cv} \rho u \, d\mathcal{V} + \int_{cs} \rho u \mathbf{v} \cdot d\mathbf{A}$$

It will be applied to the case of incompressible steady flow. The only force

[1] T. von Kármán, On Laminar and Turbulent Friction, *Z. Angew. Math. Mech.*, vol. 1, pp. 235–236, 1921.

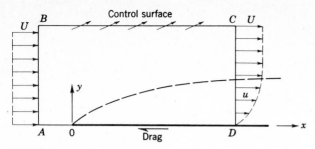

Fig. 5.15 Control volume applied to fluid flowing over one side of a flat plate.

acting is due to drag or shear at the plate, since the pressure is constant around the periphery of the control volume. For unit widths of plate normal to the paper,

$$-\text{Drag} = \rho \int_0^h u^2 \, dy - \rho U^2 h + U\rho \int_0^h (U - u) \, dy$$

The first term on the right-hand side of the equation is the efflux of x momentum from CD, and the second term is the x-momentum influx through AB. The integral in the third term is the net volume influx through AB and CD which, by continuity, must just equal the volume efflux through BC. It is multiplied by $U\rho$ to yield x-momentum efflux through BC. Combining the integrals gives

$$\text{Drag} = \rho \int_0^h u(U - u) \, dy \qquad (5.6.2)$$

The drag $D(x)$ on the plate is in the reverse direction, so that

$$D(x) = \rho \int_0^h u(U - u) \, dy \qquad (5.6.3)$$

The drag on the plate may also be expressed as an integral of the shear stress along the plate,

$$D(x) = \int_0^x \tau_0 \, dx \qquad (5.6.4)$$

Equating the last two expressions and then differentiating with respect to

x leads to

$$\tau_0 = \rho \frac{\partial}{\partial x} \int_0^h u(U - u)\, dy \tag{5.6.5}$$

which is the momentum equation for two-dimensional flow along a flat plate.

Calculations of boundary-layer growth, in general, are complex and require advanced mathematical treatment. The parallel-flow cases, laminar or turbulent, along a flat plate may be worked out approximately by use of momentum methods that do not give any detail regarding the velocity distribution—in fact a velocity distribution must be assumed. The results can be shown to agree closely with the more exact approach obtained from general viscous-flow differential equations.

For an assumed distribution which satisfies the boundary conditions $u = 0$, $y = 0$ and $u = U$, $y = \delta$, the boundary-layer thickness as well as the shear at the boundary can be determined. The velocity distribution is assumed to have the same form at each value of x,

$$\frac{u}{U} = F\left(\frac{y}{\delta}\right) = F(\eta) \qquad \eta = \frac{y}{\delta}$$

when δ is unknown. For the laminar boundary layer Prandtl assumed that

$$\frac{u}{U} = F = \tfrac{3}{2}\eta - \frac{\eta^3}{2} \qquad 0 \le y \le \delta \qquad \text{and} \qquad F = 1 \qquad \delta \le y$$

which satisfy the boundary conditions. Equation (5.6.5) may be rewritten

$$\tau_0 = \rho U^2 \frac{\partial \delta}{\partial x} \int_0^1 \left(1 - \frac{u}{U}\right)\frac{u}{U}\, d\eta$$

and

$$\tau_0 = \rho U^2 \frac{\partial \delta}{\partial x} \int_0^1 \left(1 - \tfrac{3}{2}\eta + \frac{\eta^3}{2}\right)\left(\tfrac{3}{2}\eta - \frac{\eta^3}{2}\right) d\eta = 0.139\rho U^2 \frac{\partial \delta}{\partial x}$$

At the boundary

$$\tau_0 = \mu \left.\frac{\partial u}{\partial y}\right|_{y=0} = \mu \left.\frac{U}{\delta}\frac{\partial F}{\partial \eta}\right|_{\eta=0} = \mu \left.\frac{U}{\delta}\frac{\partial}{\partial \eta}\left(\tfrac{3}{2}\eta - \frac{\eta^3}{2}\right)\right|_{\eta=0} = \tfrac{3}{2}\mu \frac{U}{\delta} \tag{5.6.6}$$

Equating the two expressions for τ_0 yields

$$\tfrac{3}{2}\mu \frac{U}{\delta} = 0.139\rho U^2 \frac{\partial \delta}{\partial x}$$

and rearranging gives

$$\delta \, d\delta = 10.78 \frac{\mu \, dx}{\rho U}$$

since δ is a function of x only in this equation. Integrating gives

$$\frac{\delta^2}{2} = 10.78 \frac{\nu}{U} x + \text{const}$$

If $\delta = 0$, for $x = 0$, the constant of integration is zero. Solving for δ/x leads to

$$\frac{\delta}{x} = 4.65 \sqrt{\frac{\nu}{Ux}} = \frac{4.65}{\sqrt{\mathbf{R}_x}} \tag{5.6.7}$$

in which $\mathbf{R}_x = Ux/\nu$ is a Reynolds number based on the distance x from the leading edge of the plate. This equation for boundary-layer thickness in laminar flow shows that δ increases as the square root of the distance from the leading edge.

Substituting the value of δ into Eq. (5.6.6) yields

$$\tau_0 = 0.322 \sqrt{\frac{\mu \rho U^3}{x}} \tag{5.6.8}$$

The shear stress varies inversely as the square root of x and directly as the three-halves power of the velocity. The drag on one side of the plate, of unit width, is

$$\text{Drag} = \int_0^l \tau_0 \, dx = 0.644 \sqrt{\mu \rho U^3 l} \tag{5.6.9}$$

Selecting other velocity distributions does not radically alter these results. The exact solution, worked out by Blasius from the general equations of viscous motion, yields the coefficients 0.332 and 0.664 for Eqs. (5.6.8) and (5.6.9), respectively.

The drag can be expressed in terms of a drag coefficient C_D times the

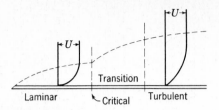

Fig. 5.16 Boundary-layer growth. (The vertical scale is greatly enlarged.)

stagnation pressure $\rho U^2/2$ and the area of plate l (per unit breadth),

$$\text{Drag} = C_D \frac{\rho U^2}{2} l$$

in which, for the laminar boundary layer,

$$C_D = \frac{1.328}{\sqrt{\mathbf{R}_l}} \tag{5.6.10}$$

and $\mathbf{R}_l = Ul/\nu$.

When the Reynolds number for the plate reaches a value between 500,000 and 1,000,000, the boundary layer becomes turbulent. Figure 5.16 indicates the growth and transition from laminar to turbulent boundary layer. The critical Reynolds number depends upon the initial turbulence of the fluid stream, the upstream edge of the plate, and the plate roughness.

Turbulent boundary layer

The momentum equation can be used to determine turbulent boundary-layer growth and shear stress along a smooth plate in a manner analogous to the treatment of the laminar boundary layer. The universal velocity-distribution law for smooth pipes, Eq. (5.4.11), provides the best basis, but the calculations are involved. A simpler approach is to use Prandtl's *one-seventh-power* law. It is $u/u_{\max} = (y/r_0)^{1/7}$, in which y is measured from the wall of the pipe and r_0 is the pipe radius. Applying it to flat plates produces

$$F = \frac{u}{U} = \left(\frac{y}{\delta}\right)^{1/7} = \eta^{1/7}$$

and

$$\tau_0 = 0.0228\rho U^2 \left(\frac{\nu}{U\delta}\right)^{1/4}$$

(5.6.11)

in which the latter expression is the shear stress at the wall of a smooth plate with a turbulent boundary layer.[1] The method used to calculate the laminar boundary layer gives

$$\tau_0 = \rho U^2 \frac{d\delta}{dx} \int_0^1 (1 - \eta^{1/7})\eta^{1/7}\, d\eta = \tfrac{7}{72}\rho U^2 \frac{d\delta}{dx}$$

(5.6.12)

By equating the expressions for shear stress, the differential equation for boundary-layer thickness δ is obtained,

$$\delta^{1/4}\, d\delta = 0.234 \left(\frac{\nu}{U}\right)^{1/4} dx$$

After integrating, and then by assuming that the boundary layer is turbulent over the whole length of the plate so that the initial conditions $x = 0$, $\delta = 0$ can be used,

$$\delta^{5/4} = 0.292 \left(\frac{\nu}{U}\right)^{1/4} x$$

Solving for δ gives

$$\delta = 0.37 \left(\frac{\nu}{U}\right)^{1/5} x^{4/5} = \frac{0.37x}{(Ux/\nu)^{1/5}} = \frac{0.37x}{\mathbf{R}_x^{1/5}}$$

(5.6.13)

The thickness increases more rapidly in the turbulent boundary layer. In it the thickness increases as $x^{4/5}$, but in the laminar boundary layer δ varies as $x^{1/2}$.

To determine the drag on a smooth, flat plate, δ is eliminated in Eqs. (5.6.11) and (5.6.13), and

$$\tau_0 = 0.029\rho U^2 \left(\frac{\nu}{Ux}\right)^{1/5}$$

(5.6.14)

[1] Equation (5.6.11) is obtained from the following pipe equations: $\tau_0 = \rho f V^2/8$, $f = 0.316/\mathbf{R}^{1/4}$ (Blasius eq.), $\mathbf{R} = V2r_0\rho/\mu$, and $V = u_m/1.235$. To transfer to the flat plate $r_0 \sim \delta$, $u_m \sim U$.

The drag for unit width on one side of the plate is

$$\text{Drag} = \int_0^l \tau_0 \, dx = 0.036\rho U^2 l \left(\frac{\nu}{Ul}\right)^{1/5} = \frac{0.036\rho U^2 l}{\mathbf{R}_l^{1/5}} \tag{5.6.15}$$

In terms of the drag coefficient,

$$C_D = 0.072\mathbf{R}_l^{-1/5} \tag{5.6.16}$$

in which $\mathbf{R}_l$ is the Reynolds number based on the length of plate.

The above equations are valid only for the range in which the Blasius resistance equation holds. For larger Reynolds numbers in smooth-pipe flow, the exponent in the velocity-distribution law is reduced. For $\mathbf{R} = 400{,}000$, $n = \frac{1}{8}$, and for $\mathbf{R} = 4{,}000{,}000$, $n = \frac{1}{10}$. The drag law, Eq. (5.6.15), is valid for a range

$$5 \times 10^5 < \mathbf{R}_l < 10^7$$

Experiment shows that the drag is slightly higher than is predicted by Eq. (5.6.16),

$$C_D = 0.074\mathbf{R}_l^{-1/5} \tag{5.6.17}$$

The boundary layer is actually laminar along the upstream part of the plate. Prandtl[1] has subtracted the drag from the equation for the upstream end of the plate up to the critical Reynolds number and then added the drag as given by the laminar equation for this portion of the plate, producing the equation

$$C_D = 0.074\mathbf{R}_l^{-1/5} - \frac{1700}{\mathbf{R}_l} \qquad 5 \times 10^5 < \mathbf{R}_l < 10^7 \tag{5.6.18}$$

In Fig. 5.17 a log-log plot of C_D vs. $\mathbf{R}_l$ shows the trend of the drag coefficients. Use of the logarithmic velocity distribution for pipes produces

$$C_D = \frac{0.455}{(\log \mathbf{R}_l)^{2.58}} \qquad 10^6 < \mathbf{R}_l < 10^9 \tag{5.6.19}$$

in which the constant term has been selected for best agreement with experimental results.

[1] L. Prandtl, Über den Reibungswiderstand strömender Luft, *Result. Aerodyn. Test Inst.*, *Goett.* III Lieferung, 1927.

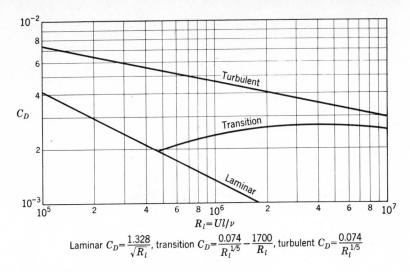

Laminar $C_D = \dfrac{1.328}{\sqrt{R_l}}$, transition $C_D = \dfrac{0.074}{R_l^{1/5}} - \dfrac{1700}{R_l}$, turbulent $C_D = \dfrac{0.074}{R_l^{1/5}}$

Fig. 5.17 The drag law for smooth plates.

EXAMPLE 5.7 A smooth, flat plate 10 ft wide and 100 ft long is towed through still water at 68°F with a speed of 20 ft/s. Determine the drag on one side of the plate and the drag on the first 10 ft of the plate.

For the whole plate

$$\mathbf{R}_l = \frac{(100 \text{ ft})\,(20 \text{ ft/s})\,(1.935 \text{ slugs/ft}^3)}{0.01 \text{ P}\,\dfrac{1 \text{ slug/ft}\cdot\text{s}}{479 \text{ P}}} = 1.85 \times 10^8$$

From Eq. (5.6.19)

$$C_D = \frac{0.455}{[\log\,(1.85 \times 10^8)]^{2.58}} = \frac{0.455}{8.2675^{2.58}} = 0.00196$$

The drag on one side is

$$\text{Drag} = C_D b l \rho \frac{U^2}{2} = 0.00196\,(10 \text{ ft})\,(100 \text{ ft})\,\frac{1.935 \text{ slugs/ft}^3}{2}\,(20 \text{ ft/s})^2 = 760 \text{ lb}$$

in which b is the plate width. If the critical Reynolds number occurs at 5×10^5, the length l_0 to the transition is

$$\frac{(l_0 \text{ ft})\,(20 \text{ ft/s})\,(1.935 \text{ slugs/ft}^3)}{0.01 \text{ P}\,\dfrac{1 \text{ slug/ft}\cdot\text{s}}{479 \text{ P}}} = 5 \times 10^5 \qquad l_0 = 0.27 \text{ ft}$$

For the first 10 ft of the plate, $\mathbf{R}_l = 1.85 \times 10^7$, $C_D = 0.00274$, and

$$\text{Drag} = 0.00274 \times 10 \times 10 \times \tfrac{1.935}{2} \times 20^2 = 106 \text{ lb}$$

Calculation of the turbulent boundary layer over rough plates proceeds in similar fashion, starting with the rough-pipe tests using sand roughnesses. At the upstream end of the flat plate, the flow may be laminar; then, in the turbulent boundary layer, where the boundary layer is still thin and the ratio of roughness height to boundary-layer thickness ϵ/δ is significant, the region of fully developed roughness occurs, and the drag is proportional to the square of the velocity. For long plates, this region is followed by a transition region where ϵ/δ becomes increasingly smaller, and eventually the plate becomes hydraulically smooth; i.e., the loss would not be reduced by reducing the roughness. Prandtl and Schlichting[1] have carried through these calculations, which are too complicated for reproduction here.

Separation; wake

Along a flat plate the boundary layer continues to grow in the downstream direction, regardless of the length of the plate, when the pressure gradient remains zero. With the pressure decreasing in the downstream direction, as in a conical reducing section, the boundary layer tends to be reduced in thickness.

For *adverse* pressure gradients, i.e., with pressure increasing in the downstream direction, the boundary layer thickens rapidly. The adverse gradient and the boundary shear decrease the momentum in the boundary layer, and if they both act over a sufficient distance, they cause the boundary layer to come to rest. This phenomenon is called *separation*. Figure 5.18a illustrates this case. The boundary streamline must leave the boundary at the separation point, and downstream from this point the adverse pressure gradient causes backflow near the wall. This region downstream from the streamline that separates from the boundary is known as the *wake*. The effect of separation is to decrease the net amount of flow work that can be done by a fluid element on the surrounding fluid at the expense of its kinetic energy, with the net result that pressure recovery is incomplete and flow losses (drag) increase. Figures 5.18b and 5.18c illustrate actual flow cases, the first with a very small adverse pressure gradient, which causes thickening of the boundary layer, and the second with a large diffuser angle, causing separation and backflow near the boundaries.

[1] Prandtl and H. Schlichting, Das Widerstandsgesetz rauher Platten, *Werft, Reederei, Hafen*, p. 1, 1934; see also *NACA Tech. Mem.* 1218, pt. II.

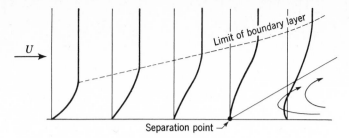

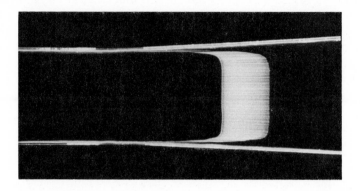

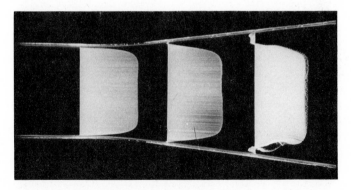

Fig. 5.18 (a) Effect of adverse pressure gradient on boundary layer. Separation. (b) Boundary-layer growth in a small-angle diffuser. (c) Boundary-layer separation in a large-angle diffuser. [*Parts (b) and (c) from the film "Fundamentals of Boundary Layers," by the National Committee for Fluid Mechanics Films and the Education Development Center.*]

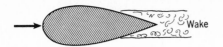

Fig. 5.19 Streamlined body.

Streamlined bodies (Fig. 5.19) are designed so that the separation point occurs as far downstream along the body as possible. If separation can be avoided, the boundary layer remains thin and the pressure is almost recovered downstream along the body. The only loss or drag is due to shear stress in the boundary layer, called *skin friction*. In the wake, the pressure is not recovered, and a *pressure drag* results. Reduction of wake size reduces the pressure drag on a body. In general, the drag is caused by both skin friction and pressure drag.

Flow around a sphere is an excellent example of the effect of separation on drag. For very small Reynolds numbers, $VD/\nu < 1$, the flow is everywhere nonturbulent, and the drag is referred to as *deformation drag*. Stokes' law (Sec. 5.7) gives the drag force for this case. For large Reynolds numbers, the flow may be considered potential flow except in the boundary layer and the wake. The boundary layer forms at the forward stagnation point and is generally laminar. In the laminar boundary layer, an adverse pressure

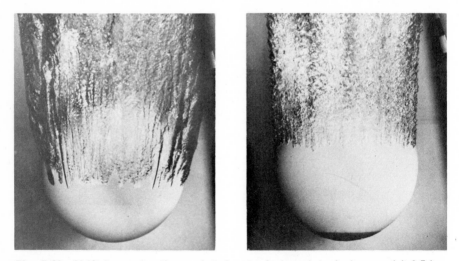

Fig. 5.20 Shift in separation point due to induced turbulence: (*a*) 8.5-in bowling ball, smooth surface, 25 ft/s entry velocity into water; (*b*) same except for 4-in diameter patch of sand on nose. (*Official U.S. Navy photograph made at Navy Ordnance Test Station, Pasadena Annex.*)

gradient causes separation more readily than in a turbulent boundary layer, because of the small amount of momentum brought into the laminar layer. If separation occurs in the laminar boundary layer, the location is farther upstream on the sphere than it is when the boundary layer becomes turbulent first and then separation occurs.

In Fig. 5.20 this is graphically portrayed by the photographs of the two spheres dropped into water at 25 ft/s. In *a*, separation occurs in the laminar boundary layer that forms along the smooth surface and causes a very large wake with a resulting large pressure drag. In *b*, the nose of the sphere, roughened by sand glued to it, induced an early transition to turbulent boundary layer before separation occurred. The high momentum transfer in the turbulent boundary layer delayed the separation so that the wake is substantially reduced, resulting in a total drag on the sphere less than half that occurring in *a*.

A plot of drag coefficient against Reynolds number (Fig. 5.21) for smooth spheres shows that the shift to turbulent boundary layer (before separation) occurs by itself at a sufficiently high Reynolds number, as evidenced by the sudden drop in drag coefficient. The exact Reynolds number for the sudden shift depends upon the smoothness of the sphere and the turbulence in the fluid stream. In fact, the sphere is frequently used as a turbulence meter by determining the Reynolds number at which the drag coefficient is 0.30, a point located in the center of the sudden drop (Fig. 5.21). By use of the hot-wire

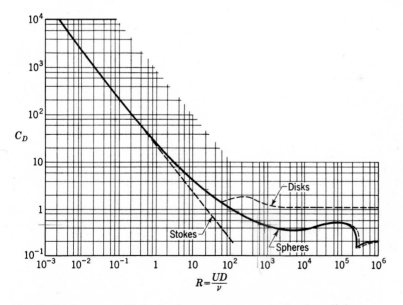

Fig. 5.21 Drag coefficients for spheres and circular disks.

anemometer, Dryden[1] has correlated the turbulence level of the fluid stream to the Reynolds number for the sphere at $C_D = 0.30$. The greater the turbulence of the fluid stream, the smaller the Reynolds number for shift in separation point.

In Sec. 7.8, for ideal-fluid flow, equations are developed that permit the velocity and pressure to be found at any point in the fluid for flow around a sphere. In ideal-fluid flow, slip is permitted at the boundary; in addition, the boundary condition states that the normal component of the velocity at a boundary is zero in steady flow. Therefore separation is ruled out of consideration. In ideal-fluid flow, which is for constant energy, there is no drag on a body and no boundary layer. A comparison of Figs. 5.20 and 7.23 shows the great contrast between ideal- and real-fluid flow around a bluff body. The ideal-fluid flow, however, does yield a good representation of velocity and pressure for the upstream portion of the flow, away from the effects of boundary-layer separation and wake formation.

5.7 DRAG ON IMMERSED BODIES

The principles of potential flow around bodies are developed in Chap. 7, and principles of the boundary layer, separation, and wake in Sec. 5.6. In this section drag is defined, some experimental drag coefficients are listed, the effect of compressibility on drag is discussed, and Stokes' law is presented. Lift is defined, and the lift and drag coefficients for an airfoil are given.

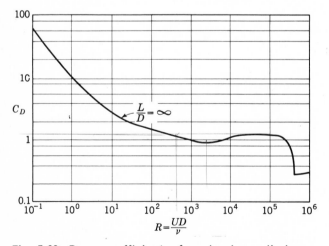

Fig. 5.22 Drag coefficients for circular cylinders.

[1] H. Dryden, Reduction of Turbulence in Wind Tunnels, *NACA Tech. Rep.* 392, 1931.

Table 5.1 *Typical drag coefficients for various cylinders in two-dimensional flow*†

Body shape		C_D	Reynolds number
Circular cylinder → ○		1.2	10^4 to 1.5×10^5
Elliptical cylinder → ⬯		0.6	4×10^4
2:1		0.46	10^5
→ 4:1		0.32	2.5×10^4 to 10^5
→ 8:1		0.29	2.5×10^4
		0.20	2×10^5
Square cylinder → □		2.0	3.5×10^4
→ ◇		1.6	10^4 to 10^5
Triangular cylinders → 120° ▷		2.0	10^4
→ ◁120°		1.72	10^4
→ 90° ▷		2.15	10^4
→ ◁90°		1.60	10^4
→ 60° ▷		2.20	10^4
→ ◁60°		1.39	10^4
→ 30° ▷		1.8	10^5
→ ◁30°		1.0	10^5
Semitubular →)		2.3	4×10^4
→ (		1.12	4×10^4

† Data from W. F. Lindsey, *NACA Tech. Rep.* 619, 1938.

Drag is defined as the force component, parallel to the relative approach velocity, exerted on the body by the moving fluid. The drag-coefficient curves for spheres and circular disks are shown in Fig. 5.21. In Fig. 5.22 the drag coefficient for an infinitely long circular cylinder (two-dimensional case) is plotted against the Reynolds number. Like the sphere, this case also has the sudden shift in separation point. In each case, the drag coefficient C_D is defined by

$$\text{Drag} = C_D A \frac{\rho U^2}{2}$$

in which A is the projected area of the body on a plane normal to the flow.

In Table 5.1 typical drag coefficients are shown for several cylinders. In general, the values given are for the range of Reynolds numbers in which the coefficient changes little with the Reynolds number.

A typical lift and drag curve for an airfoil section is shown in Fig. 5.23. Lift is the fluid-force component on a body at right angles to the relative approach velocity. The lift coefficient C_L is defined by

$$\text{Lift} = C_L A \frac{\rho U^2}{2}$$

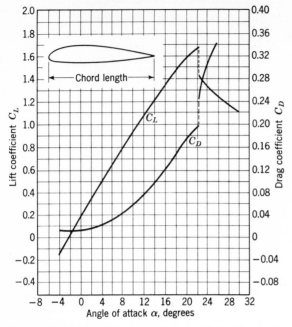

Fig. 5.23 Typical lift and drag coefficients for an airfoil.

in which A refers to the chord length times the wing length for lift and drag for airfoil sections.

Effect of compressibility on drag

To determine drag in high-speed gas flow the effects of compressibility, as expressed by the *Mach* number, are more important than the Reynolds number. The Mach number **M** is defined as the ratio of fluid velocity to velocity of sound in the fluid medium. When flow is at the critical velocity c, it has exactly the speed of the sound wave, so that small pressure waves cannot travel upstream. For this condition **M** $= 1$. When **M** is greater than unity, the flow is supersonic; and when **M** is less than unity, it is subsonic.

Any small disturbance is propagated with the speed of sound (Sec. 6.2). For example, a disturbance in still air travels outward as a spherical pressure wave. When the source of the disturbance moves with a velocity less than c, as in Fig. 5.24a, the wave travels ahead of the disturbing body and gives the fluid a chance to adjust itself to the oncoming body. By the time the particle has moved a distance Vt, the disturbance wave has moved out as far as $r = ct$

from the point O. As the disturbing body moves along, new spherical waves are sent out, but in all subsonic cases they are contained within the initial spherical wave shown. In supersonic motion of a particle (Fig. 5.24b) the body moves faster than the spherical wave emitted from it, yielding a cone-shaped wavefront with vertex at the body, as shown. The half angle of cone α is called the *Mach angle*,

$$\alpha = \sin^{-1}\frac{ct}{Vt} = \sin^{-1}\frac{c}{V}$$

The conical pressure front extends out behind the body and is called a *Mach wave* (Sec. 6.4). There is a sudden small change in velocity and pressure across a Mach wave.

The drag on bodies varies greatly with the Mach number and becomes relatively independent of the Reynolds number when compressibility effects become important. In Fig. 5.25 the drag coefficients for four projectiles are plotted against the Mach number.

For low Mach numbers, a body should be rounded in front, with a blunt nose and a long-tapering afterbody for minimum drag. For high Mach numbers (0.7 and over), the drag rises very rapidly owing to formation of the vortices behind the projectile and to formation of the shock waves; the body should have a tapered nose or thin forward edge. As the Mach numbers increase, the curves tend to drop and to approach a constant value asymptotically. This appears to be due to the fact that the reduction of pressure behind the projectile is limited to absolute zero, and hence its contribution to the total drag tends to become constant. The pointed projectile creates a narrower shock front that tends to reduce the limiting value of the drag coefficient.

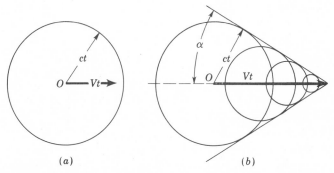

Fig. 5.24 Wave propagation produced by a particle moving at (*a*) subsonic velocity and (*b*) supersonic velocity.

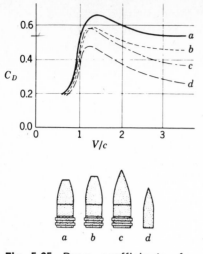

Fig. 5.25 Drag coefficients for projectiles as a function of the Mach number. (*From L. Prandtl, "Abriss der Strömungslehre," Friedrich Vieweg und Sohn, Brunswick, Germany, 1935.*)

Stokes' law

The flow of a viscous incompressible fluid around a sphere has been solved by Stokes[1] for values of the Reynolds number UD/ν below 1. The derivation is beyond the scope of this treatment; the results, however, are of value in such problems as the settling of dust particles. Stokes found the drag (force exerted on the sphere by flow of fluid around it) to be

$$\text{Drag} = 6\pi a\mu U$$

in which a is the radius of sphere and U the velocity of sphere relative to the fluid at a great distance. To find the terminal velocity for a sphere dropping through a fluid that is otherwise at rest, the buoyant force plus the drag force must just equal its weight, or

$$\tfrac{4}{3}\pi a^3\gamma + 6\pi a\mu U = \tfrac{4}{3}\pi a^3\gamma_s$$

[1] G. Stokes, *Trans. Camb. Phil. Soc.*, vol. 8, 1845; vol. 9, 1851.

in which γ is the specific weight of liquid and γ_s is the specific weight of the sphere. By solving for U, the terminal velocity is found to be

$$U = \frac{2}{9} \frac{a^2}{\mu} (\gamma_s - \gamma)$$ (5.7.1)

The straight-line portion of Fig. 5.21 represents Stokes' law.

The drag relations on particles, as given by Stokes' law and by the experimental results of Fig. 5.21, are useful in the design of settling basins for separating small solid particles from fluids. Applications include separating coolants from metal chips in machining operations, desilting river flow, and sanitary engineering applications to treatment of raw water and sewage.

EXAMPLE 5.8 A jet aircraft discharges solid particles of matter 10 μm in diameter, $S = 2.5$, at the base of the stratosphere at 11,000 m. Assume the viscosity μ of air, in poises, to be expressed by the relationship

$$\mu = 1.78 \times 10^{-4} - 3.06 \times 10^{-9}y$$

where y in meters is measured from sea level. Estimate the time for these particles to reach sea level. Neglect air currents and wind effects.

Writing $U = -dy/dt$ in Eq. (5.7.1) and recognizing the unit weight of air to be much smaller than the unit weight of the solid particles, one has

$$-\frac{dy}{dt} = \frac{2}{9} \frac{a^2\gamma_s}{\mu}$$

$$\int_0^T dt = -\int_{11,000}^0 \tfrac{9}{2}(1.78 \times 10^{-4} - 3.06 \times 10^{-9}y \text{ P}) \frac{0.1 \text{ N}\cdot\text{s/m}^2}{1 \text{ P}}$$

$$\times \frac{1}{(5 \times 10^{-6} \text{ m})^2} \frac{1}{2.5 \times 9802 \text{ N/m}^3} dy \text{ m}$$

$$T = \frac{1 \text{ d}}{86,400 \text{ s}} \left[1.78y - \frac{3.06 \times 10^{-5}y^2}{2} \right]_0^{11,000} \times 73.45 \text{ s}$$

$$= 15.07 \text{ d}$$

where d is the abbreviation for day.

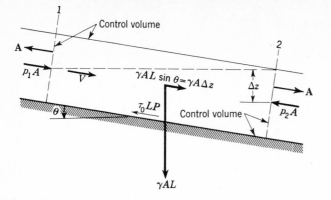

Fig. 5.26 Axial forces on free body of fluid in a conduit.

5.8 RESISTANCE TO TURBULENT FLOW IN OPEN AND CLOSED CONDUITS

In steady turbulent incompressible flow in conduits of constant cross section (steady uniform flow) the wall shear stress varies closely proportional to the square of the velocity,

$$\tau_0 = \lambda \frac{\rho}{2} V^2 \tag{5.8.1}$$

in which λ is a dimensionless coefficient. For open channels and noncircular closed conduits the shear stress is not constant over the surface. In these cases, τ_0 is used as the average wall shear stress. Secondary flows[1] occurring in noncircular conduits act to equalize the wall shear stress.

The momentum equation applied to the control volume (Fig. 5.26) comprising the liquid between sections 1 and 2, under the assumptions made, shows no net efflux of momentum, and hence shows equilibrium of forces on the control volume in the direction of motion

$$(p_1 - p_2)A + \gamma A \, \Delta z = \tau_0 LP$$

in which $\Delta z = L \sin \theta$ and P is the *wetted perimeter* of the conduit, i.e., the portion of the perimeter where the wall is in contact with the fluid (free-liquid surface excluded). The ratio A/P is called the *hydraulic radius R* of

[1] Secondary flows, not wholly understood, are transverse components that cause the main central flow to spread out into corners or near walls.

the conduit. If $p_1 - p_2 = \Delta p$,

$$\frac{\Delta p + \gamma\,\Delta z}{L} = \frac{\tau_0}{R} = \frac{\lambda \rho V^2}{2R} \qquad (5.8.2)$$

or, when divided through by γ, if $h_f = (\Delta p + \gamma\,\Delta z)/\gamma$ are the losses per unit weight,

$$\frac{h_f}{L} = S = \frac{\lambda}{R}\frac{V^2}{2g}$$

in which S represents the losses per unit weight per unit length. After solving for V,

$$V = \sqrt{\frac{2g}{\lambda}}\ \sqrt{RS} = C\ \sqrt{RS} \qquad (5.8.3)$$

This is the Chézy formula, in which originally the Chézy coefficient C was thought to be a constant for any size conduit or wall-surface condition. Various formulas for C are now generally used.

For pipes, when $\lambda = f/4$ and $R = D/4$, the Darcy-Weisbach equation is obtained,

$$h_f = f\frac{L}{D}\frac{V^2}{2g} \qquad (5.8.4)$$

in which D is the inside pipe diameter. This equation may be applied to open channels in the form

$$V = \sqrt{\frac{8g}{f}}\ \sqrt{RS} \qquad (5.8.5)$$

with values of f determined from pipe experiments.

5.9 STEADY UNIFORM FLOW IN OPEN CHANNELS

For incompressible, steady flow at constant depth in a prismatic open channel, the *Manning* formula is widely used. It can be obtained from the Chézy formula [Eq. (5.8.3)] by setting

$$C = \frac{C_m}{n}R^{1/6} \qquad (5.9.1)$$

so that

$$V = \frac{C_m}{n} R^{2/3} S^{1/2} \qquad (5.9.2)$$

which is the Manning formula.

The value of C_m is 1.49 and 1.0 for English and SI units, respectively, V is the average velocity at a cross section, R the hydraulic radius (Sec. 5.8), and S the losses per unit weight per unit length of channel or the slope of the bottom of the channel. It is also the slope of the water surface, which is parallel to the channel bottom. The coefficient n was thought to be an absolute roughness coefficient, i.e., dependent upon surface roughness only, but it actually depends upon the size and shape of channel cross section in some unknown manner. Values of the coefficient n, determined by many tests on actual canals, are given in Table 5.2. Equation (5.9.2) must have consistent English or SI units as indicated for use with the values in Table 5.2.[1]

Table 5.2 *Average values of the Manning roughness factor for various boundary materials*†

Boundary material	Manning n
Planed wood	0.012
Unplaned wood	0.013
Finished concrete	0.012
Unfinished concrete	0.014
Cast iron	0.015
Brick	0.016
Riveted steel	0.018
Corrugated metal	0.022
Rubble	0.025
Earth	0.025
Earth, with stones or weeds	0.035
Gravel	0.029

† Work by the U.S. Bureau of Reclamation and other government agencies indicates that the Manning roughness factor should be increased (say, 10 to 15 percent) for hydraulic radii greater than about 10 ft. The loss in capacity of large channels is due to the roughening of the surfaces with age, marine and plant growths, deposits, and the addition of bridge piers as the highway system is expanded.

[1] To convert the empirical equation in English units to SI units, n is taken to be dimensionless; then the constant has dimensions, and $(1.49 \text{ ft}^{1/3}/\text{s})(0.3048 \text{ m/ft})^{1/3} = 1.0 \text{ m}^{1/3}/\text{s}$.

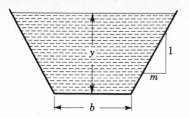

Fig. 5.27 Notation for trape-zoidal cross section.

When Eq. (5.9.2) is multiplied by the cross-sectional area A, the Manning formula takes the form

$$Q = \frac{C_m}{n} A R^{2/3} S^{1/2} \qquad (5.9.3)$$

When the cross-sectional area is known, any one of the other quantities can be obtained from Eq. (5.9.3) by direct solution.

EXAMPLE 5.9 Determine the discharge for a trapezoidal channel (Fig. 5.27) with a bottom width $b = 8$ ft and side slopes 1 on 1. The depth is 6 ft, and the slope of the bottom is 0.0009. The channel has a finished concrete lining.

From Table 5.2, $n = 0.012$. The area is

$$A = 8 \times 6 + 6 \times 6 = 84 \text{ ft}^2$$

and the wetted perimeter is

$$P = 8 + 2 \times 6\sqrt{2} = 24.96$$

By substituting into Eq. (5.9.3),

$$Q = \frac{1.49}{0.012} 84 \left(\frac{84}{24.96}\right)^{2/3} (0.0009^{1/2}) = 703 \text{ cfs}$$

Trial solutions are required in some instances when the cross-sectional area is unknown. Expressions for both the hydraulic radius and the area contain the depth in a form that cannot be solved explicitly.

EXAMPLE 5.10 What depth is required for 4 m³/s flow in a rectangular planed-wood channel 2 m wide with a bottom slope of 0.002?

If the depth is y, $A = 2y$, $P = 2 + 2y$, and $n = 0.012$. By substituting in Eq. (5.9.3),

$$4 \text{ m}^3/\text{s} = \frac{1.00}{0.012} 2y \left(\frac{2y}{2 + 2y}\right)^{2/3} 0.002^{1/2}$$

Simplifying gives

$$f(y) = y \left(\frac{y}{1 + y}\right)^{2/3} = 0.536$$

Assume $y = 1$ m; then $f(y) = 0.63$. Assume $y = 0.89$ m, then $f(y) = 0.538$. The correct depth then is about 0.89 m.

EXAMPLE 5.11 Riprap problem. A developer has been required by environmental regulatory authorities to line an open channel to prevent erosion. The channel is trapezoidal in cross section and has a slope of 0.0009. The bottom width is 10 ft and side slopes are 2:1 (horizontal to vertical). If he uses roughly spherical rubble ($\gamma_s = 135$ lb/ft^3) for the lining, what is the minimum D_{50} of the rubble that can be used? The design flow is 1000 cfs. Assume the shear that rubble can withstand is described by

$$\tau = 0.040 \, (\gamma_s - \gamma) D_{50} \quad \text{lb/ft}^2$$

in which γ_s is the unit weight of rock and D_{50} is the average rock diameter in feet.

A Manning n of 0.03 is appropriate for the rubble. To find the size of channel, from Eq. (5.9.3)

$$1000 = \frac{1.49}{0.03} \frac{[y(10 + 2y)]^{5/3}}{(10 + 2\sqrt{5}\,y)^{2/3}} 0.03$$

By trial solution the depth is $y = 8.62$ ft, and the hydraulic radius $R = 4.84$ ft. From Eq. (5.8.2)

$$\tau_0 = \gamma R S = 62.4 \times 4.84 \times 0.0009 = 0.272 \text{ lb/ft}^2$$

To find the D_{50} size for incipient movement $\tau = \tau_0$, and

$$0.040(135 - 62.4)D_{50} = 0.272$$

Hence $D_{50} = 0.0936$ ft.

More general cases of open-channel flow are considered in Chap. 11.

5.10 STEADY INCOMPRESSIBLE FLOW THROUGH SIMPLE PIPE SYSTEMS

Colebrook formula

A force balance for steady flow (no acceleration) in a pipe (Fig. 5.28) yields

$$\Delta p \, \pi r_0^2 = \tau_0 2\pi r_0 \, \Delta L$$

This simplifies to

$$\tau_0 = \frac{\Delta p}{\Delta L} \frac{r_0}{2} \tag{5.10.1}$$

which holds for laminar or turbulent flow. The Darcy-Weisbach equation (5.8.4) may be written

$$\Delta p = \gamma h_f = f \frac{\Delta L}{2r_0} \rho \frac{V^2}{2}$$

Eliminating Δp in the two equations and simplifying gives

$$\sqrt{\frac{\tau_0}{\rho}} = \sqrt{\frac{f}{8}} V \tag{5.10.2}$$

which relates wall shear stress, friction factor, and average velocity. The average velocity V may be obtained from Eq. (5.4.11) by integrating over the cross section. Substituting for V in Eq. (5.10.2) and simplifying produces the equation for friction factor in smooth-pipe flow,

$$\frac{1}{\sqrt{f}} = A_s + B_s \ln \left(\mathbf{R} \sqrt{f} \right) \tag{5.10.3}$$

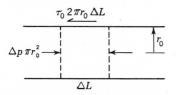

Fig. 5.28 Equilibrium conditions for steady flow in a pipe.

With the Nikuradse[1] data for smooth pipes, the equation becomes

$$\frac{1}{\sqrt{f}} = 0.86 \ln (\mathbf{R} \sqrt{f}) - 0.8 \tag{5.10.4}$$

For rough pipes in the complete turbulence zone,

$$\frac{1}{\sqrt{f}} = F_2 \left(m, \frac{\epsilon'}{D} \right) + B_r \ln \frac{\epsilon}{D} \tag{5.10.5}$$

in which F_2 is, in general, a constant for a given form and spacing of the roughness elements. For the Nikuradse sand-grain roughness (Fig. 5.31) Eq. (5.10.5) becomes

$$\frac{1}{\sqrt{f}} = 1.14 - 0.86 \ln \frac{\epsilon}{D} \tag{5.10.6}$$

The roughness height ϵ for sand-roughened pipes may be used as a measure of the roughness of commercial pipes. If the value of f is known for a commercial pipe in the fully developed wall turbulence zone, i.e., large Reynolds numbers and the loss proportional to the square of the velocity, the value of ϵ can be computed by Eq. (5.10.6). In the transition region, where f depends upon both ϵ/D and $\mathbf{R}$, sand-roughened pipes produce different results from commercial pipes. This is made evident by a graph based on Eqs. (5.10.4) and (5.10.6) with both sand-roughened and commercial-pipe-test results shown. Rearranging Eq. (5.10.6) gives

$$\frac{1}{\sqrt{f}} + 0.86 \ln \frac{\epsilon}{D} = 1.14$$

and adding $0.86 \ln (\epsilon/D)$ to each side of Eq. (5.10.4) leads to

$$\frac{1}{\sqrt{f}} + 0.86 \ln \frac{\epsilon}{D} = 0.86 \ln \left(\mathbf{R} \sqrt{f} \frac{\epsilon}{D} \right) - 0.8$$

By selecting $1/\sqrt{f} + 0.86 \ln (\epsilon/D)$ as ordinate and $\ln (\mathbf{R} \sqrt{f} \, \epsilon/D)$ as abscissa (Fig. 5.29), smooth-pipe-test results plot as a straight line with slope $+0.86$ and rough-pipe-test results in the complete turbulence zone plot as the horizontal line. Nikuradse sand-roughness-test results plot along the dashed

[1] J. Nikuradse, Gesetzmässigkeiten der turbulenten Strömung in glatten Rohren, *Ver. Dtsch. Ing. Forschungsh.*, vol. 356, 1932.

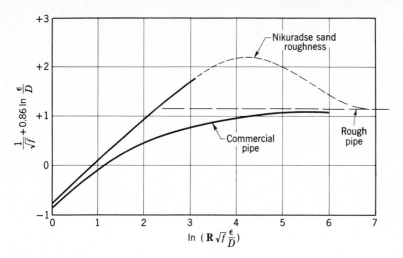

Fig. 5.29   Colebrook transition function.

line in the transition region, and commercial-pipe-test results plot along the lower curved line.

The explanation of the difference in shape of the artificial roughness curve of Nikuradse and the commercial roughness curve is that the laminar sublayer, or laminar film, covers all the artificial roughness or allows it to protrude uniformly as the film thickness decreases. With commercial roughness, which varies greatly in uniformity, small portions extend beyond the film first, as the film decreases in thickness with increasing Reynolds number. An empirical transition function for commercial pipes for the region between smooth pipes and the complete turbulence zone has been developed by Colebrook,[1]

$$\frac{1}{\sqrt{f}} = -0.86 \ln \left(\frac{\epsilon/D}{3.7} + \frac{2.51}{\mathbf{R}\,\sqrt{f}} \right) \qquad (5.10.7)$$

which is the basis for the Moody diagram (Fig. 5.32).

Pipe flow

In steady incompressible flow in a pipe the irreversibilities are expressed in terms of a head loss, or drop in *hydraulic grade line* (Sec. 10.1). The hydraulic

[1] C. F. Colebrook, Turbulent Flow in Pipes, with Particular Reference to the Transition Region Between the Smooth and Rough Pipe Laws, *J. Inst. Civ. Eng. Lond.*, vol. 11, pp. 133–156, 1938–1939.

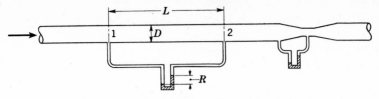

Fig. 5.30 Experimental arrangement for determining head loss in a pipe.

grade line is p/γ above the center of the pipe, and if z is the elevation of the center of the pipe, then $z + p/\gamma$ is the elevation of a point on the hydraulic grade line. The locus of values of $z + p/\gamma$ along the pipeline gives the hydraulic grade line. Losses, or irreversibilities, cause this line to drop in the direction of flow. The Darcy-Weisbach equation

$$h_f = f \frac{L}{D} \frac{V^2}{2g} \tag{5.8.4}$$

is generally adopted for pipe-flow calculations.[1] h_f is the head loss, or drop in hydraulic grade line, in the pipe length L, having an inside diameter D and an average velocity V. h_f has the dimension length and is expressed in terms of foot-pounds per pound or meter-newtons per newton. The friction factor f is a dimensionless factor that is required to make the equation produce the correct value for losses. All quantities in Eq. (5.8.4) except f can be measured experimentally. A typical setup is shown in Fig. 5.30. By measuring the discharge and inside diameter, the average velocity can be computed. The head loss h_f is measured by a differential manometer attached to piezometer openings at sections 1 and 2, distance L apart.

Experimentation shows the following to be true in turbulent flow:

1. The head loss varies directly as the length of the pipe.
2. The head loss varies almost as the square of the velocity.
3. The head loss varies almost inversely as the diameter.
4. The head loss depends upon the surface roughness of the interior pipe wall.
5. The head loss depends upon the fluid properties of density and viscosity.
6. The head loss is independent of the pressure.

The friction factor f must be selected so that Eq. (5.8.4) correctly yields the head loss; hence, f cannot be a constant but must depend upon velocity V, diameter D, density ρ, viscosity μ, and certain characteristics of the wall

---

[1] See Sec. 10.1 for development of empirical pipe-flow formulas for special uses.

roughness signified by ϵ, ϵ', and m, where ϵ is a measure of the *size* of the roughness projections and has the dimensions of a length, ϵ' is a measure of the *arrangement* or *spacing* of the roughness elements and also has the dimensions of a length, and m is a form factor, depending upon the *shape* of the individual roughness elements and is dimensionless. The term f, instead of being a simple constant, turns out to depend upon seven quantities,

$$f = f(V,D,\rho,\mu,\epsilon,\epsilon',m) \tag{5.10.8}$$

Since f is a dimensionless factor, it must depend upon the grouping of these quantities into dimensionless parameters. For *smooth* pipe $\epsilon = \epsilon' = m = 0$, leaving f dependent upon the first four quantities. They can be arranged in only one way to make them dimensionless, namely, $VD\rho/\mu$, which is the Reynolds number. For *rough* pipes the terms ϵ, ϵ' may be made dimensionless by dividing by D. Therefore, in general,

$$f = f\left(\frac{VD\rho}{\mu}, \frac{\epsilon}{D}, \frac{\epsilon'}{D}, m\right) \tag{5.10.9}$$

The proof of this relationship is left to experimentation. For smooth pipes a plot of all experimental results shows the functional relationship, subject to a scattering of ± 5 percent. The plot of friction factor against the Reynolds number on a log-log chart is called a *Stanton diagram*. Blasius[1] was the first to correlate the smooth-pipe experiments in turbulent flow. He presented the results by an empirical formula that is valid up to about $\mathbf{R} = 100,000$. The Blasius formula is

$$f = \frac{0.316}{\mathbf{R}^{1/4}} \tag{5.10.10}$$

In rough pipes the term ϵ/D is called the *relative roughness*. Nikuradse[2] proved the validity of the relative-roughness concept by his tests on sand-roughened pipes. He used three sizes of pipes and glued sand grains ($\epsilon = $ diameter of the sand grains) of practically constant size to the interior walls so that he had the same values of ϵ/D for different pipes. These experiments (Fig. 5.31) show that for one value of ϵ/D the f, $\mathbf{R}$ curve is smoothly connected regardless of the actual pipe diameter. These tests did not permit variation

[1] H. Blasius, Das Ähnlichkeitsgesetz bei Reibungsvorgängen in Flüssigkeiten, *Ver. Dtsch. Ing. Forschungsh.*, vol. 131, 1913.
[2] J. Nikuradse, Strömungsgesetze in rauhen Rohren, *Ver. Dtsch. Ing. Forschungsh.*, vol. 361, 1933.

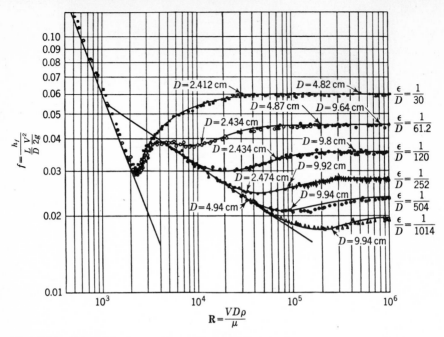

Fig. 5.31 Nikuradse's sand-roughened-pipe tests.

of ϵ'/D or m but proved the validity of the equation

$$f = f\left(\mathbf{R}, \frac{\epsilon}{D}\right)$$

for one type of roughness.

Because of the extreme complexity of naturally rough surfaces, most of the advances in understanding the basic relationships have been developed around experiments on artificially roughened pipes. Moody[1] has constructed one of the most convenient charts for determining friction factors in clean, commercial pipes. This chart, presented in Fig. 5.32, is the basis for pipe-flow calculations in this chapter. The chart is a Stanton diagram that expresses f as a function of relative roughness and the Reynolds number. The values of absolute roughness of the commercial pipes are determined by experiment in which f and $\mathbf{R}$ are found and substituted into the Colebrook formula, Eq. (5.10.7), which closely represents natural pipe trends. These are listed in the table in the lower left-hand corner of Fig. 5.32. The Colebrook formula provides the shape of the ϵ/D = const curves in the transition region.

[1] L. F. Moody, Friction Factors for Pipe Flow, *Trans. ASME*, November 1944.

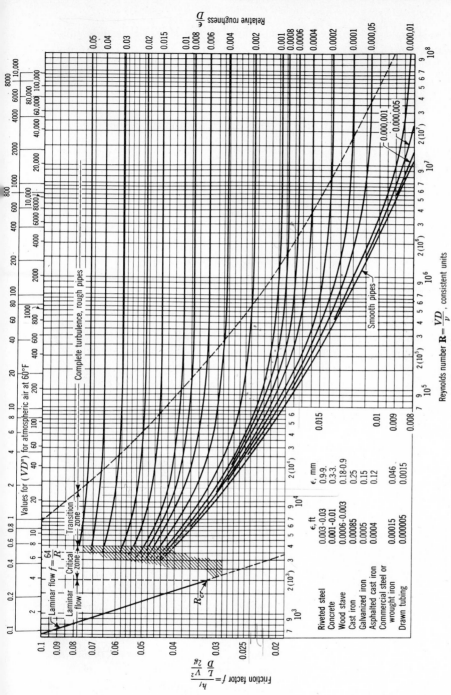

Fig. 5.32 Moody diagram.

The straight line marked "laminar flow" is the Hagen-Poiseuille equation. Equation (5.2.10b),

$$V = \frac{\Delta p r_0^2}{8\mu L}$$

may be transformed into Eq. (5.8.4) with $\Delta p = \gamma h_f$ and by solving for h_f,

$$h_f = \frac{V 8\mu L}{\gamma r_0^2} = \frac{64\mu}{\rho D} \frac{L}{D} \frac{V}{2g} = \frac{64}{\rho D V/\mu} \frac{L}{D} \frac{V^2}{2g}$$

or

$$h_f = f \frac{L}{D} \frac{V^2}{2g} = \frac{64}{\mathbf{R}} \frac{L}{D} \frac{V^2}{2g} \tag{5.10.11}$$

from which

$$f = \frac{64}{\mathbf{R}} \tag{5.10.12}$$

This equation, which plots as a straight line with slope -1 on a log-log chart, may be used for the solution of laminar-flow problems in pipes. It applies to all roughnesses, as the head loss in laminar flow is independent of wall roughness. The Reynolds critical number is about 2000, and the *critical zone*, where the flow may be either laminar or turbulent, is about 2000 to 4000.

It should be noted that the relative-roughness curves $\epsilon/D = 0.001$ and smaller approach the smooth-pipe curve for decreasing Reynolds numbers. This can be explained by the presence of a laminar film at the wall of the pipe that decreases in thickness as the Reynolds number increases. For certain ranges of Reynolds numbers in the transition zone, the film completely covers small roughness projections, and the pipe has a friction factor the same as that of a smooth pipe. For larger Reynolds numbers, projections protrude through laminar film, and each projection causes extra turbulence that increases the head loss. For the zone marked "complete turbulence, rough pipes," the film thickness is negligible compared with the height of roughness projections, and each projection contributes fully to the turbulence. Viscosity does not affect the head loss in this zone, as evidenced by the fact that the friction factor does not change with the Reynolds number. In this zone the loss follows the V^2 law; i.e., it varies directly as the square of the velocity.

Two auxiliary scales are given along the top of the Moody diagram. One is for water at 60°F, and the other is for air at standard atmospheric pressure and 60°F. Since the kinematic viscosity is constant in each case, the Reynolds number is a function of VD. For these two scales only, D must be expressed in inches and V in feet per second.

Simple pipe problems

The three simple pipe-flow cases that are basic to solutions of the more complex problems are

	Given	*To find*
I.	Q, L, D, ν, ϵ	h_f
II.	h_f, L, D, ν, ϵ	Q
III.	h_f, Q, L, ν, ϵ	D

In each of these cases the Darcy-Weisbach equation, the continuity equation, and the Moody diagram are used to determine the unknown quantity.

In the first case the Reynolds number and the relative roughness are readily determined from the data given, and h_f is found by determining f from the Moody diagram and substituting into the Darcy-Weisbach equation.

EXAMPLE 5.12 Determine the head loss due to the flow of 2000 gal/min of oil, $\nu = 0.0001$ ft²/s, through 1000 ft of 8-in-diameter cast-iron pipe.

$$V = \frac{2000 \text{ gal/min}}{\dfrac{448.8 \text{ gal/min}}{1 \text{ ft}^3/\text{s}} \dfrac{\pi}{3^2} \text{ ft}^2} = 12.8 \text{ ft/s}$$

$$\mathbf{R} = \frac{VD}{\nu} = \frac{12.8 \text{ ft/s}}{0.0001 \text{ ft}^2/\text{s}} \left(\tfrac{2}{3} \text{ ft}\right) = 85{,}500$$

The relative roughness is $\epsilon/D = 0.00085/0.667 = 0.0013$. From Fig. 5.32, by interpolation, $f = 0.024$; hence

$$h_f = f \frac{L}{D} \frac{V^2}{2g} = 0.024 \times \frac{1000 \text{ ft}}{\tfrac{2}{3} \text{ ft}} \frac{(12.8 \text{ ft/s})^2}{64.4 \text{ ft/s}^2} = 91.8 \text{ ft} \cdot \text{lb/lb}$$

In the second case, V and f are unknowns, and the Darcy-Weisbach equation and Moody diagram must be used simultaneously to find their values. Since ϵ/D is known, a value of f may be assumed by inspection of the Moody diagram. Substitution of this trial f into the Darcy-Weisbach equation produces a trial value of V, from which a trial Reynolds number is computed. With the Reynolds number an improved value of f is found from the Moody diagram. When f has been found correct to two significant figures, the corresponding V is the value sought and Q is determined by multiplying by the area.

EXAMPLE 5.13 Water at 15°C flows through a 30-cm-diameter riveted steel pipe, $\epsilon = 3$ mm, with a head loss of 6 m in 300 m. Determine the flow.

The relative roughness is $\epsilon/D = 0.003/0.3 = 0.01$, and from Fig. 5.32 a trial f is taken as 0.04. By substituting into Eq. (5.8.4),

$$6 \text{ m} = 0.04 \times \frac{300 \text{ m}}{0.3 \text{ m}} \frac{(V \text{ m/s})^2}{2 \times 9.806 \text{ m/s}^2}$$

from which $V = 1.715$ m/s. From Appendix C, $\nu = 1.13 \times 10^{-6}$ m²/s, and so

$$\mathbf{R} = \frac{VD}{\nu} = \frac{(1.715 \text{ m/s})(0.30 \text{ m})}{1.13 \times 10^{-6} \text{ m}^2/\text{s}} = 455{,}000$$

From the Moody diagram $f = 0.038$, and

$$Q = AV = \pi(0.15 \text{ m})^2 \sqrt{\frac{(6 \text{ m} \times 0.3 \text{ m})(2)(9.806 \text{ m/s}^2)}{0.038(300 \text{ m})}} = 0.1245 \text{ m}^3/\text{s}$$

In the third case, with D unknown, there are three unknowns in Eq. (5.8.4), f, V, D; two in the continuity equation, V, D; and three in the Reynolds number equation, $V, D, \mathbf{R}$. The relative roughness is also unknown. Using the continuity equation to eliminate the velocity in Eq. (5.8.4) and in the expression for $\mathbf{R}$ simplifies the problem. Equation (5.8.4) becomes

$$h_f = f \frac{L}{D} \frac{Q^2}{2g(D^2\pi/4)^2}$$

or

$$D^5 = \frac{8LQ^2}{h_f g \pi^2} f = C_1 f \tag{5.10.13}$$

in which C_1 is the known quantity $8LQ^2/h_f g\pi^2$. As $VD^2 = 4Q/\pi$ from continuity,

$$\mathbf{R} = \frac{VD}{\nu} = \frac{4Q}{\pi\nu}\frac{1}{D} = \frac{C_2}{D} \tag{5.10.14}$$

in which C_2 is the known quantity $4Q/\pi\nu$. The solution is now effected by the following procedure:
1. Assume a value of f.
2. Solve Eq. (5.10.13) for D.
3. Solve Eq. (5.10.14) for $\mathbf{R}$.
4. Find the relative roughness ϵ/D.

5. With **R** and ϵ/D, look up a new f from Fig. 5.32.
6. Use the new f, and repeat the procedure.
7. When the value of f does not change in the first two significant figures, all equations are satisfied and the problem is solved.

Normally only one or two trials are required. Since standard pipe sizes are usually selected, the next larger size of pipe than that given by the computation is taken. Nominal standard pipe sizes are $\frac{1}{8}$, $\frac{1}{4}$, $\frac{3}{8}$, $\frac{1}{2}$, $\frac{3}{4}$, 1, $1\frac{1}{4}$, $1\frac{1}{2}$, 2, $2\frac{1}{2}$, 3, $3\frac{1}{2}$, 4, 5, 6, 8, 10, 12, 14, 16, 18, 24, and 30 in. The inside diameters are larger than the nominal up to 12 in. Above the 12-in size the actual inside diameter depends upon the "schedule" of the pipe, and manufacturers' tables should be consulted. Throughout this chapter the nominal size is taken as the actual inside diameter.

EXAMPLE 5.14 Determine the size of clean wrought-iron pipe required to convey 4000 gpm oil, $\nu = 0.0001$ ft^2/s, 10,000 ft with a head loss of 75 ft·lb/lb.
The discharge is

$$Q = \frac{4000}{448.8} = 8.93 \text{ cfs}$$

From Eq. (5.10.13)

$$D^5 = \frac{8 \times 10,000 \times 8.93^2}{75 \times 32.2 \times \pi^2} f = 267.0f$$

and from Eq. (5.10.14)

$$\mathbf{R} = \frac{4 \times 8.93}{\pi 0.0001} \frac{1}{D} = \frac{113,800}{D}$$

and from Fig. 5.32, $\epsilon = 0.00015$ ft.
If $f = 0.02$, $D = 1.398$ ft, $\mathbf{R} = 81,400$, $\epsilon/D = 0.00011$, and from Fig. 5.32, $f = 0.019$. In repeating the procedure, $D = 1.382$, $\mathbf{R} = 82,300$, $f = 0.019$. Therefore, $D = 1.382 \times 12 = 16.6$ in. If a 75-ft head loss is the maximum allowable, an 18-in pipe is required.

In each of the cases considered, the loss has been expressed in feet of head or in foot-pounds per pound. For horizontal pipes, this loss shows up as a gradual reduction in pressure along the line. For nonhorizontal cases, the energy equation (3.10.1) is applied to the two end sections of the pipe, and the loss term is included; thus

$$\frac{V_1^2}{2g} + \frac{p_1}{\gamma} + z_1 = \frac{V_2^2}{2g} + \frac{p_2}{\gamma} + z_2 + h_f \qquad (5.10.15)$$

in which the kinetic-energy correction factors have been taken as unity. The upstream section is given the subscript 1 and the downstream section the subscript 2. The total head at section 1 is equal to the sum of the total head at section 2 and all the head losses between the two sections.

EXAMPLE 5.15 In the preceding example, for $D = 16.6$ in, if the specific gravity is 0.85, $p_1 = 40$ psi, $z_1 = 200$ ft, and $z_2 = 50$ ft, determine the pressure at section 2.

In Eq. (5.10.15) $V_1 = V_2$; hence

$$\frac{40 \text{ psi}}{0.85 \times 0.433 \text{ psi/ft}} + 200 \text{ ft} = \frac{p_2 \text{ psi}}{0.85 \times 0.433 \text{ psi/ft}} + 50 \text{ ft} + 75 \text{ ft}$$

and

$$p_2 = 67.6 \text{ psi}$$

Digital-computer solution of simple pipe problems

In solving simple pipe problems by computer, the Colebrook equation (5.10.7) would be used in place of the Moody diagram, which is its graphical repre-

```
C       SCLUTICN CF SIMPLE PIPE PRCBLEM TC FINU HEAC LOSS.   DATA GIVEN IN
C       CCNSISTENT UNITS.
        REAL K,MU,L
      2 FURMAT('0       EPS=',F8.6,3H D=,F7.4,3H Q=,F7.3,3H L=,F8.2,4H MU=,
     2F9.7,5H RHC=,F8.3,3H G=,F7.3)
      3 FORMAT('0       HEAD=',F10.3,3H R=,F11.1,7H EPS/C=,F8.6,3H F=,F7.4)
        NAMELIST/CATA/EPS,D,Q,L,MU,RHC,G
      1 REAC(5,CATA,END=99)
        WRITE(6,2) EPS,D,Q,L,MU,RHC,G
        V=Q/(.7854*D*D)
        R=V*D*RHC/MU
        K=EPS/D
        A=.094*K**.225+.53*K
        B=88.*K**.44
        C=1.62*K**.134
        F=A+B/R**C
        HEAD=F*L*V*V/(2.*G*D)
        WRITE(6,3) HEAD,R,K,F
        GC TO 1
        CALL SYSTEM
        ENC
     ECATA EPS=8.5E-4,D=1.333,Q=6.62,L=12346.,MU=2.E-5,RHO=1.935,G=32.2 EEND
     EDATA EPS=.0015,D=0.6,Q=.35,L=500.,MU=.062,RHO=860,G=9.806, EEND
        EPS=C.CCC850 D= 1.3330 Q=  6.620 L=12346.00 ML=C.0000200 RHO=   1.935 G= 32.200
       HEAC=    62.752 R=   611769.9 EPS/D=0.000638 F= 0.0194
        EPS=0.CC1500 D= 0.6000 Q=  0.350 L=  500.00 ML=0.0620000 RHO= 860.000 G=  9.806
       HEAD=     2.178 R=   10302.3 EPS/D=0.002500 F= 0.0334
```

Fig. 5.33 Computer program, data, and results for determination of head loss in a single pipe.

```
C      SCLUTICN OF SIMPLE PIPE PRCBLEM TC FIND UISCHARGE.
C      CATA IN CCNSISTENT LNITS
       REAL K,MU,L
   1 FORMAT('0  ANSWER UCES NOT CONVERGE')
   2 FORMAT('0      EPS=',F8.6,3H D=,F7.4,6H HEAU=,F10.3,3H L=,F8.2,4H M
  2U=,F9.7,5H RHO=,F8.3,3H G=,F7.3,3H F=,F7.4)
   3 FORMAT('0      C=',F7.3,3H V=,F7.3,3H F=,F7.4,3H R=,F11.1)
       NAMELIST/DATA/EPS,D,HEAD,L,MU,RHO,G,F
   5 REAC(5,CATA,END=99)
       WRITE(6,2) EPS,D,HEAD,L,MU,RHC,G,F
       I=0
       K=EPS/C
       A=.C94*K**.225+.53*K
       B=88.*K**.44
       C=1.62*K**.134
   6 V=SQRT(2.*G*D*HEAD/(F*L))
       R=V*D*RHO/MU
       F1=A+B/R**C
       IF(ABS(F1-F).LT..00C5) GU TO 7
       I=I+1
       IF(I.EC.8) GO TO 8
       F=F1
       GO TJ 6
   7 Q=.7854*C*D*V
       WRITE(6,3) Q,V,F,R
       GO TO 5
   8 WRITE(6,1)
       GO TO 5
  99 CALL SYSTEM
       END
&CATA EPS=5.E-4,C=3.33,L=1650.,HEAD=18.,MU=1.33E-5,RHO=1.92,G=32.2,F=.02 &END
&CATA EPS=.0C18,D=1.,HEAD=4.,L=200.,MU=.001,RHO=997.3,G=9.806, &END
   EPS=0.0CC500 D= 3.3300 HEAD=   18.000 L= 165C.00 MU=0.C000133 RHO=   1.920 G= 32.200 F= 0.0200
   Q=113.C50 V= 12.985 F= 0.C139 R= 6242214.0
   EPS=0.CC1800 D= 1.0000 HEAD=   4.000 L=  200.00 MU=0.0010000 RHO= 997.300 G=  9.806 F= 0.0139
   C=  3.192 V=  4.064 F= 0.C237 R=  4053252.0
```

Fig. 5.34 Computer program, data, and results for determination of discharge in a single pipe.

sentation. Wood[1] has developed an empirical, explicit form of the Colebrook equation which closely approximates it for values of $\mathbf{R} > 10{,}000$ and $1 \times 10^{-5} < \epsilon/D < 0.04$. It is (for $k = \epsilon/D$)

$$f = a + b\mathbf{R}^{-c}$$

$$a = 0.094k^{0.225} + 0.53k \qquad b = 88k^{0.44} \qquad c = 1.62k^{0.134}$$

The first type, for solution of head loss (HEAD), is direct, as given in Fig. 5.33. In the programs EPS = ϵ, MU = μ, RHO = ρ. The second type, for solution for discharge, starts with an assumption for f, Fig. 5.34. The value of F (F1) is improved until the criterion (F does not change by 0.0005 in an iteration) is satisfied. The third type, for determination of diameter, also starts with an assumed value of F, Fig. 5.35. If the criterion is not met in eight iterations, the program moves to the next set of data. The programs may be modified to include minor losses, which are discussed in the remaining portion of this section.

[1] Don J. Wood, An Explicit Friction Factor Relationship, *Civ. Eng.*, December 1966, pp. 60, 61.

```
C      SOLUTICN CF SIMPLE PIPE PROBLEM FCR DIAMETER.  CATA IN CONSISTENT
C      LNITS.
       REAL K,MU,L
     1 FORMAT('0   ANSWER DCES NOT CONVERGE')
     2 FORMAT('0      EPS=',F8.6,6H HEAC=,F10.3,3H C=,F7.3,3H L=,F8.2,4H M
      2U=,F9.7,5H RHO=,F8.3,3H G=,F7.3,3H F=,F7.4)
     3 FORMAT('0      C=',F7.4,3H F=,F7.4,3H R=,F11.1)
       NAMELIST/CATA/EPS,HEAD,Q,L,MU,RHO,G,F
     5 READ(5,CATA,END=99)
       WRITE(6,2) EPS,HEAC,Q,L,MU,RHO,G,F
       I=0
     6 D=(F*L*C*C/(2.*G*HEAC*.7854**2))**.2
       K=EPS/D
       V=C/(.7E54*D*D)
       R=V*C*RHO/MU
       A=.C94*K**.225+.53*K
       B=88.*K**.44
       C=1.62*K**.134
       F1=A+B/R**C
       IF(ABS(F1-F).LT..0005) GO TO 7
       I=I+1
       IF(I.EC.8) GO TO 8
       F=F1
       GC TO 6
     7 WRITE(6,3) D,F,R
       CC TO 5
     8 WRITE(6,1)
       GO TO 5
    99 CALL SYSTEM
       END
 &CATA EPS=.CCC85,HEAD=23.,Q=7.5,L=3200.,MU=4.E-4,RHC=1.69,G=32.2,F=.02 &END
 &CATA EPS=.CC15,HEAD=7.,C=.3,L=1000.,MU=.02,RHO=87C.,G=9.8C6,F=.C2, &END
       EPS=C.CCC850 HEAD=    23.000 Q=   7.500 L= 3200.00 MU=0.CCC4000 RHO=    1.690 G= 32.200 F= 0.0200
       D= 1.3758 F= 0.0250 R=    29324.4
       EPS=0.CC1500 HEAD=    7.000 Q=   0.300 L= 1COC.00 MU=0.C200000 RHO=  87C.000 G=  9.806 F= 0.0200
       D= C.5029 F= 0.0303 R=    33039.8
```

Fig. 5.35 Computer program, data, and results for determination of diameter in a single pipe.

Minor losses

Those losses which occur in pipelines due to bends, elbows, joints, valves, etc., are called *minor losses*. This is a misnomer, because in many situations they are more important than the losses due to pipe friction considered in the preceding section, but it is the conventional name. In almost all cases the minor loss is determined by experiment. However, one important exception is the head loss due to a *sudden expansion* in a pipeline (Sec. 3.11).

Equation (3.11.22) may also be written

$$h_e = K \frac{V_1{}^2}{2g} = \left[1 - \left(\frac{D_1}{D_2} \right)^2 \right]^2 \frac{V_1{}^2}{2g} \tag{5.10.16}$$

in which

$$K = \left[1 - \left(\frac{D_1}{D_2} \right)^2 \right]^2 \tag{5.10.17}$$

From Eq. (5.10.16) it is obvious that the head loss varies as the square of the velocity. This is substantially true for all minor losses in turbulent flow. A convenient method of expressing the minor losses in flow is by means of the coefficient K, usually determined by experiment.

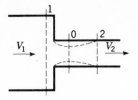

Fig. 5.36 Sudden contraction in a pipeline.

If the sudden expansion is from a pipe to a reservoir, $D_1/D_2 = 0$ and the loss becomes $V_1^2/2g$; that is, the complete kinetic energy in the flow is converted into thermal energy.

The head loss h_c due to a *sudden contraction* in the pipe cross section, illustrated in Fig. 5.36, is subject to the same analysis as the sudden expansion, provided that the amount of contraction of the jet is known. The process of converting pressure head into velocity head is very efficient; hence, the head loss from section 1 to the *vena contracta*[1] is small compared with the loss from section 0 to section 2, where velocity head is being reconverted into pressure head. By applying Eq. (3.11.22) to this expansion the head loss is computed to be

$$h_c = \frac{(V_0 - V_2)^2}{2g}$$

With the continuity equation $V_0 C_c A_2 = V_2 A_2$, in which C_c is the contraction coefficient, i.e., the area of jet at section 0 divided by the area of section 2, the head loss is computed to be

$$h_c = \left(\frac{1}{C_c} - 1\right)^2 \frac{V_2^2}{2g} \qquad (5.10.18)$$

The contraction coefficient C_c for water, determined by Weisbach,[2] is presented in the tabulation.

A_2/A_1	0.1	0.2	0.3	0.4	0.5	0.6	0.7	0.8	0.9	1.0
C_c	0.624	0.632	0.643	0.659	0.681	0.712	0.755	0.813	0.892	1.00

[1] The *vena contracta* is the section of greatest contraction of the jet.
[2] Julius Weisbach, "Die Experimental-Hydraulik," p. 133, Englehardt, Freiburg, 1855.

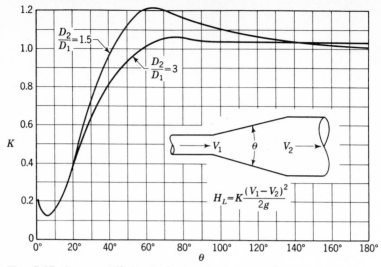

Fig. 5.37 Loss coefficients for conical expansions.

The head loss at the entrance to a pipeline from a reservoir is usually taken as $0.5V^2/2g$ if the opening is square-edged. For well-rounded entrances, the loss is between $0.01V^2/2g$ and $0.05V^2/2g$ and may usually be neglected. For re-entrant openings, as with the pipe extending into the reservoir beyond the wall, the loss is taken as $1.0V^2/2g$, for thin pipe walls.

The head loss due to gradual expansions (including pipe friction over the length of the expansion) has been investigated experimentally by Gibson,[1] whose results are given in Fig. 5.37.

Table 5.3 *Head-loss coefficients K for various fittings*

Fitting	K
Globe valve (fully open)	10.0
Angle valve (fully open)	5.0
Swing check valve (fully open)	2.5
Gate valve (fully open)	0.19
Close return bend	2.2
Standard tee	1.8
Standard elbow	0.9
Medium sweep elbow	0.75
Long sweep elbow	0.60

[1] A. H. Gibson, The Conversion of Kinetic to Pressure Energy in the Flow of Water Through Passages Having Divergent Boundaries, *Engineering*, vol. 93, p. 205, 1912.

A summary of representative head loss coefficients K for typical fittings, published by the Crane Company,[1] is given in Table 5.3.

Minor losses may be expressed in terms of the equivalent length L_e of pipe that has the same head loss in foot-pounds per pound (meter-newtons per newton) for the same discharge; thus

$$f \frac{L_e}{D} \frac{V^2}{2g} = K \frac{V^2}{2g}$$

in which K may refer to one minor head loss or to the sum of several losses. Solving for L_e gives

$$L_e = \frac{KD}{f} \qquad\qquad (5.10.19)$$

For example, if the minor losses in a 12-in pipeline add to $K = 20$, and if $f = 0.020$ for the line, then to the actual length of line may be added $20 \times 1/0.020 = 1000$ ft, and this additional or equivalent length causes the same resistance to flow as the minor losses.

EXAMPLE 5.16 Find the discharge through the pipeline in Fig. 5.38 for $H = 10$ m, and determine the head loss H for $Q = 60\,\text{l/s}$.

The energy equation applied to points 1 and 2, including all the losses, may be written

$$H_1 + 0 + 0 = \frac{V_2^2}{2g} + 0 + 0 + \frac{1}{2}\frac{V_2^2}{2g} + f\frac{102\,\text{m}}{0.15\,\text{m}}\frac{V_2^2}{2g} + 2 \times 0.9\frac{V_2^2}{2g} + 10\frac{V_2^2}{2g}$$

or

$$H_1 = \frac{V_2^2}{2g}(13.3 + 680f)$$

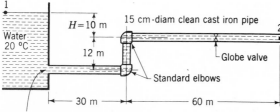

Fig. 5.38 Pipeline with minor losses.

[1] Crane Company, Flow of Fluids, *Tech. Pap.* 409, May, 1942.

When the head is given, this problem is solved as the second type of simple pipe problem. If $f = 0.022$,

$$10 = \frac{V_2^2}{2g} (13.3 + 680 \times 0.022)$$

and $V_2 = 2.63$ m/s.   From Appendix C, $\nu = 1.01 \ \mu m^2/s$, $\epsilon/D = 0.0017$, $\mathbf{R} = (2.63$ m/s$)(0.15$ m/$1.01 \ \mu m^2/s) = 391{,}000$. From Fig. 5.32, $f = 0.023$. Repeating the procedure gives $V_2 = 2.60$ m/s, $\mathbf{R} = 380{,}000$, and $f = 0.023$. The discharge is

$$Q = V_2 A_2 = (2.60 \text{ m/s}) (\pi/4) (0.15 \text{ m})^2 = 46 \text{ l/s}$$

For the second part, with Q known, the solution is straightforward:

$$V_2 = \frac{Q}{A} = \frac{0.06 \text{ m}^3/\text{s}}{(\pi/4)(0.15 \text{ m})^2} = 3.40 \text{ m/s} \qquad \mathbf{R} = 505{,}000 \qquad f = 0.023$$

and

$$H_1 = \frac{(3.4 \text{ m/s})^2}{2 \times 9.806 \text{ m/s}^2} (13.3 + 680 \times 0.023) = 17.06 \text{ m}$$

With equivalent lengths [Eq. (5.10.19)] the value of f is approximated, say $f = 0.022$. The sum of minor losses is $K = 13.3$, in which the kinetic energy at 2 is considered a minor loss,

$$L_e = \frac{13.3 \times 0.15}{0.022} = 90.7 \text{ m}$$

Hence the total length of pipe is $90.7 + 102 = 192.7$ m. The first part of the problem is solved by this method,

$$10 \text{ m} = f \frac{L + L_e}{D} \frac{V_2^2}{2g} = f \frac{192.7 \text{ m}}{0.15 \text{ m}} \frac{(V_2 \text{ m/s})^2}{2g \text{ m/s}^2}$$

If $f = 0.022$, $V_2 = 2.63$ m/s, $\mathbf{R} = 391{,}000$, and $f = 0.023$, then $V_2 = 2.58$ m/s and $Q = 45.6$ l/s.   Normally it is not necessary to use the new f to improve L_e.

Minor losses may be neglected in those situations where they comprise only 5 percent or less of the head losses due to pipe friction.   The friction factor, at best, is subject to about 5 percent error, and it is meaningless to select values to more than two significant figures. In general, minor losses may be neglected when, on the average, there is a length of 1000 diameters between each minor loss.

Compressible flow in pipes is treated in Chap. 6. Complex pipe-flow situations are treated in Chap. 10.

5.11 LUBRICATION MECHANICS

The effect of viscosity on flow and its effects on head losses have been examined in the preceding sections of this chapter. A laminar-flow case of great practical importance is the hydrodynamic theory of lubrication. Simple aspects of this theory are developed in this section.

Large forces are developed in small clearances when the surfaces are slightly inclined and one is in motion so that fluid is "wedged" into the decreasing space. The slipper bearing, which operates on this principle, is illustrated in Fig. 5.39. The journal bearing (Fig. 5.40) develops its force by the same action, except that the surfaces are curved.

The laminar-flow equations may be used to develop the theory of lubrication. The assumption is made that there is no flow out of the ends of the bearing normal to the plane of Fig. 5.39. From Eq. (5.1.4), which relates pressure drop and shear stress, the equation for the force P that the bearing will support is worked out, and the drag on the bearing is computed.

Substituting Newton's law of viscosity into Eq. (5.1.4) produces

$$\frac{dp}{dx} = \mu \frac{d^2u}{dy^2} \qquad\qquad (5.11.1)$$

Since the inclination of the upper portion of the bearing (Fig. 5.39) is very

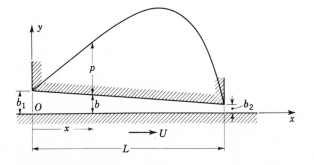

Fig. 5.39 Sliding bearing.

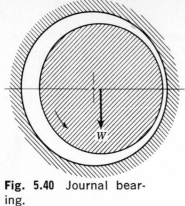

Fig. 5.40 Journal bearing.

slight, it may be assumed that the velocity distribution is the same as if the plates were parallel and that p is independent of y. Integrating Eq. (5.11.1) twice with respect to y, with dp/dx constant, produces

$$\frac{dp}{dx} \int dy = \mu \int \frac{d^2u}{dy^2} \, dy + A \qquad \text{or} \qquad \frac{dp}{dx} y = \mu \frac{du}{dy} + A$$

and the second time

$$\frac{dp}{dx} \int y \, dy = \mu \int \frac{du}{dy} \, dy + A \int dy + B \qquad \text{or} \qquad \frac{dp}{dx} \frac{y^2}{2} = \mu u + Ay + B$$

The constants of integration A, B are determined from the conditions $u = 0$, $y = b$; $u = U$, $y = 0$. Substituting in turn produces

$$\frac{dp}{dx} \frac{b^2}{2} = Ab + B \qquad \mu U + B = 0$$

Eliminating A and B and solving for u results in

$$u = \frac{y}{2\mu} \frac{dp}{dx} (y - b) + U \left(1 - \frac{y}{b}\right) \tag{5.11.2}$$

The discharge Q must be the same at each cross section. By integrating over a typical section, again with dp/dx constant,

$$Q = \int_0^b u \, dy = \frac{Ub}{2} - \frac{b^3}{12\mu} \frac{dp}{dx} \tag{5.11.3}$$

Now, since Q cannot vary with x, b may be expressed in terms of x, $b = b_1 - \alpha x$, in which $\alpha = (b_1 - b_2)/L$, and the equation is integrated with respect to x to determine the pressure distribution. Solving Eq. (5.11.3) for dp/dx produces

$$\frac{dp}{dx} = \frac{6\mu U}{(b_1 - \alpha x)^2} - \frac{12\mu Q}{(b_1 - \alpha x)^3} \tag{5.11.4}$$

Integrating gives

$$\int \frac{dp}{dx}\, dx = 6\mu U \int \frac{dx}{(b_1 - \alpha x)^2} - 12\mu Q \int \frac{dx}{(b_1 - \alpha x)^3} + C$$

or

$$p = \frac{6\mu U}{\alpha(b_1 - \alpha x)} - \frac{6\mu Q}{\alpha(b_1 - \alpha x)^2} + C$$

In this equation Q and C are unknowns. Since the pressure must be the same, say zero, at the ends of the bearing, namely, $p = 0$, $x = 0$; $p = 0$, $x = L$, the constants can be determined,

$$Q = \frac{Ub_1 b_2}{b_1 + b_2} \qquad C = -\frac{6\mu U}{\alpha(b_1 + b_2)}$$

With these values inserted, the equation for pressure distribution becomes

$$p = \frac{6\mu U x (b - b_2)}{b^2(b_1 + b_2)} \tag{5.11.5}$$

This equation shows that p is positive between $x = 0$ and $x = L$ if $b > b_2$. It is plotted in Fig. 5.39 to show the distribution of pressure throughout the bearing. With this one-dimensional method of analysis the very slight change in pressure along a vertical line $x = \text{const}$ is neglected.

The total force P that the bearing will sustain, per unit width, is

$$P = \int_0^L p\, dx = \frac{6\mu U}{b_1 + b_2} \int_0^L \frac{x(b - b_2)\, dx}{b^2}$$

After substituting the value of b in terms of x and performing the integration,

$$P = \frac{6\mu U L^2}{(b_1 - b_2)^2} \left(\ln \frac{b_1}{b_2} - 2\frac{b_1 - b_2}{b_1 + b_2} \right) \tag{5.11.6}$$

The drag force D required to move the lower surface at speed U is expressed by

$$D = \int_0^L \tau \bigg|_{y=0} dx = - \int_0^L \mu \frac{du}{dy} \bigg|_{y=0} dx$$

By evaluating du/dy from Eq. (5.11.2), for $y = 0$,

$$\frac{du}{dy}\bigg|_{y=0} = - \frac{b}{2\mu}\frac{dp}{dx} - \frac{U}{b}$$

This value in the integral, along with the value of dp/dx from Eq. (5.11.4), gives

$$D = \frac{2\mu U L}{b_1 - b_2}\left(2 \ln \frac{b_1}{b_2} - 3\frac{b_1 - b_2}{b_1 + b_2}\right) \tag{5.11.7}$$

The maximum load P is computed with Eq. (5.11.6) when $b_1 = 2.2b_2$. With this ratio,

$$P = 0.16\frac{\mu U L^2}{b_2{}^2} \qquad D = 0.75\frac{\mu U L}{b_2} \tag{5.11.8}$$

The ratio of load to drag for optimum load is

$$\frac{P}{D} = 0.21\frac{L}{b_2} \tag{5.11.9}$$

which can be very large since b_2 can be very small.

EXAMPLE 5.17   A vertical turbine shaft carries a load of 80,000 lb on a thrust bearing consisting of 16 flat rocker plates, 3 by 9 in, arranged with their long dimensions radial from the shaft and with their centers on a circle of radius 1.5 ft. The shaft turns at 120 rpm; $\mu = 0.002$ lb·s/ft². If the plates take the angle for maximum load, neglecting effects of curvature of path and radial lubricant flow, find (a) the clearance between rocker plate and fixed plate; (b) the torque loss due to the bearing.

(a)   Since the motion is considered straight-line,

$$U = 1.5 \left(\tfrac{120}{60}\right)(2\pi) = 18.85 \text{ ft/s} \qquad L = 0.25 \text{ ft}$$

The load is 5000 lb for each plate, which is $5000/0.75 = 6667$ lb for unit

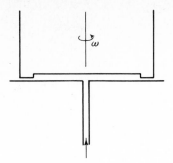

Fig. 5.41 Hydrostatic lubrication by high-pressure pumping of oil.

width. By solving for the clearance b_2, from Eq. (5.11.8),

$$b_2 = \sqrt{\frac{0.16\mu U L^2}{P}} = 0.4 \times 0.25 \sqrt{\frac{0.002 \times 18.85}{6667}} = 2.38 \times 10^{-4} = 0.0029 \text{ in}$$

(*b*)   The drag due to one rocker plate is, per foot of width,

$$D = 0.75 \frac{\mu U L}{b_2} = \frac{0.75 \times 0.002 \times 18.85 \times 0.25}{2.38 \times 10^{-4}} = 29.6 \text{ lb}$$

For a 9-in plate, $D = 29.6 \times 0.75 = 22.2$ lb. The torque loss due to the 16 rocker plates is

$$16 \times 22.2 \times 1.5 = 533 \text{ ft·lb}$$

Another form of lubrication, called *hydrostatic lubrication*,[1] has many important applications. It involves the continuous high-pressure pumping of oil under a step bearing, as illustrated in Fig. 5.41. The load may be lifted by the lubrication before rotation starts, which greatly reduces starting friction.

PROBLEMS

5.1   Determine the formulas for shear stress on each plate and for the velocity distribution for flow in Fig. 5.1 when an adverse pressure gradient exists such that $Q = 0$.

[1] For further information on hydrostatic lubrication see D. D. Fuller, Lubrication Mechanics, in V. L. Streeter (ed.), "Handbook of Fluid Dynamics," pp. **22**–21 to **22**–30, McGraw-Hill, New York, 1961.

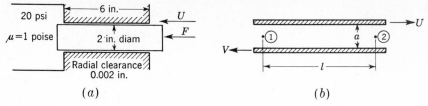

(a)   (b)

Fig. 5.42

5.2 In Fig. 5.1, with U positive as shown, find the expression for $d(p + \gamma h)/dl$ such that the shear is zero at the fixed plate. What is the discharge for this case?

5.3 In Fig. 5.42a, $U = 2$ ft/s. Find the rate at which oil is carried into the pressure chamber by the piston and the shear force and total force F acting on the piston.

5.4 Determine the force on the piston of Fig. 5.42a due to shear and the leakage from the pressure chamber for $U = 0$.

5.5 Find F and U in Fig. 5.42a such that no oil is lost through the clearance from the pressure chamber.

5.6 Derive an expression for the flow past a fixed cross section of Fig. 5.42b for laminar flow between the two moving plates.

5.7 In Fig. 5.42b, for $p_1 = p_2 = 1$ kg$_f$/cm^2, $U = 2V = 2$ m/s, $a = 1.5$ mm, $\mu = 0.5$ P, find the shear stress at each plate.

5.8 Compute the kinetic-energy and momentum correction factors for laminar flow between fixed parallel plates.

5.9 Determine the formula for angle θ for fixed parallel plates so that laminar flow at constant pressure takes place.

5.10 With a free body, as in Fig. 5.43, for uniform flow of a thin lamina of liquid down an inclined plane, show that the velocity distribution is

$$u = \frac{\gamma}{2\mu} (b^2 - s^2) \sin \theta$$

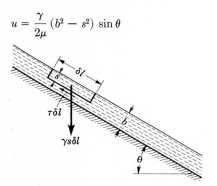

Fig. 5.43

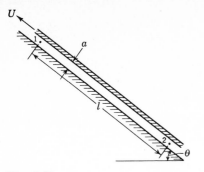

Fig. 5.44

and that the discharge per unit width is

$$Q = \frac{\gamma}{3\mu} b^3 \sin \theta$$

5.11 Derive the velocity distribution of Prob. 5.10 by inserting into the appropriate equation prior to Eq. (5.1.2) the condition that the shear at the free surface must be zero.

5.12 In Fig. 5.44, $p_1 = 6$ psi, $p_2 = 8$ psi, $l = 4$ ft, $a = 0.006$ ft, $\theta = 30°$, $U = 3$ ft/s, $\gamma = 50$ lb/ft^3, and $\mu = 0.8$ P. Determine the tangential force per square foot exerted on the upper plate and its direction.

5.13 For $\theta = 90°$ in Fig. 5.44, what speed U is required for no discharge? $S = 0.83$, $a = 3$ mm, $p_1 = p_2$, and $\mu = 0.2$ kg/m·s.

5.14 The belt conveyer (Fig. 5.45) delivers fluid to a reservoir of such a depth that the velocity on the free-liquid surface on the belt is zero. By considering only the work done by the belt on the fluid in shear, find how efficient this device is in transferring energy to the fluid.

5.15 What is the velocity distribution of the fluid on the belt and the volume rate of fluid being transported in Prob. 5.14?

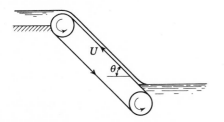

Fig. 5.45

5.16 What is the time rate of momentum and kinetic energy passing through a cross section that is normal to the flow if in Eq. (5.1.3) $Q = 0$?

5.17 A film of fluid 0.005 ft thick flows down a fixed vertical surface with a surface velocity of 2 ft/s. Determine the fluid viscosity. $\gamma = 55$ lb/ft³.

5.18 Determine the momentum correction factor for laminar flow in a round tube.

5.19 Water at standard conditions is flowing laminarly in a tube at pressure p_1 and diameter d_1. This tube expands to a diameter of $2d_1$ and pressure p_2, and the flow is again described by Eq. (5.2.6) some distance downstream of the expansion. Determine the force on the tube which results from the expansion.

5.20 At what distance r from the center of a tube of radius r_0 does the average velocity occur in laminar flow?

5.21 Determine the maximum wall shear stress for laminar flow in a tube of diameter D with fluid properties μ and ρ given.

5.22 Show that laminar flow between parallel plates may be used in place of flow through an annulus for 2 percent accuracy if the clearance is no more than 4 percent of the inner radius.

5.23 What are the losses per kilogram per meter of tubing for flow of mercury at 35°C through 0.6-mm-diameter tube at $\mathbf{R} = 1800$?

5.24 Determine the shear stress at the wall of a $\frac{1}{16}$-in-diameter tube when water at 80°F flows through it with a velocity of 1 ft/s.

5.25 Determine the pressure drop per meter of 3-mm-ID tubing for flow of liquid, $\mu = 60$ cP, sp gr $= 0.83$, at $\mathbf{R} = 100$.

5.26 Glycerin at 80°F flows through a $\frac{3}{8}$-in-diameter pipe with a pressure drop of 5 psi/ft. Find the discharge and the Reynolds number.

5.27 Calculate the diameter of vertical pipe needed for flow of liquid at $\mathbf{R} = 1800$ when the pressure remains constant. $\nu = 1.5\ \mu\text{m}^2/\text{s}$.

5.28 Calculate the discharge of the system in Fig. 5.46, neglecting all losses except through the pipe.

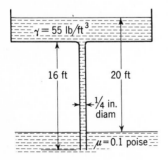

Fig. 5.46

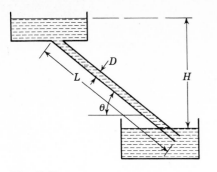

Fig. 5.47

5.29 In Fig. 5.47, $H = 10$ m, $L = 20$ m, $\theta = 30°$, $D = 8$ mm, $\gamma = 10$ kN/m^3, and $\mu = 0.08$ kg/m·s. Find the head loss per unit length of pipe and the discharge in liters per minute.

5.30 In Fig. 5.47 and Prob. 5.29, find H if the velocity is 3 m/s.

5.31 Oil, sp gr 0.85, $\mu = 0.50$ P, flows through an annulus $a = 0.60$ in, $b = 0.30$ in. When the shear stress at the outer wall is 0.25 lb/ft^2, calculate (a) the pressure drop per foot for a horizontal system, (b) the discharge in gallons per hour, and (c) the axial force exerted on the inner tube per foot of length.

5.32 What is the Reynolds number of flow of 0.3 m^3/s oil, sp gr 0.86, $\mu = 0.27$ P, through an 45-cm-diameter pipe?

5.33 Calculate the flow of crude oil, sp gr 0.86, at 80°F in a $\frac{3}{8}$-in-diameter tube to yield a Reynolds number of 1200.

5.34 Determine the velocity of kerosene at 90°F in a 3-in pipe to be dynamically similar to the flow of 3 m^3/s air at 1.4 kg$_f$/cm^2 abs and 15°C through a 75-cm duct.

5.35 What is the Reynolds number for a sphere 0.004 ft in diameter falling through water at 80°F at 0.5 ft/s?

5.36 Show that the power input for laminar flow in a round tube is $Q \, \Delta p$ by integration of Eq. (5.1.7).

5.37 By use of the one-seventh-power law of velocity distribution $u/u_{\max} = (y/r_0)^{1/7}$, determine the mixing-length distribution l/r_0 in terms of y/r_0 from Eq. (5.4.4).

5.38 A fluid is agitated so that the kinematic eddy viscosity increases linearly from zero ($y = 0$) at the bottom of the tank to 0.2 m^2/s at $y = 60$ cm. For uniform particles with fall velocities of 30 cm/s in still fluid, find the concentration at $y = 30$ cm if it is 10 per liter at $y = 60$ cm.

5.39 Plot a curve of $\epsilon/u_* r_0$ as a function of y/r_0 using Eq. (5.4.9) for velocity distribution in a pipe.

5.40   Find the value of y/r_0 in a pipe where the velocity equals the average velocity for turbulent flow.

5.41   Plot the velocity profiles for Prandtl's exponential velocity formula for values of n of $\frac{1}{7}$, $\frac{1}{8}$, and $\frac{1}{9}$.

5.42   Estimate the skin-friction drag on an airship 300 ft long, average diameter 60 ft, with velocity of 80 mph traveling through air at 13 psia and 80°F.

5.43   The velocity distribution in a boundary layer is given by $u/U = 3(y/\delta) - 2(y/\delta)^2$. Show that the displacement thickness of the boundary layer is $\delta_1 = \delta/6$.

5.44   Using the velocity distribution $u/U = \sin(\pi y/2\delta)$, determine the equation for growth of the laminar boundary layer and for shear stress along a smooth flat plate in two-dimensional flow.

5.45   Compare the drag coefficients that are obtained with the velocity distributions given in Probs. 5.43 and 5.44.

5.46   Work out the equations for growth of the turbulent boundary layer, based on the exponential law $u/U = (y/\delta)^{1/9}$ and $f = 0.185/\mathbf{R}^{1/5}$. $(\tau_0 = \rho f\, V^2/8.)$

5.47   Air at 20°C, 1 $\mathrm{kg}_f/\mathrm{cm}^2$ abs flows along a smooth plate with a velocity of 150 km/h. How long does the plate have to be to obtain a boundary-layer thickness of 8 mm?

5.48   The walls of a wind tunnel are sometimes made divergent to offset the effect of the boundary layer in reducing the portion of the cross section in which the flow is of constant speed. At what angle must plane walls be set so that the displacement thickness does not encroach upon the tunnel's constant-speed cross section at distances greater than 0.8 ft from the leading edge of the wall? Use the data of Prob. 5.47.

5.49   What is the terminal velocity of a 2-in-diameter metal ball, sp gr 3.5, dropped in oil, sp gr 0.80, $\mu = 1$ P? What would be the terminal velocity for the same-size ball but with a 7.0 sp gr? How do these results agree with the experiments attributed to Galileo at the Leaning Tower of Pisa?

5.50   At what speed must a 15-cm sphere travel through water at 10°C to have a drag of 5 N?

5.51   A spherical balloon contains helium and ascends through air at 14 psia, 40°F. Balloon and payload weigh 300 lb. What diameter permits ascension at 10 ft/s? $C_D = 0.21$. If the balloon is tethered to the ground in a 10-mph wind, what is the angle of inclination of the retaining cable?

5.52   How many 30-m-diameter parachutes $(C_D = 1.2)$ should be used to drop a bulldozer weighing 45 kN at a terminal speed of 10 m/s through air at 100,000 Pa abs at 20°C?

5.53   An object weighing 300 lb is attached to a circular disk and dropped from a plane. What diameter should the disk be to have the object strike the ground at 72 ft/s? The disk is attached so that it is normal to direction of motion. $p = 14.7$ psia; $t = 70$°F.

5.54 A circular disk 3 m in diameter is held normal to a 100 km/h airstream ($\rho =$ 0.0024 slug/ft^3). What force is required to hold it at rest?

5.55 Discuss the origin of the drag on a disk when its plane is parallel to the flow and when it is normal to it.

5.56 A semitubular cylinder of 6-in radius with concave side upstream is submerged in water flowing 2 ft/s. Calculate the drag for a cylinder 24 ft long.

5.57 A projectile of the form of (a), Fig. 5.25, is 108 mm in diameter and travels at 1 km/s through air, $\rho = 1$ kg/m^3; $c = 300$ m/s. What is its drag?

5.58 On the basis of the discussion of the Mach angle explain why a supersonic airplane is often seen before it is heard.

5.59 If an airplane 1 mi above the earth passes over an observer and the observer does not hear the plane until it has traveled 1.6 mi farther, what is its speed? Sound velocity is 1080 ft/s. What is its Mach angle?

5.60 Give some reason for the discontinuity in the curves of Fig. 5.23 at the angle of attack of 22°.

5.61 What is the ratio of lift to drag for the airfoil section of Fig. 5.23 for an angle of attack of 2°?

5.62 Determine the settling velocity of small metal spheres, sp gr 4.5, diameter 0.1 mm, in crude oil at 25°C.

5.63 A spherical dust particle at an altitude of 50 mi is radioactive as a result of an atomic explosion. Determine the time it will take to settle to earth if it falls in accordance with Stokes' law. Its size and sp gr are 25 μm and 2.5. Neglect wind effects. Use isothermal atmosphere at 0°F.

5.64 How large a spherical particle of dust, sp gr 2.5, will settle in atmospheric air at 20°C in obedience to Stokes' law? What is the settling velocity?

5.65 The Chézy coefficient is 127 for flow in a rectangular channel 6 ft wide, 2 ft deep, with bottom slope of 0.0016. What is the discharge?

5.66 A rectangular channel 1 m wide, Chézy $C = 60$, $S = 0.0064$, carries 1 m^3/s. Determine the velocity.

5.67 What is the value of the Manning roughness factor n in Prob. 5.66?

5.68 A rectangular, brick-lined channel 6 ft wide and 5 ft deep carries 210 cfs. What slope is required for the channel?

5.69 The channel cross section shown in Fig. 5.48 is made of unplaned wood and has a slope of 0.0009. What is the discharge?

5.70 A trapezoidal, unfinished concrete channel carries water at a depth of 6 ft. Its bottom width is 8 ft and side slope 1 horizontal to 1$\frac{1}{2}$ vertical. For a bottom slope of 0.004 what is the discharge?

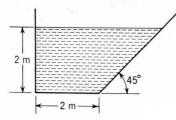

Fig. 5.48

5.71 A trapezoidal channel with bottom slope 0.003, bottom width 1.2 m, and side slopes 2 horizontal to 1 vertical carries 6 m³/s at a depth of 1.2 m. What is the Manning roughness factor?

5.72 A trapezoidal earth canal, bottom width 8 ft and side slope 2 on 1 (2 horizontal to 1 vertical), is to be constructed to carry 280 cfs. The best velocity for nonscouring is 2.8 ft/s with this material. What is the bottom slope required?

5.73 What diameter is required of a semicircular corrugated-metal channel to carry 2 m³/s when its slope is 0.01?

5.74 A semicircular corrugated-metal channel 10 ft in diameter has a bottom slope of 0.004. What is its capacity when flowing full?

5.75 Calculate the depth of flow of 60 m³/s in a gravel trapezoidal channel with bottom width of 4 m, side slopes of 3 horizontal to 1 vertical, and bottom slope of 0.001.

5.76 What is the velocity of flow of 260 cfs in a rectangular channel 12 ft wide? $S = 0.0049; n = 0.016$.

5.77 A trapezoidal channel, brick-lined, is to be constructed to carry 35 m³/s a distance of 8 km with a head loss of 5 m. The bottom width is 4 m, the side slopes 1 on 1. What is the velocity?

5.78 How does the discharge vary with depth in Fig. 5.49?

5.79 How does the velocity vary with depth in Fig. 5.49?

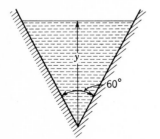

Fig. 5.49

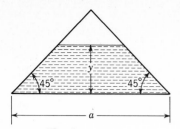

Fig. 5.50

5.80 Determine the depth of flow in Fig. 5.49 for discharge of 12 cfs. It is made of riveted steel with bottom slope 0.02.

5.81 Determine the depth y (Fig. 5.50) for maximum velocity for given n and S.

5.82 Determine the depth y (Fig. 5.50) for maximum discharge for given n and S.

5.83 A test on 30-cm-diameter pipe with water showed a gage difference of 33 cm on a mercury-water manometer connected to two piezometer rings 120 m apart. The flow was 0.23 m³/s. What is the friction factor?

5.84 By using the Blasius equation for determination of friction factor, determine the horsepower per mile required to pump 3.0 cfs liquid, $\nu = 3.3 \times 10^{-4}$ ft²/s, $\gamma = 55$ lb/ft³, through an 18 in pipeline.

5.85 Determine the head loss per kilometer required to maintain a velocity of 4 m/s in a 1-cm-diameter pipe. $\nu = 4 \times 10^{-5}$ m²/s.

5.86 Fluid flows through a $\frac{1}{2}$-in diameter tube at a Reynolds number of 1800. The head loss is 30 ft in 100 ft of tubing. Calculate the discharge in gallons per minute.

5.87 What size galvanized-iron pipe is needed to be "hydraulically smooth" at $\mathbf{R} = 3.5 \times 10^5$? (A pipe is said to be hydraulically smooth when it has the same losses as a smoother pipe under the same conditions.)

5.88 Above what Reynolds number is the flow through a 3-m-diameter riveted steel pipe, $\epsilon = 3$ mm, independent of the viscosity of the fluid?

5.89 Determine the absolute roughness of a 1-ft-diameter pipe that has a friction factor $f = 0.03$ for $\mathbf{R} = 1,000,000$.

5.90 What diameter clean galvanized-iron pipe has the same friction factor for $\mathbf{R} = 100,000$ as a 30-cm-diameter cast-iron pipe?

5.91 Under what conditions do the losses in an artificially roughened pipe vary as some power of the velocity greater than the second?

5.92 Why does the friction factor increase as the velocity decreases in laminar flow in a pipe?

5.93 Look up the friction factor for atmospheric air at 60°F traveling 60 ft/s through a 3-ft-diameter galvanized pipe.

5.94 Water at 20°C is to be pumped through a kilometer of 20 cm diameter wrought-iron pipe at the rate of 60 l/s. Compute the head loss and power required.

5.95 16,000 ft³/min atmospheric air at 90°F is conveyed 1000 ft through a 4-ft-diameter wrought-iron pipe. What is the head loss in inches of water?

5.96 What power motor for a fan must be purchased to circulate standard air in a wind tunnel at 500 km/h? The tunnel is a closed loop, 60 m long, and it can be assumed to have a constant circular cross section with a 2 m diameter. Assume smooth pipe.

5.97 Must there be a provision made to cool the air at some section of the tunnel described in Prob. 5.96? To what extent?

5.98 2.0 cfs oil, $\mu = 0.16$ P, $\gamma = 54$ lb/ft³, is pumped through a 12-in pipeline of cast iron. If each pump produces 80 psi, how far apart may they be placed?

5.99 A 6-cm diameter smooth pipe 150 m long conveys 10 l/s water at 25°C from a water main, $p = 1.6$ MN/m², to the top of a building 25 m above the main. What pressure can be maintained at the top of the building?

5.100 For water at 150°F calculate the discharge for the pipe of Fig. 5.51.

5.101 In Fig. 5.51, how much power would be required to pump 160 gpm from a reservoir at the bottom of the pipe to the reservoir shown?

5.102 A 12-mm-diameter commercial steel pipe 15 m long is used to drain an oil tank. Determine the discharge when the oil level in the tank is 2 m above the exit end of the pipe. $\mu = 0.10$ P; $\gamma = 8$ kN/m³.

5.103 Two liquid reservoirs are connected by 200 ft of 2-in-diameter smooth tubing. What is the flow rate when the difference in elevation is 50 ft? $\nu = 0.001$ ft²/s.

5.104 For a head loss of 8 cm water in a length of 200 m for flow of atmospheric air

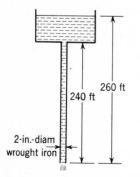

260 ft

240 ft

2-in.-diam
wrought iron

Fig. 5.51

at 15°C through a 1.25-m-diameter duct, $\epsilon = 1$ mm, calculate the flow in cubic meters per minute.

5.105 A gas of molecular weight 37 flows through a 24-in-diameter galvanized duct at 90 psia and 100°F. The head loss per 100 ft of duct is 2 in H_2O. What is the flow in slugs per hour? $\mu = 0.194$ mP.

5.106 What is the power per kilometer required for a 70 percent efficient blower to maintain the flow of Prob. 5.105?

5.107 The 100 lb_m/min air required to ventilate a mine is admitted through 2000 ft of 12-in-diameter galvanized pipe. Neglecting minor losses, what head in inches of water does a blower have to produce to furnish this flow? $p = 14$ psia; $t = 90°F$.

5.108 In Fig. 5.47 $H = 20$ m, $L = 150$ m, $D = 5$ cm, $S = 0.85, \mu = 4$ cP, $\epsilon = 1$ mm. Find the newtons per second flowing.

5.109 In a process 10,000 lb/h of distilled water at 70°F is conducted through a smooth tube between two reservoirs having a distance between them of 30 ft and a difference in elevation of 4 ft. What size tubing is needed?

5.110 What size of new cast-iron pipe is needed to transport 300 l/s water at 25°C for 1 km with head loss of 2 m?

5.111 Two types of steel plate, having surface roughnesses of $\epsilon_1 = 0.0003$ ft and $\epsilon_2 = 0.0001$ ft, have a cost differential of 10 percent more for the smoother plate. With an allowable stress in each of 10,000 psi, which plate should be selected to convey 100 cfs water at 200 psi with a head loss of 6 ft/mi?

5.112 An old pipe 2 m in diameter has a roughness of $\epsilon = 30$ mm. A 12-mm-thick lining would reduce the roughness to $\epsilon = 1$ mm. How much in pumping costs would be saved per year per kilometer of pipe for water at 20°C with discharge of 6 m^3/s? The pumps and motors are 80 percent efficient, and power costs 1 cent per kilowatthour.

5.113 Calculate the diameter of new wood-stave pipe in excellent condition needed to convey 300 cfs water at 60°F with a head loss of 1 ft per 1000 ft of pipe.

5.114 Two oil reservoirs with difference in elevation of 5 m are connected by 300 m of commercial steel pipe. What size must the pipe be to convey 50 l/s? $\mu = 0.05$ kg/m·s, $\gamma = 8$ kN/m^3.

5.115 200 cfs air, $p = 16$ psia, $t = 70°F$, is to be delivered to a mine with a head loss of 3 in H_2O per 1000 ft. What size galvanized pipe is needed?

5.116 Compute the losses in foot-pounds per pound due to flow of 25 m^3/min air, $p = 1$ kg$_f$/cm^2, $t = 20°C$, through a sudden expansion from 30- to 90-cm pipe. How much head would be saved by using a 10° conical diffuser?

5.117 Calculate the value of H in Fig. 5.52 for 125 l/s water at 15°C through commercial steel pipe. Include minor losses.

5.118 In Prob. 5.28 what would be the discharge if a globe valve were inserted into the line? Assume a smooth pipe and a well-rounded inlet, with $\mu = 1$ cP.

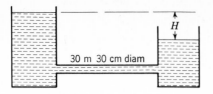

Fig. 5.52

5.119 In Fig. 5.52 for $H = 3$ m, calculate the discharge of oil, $S = 0.8$, $\mu = 7$ cP, through smooth pipe. Include minor losses.

5.120 If a valve is placed in the line in Prob. 5.119 and adjusted to reduce the discharge by one-half, what is K for the valve and what is its equivalent length of pipe at this setting?

5.121 A water line connecting two reservoirs at 70°F has 5000 ft of 24-in-diameter steel pipe, three standard elbows, a globe valve, and a re-entrant pipe entrance. What is the difference in reservoir elevations for 20 cfs?

5.122 Determine the discharge in Prob. 5.121 if the difference in elevation is 40 ft.

5.123 Compute the losses in power due to flow of 3 m³/s water through a sudden contraction from 2- to 1.3-m-diameter pipe.

5.124 What is the equivalent length of 2-in-diameter pipe, $f = 0.022$, for (a) a re-entrant pipe entrance, (b) a sudden expansion from 2 to 4 in diameter, (c) a globe valve and a standard tee?

5.125 Find H in Fig. 5.53 for 200 gpm oil flow, $\mu = 0.1$ P, $\gamma = 60$ lb/ft³, for the angle valve wide open.

5.126 Find K for the angle valve in Prob. 5.125 for flow of 10 l/s at the same H.

5.127 What is the discharge through the system of Fig. 5.53 for water at 25°C when $H = 8$ m?

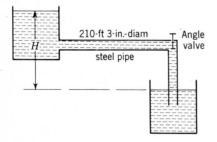

Fig. 5.53

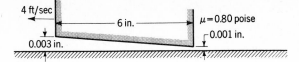

Fig. 5.54

5.128 Compare the smooth-pipe curve on the Moody diagram with Eq. (5.10.4) for $\mathbf{R} = 10^5, 10^6, 10^7$.

5.129 Check the location of line $\epsilon/D = 0.0002$ on the Moody diagram with Eq. (5.10.7).

5.130 In Eq. (5.10.7) show that when $\epsilon = 0$ it reduces to Eq. (5.10.4) and that, when $\mathbf{R}$ is very large, it reduces to Eq. (5.10.6).

√5.131 In Fig. 5.54 the rocker plate has a width of 1 ft. Calculate (*a*) the load the bearing will sustain, (*b*) the drag on the bearing. Assume no flow normal to the paper.

√5.132 Find the maximum pressure in the fluid of Prob. 5.131, and determine its location.

5.133 Determine the pressure center for the rocker plate of Prob. 5.131.

5.134 Show that a shaft concentric with a bearing can sustain no load.

5.135 The shear stress in a fluid flowing between two fixed parallel plates

(*a*) is constant over the cross section
(*b*) is zero at the plates and increases linearly to the midpoint
(*c*) varies parabolically across the section
(*d*) is zero at the midplane and varies linearly with distance from the midplane
(*e*) is none of these answers

5.136 The velocity distribution for flow between two fixed parallel plates

(*a*) is constant over the cross section
(*b*) is zero at the plates and increases linearly to the midplane
(*c*) varies parabolically across the section
(*d*) varies as the three-halves power of the distance from the midpoint
(*e*) is none of these answers

5.137 The discharge between two parallel plates, distance a apart, when one has the velocity U and the shear stress is zero at the fixed plate, is

(*a*) $Ua/3$ (*b*) $Ua/2$ (*c*) $2Ua/3$ (*d*) Ua (*e*) none of these answers

5.138 Fluid is in laminar motion between two parallel plates, with one plate in motion, and is under the action of a pressure gradient so that the discharge through any fixed cross section is zero. The minimum velocity occurs at a point which is distant from the fixed plate

(a) $a/6$ (b) $a/3$ (c) $a/2$ (d) $2a/3$ (e) none of these answers

5.139   In Prob. 5.138 the value of the minimum velocity is

(a) $-3U/4$ (b) $-2U/3$ (c) $-U/2$ (d) $-U/3$ (e) $-U/6$

5.140   The relation between pressure and shear stress in one-dimensional laminar flow in the x direction is given by

(a) $dp/dx = \mu\,d\tau/dy$ (b) $dp/dy = d\tau/dx$ (c) $dp/dy = \mu\,d\tau/dx$
(d) $dp/dx = d\tau/dy$ (e) none of these answers

5.141   The expression for power input per unit volume to a fluid in one-dimensional laminar motion in the x direction is

(a) $\tau\,du/dy$ (b) τ/μ^2 (c) $\mu\,du/dy$ (d) $\tau(du/dy)^2$ (e) none of these answers

5.142   When liquid is in laminar motion at constant depth and flowing down an inclined plate (y measured normal to surface),

(a) the shear is zero throughout the liquid   (b) $d\tau/dy = 0$ at the plate
(c) $\tau = 0$ at the surface of the liquid
(d) the velocity is constant throughout the liquid   (e) there are no losses

5.143   The shear stress in a fluid flowing in a round pipe

(a) is constant over the cross section
(b) is zero at the wall and increases linearly to the center
(c) varies parabolically across the section
(d) is zero at the center and varies linearly with the radius
(e) is none of these answers

5.144   When the pressure drop in a 24-in-diameter pipeline is 10 psi in 100 ft, the wall shear stress in pounds per square foot is

(a) 0   (b) 7.2   (c) 14.4   (d) 720   (e) none of these answers

5.145   In laminar flow through a round tube the discharge varies

(a) linearly as the viscosity
(b) as the square of the radius
(c) inversely as the pressure drop
(d) inversely as the viscosity
(e) as the cube of the diameter

5.146   When a tube is inclined, the term $-dp/dl$ is replaced by

(a) $-dz/dl$ (b) $-\gamma\,dz/dl$ (c) $-d(p+z)/dl$
(d) $-d(p+\rho z)/dl$ (e) $-d(p+\gamma z)/dl$

5.147 The upper critical Reynolds number is

(*a*) important from a design viewpoint
(*b*) the number at which turbulent flow changes to laminar flow
(*c*) about 2000
(*d*) not more than 2000
(*e*) of no practical importance in pipe-flow problems

5.148 The Reynolds number for pipe flow is given by

(*a*) VD/ν (*b*) $VD\mu/\rho$ (*c*) $VD\rho/\nu$ (*d*) VD/μ
(*e*) none of these answers

5.149 The lower critical Reynolds number has the value

(*a*) 200   (*b*) 1200   (*c*) 12,000   (*d*) 40,000   (*e*) none of these answers

5.150 The Reynolds number for a 3-cm-diameter sphere moving 3 m/s through oil, sp gr 0.90, $\mu = 0.10$ kg/m·s, is

(*a*) 404   (*b*) 808   (*c*) 900   (*d*) 8080   (*e*) none of these answers

5.151 The Reynolds number for 10 cfs discharge of water at 68°F through a 12-in-diameter pipe is

(*a*) 2460   (*b*) 980,000   (*c*) 1,178,000   (*d*) 14,120,000
(*e*) none of these answers

5.152 The Prandtl mixing length is

(*a*) independent of radial distance from pipe axis
(*b*) independent of the shear stress
(*c*) zero at the pipe wall   (*d*) a universal constant
(*e*) useful for computing laminar-flow problems

5.153 In a fluid stream of low viscosity

(*a*) the effect of viscosity does not appreciably increase the drag on a body
(*b*) the potential theory yields the drag force on a body
(*c*) the effect of viscosity is limited to a narrow region surrounding a body
(*d*) the deformation drag on a body always predominates
(*e*) the potential theory contributes nothing of value regarding flow around bodies

5.154 The lift on a body immersed in a fluid stream is

(*a*) due to buoyant force
(*b*) always in the opposite direction to gravity
(*c*) the resultant fluid force on the body
(*d*) the dynamic fluid-force component exerted on the body normal to the approach velocity
(*e*) the dynamic fluid-force component exerted on the body parallel to the approach velocity

5.155   The displacement thickness of the boundary layer is

(a) the distance from the boundary affected by boundary shear
(b) one-half the actual thickness of the boundary layer
(c) the distance to the point where $u/U = 0.99$
(d) the distance the main flow is shifted
(e) none of these answers

5.156   The shear stress at the boundary of a flat plate is

(a) $\partial p/\partial x$ (b) $\mu\, \partial u/\partial y\,|_{y=0}$ (c) $\rho\, \partial u/\partial y\,|_{y=0}$ (d) $\mu\, \partial u/\partial y\,|_{y=\delta}$
(e) none of these answers

5.157   Which of the following velocity distributions u/U satisfies the boundary conditions for flow along a flat plate? $\eta = y/\delta$.

(a) e^{η} (b) $\cos \pi\eta/2$ (c) $\eta - \eta^2$ (d) $2\eta - \eta^3$ (e) none of these answers

5.158   The drag coefficient for a flat plate is ($D = $ drag)

(a) $2D/\rho U^2 l$ (b) $\rho U l/D$ (c) $\rho U l/2D$ (d) $\rho U^2 l/2D$
(e) none of these answers

5.159   The average velocity divided by the maximum velocity, as given by the one-seventh-power law, is

(a) $\frac{49}{120}$ (b) $\frac{1}{2}$ (c) $\frac{6}{7}$ (d) $\frac{98}{120}$ (e) none of these answers

5.160   The laminar-boundary-layer thickness varies as

(a) $1/x^{1/2}$ (b) $x^{1/7}$ (c) $x^{1/2}$ (d) $x^{6/7}$ (e) none of these answers

5.161   The turbulent-boundary-layer thickness varies as

(a) $1/x^{1/5}$ (b) $x^{1/5}$ (c) $x^{1/2}$ (d) $x^{4/5}$ (e) none of these answers

5.162   In flow along a rough plate, the order of flow type from upstream to downstream is

(a) laminar, fully developed wall roughness, transition region, hydraulically smooth
(b) laminar, transition region, hydraulically smooth, fully developed wall roughness
(c) laminar, hydraulically smooth, transition region, fully developed wall roughness
(d) laminar, hydraulically smooth, fully developed wall roughness, transition region
(e) laminar, fully developed wall roughness, hydraulically smooth, transition region

5.163   Separation is caused by

(a) reduction of pressure to vapor pressure
(b) reduction of pressure gradient to zero
(c) an adverse pressure gradient
(d) the boundary-layer thickness reducing to zero
(e) none of these answers

5.164 Separation occurs when

(a) the cross section of a channel is reduced
(b) the boundary layer comes to rest
(c) the velocity of sound is reached
(d) the pressure reaches a minimum
(e) a valve is closed

5.165 The wake

(a) is a region of high pressure
(b) is the principal cause of skin friction
(c) always occurs when deformation drag predominates
(d) always occurs after a separation point
(e) is none of these answers

5.166 Pressure drag results from

(a) skin friction
(b) deformation drag
(c) breakdown of potential flow near the forward stagnation point
(d) occurrence of a wake
(e) none of these answers

5.167 A body with a rounded nose and long, tapering tail is usually best suited for

(a) laminar flow
(b) turbulent subsonic flow
(c) supersonic flow
(d) flow at speed of sound
(e) none of these answers

5.168 A sudden change in position of the separation point in flow around a sphere occurs at a Reynolds number of about

(a) 1   (b) 300   (c) 30,000   (d) 3,000,000   (e) none of these answers

5.169 The effect of compressibility on the drag force is to

(a) increase it greatly near the speed of sound
(b) decrease it near the speed of sound
(c) cause it asymptotically to approach a constant value for large Mach numbers
(d) cause it to increase more rapidly than the square of the speed at high Mach numbers
(e) reduce it throughout the whole flow range

5.170 The terminal velocity of a small sphere settling in a viscous fluid varies as the

(a) first power of its diameter
(b) inverse of the fluid viscosity
(c) inverse square of the diameter
(d) inverse of the diameter
(e) square of the difference in specific weights of solid and fluid

5.171   The losses in open-channel flow generally vary as the

(a) first power of the roughness
(b) inverse of the roughness
(c) square of the velocity
(d) inverse square of the hydraulic radius
(e) velocity

5.172   The most simple form of open-channel-flow computation is

(a) steady uniform
(b) steady nonuniform
(c) unsteady uniform
(d) unsteady nonuniform
(e) gradually varied

5.173   In an open channel of great width the hydraulic radius equals

(a) $y/3$ (b) $y/2$ (c) $2y/3$ (d) y (e) none of these answers

5.174   The Manning roughness coefficient for finished concrete is

(a) 0.002   (b) 0.020   (c) 0.20   (d) dependent upon hydraulic radius
(e) none of these answers

5.175   In turbulent flow a rough pipe has the same friction factor as a smooth pipe

(a) in the zone of complete turbulence, rough pipes
(b) when the friction factor is independent of the Reynolds number
(c) when the roughness projections are much smaller than the thickness of the laminar film
(d) everywhere in the transition zone
(e) when the friction factor is constant

5.176   The friction factor in turbulent flow in smooth pipes depends upon the following:

(a) V, D, ρ, L, μ (b) Q, L, μ, ρ (c) V, D, ρ, p, μ (d) V, D, μ, ρ
(e) p, L, D, Q, V

5.177   In a *given* rough pipe, the losses depend upon

(a) f, V (b) μ, ρ (c) **R**   (d) Q only   (e) none of these answers

5.178   In the complete-turbulence zone, rough pipes,

(a) rough and smooth pipes have the same friction factor
(b) the laminar film covers the roughness projections
(c) the friction factor depends upon Reynolds number only
(d) the head loss varies as the square of the velocity
(e) the friction factor is independent of the relative roughness

5.179   The friction factor for flow of water at 60°F through a 2-ft-diameter cast-iron pipe with a velocity of 5 ft/s is

(*a*) 0.013   (*b*) 0.017   (*c*) 0.019   (*d*) 0.021   (*e*) none of these answers

5.180   The procedure to follow in solving for losses when $Q, L, D, \nu,$ and ϵ are given is to

(*a*) assume an f, look up **R** on Moody diagram, etc.
(*b*) assume an h_f, solve for f, check against **R** on Moody diagram
(*c*) assume an f, solve for h_f, compute **R**, etc.
(*d*) compute **R**, look up f for ϵ/D, solve for h_f
(*e*) assume an **R**, compute V, look up f, solve for h_f

5.181   The procedure to follow in solving for discharge when $h_f, L, D, \nu,$ and ϵ are given is to

(*a*) assume an f, compute V, **R**, ϵ/D, look up f, and repeat if necessary
(*b*) assume an **R**, compute f, check ϵ/D, etc.
(*c*) assume a V, compute **R**, look up f, compute V again, etc.
(*d*) solve Darcy-Weisbach for V, compute Q
(*e*) assume a Q, compute V, **R**, look up f, etc.

5.182   The procedure to follow in solving for pipe diameter when $h_f, Q, L, \nu,$ and ϵ are given is to

(*a*) assume a D, compute V, **R**, ϵ/D, look up f, and repeat
(*b*) compute V from continuity, assume an f, solve for D
(*c*) eliminate V in **R** and Darcy-Weisbach, using continuity, assume an f, solve for D, **R**, look up f, and repeat
(*d*) assume an **R** and an ϵ/D, look up f, solve Darcy-Weisbach for V^2/D, and solve simultaneously with continuity for V and D, compute new **R**, etc.
(*e*) assume a V, solve for D, **R**, ϵ/D, look up f, and repeat

5.183   The losses due to a sudden contraction are given by

(*a*) $\left(\dfrac{1}{C_c^2} - 1\right)\dfrac{V_2^2}{2g}$ (*b*) $(1 - C_c^2)\dfrac{V_2^2}{2g}$ (*c*) $\left(\dfrac{1}{C_c} - 1\right)^2 \dfrac{V_2^2}{2g}$

(*d*) $(C_c - 1)^2 \dfrac{V_2^2}{2g}$ (*e*) none of these answers

5.184   The losses at the exit of a submerged pipe in a reservoir are

(*a*) negligible   (*b*) $0.05(V^2/2g)$ (*c*) $0.5(V^2/2g)$ (*d*) $V^2/2g$
(*e*) none of these answers

5.185   Minor losses usually may be neglected when

(*a*) there are 100 ft of pipe between special fittings
(*b*) their loss is 5 percent or less of the friction loss

(c) there are 500 diameters of pipe between minor losses
(d) there are no globe valves in the line
(e) rough pipe is used

5.186   The length of pipe ($f = 0.025$) in diameters, equivalent to a globe valve, is
(a) 40   (b) 200   (c) 300   (d) 400   (e) not determinable; insufficient data

5.187   The hydraulic radius is given by

(a) wetted perimeter divided by area
(b) area divided by square of wetted perimeter
(c) square root of area
(d) area divided by wetted perimeter
(e) none of these answers

5.188   The hydraulic radius of a 6 by 12 cm cross section is, in centimeters,

(a) $\frac{3}{2}$ (b) 2   (c) 3   (d) 6   (e) none of these answers

5.189   In the theory of lubrication the assumption is made that

(a) the velocity distribution is the same at all cross sections
(b) the velocity distribution at any section is the same as if the plates were parallel
(c) the pressure variation along the bearing is the same as if the plates were parallel
(d) the shear stress varies linearly between the two surfaces
(e) the velocity varies linearly between the two surfaces

5.190   A 4-in-diameter shaft rotates at 240 rpm in a bearing with a radial clearance of 0.006 in. The shear stress in an oil film, $\mu = 0.1$ P, is, in pounds per square foot,

(a) 0.15   (b) 1.75   (c) 3.50   (d) 16.70   (e) none of these answers

COMPRESSIBLE FLOW

In Chap. 5, viscous incompressible-fluid-flow situations were mainly considered. In this chapter, on compressible flow, one new variable enters, the density, and one extra equation is available, the equation of state, which relates pressure and density. The other equations—continuity, momentum, and the first and second laws of thermodynamics—are also needed in the analysis of compressible-fluid-flow situations. In this chapter topics in steady one-dimensional flow of a perfect gas are discussed. The one-dimensional approach is limited to those applications in which the velocity and density may be considered constant over any cross section. When density changes are gradual and do not change by more than a few percent, the flow may be treated as incompressible with the use of an average density.

The following topics are discussed in this chapter: perfect-gas relationships, speed of a sound wave, Mach number, isentropic flow, shock waves, Fanno and Rayleigh lines, adiabatic flow, flow with heat transfer, isothermal flow, and the analogy between shock waves and open-channel waves.

6.1 PERFECT-GAS RELATIONSHIPS

In Sec. 1.6 [Eq. (1.6.2)] a perfect gas is defined as a fluid that has constant specific heats and follows the law

$$p = \rho RT \tag{6.1.1}$$

in which p and T are the absolute pressure and absolute temperature, respectively, ρ is the density, and R the gas constant. In this section specific heats are defined; the specific heat ratio is introduced and related to specific heats and the gas constant; internal energy and enthalpy are related to temperature; entropy relations are established; and the isentropic and reversible polytropic processes are introduced.

In general, the specific heat c_v at constant volume is defined by

$$c_v = \left(\frac{\partial u}{\partial T}\right)_v \qquad (6.1.2)$$

in which u is the internal energy[1] per unit mass. In words, c_v is the amount of internal-energy increase required by a unit mass of gas to increase its temperature by one degree when its volume is held constant. In thermodynamic theory it is proved that u is a function only of temperature for a perfect gas.

The specific heat c_p at constant pressure is defined by

$$c_p = \left(\frac{\partial h}{\partial T}\right)_p \qquad (6.1.3)$$

in which h is the enthalpy per unit mass given by $h = u + p/\rho$. Since p/ρ is equal to RT and u is a function only of temperature for a perfect gas, h depends only on temperature. Many of the common gases, such as water vapor, hydrogen, oxygen, carbon monoxide, and air, have a fairly small change in specific heats over the temperature range 500 to 1000°R, and an intermediate value is taken for their use as perfect gases. Table C.3 of Appendix C lists some common gases with values of specific heats at 80°F.

For perfect gases Eq. (6.1.2) becomes

$$du = c_v \, dT \qquad (6.1.4)$$

and Eq. (6.1.3) becomes

$$dh = c_p \, dT \qquad (6.1.5)$$

Then, from

$$h = u + \frac{p}{\rho} = u + RT$$

---

[1] The definitions for c_v and c_p are for equilibrium conditions; hence the internal energy e of Eq. (3.2.7) is u.

differentiating gives

$$dh = du + R \, dT$$

and substitution of Eqs. (6.1.4) and (6.1.5) leads to

$$c_p = c_v + R \tag{6.1.6}$$

which is valid for any gas obeying Eq. (6.1.1) (even when c_p and c_v are changing with temperature). If c_p and c_v are given in heat units per unit mass, i.e., kilocalorie per kilogram per kelvin or Btu per slug per degree Rankine, then R must have the same units. The conversion factor is 1 Btu = 778 ft·lb or 1 kcal = 4187 J.

The *specific-heat ratio* k is defined as the ratio

$$k = \frac{c_p}{c_v} \tag{6.1.7}$$

Solving with Eq. (6.1.6) gives

$$c_p = \frac{k}{k-1} R \qquad c_v = \frac{R}{k-1} \tag{6.1.8}$$

Entropy relationships

The first law of thermodynamics for a system states that the heat added to a system is equal to the work done by the system plus its increase in internal energy [Eq. (3.7.4)]. In terms of the entropy s the equation takes the form

$$T \, ds = du + p \, d\frac{1}{\rho} \tag{3.7.6}$$

which is a relationship between thermodynamic properties and must hold for all pure substances.

The internal energy change for a perfect gas is

$$u_2 - u_1 = c_v(T_2 - T_1) \tag{6.1.9}$$

and the enthalpy change is

$$h_2 - h_1 = c_p(T_2 - T_1) \tag{6.1.10}$$

The change in entropy

$$ds = \frac{du}{T} + \frac{p}{T} d \frac{1}{\rho} = c_v \frac{dT}{T} + R\rho \, d \frac{1}{\rho} \tag{6.1.11}$$

may be obtained from Eqs. (6.1.4) and (6.1.1). After integrating,

$$s_2 - s_1 = c_v \ln \frac{T_2}{T_1} + R \ln \frac{\rho_1}{\rho_2} \tag{6.1.12}$$

By use of Eqs. (6.1.8) and (6.1.1), Eq. (6.1.12) becomes

$$s_2 - s_1 = c_v \ln \left[\frac{T_2}{T_1} \left(\frac{\rho_1}{\rho_2} \right)^{k-1} \right] \tag{6.1.13}$$

or

$$s_2 - s_1 = c_v \ln \left[\frac{p_2}{p_1} \left(\frac{\rho_1}{\rho_2} \right)^{k} \right] \tag{6.1.14}$$

and

$$s_2 - s_1 = c_v \ln \left[\left(\frac{T_2}{T_1} \right)^{k} \left(\frac{p_2}{p_1} \right)^{1-k} \right] \tag{6.1.15}$$

These equations are forms of the second law of thermodynamics.

If the process is reversible, $ds = dq_H/T$, or $T \, ds = dq_H$; further, if the process should also be adiabatic, $dq_H = 0$. Thus $ds = 0$ for a reversible, adiabatic process, or $s = $ const; the *reversible, adiabatic* process is therefore *isentropic*. Then, from Eq. (6.1.14) for $s_2 = s_1$,

$$\frac{p_1}{\rho_1^{\,k}} = \frac{p_2}{\rho_2^{\,k}} \tag{6.1.16}$$

Equation (6.1.16) combined with the general gas law yields

$$\frac{T_2}{T_1} = \left(\frac{p_2}{p_1} \right)^{(k-1)/k} = \left(\frac{\rho_2}{\rho_1} \right)^{k-1} \tag{6.1.17}$$

The enthalpy change for an isentropic process is

$$h_2 - h_1 = c_p(T_2 - T_1) = c_p T_1 \left(\frac{T_2}{T_1} - 1 \right) = c_p T_1 \left[\left(\frac{p_2}{p_1} \right)^{(k-1)/k} - 1 \right] \tag{6.1.18}$$

The *polytropic* process is defined by

$$\frac{p}{\rho^n} = \text{const} \tag{6.1.19}$$

and is an approximation to certain actual processes in which p would plot substantially as a straight line against ρ on log-log paper. This relationship is frequently used to calculate the work when the polytropic process is reversible, by substitution into the relation $W = \int p \, d\mathcal{v}$. Heat transfer occurs in a reversible polytropic process except when $n = k$, the isentropic case.

EXAMPLE 6.1 Express R in kilocalories per kilogram per kelvin for helium.
From Table C.3, $R = 2077$ m·N/kg·K; therefore

$$R = (2077 \text{ m} \cdot \text{N/kg} \cdot \text{K}) \frac{1 \text{ kcal}}{4187 \text{ m} \cdot \text{N}} = 0.496 \text{ kcal/kg} \cdot \text{K}$$

EXAMPLE 6.2 Compute the value of R from the values of k and c_p for air and check in Table C.3.
From Eq. (6.1.8)

$$R = \frac{k-1}{k} c_p = \frac{1.40 - 1.0}{1.40} \times 0.240 \text{ Btu/lb}_m \cdot {}^\circ\text{R} = 0.0686 \text{ Btu/lb}_m \cdot {}^\circ\text{R}$$

By converting from Btu to foot-pounds

$$R = 0.0686 \text{ Btu/lb}_m \cdot {}^\circ\text{R} \times 778 \text{ ft} \cdot \text{lb/Btu} = 53.3 \text{ ft} \cdot \text{lb/lb}_m \cdot {}^\circ\text{R}$$

which checks with the value in Table C.3.

EXAMPLE 6.3 Compute the enthalpy change in 5 kg of oxygen when the initial conditions are $p_1 = 130$ kPa abs, $t_1 = 10°C$, and the final conditions are $p_2 = 500$ kPa abs, $t_2 = 95°C$.
Enthalpy is a function of temperature only. By Eq. (6.1.10) the enthalpy change for 5 kg oxygen is

$$H_2 - H_1 = 5 \text{ kg} \times c_p(T_2 - T_1)$$

$$= (5 \text{ kg})(0.219 \text{ kcal/kg} \cdot \text{K})(95 - 10 \text{ K}) = 93.08 \text{ kcal}$$

EXAMPLE 6.4 Determine the entropy change in 4.0 slugs of water vapor when the initial conditions are $p_1 = 6$ psia, $t_1 = 110°F$, and the final conditions are $p_2 = 40$ psia, $t_2 = 38°F$.

From Eq. (6.1.15) and Table C.3

$$s_2 - s_1 = 0.335 \ln \left[\left(\frac{460 + 38}{460 + 110} \right)^{1.33} \left(\frac{40}{6} \right)^{-0.33} \right] = -0.270 \text{ Btu/lb}_m \cdot °R$$

or

$$S_2 - S_1 = -(0.271 \text{ Btu/lb}_m \cdot °R)(4.0 \text{ slugs})(32.17 \text{ lb}_m/\text{slug})$$

$$= -34.7 \text{ Btu/}°R$$

EXAMPLE 6.5 A cylinder containing 2 kg nitrogen at 1.4 kg_f/cm^2 abs and 5°C is compressed isentropically to 3 kg_f/cm^2 abs. Find the final temperature and the work required.

By Eq. (6.1.17)

$$T_2 = T_1 \left(\frac{p_2}{p_1} \right)^{(k-1)/k} = (273 + 5 \text{ K}) \left(\frac{3}{1.4} \right)^{(1.4-1)/1.4} = 345.6 \text{ K} = 72.6°C$$

From the principle of conservation of energy, the work done on the gas must equal its increase in internal energy, since there is no heat transfer in an isentropic process; i.e.,

$$u_2 - u_1 = c_v(T_2 - T_1) \text{ kcal/kg} = \text{work/kg}$$

or

$$\text{Work} = (2 \text{ kg})(0.177 \text{ kcal/kg} \cdot \text{K})(345.6 - 278 \text{ K}) = 23.93 \text{ kcal}$$

EXAMPLE 6.6 3.0 slugs of air are involved in a reversible polytropic process in which the initial conditions $p_1 = 12$ psia, $t_1 = 60°F$ change to $p_2 = 20$ psia, and volume $\mathcal{U} = 1011$ ft³. Determine (a) the formula for the process, (b) the work done on the air, (c) the amount of heat transfer, and (d) the entropy change.

$$(a) \quad \rho_1 = \frac{p_1}{RT_1} = \frac{12 \times 144}{53.3 \times 32.17(460 + 60)} = 0.00194 \text{ slugs/ft}^3$$

R was converted to foot-pounds per slug per degree Rankine by multiplying by

32.17. Also

$$\rho_2 = \tfrac{3}{1011} = 0.002967 \text{ slugs/ft}^3$$

From Eq. (6.1.19)

$$\frac{p_1}{\rho_1{}^n} = \frac{p_2}{\rho_2{}^n}$$

$$n = \frac{\ln{(p_2/p_1)}}{\ln{(\rho_2/\rho_1)}} = \frac{\ln{\tfrac{20}{12}}}{\ln{(0.002967/0.00194)}} = 1.20$$

hence

$$\frac{p}{\rho^{1.2}} = \text{const}$$

describes the polytropic process.

    (*b*)   Work of expansion is

$$W = \int_{\mathcal{V}_1}^{\mathcal{V}_2} p \, d\mathcal{V}$$

This is the work done by the gas on its surroundings. Since

$$p_1\mathcal{V}_1{}^n = p_2\mathcal{V}_2{}^n = p\mathcal{V}^n$$

by substituting into the integral,

$$W = p_1\mathcal{V}_1{}^n \int_{\mathcal{V}_1}^{\mathcal{V}_2} \frac{d\mathcal{V}}{\mathcal{V}^n} = \frac{p_2\mathcal{V}_2 - p_1\mathcal{V}_1}{1 - n} = \frac{mR}{1 - n}(T_2 - T_1)$$

if m is the mass of gas. $\mathcal{V}_2 = 1011 \text{ ft}^3$ and

$$\mathcal{V}_1 = \mathcal{V}_2 \left(\frac{p_2}{p_1}\right)^{1/n} = 1011(\tfrac{20}{12})^{1/1.2} = 1547 \text{ ft}^3$$

Then

$$W = \frac{20 \times 144 \times 1011 - 12 \times 144 \times 1548}{1 - 1.2} = -1{,}184{,}000 \text{ ft} \cdot \text{lb}$$

Hence the work done on the gas is 1,184,000 ft·lb.

(c)   From the first law of thermodynamics the heat added minus the work done by the gas must equal the increase in internal energy; i.e.,

$$Q_H - W = U_2 - U_1 = c_v m (T_2 - T_1)$$

First

$$T_2 = \frac{p_2}{p_2 R} = \frac{20 \times 144}{0.002965 \times 53.3 \times 32.17} = 566°R$$

Then

$$Q_H = -\frac{1,184,000}{778} + 0.171 \times 32.17 \times 3 (566 - 520)$$

$$= -761 \text{ Btu}$$

761 Btu was transferred from the mass of air.

(d)   From Eq. (6.1.14) the entropy change is computed:

$$s_2 - s_1 = 0.171 \ln \left[\frac{20}{12} \left(\frac{0.00194}{0.002965} \right)^{1.4} \right] = -0.01420 \text{ Btu/lb}_m \cdot °R$$

and

$$S_2 - S_1 = -0.01420 \times 3 \times 32.17 = -1.370 \text{ Btu/°R}$$

A rough check on the heat transfer can be made by using Eq. (3.7.5), by using an average temperature $T = (520 + 566)/2 = 543$, and by remembering that the losses are zero in a reversible process.

$$Q_H = T(S_2 - S_1) = 543(-1.386) = -753 \text{ Btu}$$

6.2 SPEED OF A SOUND WAVE; MACH NUMBER

The speed of a small disturbance in a conduit can be determined by application of the momentum equation and the continuity equation.   The question is first raised whether a *stationary* small change in velocity, pressure, and density can occur in a channel.   By referring to Fig. 6.1, the continuity equation can be written

$$\rho V A = (\rho + d\rho)(V + dV)A$$

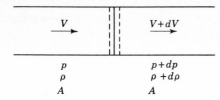

Fig. 6.1 Steady flow in prismatic channel with sudden small change in velocity, pressure, and density.

in which A is the cross-sectional area of channel. The equation can be reduced to

$$\rho \, dV + V \, d\rho = 0$$

When the momentum equation [Eq. (3.11.2)] is applied to the control volume within the dotted lines,

$$pA - (p + dp)A = \rho VA(V + dV - V)$$

or

$$dp = -\rho V \, dV$$

If $\rho \, dV$ is eliminated between the two equations,

$$V^2 = \frac{dp}{d\rho} \tag{6.2.1}$$

Hence, a small disturbance or sudden change in conditions in steady flow can occur only when the particular velocity $V = \sqrt{dp/d\rho}$ exists in the conduit. Now this problem can be converted to the unsteady flow of a small disturbance through still fluid by superposing on the whole system and its surroundings the velocity V to the left, since this in no way affects the dynamics of the system. This is called the speed of sound c in the medium. The disturbance from a point source would cause a spherical wave to emanate, but at some distance from the source the wavefront would be essentially linear or one-dimensional. Large disturbances may travel faster than the speed of sound, e.g., a bomb explosion. The equation for speed of sound

$$c = \sqrt{\frac{dp}{d\rho}} \tag{6.2.2}$$

may be expressed in several useful forms. The bulk modulus of elasticity can be introduced:

$$K = -\frac{dp}{d\upsilon/\upsilon}$$

in which υ is the volume of fluid subjected to the pressure change dp. Since

$$\frac{d\upsilon}{\upsilon} = \frac{dv_s}{v_s} = -\frac{d\rho}{\rho}$$

K may be expressed as

$$K = \frac{\rho\, dp}{d\rho}$$

Then, from Eq. (6.2.2),

$$c = \sqrt{\frac{K}{\rho}} \tag{6.2.3}$$

This equation applies to liquids as well as gases.

EXAMPLE 6.7   Carbon tetrachloride has a bulk modulus of elasticity of 11,460 kg_f/cm^2 and a density of 1593 kg/m^3. What is the speed of sound in the medium?

$$c = \sqrt{\frac{K}{\rho}} = \sqrt{\frac{(11{,}460 \text{ kg}_f/\text{cm}^2)\,(9.806 \text{ N/kg}_f)}{1593 \text{ kg/m}^3}} \times 10^4 \text{ cm}^2/\text{m}^2 = 840 \text{ m/s}$$

Since the pressure and temperature changes due to passage of a sound wave are extremely small, the process is almost reversible. Also, the relatively rapid process of passage of the wave, together with the minute temperature changes, makes the process almost adiabatic. In the limit, the process may be considered to be isentropic,

$$p\rho^{-k} = \text{const} \qquad \frac{dp}{d\rho} = \frac{kp}{\rho}$$

and

$$c = \sqrt{\frac{kp}{\rho}} \tag{6.2.4}$$

or, from the perfect-gas law $p = \rho RT$,

$$c = \sqrt{kRT} \tag{6.2.5}$$

which shows that the speed of sound in a perfect gas is a function of its absolute temperature only. In flow of gas through a conduit, the speed of sound generally changes from section to section as the temperature is changed by density changes and friction effects. In isothermal flow the speed of sound remains constant.

The Mach number has been defined as the ratio of velocity of a fluid to the local velocity of sound in the medium,

$$\mathbf{M} = \frac{V}{c} \tag{6.2.6}$$

Squaring the Mach number produces V^2/c^2, which may be interpreted as the ratio of kinetic energy of the fluid to its thermal energy, since kinetic energy is proportional to V^2 and thermal energy is proportional to T. The Mach number is a measure of the importance of compressibility. In an incompressible fluid K is infinite and $\mathbf{M} = 0$. For perfect gases

$$K = kp \tag{6.2.7}$$

when the compression is isentropic.

EXAMPLE 6.8 What is the speed of sound in dry air at sea level when $t = 68°F$ and in the stratosphere when $t = -67°F$?

At sea level, from Eq. (6.2.5),

$$c = \sqrt{1.4 \times 32.17 \times 53.3(460 + 68)} = 1126 \text{ ft/s}$$

and in the stratosphere

$$c = \sqrt{1.4 \times 32.17 \times 53.3(460 - 67)} = 971 \text{ ft/s}$$

6.3 ISENTROPIC FLOW

Frictionless adiabatic, or isentropic, flow is an ideal that cannot be reached in the flow of real gases. It is approached, however, in flow through transitions, nozzles, and venturi meters where friction effects are minor, owing to the short distances traveled, and heat transfer is minor because the changes that a particle undergoes are slow enough to keep the velocity and temperature gradients small.[1] The performance of fluid machines is frequently compared with the performance assuming isentropic flow. In this section one-dimensional steady flow of a perfect gas through converging and converging-diverging ducts is studied.

Some very general results can be obtained by use of Euler's equation (3.5.4), neglecting elevation changes,

$$V \, dV + \frac{dp}{\rho} = 0 \tag{6.3.1}$$

and the continuity equation

$$\rho A V = \text{const} \tag{6.3.2}$$

Differentiating $\rho A V$ and then dividing through by $\rho A V$ gives

$$\frac{d\rho}{\rho} + \frac{dV}{V} + \frac{dA}{A} = 0 \tag{6.3.3}$$

From Eq. (6.2.2) dp is obtained and substituted into Eq. (6.3.1), yielding

$$V \, dV + c^2 \frac{d\rho}{\rho} = 0 \tag{6.3.4}$$

Eliminating $d\rho/\rho$ in the last two equations and rearranging give

$$\frac{dA}{dV} = \frac{A}{V} \left(\frac{V^2}{c^2} - 1 \right) = \frac{A}{V} (\mathbf{M}^2 - 1) \tag{6.3.5}$$

The assumptions underlying this equation are that the flow is steady and frictionless. No restrictions as to heat transfer have been imposed. Equation (6.3.5) shows that for subsonic flow ($\mathbf{M} < 1$), dA/dV is always negative; i.e., the channel area must decrease for increasing velocity. As dA/dV is zero for $\mathbf{M} = 1$ only, the velocity keeps increasing until the minimum section or throat is reached, and that is the only section at which sonic flow may occur. Also, for Mach numbers greater than unity (supersonic flow) dA/dV is positive and the area must increase for an increase in velocity. Hence to obtain supersonic steady flow from a fluid at rest in a reservoir, it must first pass through a converging duct and then a diverging duct.

[1] H. W. Liepmann and A. Roshko, "Elements of Gas Dynamics," p. 51, Wiley, New York, 1957.

When the analysis is restricted to isentropic flow, Eq. (6.1.16) may be written

$$p = p_1 \rho_1^{-k} \rho^k \qquad (6.3.6)$$

Differentiating and substituting for dp in Eq. (6.3.1) give

$$V \, dV + k \frac{p_1}{\rho_1^{\,k}} \rho^{k-2} \, d\rho = 0$$

Integration yields

$$\frac{V^2}{2} + \frac{k}{k-1} \frac{p_1}{\rho_1^{\,k}} \rho^{k-1} = \text{const}$$

or

$$\frac{V_1^{\,2}}{2} + \frac{k}{k-1} \frac{p_1}{\rho_1} = \frac{V_2^{\,2}}{2} + \frac{k}{k-1} \frac{p_2}{\rho_2} \qquad (6.3.7)$$

Equation (6.3.7) can be derived from Eq. (3.7.2) for adiabatic flow ($dq_H = 0$) using Eq. (6.1.8). This avoids the restriction to isentropic flow. This equation is useful when expressed in terms of temperature; from $p = \rho RT$

$$\frac{V_1^{\,2}}{2} + \frac{k}{k-1} RT_1 = \frac{V_2^{\,2}}{2} + \frac{k}{k-1} RT_2 \qquad (6.3.8)$$

For adiabatic flow from a reservoir where conditions are given by p_0, ρ_0, T_0, at any other section

$$\frac{V^2}{2} = \frac{kR}{k-1} (T_0 - T) \qquad (6.3.9)$$

In terms of the local Mach number V/c, with $c^2 = kRT$,

$$\mathbf{M}^2 = \frac{V^2}{c^2} = \frac{2kR(T_0 - T)}{(k-1)kRT} = \frac{2}{k-1}\left(\frac{T_0}{T} - 1\right)$$

or

$$\frac{T_0}{T} = 1 + \frac{k-1}{2} \mathbf{M}^2 \qquad (6.3.10)$$

From Eqs. (6.3.10) and (6.1.17), which now restrict the following equations to isentropic flow,

$$\frac{p_0}{p} = \left(1 + \frac{k-1}{2}\mathbf{M}^2\right)^{k/(k-1)} \tag{6.3.11}$$

and

$$\frac{\rho_0}{\rho} = \left(1 + \frac{k-1}{2}\mathbf{M}^2\right)^{1/(k-1)} \tag{6.3.12}$$

Flow conditions are termed critical at the throat section when the velocity there is sonic. Sonic conditions are marked with an asterisk. $\mathbf{M} = 1$; $c* = V* = \sqrt{kRT^*}$. By applying Eqs. (6.3.10) to (6.3.12) to the throat section for critical conditions (for $k = 1.4$ in the numerical portion),

$$\frac{T^*}{T_0} = \frac{2}{k+1} = 0.833 \qquad k = 1.40 \tag{6.3.13}$$

$$\frac{p^*}{p_0} = \left(\frac{2}{k+1}\right)^{k/(k-1)} = 0.528 \qquad k = 1.40 \tag{6.3.14}$$

$$\frac{\rho^*}{\rho_0} = \left(\frac{2}{k+1}\right)^{1/(k-1)} = 0.634 \qquad k = 1.40 \tag{6.3.15}$$

These relations show that for airflow, the absolute temperature drops about 17 percent from reservoir to throat, the critical pressure is 52.8 percent of the reservoir pressure, and the density is reduced by about 37 percent.

The variation of area with the Mach number for the critical case is obtained by use of the continuity equation and Eqs. (6.3.10) to (6.3.15). First

$$\rho A V = \rho^* A^* V^* \tag{6.3.16}$$

in which A^* is the minimum, or throat, area. Then

$$\frac{A}{A^*} = \frac{\rho^*}{\rho}\frac{V^*}{V} \tag{6.3.17}$$

Now, $V^* = c^* = \sqrt{kRT^*}$, and $V = c\mathbf{M} = \mathbf{M}\sqrt{kRT}$, so that

$$\frac{V^*}{V} = \frac{1}{\mathbf{M}}\sqrt{\frac{T^*}{T}} = \frac{1}{\mathbf{M}}\sqrt{\frac{T^*}{T_0}}\sqrt{\frac{T_0}{T}} = \frac{1}{\mathbf{M}}\left\{\frac{1+[(k-1)/2]\mathbf{M}^2}{(k+1)/2}\right\}^{1/2} \tag{6.3.18}$$

by use of Eqs. (6.3.13) and (6.3.10). In a similar manner

$$\frac{\rho^*}{\rho} = \frac{\rho^*}{\rho_0}\frac{\rho_0}{\rho} = \left\{\frac{1 + [(k-1)/2]\mathbf{M}^2}{(k+1)/2}\right\}^{1/(k-1)} \tag{6.3.19}$$

By substituting the last two equations into Eq. (6.3.17),

$$\frac{A}{A^*} = \frac{1}{\mathbf{M}}\left\{\frac{1 + [(k-1)/2]\mathbf{M}^2}{(k+1)/2}\right\}^{(k+1)/2(k-1)} \tag{6.3.20}$$

which yields the variation of area of duct in terms of Mach number. A/A^* is never less than unity, and for any value greater than unity there will be two values of Mach number, one less than and one greater than unity. For gases with $k = 1.40$, Eq. (6.3.20) reduces to

$$\frac{A}{A^*} = \frac{1}{\mathbf{M}}\left(\frac{5 + \mathbf{M}^2}{6}\right)^3 \qquad k = 1.40 \tag{6.3.21}$$

The maximum mass flow rate $\dot{m}_{\max}$ can be expressed in terms of the throat area and reservoir conditions:

$$\dot{m}_{\max} = \rho^* A^* V^* = \rho_0 \left(\frac{2}{k+1}\right)^{1/(k-1)} A^* \sqrt{\frac{kR2T_0}{k+1}}$$

by Eqs. (6.3.15) and (6.3.13). Replacing ρ_0 by p_0/RT_0 gives

$$\dot{m}_{\max} = \frac{A^* p_0}{\sqrt{T_0}} \sqrt{\frac{k}{R}\left(\frac{2}{k+1}\right)^{(k+1)/(k-1)}} \tag{6.3.22}$$

For $k = 1.40$ this reduces to

$$\dot{m}_{\max} = 0.686 \frac{A^* p_0}{\sqrt{RT_0}} \tag{6.3.23}$$

which shows that the mass flow rate varies linearly as A^* and p_0 and varies inversely as the square root of the absolute temperature.

For subsonic flow throughout a converging-diverging duct, the velocity at the throat must be less than sonic velocity, or $\mathbf{M}_t < 1$ with subscript t indicating the throat section. The mass rate of flow $\dot{m}$ is obtained from

$$\dot{m} = \rho V A = A \sqrt{2p_0\rho_0 \frac{k}{k-1}\left(\frac{p}{p_0}\right)^{2/k}\left[1 - \left(\frac{p}{p_0}\right)^{(k-1)/k}\right]} \tag{6.3.24}$$

which is derived from Eqs. (6.3.9) and (6.3.6) and the perfect-gas law. This equation holds for any section and is applicable as long as the velocity at the throat is subsonic. It may be applied to the throat section, and for this section, from Eq. (6.3.14),

$$\frac{p_t}{p_0} \geq \left(\frac{2}{k+1}\right)^{k/(k-1)}$$

p_t is the throat pressure. When the equals sign is used in the expression, Eq. (6.3.24) reduces to Eq. (6.3.22).

For maximum mass flow rate, the flow downstream from the throat may be either supersonic or subsonic, depending upon the downstream pressure. Substituting Eq. (6.3.22) for $\dot{m}$ in Eq. (6.3.24) and simplifying gives

$$\left(\frac{p}{p_0}\right)^{2/k}\left[1-\left(\frac{p}{p_0}\right)^{(k-1)/k}\right] = \frac{k-1}{2}\left(\frac{2}{k+1}\right)^{(k+1)/(k-1)}\left(\frac{A^*}{A}\right)^2 \qquad (6.3.25)$$

A may be taken as the outlet area and p as the outlet pressure. For a given A^*/A (less than unity) there will be two values of p/p_0 between zero and unity, the upper value for subsonic flow through the diverging duct and the lower value for supersonic flow through the diverging duct. For all other pressure ratios less than the upper value complete isentropic flow is impossible and shock waves form in or just downstream from the diverging duct. They are briefly discussed in the following section.

Appendix Table C.4 is quite helpful in solving isentropic flow problems for $k = 1.4$. Equations (6.3.10), (6.3.11), (6.3.12), and (6.3.21) are presented in tabular form.

EXAMPLE 6.9   A preliminary design of a wind tunnel to produce Mach number 3.0 at the exit is desired. The mass flow rate is 1 kg/s for $p_0 = 0.9$ kg$_f$/cm^2, $t_0 = 25°$C. Determine (a) the throat area, (b) the outlet area, and (c) the velocity, pressure, temperature, and density at the outlet.

(a)   The throat area can be determined from Eq. (6.3.23):

$$A^* = \frac{\dot{m}_{\max} \sqrt{RT_0}}{0.686p_0} = \frac{1 \text{ kg/s} \sqrt{(287 \text{ m} \cdot N/\text{kg} \cdot K)(273 + 25 \text{ K})}}{(0.686 \times 0.9 \text{ kg}_f/\text{cm}^2)(9.806 \text{ N/kg}_f)(10^4 \text{ cm}^2/\text{m}^2)}$$

$$= 0.00483 \text{ m}^2$$

(b)   The area of outlet is determined from Table C.4:

$$\frac{A}{A^*} = 4.23 \qquad A = 4.23 \times 0.00483 \text{ m}^2 = 0.0204 \text{ m}^2$$

(c)   From Table C.4

$$\frac{p}{p_0} = 0.027 \qquad \frac{\rho}{\rho_0} = 0.076 \qquad \frac{T}{T_0} = 0.357$$

From the gas law

$$\rho_0 = \frac{p_0}{RT_0} = \frac{(0.9 \text{ kg}_f/\text{cm}^2)\,(9.806 \text{ N/kg}_f)\,(10^4 \text{ cm}^2/\text{m}^2)}{(287 \text{ m} \cdot \text{N/kg} \cdot \text{K})\,(273 + 25 \text{ K})} = 1.0319 \text{ kg/m}^3$$

hence at the exit

$$p = 0.027 \times 0.9 \text{ kg}_f/\text{cm}^2 = 0.0243 \text{ kg}_f/\text{cm}^2$$

$$T = 0.357(273 + 25 \text{ K}) = 106.39 \text{ K} = -166.6°\text{C}$$

$$\rho = 0.076 \times 1.0319 \text{ kg/m}^3 = 0.0784 \text{ kg/m}^3$$

From the continuity equation

$$V = \frac{\dot{m}_{\max}}{\rho A} = \frac{1 \text{ kg/s}}{(0.078 \text{ kg/m}^3)\,(0.0204 \text{ m}^2)} = 628.5 \text{ m/s}$$

EXAMPLE 6.10 A converging-diverging air duct has a throat cross section of 0.40 ft² and an exit cross section of 1.0 ft². Reservoir pressure is 30 psia, and temperature is 60°F. Determine the range of Mach numbers and the pressure range at the exit for isentropic flow. Find the maximum flow rate.

From Table C.4 [Eq. (6.3.21)] **M** = 2.44 and 0.24. Each of these values of Mach number at the exit is for critical conditions; hence the Mach number range for isentropic flow is 0 to 0.24 and the one value 2.44.

From Table C.4 [Eq. (6.3.11)] for **M** = 2.44, p = 30 × 0.064 = 1.92 psia, and for **M** = 0.24, p = 30 × 0.961 = 28.83 psia. The downstream pressure range is then from 28.83 to 30 psia, and the isolated point is 1.92 psia. The maximum mass flow rate is determined from Eq. (6.3.23):

$$\dot{m}_{\max} = \frac{0.686 \times 0.40 \times 30 \times 144}{\sqrt{53.3 \times 32.17(460 + 60)}} = 1.255 \text{ slugs/s} = 40.4 \text{ lb}_m/\text{s}$$

EXAMPLE 6.11 A converging-diverging duct in an air line downstream from a reservoir has a 5-cm-diameter throat. Determine the mass rate of flow when

$p_0 = 8 \text{ kg}_f/\text{cm}^2 \text{ abs}$, $t_0 = 33°\text{C}$, and $p = 5 \text{ kg}_f/\text{cm}^2 \text{ abs}$ at the throat.

$$\rho_0 = \frac{p_0}{RT_0} = \frac{(8 \text{ kg}_f/\text{cm}^2)\,(9.806 \text{ N/kg}_f)\,(10^4 \text{ cm}^2/\text{m}^2)}{(287 \text{ m}\cdot\text{N/kg}\cdot\text{K})\,(273 + 33 \text{ K})} = 8.933 \text{ kg/m}^3$$

From Eq. (6.3.24)

$$\dot{m} = \frac{\pi}{4}\,(0.05 \text{ m})^2$$

$$\times \sqrt{(2 \times 8 \text{ kg}_f/\text{cm}^2)(9.806 \text{ N/kg}_f)(10^4\,\text{cm}^2/\text{m}^2)\,\frac{1.4}{1.4-1}\left(\frac{5}{8}\right)^{2/1.4}\left[1 - \left(\tfrac{5}{8}\right)^{0.4/1.4}\right]}$$

$$= 1.166 \text{ kg/s}$$

6.4 SHOCK WAVES

In one-dimensional flow the only type of shock wave that can occur is a normal compression shock wave, as illustrated in Fig. 6.2. For a complete discussion of converging-diverging flow for all downstream pressure ranges,[1] oblique shock waves must be taken into account as they occur at the exit. In the preceding section isentropic flow was shown to occur throughout a converging-diverging tube for a range of downstream pressures in which the flow was subsonic throughout and for one downstream pressure for supersonic flow through the diffuser (diverging portion). In this section the normal shock wave in a diffuser is studied, with isentropic flow throughout the tube, except for the shock-wave surface. The shock wave occurs in supersonic flow and reduces the flow to subsonic flow, as proved in the following section. It has very little thickness, of the order of the molecular mean free path of the gas. The controlling equations for adiabatic flow are (Fig. 6.2)

Continuity: $$G = \frac{\dot{m}}{A} = \rho_1 V_1 = \rho_2 V_2 \tag{6.4.1}$$

Energy: $$\frac{V_1^2}{2} + h_1 = \frac{V_2^2}{2} + h_2 = h_0 = \frac{V^2}{2} + \frac{k}{k-1}\frac{p}{\rho} \tag{6.4.2}$$

which are obtained from Eq. (3.7.1) for no change in elevation, no heat trans-

[1] H. W. Liepmann and A. Roshko, "Elements of Gas Dynamics," Wiley, New York, 1957.

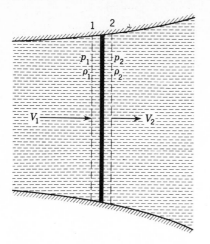

Fig. 6.2 Normal compression shock wave.

fer, and no work done. $h = u + p/\rho = c_p T$ is the enthalpy, and h_0 is the value of stagnation enthalpy, i.e., its value in the reservoir or where the fluid is at rest. Equation (6.4.2) holds for real fluids and is valid both upstream and downstream from a shock wave. The momentum equation (3.11.2) for a control volume between sections 1 and 2 becomes

$$(p_1 - p_2)A = \rho_2 A V_2{}^2 - \rho_1 A V_1{}^2$$

or

$$p_1 + \rho_1 V_1{}^2 = p_2 + \rho_2 V_2{}^2 \tag{6.4.3}$$

For given upstream conditions h_1, p_1, V_1, ρ_1, the three equations are to be solved for p_2, ρ_2, V_2. The equation of state for a perfect gas is also available for use, $p = \rho R T$. The value of p_2 is

$$p_2 = \frac{1}{k+1}\left[2\rho_1 V_1{}^2 - (k-1)p_1\right] \tag{6.4.4}$$

Once p_2 is determined, by combination of the continuity and momentum equations

$$p_1 + \rho_1 V_1{}^2 = p_2 + \rho_1 V_1 V_2 \tag{6.4.5}$$

V_2 is readily obtained. Finally ρ_2 is obtained from the continuity equation.

For given upstream conditions, with $M_1 > 1$, the values of p_2, V_2, ρ_2, and $M_2 = V_2/\sqrt{kp_2/\rho_2}$ exist and $M_2 < 1$. By eliminating V_1 and V_2 between Eqs. (6.4.1), (6.4.2), and (6.4.3) the Rankine-Hugoniot equations are obtained:

$$\frac{p_2}{p_1} = \frac{[(k+1)/(k-1)](\rho_2/\rho_1) - 1}{[(k+1)/(k-1)] - \rho_2/\rho_1} \tag{6.4.6}$$

and

$$\frac{\rho_2}{\rho_1} = \frac{1 + [(k+1)/(k-1)]p_2/p_1}{[(k+1)/(k-1)] + p_2/p_1} = \frac{V_1}{V_2} \tag{6.4.7}$$

These equations, relating conditions on either side of the shock wave, take the place of the isentropic relation, Eq. (6.1.16), $p\rho^{-k} = \text{const.}$

From Eq. (6.4.2), the energy equation,

$$\frac{V^2}{2} + \frac{k}{k-1}\frac{p}{\rho} = \frac{c^{*2}}{2} + \frac{c^{*2}}{k-1} = \frac{k+1}{k-1}\frac{c^{*2}}{2} \tag{6.4.8}$$

since the equation holds for all points in adiabatic flow without change in elevation, and $c^* = \sqrt{kp^*/\rho^*}$ is the velocity of sound. Dividing Eq. (6.4.3) by Eq. (6.4.1) gives

$$V_1 - V_2 = \frac{p_2}{\rho_2 V_2} - \frac{p_1}{\rho_1 V_1}$$

and by eliminating p_2/ρ_2 and p_1/ρ_1 by use of Eq. (6.4.8) leads to

$$V_1 - V_2 = (V_1 - V_2)\left[\frac{c^{*2}(k+1)}{2kV_1V_2} + \frac{k-1}{2k}\right] \tag{6.4.9}$$

which is satisfied by $V_1 = V_2$ (no shock wave) or by

$$V_1V_2 = c^{*2} \tag{6.4.10}$$

It may be written

$$\frac{V_1}{c^*}\frac{V_2}{c^*} = 1 \tag{6.4.11}$$

When V_1 is greater than c^*, the upstream Mach number is greater than unity and V_2 is less than c^*, and so the final Mach number is less than unity, and vice versa. It is shown in the following section that the process can occur only from supersonic upstream to subsonic downstream.

By use of Eq. (6.1.14), together with Eqs. (6.4.4), (6.4.6), and (6.4.7), an expression for change of entropy across a normal shock wave may be obtained in terms of M_1 and k. From Eq. (6.4.4)

$$\frac{p_2}{p_1} = \frac{1}{k+1}\left[\frac{2k\rho_1 V_1^2}{kp_1} - (k-1)\right] \tag{6.4.12}$$

Since $c_1^2 = kp_1/\rho_1$ and $M_1 = V_1/c_1$, from Eq. (6.4.12),

$$\frac{p_2}{p_1} = \frac{2kM_1^2 - (k-1)}{k+1} \tag{6.4.13}$$

Placing this value of p_2/p_1 in Eq. (6.4.7) yields

$$\frac{\rho_2}{\rho_1} = \frac{M_1^2(k+1)}{2 + M_1^2(k-1)}$$

Substituting these pressure and density ratios into Eq. (6.1.14) gives

$$s_2 - s_1 = c_v \ln\left\{\frac{2kM_1^2 - k + 1}{k+1}\left[\frac{2 + M_1^2(k-1)}{M_1^2(k+1)}\right]^k\right\} \tag{6.4.14}$$

By substitution of $M_1 > 1$ into this equation for the appropriate value of k, the entropy may be shown to increase across the shock wave, indicating that the normal shock may proceed from supersonic flow upstream to subsonic flow downstream. Substitution of values of $M_1 < 1$ into Eq. (6.4.14) has no meaning, since Eq. (6.4.13) yields a negative value of the ratio p_2/p_1. The equations for derivation of the gas tables in Table C.5 are developed in the following treatment. From Eq. (6.3.10), which holds for the adiabatic flow across the shock wave,

$$\frac{T_0}{T_1} = 1 + \frac{k-1}{2}M_1^2 \qquad \frac{T_0}{T_2} = 1 + \frac{k-1}{2}M_2^2$$

hence

$$\frac{T_2}{T_1} = \frac{1 + [(k-1)/2]M_1^2}{1 + [(k-1)/2]M_2^2} \tag{6.4.15}$$

Since $V_1 = M_1\sqrt{kRT_1}$ and $V_2 = M_2\sqrt{kRT_2}$, use of the momentum equation

(6.4.3) gives

$$p_1 + \rho_1 \mathbf{M}_1{}^2 kRT_1 = p_2 + \rho_2 \mathbf{M}_2{}^2 kRT_2$$

and

$$\frac{p_2}{p_1} = \frac{1 + k\mathbf{M}_1{}^2}{1 + k\mathbf{M}_2{}^2} \tag{6.4.16}$$

Now, from the continuity equation (6.4.1) and the gas law

$$\frac{\rho_2}{\rho_1} = \frac{p_2 T_1}{p_1 T_2} = \frac{V_1}{V_2} = \frac{\mathbf{M}_1}{\mathbf{M}_2}\sqrt{\frac{T_1}{T_2}} \quad \text{or} \quad \frac{\mathbf{M}_1}{\mathbf{M}_2} = \frac{p_2}{p_1}\sqrt{\frac{T_1}{T_2}} \tag{6.4.17}$$

Eliminating T_2/T_1 and p_2/p_1 in Eqs. (6.4.15), (6.4.16), and (6.4.17) gives

$$\frac{\mathbf{M}_1}{\mathbf{M}_2} = \frac{1 + k\mathbf{M}_1{}^2}{1 + k\mathbf{M}_2{}^2}\sqrt{\frac{1 + [(k-1)/2]\mathbf{M}_2{}^2}{1 + [(k-1)/2]\mathbf{M}_1{}^2}} \tag{6.4.18}$$

which can be solved for $\mathbf{M}_2$

$$\mathbf{M}_2{}^2 = \frac{\mathbf{M}_1{}^2 + 2/(k-1)}{[2k/(k-1)]\mathbf{M}_1{}^2 - 1} \tag{6.4.19}$$

When Eq. (6.4.19) is substituted back into Eq. (6.4.15), T_2/T_1 is determined in terms of $\mathbf{M}_1$.

To determine the ratio of stagnation pressures across the normal shock wave, one may write

$$\frac{p_{0_2}}{p_{0_1}} = \frac{p_{0_2}}{p_2}\frac{p_2}{p_1}\frac{p_1}{p_{0_1}} \tag{6.4.20}$$

Now by use of Eqs. (6.3.11) and (6.4.13)

$$\frac{p_{0_2}}{p_{0_1}} = \left\{\frac{1 + [(k-1)/2]\mathbf{M}_2{}^2}{1 + [(k-1)/2]\mathbf{M}_1{}^2}\right\}^{k/(k-1)} \frac{2k\mathbf{M}_1{}^2 - (k-1)}{k+1} \tag{6.4.21}$$

By use of Eq. (6.4.19) with Eq. (6.4.21) the stagnation-pressure ratio is expressed in terms of $\mathbf{M}_1$.

In the next section the shock wave is examined further by introduction of Fanno and Rayleigh lines.

EXAMPLE 6.12 If a normal shock wave occurs in the flow of helium, $p_1 = 1$ psia, $t_1 = 40°F$, $V_1 = 4500$ ft/s, find p_2, ρ_2, V_2, and t_2.

From Table C.3, $R = 386$, $k = 1.66$, and

$$\rho_1 = \frac{p_1}{RT_1} = \frac{1 \times 144}{386 \times 32.17(460 + 40)} = 0.0000232 \text{ slugs/ft}^3$$

From Eq. (6.4.4)

$$p_2 = \frac{1}{1.66 + 1} [2 \times 0.0000232 \times 4500^2 - (1.66 - 1) \times 144 \times 1]$$

$$= 317 \text{ lb/ft}^2 \text{ abs}$$

From Eq. (6.4.5)

$$V_2 = V_1 - \frac{p_2 - p_1}{\rho_1 V_1} = 4500 - \frac{317 - 144}{4500 \times 0.0000232} = 2843 \text{ ft/s}$$

From Eq. (6.4.1)

$$\rho_2 = \rho_1 \frac{V_1}{V_2} = \frac{0.000746}{32.17} \times \frac{4500}{2843} = 0.0000367 \text{ slugs/ft}^3$$

and

$$t_2 = T_2 - 460 = \frac{p_2}{\rho_2 R} - 460 = \frac{317}{0.0000367 \times 32.17 \times 386} - 460 = 236°F$$

6.5 FANNO AND RAYLEIGH LINES

To examine more closely the nature of the flow change in the short distance across a shock wave, where the area may be considered constant, the continuity and energy equations are combined for steady, frictional, adiabatic flow. By considering upstream conditions fixed, that is, p_1, V_1, ρ_1, a plot may be made of all possible conditions at section 2, Fig. 6.2. The lines on such a plot for constant mass flow G are called *Fanno lines*. The most revealing plot is that of enthalpy against entropy, i.e., an *hs* diagram.

The entropy equation for a perfect gas, Eq. (6.1.14), is

$$s - s_1 = c_v \ln \left[\frac{p}{p_1} \left(\frac{\rho_1}{\rho} \right)^k \right] \tag{6.5.1}$$

The energy equation for adiabatic flow with no change in elevation, from Eq. (6.4.2), is

$$h_0 = h + \frac{V^2}{2} \tag{6.5.2}$$

and the continuity equation for no change in area, from Eq. (6.4.1), is

$$G = \rho V \tag{6.5.3}$$

The equation of state, linking h, p, and ρ, is

$$h = c_p T = \frac{c_p p}{R\rho} \tag{6.5.4}$$

By eliminating p, ρ, and V from the four equations,

$$s = s_1 + c_v \ln \left[\frac{\rho_1{}^k}{p_1} \frac{R}{c_p} \left(\frac{\sqrt{2}}{G} \right)^{k-1} \right] + c_v \ln \left[h(h_0 - h)^{(k-1)/2} \right] \tag{6.5.5}$$

which is shown on Fig. 6.3 (not to scale). To find the conditions for maximum entropy, Eq. (6.5.5) is differentiated with respect to h and ds/dh set equal to zero. By indicating by subscript a values at the maximum entropy point,

$$\frac{ds}{dh} = 0 = \frac{1}{h_a} - \frac{k-1}{2} \frac{1}{h_0 - h_a} \qquad \text{or} \qquad h_a = \frac{2}{k+1} h_0$$

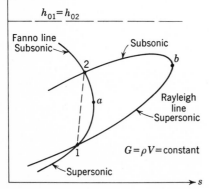

Fig. 6.3 Fanno and Rayleigh lines.

After substituting this into Eq. (6.5.2) to find V_a,

$$h_0 = \frac{k+1}{2} h_a = h_a + \frac{V_a{}^2}{2}$$

and

$$V_a{}^2 = (k-1)h_a = (k-1)c_pT_a = (k-1)\frac{kR}{k-1}T_a = kRT_a = c_a{}^2 \quad (6.5.6)$$

Hence the maximum entropy at point a is for **M** $= 1$, or sonic conditions. For $h > h_a$ the flow is subsonic, and for $h < h_a$ the flow is supersonic. The two conditions, before and after the shock, must lie on the proper Fanno line for the area at which the shock wave occurs. The momentum equation was not used to determine the Fanno line, and so the complete solution is not determined yet.

Rayleigh line

Conditions before and after the shock must also satisfy the momentum and continuity equations. Assuming constant upstream conditions and constant area, Eqs. (6.5.1), (6.5.3), (6.5.4), and (6.4.1) are used to determine the *Rayleigh line*. Eliminating V in the continuity and momentum equations gives

$$p + \frac{G^2}{\rho} = \text{const} = B \qquad (6.5.7)$$

Next eliminating p from this equation and the entropy equation gives

$$s = s_1 + c_v \ln \frac{\rho_1{}^k}{p_1} + c_v \ln \frac{B - G^2/\rho}{\rho^k} \qquad (6.5.8)$$

Enthalpy may be expressed as a function of ρ and upstream conditions, from Eq. (6.5.7):

$$h = c_p T = c_p \frac{p}{R\rho} = \frac{c_p}{R}\frac{1}{\rho}\left(B - \frac{G^2}{\rho}\right) \qquad (6.5.9)$$

The last two equations determine s and h in terms of the parameter ρ and plot on the hs diagram as indicated in Fig. 6.3. This is a *Rayleigh line*.

The value of maximum entropy is found by taking $ds/d\rho$ and $dh/d\rho$ from the equations; then by division and equating to zero, using subscript b for maximum point:

$$\frac{ds}{dh} = \frac{c_v}{c_p} R\rho_b \frac{G^2/[\rho_b(B - G^2/\rho_b)] - k}{2G^2/\rho_b - B} = 0$$

To satisfy this equation, the numerator must be zero and the denominator not zero. The numerator set equal to zero yields

$$k = \frac{G^2}{\rho_b(B - G^2/\rho_b)} = \frac{\rho_b^2 V_b^2}{\rho_b p_b} \quad \text{or} \quad V_b^2 = \frac{kp_b}{\rho_b} = c_b^2$$

that is, **M** = 1. For this value the denominator is not zero. Again, as with the Fanno line, sonic conditions occur at the point of maximum entropy. Since the flow conditions must be on both curves, just before and just after the shock wave, it must suddenly change from one point of intersection to the other. The entropy cannot decrease, as no heat is being transferred from the flow, so that the upstream point must be the intersection with least entropy. In all gases investigated the intersection in the subsonic flow has the greater entropy. Thus the shock occurs from supersonic to subsonic.

The Fanno and Rayleigh lines are of value in analyzing flow in constant-area ducts. These are treated in Secs. 6.6 and 6.7.

Converging-diverging nozzle flow

Following the presentation of Liepmann and Roshko (see references at end of chapter), the various flow situations for converging-diverging nozzles are investigated. Equation (6.3.20) gives the relation between area ratio and Mach number for isentropic flow throughout the nozzle. By use of Eq. (6.3.11) the area ratio is obtained as a function of pressure ratio

$$\frac{A^*}{A} = \frac{\rho V}{\rho^* V^*} = \frac{[1 - (p/p_0)^{(k-1)/k}]^{1/2}(p/p_0)^{1/k}}{[(k-1)/2]^{1/2}[2/(k+1)]^{(k+1)/2(k-1)}} \tag{6.5.10}$$

Figure 6.4 is a plot of area ratio vs. pressure ratio and **M**, good only for isentropic flow ($k = 1.4$).

By use of the area ratios the distribution of pressure and Mach number along a given converging-diverging nozzle can now be plotted. Figure 6.5 illustrates the various flow conditions that may occur. If the downstream pressure is p_c or greater, isentropic subsonic flow occurs throughout the

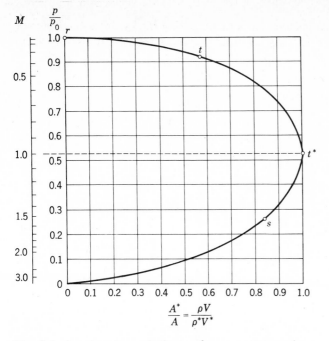

Fig. 6.4 Isentropic relations for a converging-diverging nozzle ($k = 1.4$). (*By permission, from H. W. Liepmann and A. Roshko, "Elements of Gas Dynamics," John Wiley & Sons, Inc., New York, 1957.*)

tube. If the pressure is at j, isentropic flow occurs throughout, with subsonic flow to the throat, sonic flow at the throat, and supersonic flow downstream. For downstream pressure between c and f, a shock wave occurs within the nozzle, as shown for p_d. For pressure at p_f a normal shock wave occurs at the exit, and for pressures between p_f and p_j oblique shock waves at the exit develop.

6.6 ADIABATIC FLOW WITH FRICTION IN CONDUITS

Gas flow through a pipe or constant-area duct is analyzed in this section subject to the following assumptions:

1. Perfect gas (constant specific heats).
2. Steady, one-dimensional flow.
3. Adiabatic flow (no heat transfer through walls).

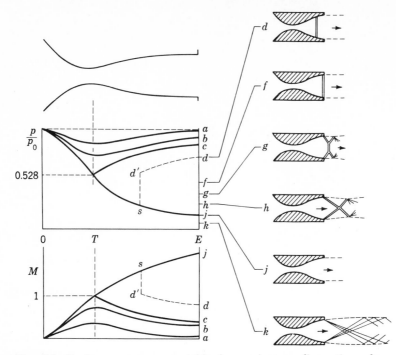

Fig. 6.5 Various pressure and Mach number configurations for flow through a nozzle. (*By permission from H. W. Liepmann and A. Roshko, "Elements of Gas Dynamics," John Wiley & Sons, Inc., New York, 1957.*)

4. Constant friction factor over length of conduit.
5. Effective conduit diameter D is four times hydraulic radius (cross-sectioned area divided by perimeter).
6. Elevation changes are unimportant compared with friction effects.
7. No work added to or extracted from the flow.

The controlling equations are continuity, energy, momentum, and the equation of state. The Fanno line, developed in Sec. 6.5 and shown in Fig. 6.3, was for constant area and used the continuity and energy equations; hence, it applies to adiabatic flow in a duct of constant area. A particle of gas at the upstream end of the duct may be represented by a point on the appropriate Fanno line for proper stagnation enthalpy h_0 and mass flow rate G per unit area. As the particle moves downstream, its properties change, owing to friction or irreversibilities such that the entropy always increases in adiabatic flow. Thus the point representing these properties moves along the Fanno

Table 6.1

Property	Subsonic flow	Supersonic flow
Velocity V	Increases	Decreases
Mach number **M**	Increases	Decreases
Pressure p	Decreases	Increases
Temperature T	Decreases	Increases
Density ρ	Decreases	Increases
Stagnation enthalpy	Constant	Constant
Entropy	Increases	Increases

line toward the maximum s point, where **M** $= 1$. If the duct is fed by a converging-diverging nozzle, the flow may originally be supersonic; the velocity must then decrease downstream. If the flow is subsonic at the upstream end, the velocity must increase in the downstream direction.

For exactly one length of pipe, depending upon upstream conditions, the flow is just sonic (**M** $= 1$) at the downstream end. For shorter lengths of pipe, the flow will not have reached sonic conditions at the outlet, but for longer lengths of pipe, there must be shock waves (and possibly *choking*) if supersonic and choking effects if subsonic. Choking means that the mass flow rate specified cannot take place in this situation and less flow will occur. Table 6.1 indicates the trends in properties of a gas in adiabatic flow through a constant-area duct, as can be shown from the equations in this section.

The gas cannot change gradually from subsonic to supersonic or vice versa in a constant-area duct.

The momentum equation must now include the effects of wall shear stress and is conveniently written for a segment of duct of length δx (Fig. 6.6):

$$pA - \left(p + \frac{dp}{dx}\delta x \right) A - \tau_0 \pi D\, \delta x = \rho V A \left(V + \frac{dV}{dx}\delta x - V \right)$$

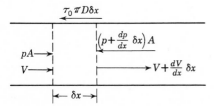

Fig. 6.6 Notation for application of momentum equation.

Upon simplification,

$$dp + \frac{4\tau_0}{D} dx + \rho V \, dV = 0 \tag{6.6.1}$$

By use of Eq. (5.10.2), $\tau_0 = \rho f V^2/8$, in which f is the Darcy-Weisbach friction factor,

$$dp + \frac{f\rho V^2}{2D} dx + \rho V \, dV = 0 \tag{6.6.2}$$

For constant f, or average value over the length of reach, this equation can be transformed into an equation for x as a function of Mach number. By dividing Eq. (6.6.2) by p,

$$\frac{dp}{p} + \frac{f}{2D} \frac{\rho V^2}{p} dx + \frac{\rho V}{p} dV = 0 \tag{6.6.3}$$

each term is now developed in terms of $\mathbf{M}$. By definition $V/c = \mathbf{M}$,

$$V^2 = \mathbf{M}^2 \frac{kp}{\rho} \tag{6.6.4}$$

or

$$\frac{\rho V^2}{p} = k\mathbf{M}^2 \tag{6.6.5}$$

for the middle term of the momentum equation. Rearranging Eq. (6.6.4) gives

$$\frac{\rho V}{p} dV = k\mathbf{M}^2 \frac{dV}{V} \tag{6.6.6}$$

Now to express dV/V in terms of $\mathbf{M}$, from the energy equation,

$$h_0 = h + \frac{V^2}{2} = c_p T + \frac{V^2}{2} \tag{6.6.7}$$

Differentiating gives

$$c_p \, dT + V \, dV = 0 \tag{6.6.8}$$

Dividing through by $V^2 = \mathbf{M}^2 kRT$ yields

$$\frac{c_p}{R} \frac{1}{k\mathbf{M}^2} \frac{dT}{T} + \frac{dV}{V} = 0$$

Since $c_p/R = k/(k-1)$,

$$\frac{dT}{T} = -\mathbf{M}^2(k-1)\frac{dV}{V} \tag{6.6.9}$$

Differentiating $V^2 = \mathbf{M}^2 kRT$ and dividing by the equation give

$$2\frac{dV}{V} = 2\frac{d\mathbf{M}}{\mathbf{M}} + \frac{dT}{T} \tag{6.6.10}$$

Eliminating dT/T in Eqs. (6.6.9) and (6.6.10) and simplifying lead to

$$\frac{dV}{V} = \frac{d\mathbf{M}/\mathbf{M}}{[(k-1)/2]\mathbf{M}^2 + 1} \tag{6.6.11}$$

which permits elimination of dV/V from Eq. (6.6.6), yielding

$$\frac{\rho V}{p} dV = \frac{k\mathbf{M}\, d\mathbf{M}}{[(k-1)/2]\mathbf{M}^2 + 1} \tag{6.6.12}$$

And finally, to express dp/p in terms of $\mathbf{M}$, from $p = \rho RT$ and $G = \rho V$,

$$pV = GRT \tag{6.6.13}$$

By differentiation

$$\frac{dp}{p} = \frac{dT}{T} - \frac{dV}{V}$$

Equations (6.6.9) and (6.6.11) are used to eliminate dT/T and dV/V:

$$\frac{dp}{p} = -\frac{(k-1)\mathbf{M}^2 + 1}{[(k-1)/2]\mathbf{M}^2 + 1}\frac{d\mathbf{M}}{\mathbf{M}} \tag{6.6.14}$$

Equations (6.6.5), (6.6.12), and (6.6.14) are now substituted into the momentum equation (6.6.3). After rearranging,

$$\frac{f}{D} dx = \frac{2(1 - \mathbf{M}^2)}{k\mathbf{M}^3\{[(k - 1)/2]\mathbf{M}^2 + 1\}} d\mathbf{M}$$

$$= \frac{2}{k} \frac{d\mathbf{M}}{\mathbf{M}^3} - \frac{k + 1}{k} \frac{d\mathbf{M}}{\mathbf{M}\{[(k - 1)/2]\mathbf{M}^2 + 1\}} \tag{6.6.15}$$

which can be integrated directly. By using the limits $x = 0$, $\mathbf{M} = \mathbf{M}_0$, $x = l$, $\mathbf{M} = \mathbf{M}$,

$$\frac{fl}{D} = -\frac{1}{k\mathbf{M}^2}\Big]_{\mathbf{M}_0}^{\mathbf{M}} - \frac{k + 1}{2k} \ln \frac{\mathbf{M}^2}{[(k - 1)/2]\mathbf{M}^2 + 1}\Big]_{\mathbf{M}_0}^{\mathbf{M}} \tag{6.6.16}$$

$$= \frac{1}{k}\left(\frac{1}{\mathbf{M}_0^2} - \frac{1}{\mathbf{M}^2}\right) + \frac{k + 1}{2k} \ln \left[\left(\frac{\mathbf{M}_0}{\mathbf{M}}\right)^2 \frac{(k - 1)\mathbf{M}^2 + 2}{(k - 1)\mathbf{M}_0^2 + 2}\right] \tag{6.6.17}$$

For $k = 1.4$, this reduces to

$$\frac{fl}{D} = \frac{5}{7}\left(\frac{1}{\mathbf{M}_0^2} - \frac{1}{\mathbf{M}^2}\right) + \tfrac{6}{7} \ln \left[\left(\frac{\mathbf{M}_0}{\mathbf{M}}\right)^2 \frac{\mathbf{M}^2 + 5}{\mathbf{M}_0^2 + 5}\right] \qquad k = 1.4 \tag{6.6.18}$$

If $\mathbf{M}_0$ is greater than 1, $\mathbf{M}$ cannot be less than 1, and if $\mathbf{M}_0$ is less than 1, $\mathbf{M}$ cannot be greater than 1. For the limiting condition $\mathbf{M} = 1$ and $k = 1.4$,

$$\frac{fL_{\max}}{D} = \frac{5}{7}\left(\frac{1}{\mathbf{M}_0^2} - 1\right) + \tfrac{6}{7} \ln \frac{6\mathbf{M}_0^2}{\mathbf{M}_0^2 + 5} \qquad k = 1.4 \tag{6.6.19}$$

There is some evidence[1] to indicate that friction factors may be smaller in supersonic flow.

EXAMPLE 6.13 Determine the maximum length of 5-cm-ID pipe, $f = 0.02$ for flow of air, when the Mach number at the entrance to the pipe is 0.30.
    From Eq. (6.6.19)

$$\frac{0.02}{0.05} L_{\max} = \frac{5}{7}\left(\frac{1}{0.3^2} - 1\right) + \tfrac{6}{7} \ln \frac{6 \times 0.30^2}{0.30^2 + 5}$$

from which $L_{\max} = 13.25$ m.

[1] J. H. Keenan and E. P. Neumann, Measurements of Friction in a Pipe for Subsonic and Supersonic Flow of Air, *J. Appl. Mech.*, vol. 13, no. 2, p. A-91, 1946.

The pressure, velocity, and temperature may also be expressed in integral form in terms of the Mach number. To simplify the equations that follow they will be integrated from upstream conditions to conditions at $\mathbf{M} = 1$, indicated by p^*, V^*, and T^*. From Eq. (6.6.14)

$$\frac{p^*}{p_1} = \mathbf{M}_0 \sqrt{\frac{(k-1)\mathbf{M}_0{}^2 + 2}{k+1}} \tag{6.6.20}$$

From Eq. (6.6.11)

$$\frac{V^*}{V_0} = \frac{1}{\mathbf{M}_0} \sqrt{\frac{(k-1)\mathbf{M}_0{}^2 + 2}{k+1}} \tag{6.6.21}$$

From Eqs. (6.6.9) and (6.6.11)

$$\frac{dT}{T} = -(k-1) \frac{\mathbf{M}\, d\mathbf{M}}{[(k-1)/2]\mathbf{M}^2 + 1}$$

which, when integrated, yields

$$\frac{T^*}{T_0} = \frac{(k-1)\mathbf{M}_0{}^2 + 2}{k+1} \tag{6.6.22}$$

EXAMPLE 6.14 A 4.0 in-ID pipe, $f = 0.020$, has air at 14.7 psia and at $t = 60°\text{F}$ flowing at the upstream end with Mach number 3.0. Determine $L_{\max}$, p^*, V^*, T^*, and values of p_0', V_0', T_0', and L at $\mathbf{M} = 2.0$.
From Eq. (6.6.19)

$$\frac{0.02}{0.333} L_{\max} = \tfrac{5}{7}(\tfrac{1}{9} - 1) + \tfrac{6}{7} \ln \frac{6 \times 3^2}{3^2 + 5}$$

from which $L_{\max} = 8.69$ ft. If the flow originated at $\mathbf{M} = 2$, the length $L_{\max}$ is given by the same equation:

$$\frac{0.02}{0.333} L_{\max} = \tfrac{5}{7}(\tfrac{1}{4} - 1) + \tfrac{6}{7} \ln \frac{6 \times 2^2}{2^2 + 5}$$

from which $L_{\max} = 5.08$ ft.
Hence the length from the upstream section at $\mathbf{M} = 3$ to the section where $\mathbf{M} = 2$ is $8.69 - 5.08 = 3.61$ ft.
The velocity at the entrance is

$$V = \sqrt{kRT}\, \mathbf{M} = \sqrt{1.4 \times 53.3 \times 32.17(460 + 60)} \times 3 = 3352 \text{ ft/s}$$

From Eqs. (6.6.20) to (6.6.22)

$$\frac{p^*}{14.7} = 3\sqrt{\frac{0.4 \times 3^2 + 2}{2.4}} = 4.583$$

$$\frac{V^*}{3354} = \frac{1}{3}\sqrt{\frac{0.4 \times 3^2 + 2}{2.4}} = 0.509$$

$$\frac{T^*}{520} = \frac{0.4 \times 3^2 + 2}{2.4} = \frac{7}{3}$$

So $p^* = 67.4$ psia, $V^* = 1707$ ft/s, $T^* = 1213°$R. For $\mathbf{M} = 2$ the same equations are now solved for p_0', V_0', and T_0':

$$\frac{67.4}{p_0'} = 2\sqrt{\frac{0.4 \times 2^2 + 2}{2.4}} = 2.45$$

$$\frac{1707}{V_0'} = \frac{1}{2}\sqrt{\frac{0.4 \times 2^2 + 2}{2.4}} = 0.6124$$

$$\frac{1213}{T_0'} = \frac{0.4 \times 2^2 + 2}{2.4} = \frac{3}{2}$$

So $p_0' = 27.5$ psia, $V_0' = 2787$ ft/s, and $T_0' = 809°$R.

6.7 FRICTIONLESS FLOW THROUGH DUCTS WITH HEAT TRANSFER

The steady flow of a perfect gas (with constant specific heats) through a constant-area duct is considered in this section. Friction is neglected, and no work is done on or by the flow.

The appropriate equations for analysis of this case are

Continuity: $G = \dfrac{\dot{m}}{A} = \rho V$ (6.7.1)

Momentum: $p + \rho V^2 = \text{const}$ (6.7.2)

Energy: $q_H = h_2 - h_1 + \dfrac{V_2{}^2 - V_1{}^2}{2}$

$$= c_p(T_2 - T_1) + \dfrac{V_2{}^2 - V_1{}^2}{2}$$

$$= c_p(T_{02} - T_{01}) \qquad\qquad (6.7.3)$$

T_{01} and T_{02} are the isentropic stagnation temperatures, i.e., the temperature produced at a section by bringing the flow isentropically to rest.

The Rayleigh line, obtained from the solution of momentum and continuity for a constant cross section by neglecting friction, is very helpful in examining the flow. First, eliminating V in Eqs. (6.7.1) and (6.7.2) gives

$$p + \dfrac{G^2}{\rho} = \text{const} \qquad\qquad (6.7.4)$$

which is Eq. (6.5.7). Equations (6.5.8) and (6.5.9) express the entropy s and enthalpy h in terms of the parameter ρ for the assumptions of this section, as in Fig. 6.7.

Since by Eq. (3.8.4), for no losses, entropy can increase only when heat is added, the properties of the gas must change as indicated in Fig. 6.7, moving toward the maximum entropy point as heat is added. At the maximum s point there is no change in entropy for a small change in h, and isentropic conditions apply to the point. The speed of sound under isentropic conditions is given by $c = \sqrt{dp/d\rho}$ as given by Eq. (6.2.2). From Eq. (6.7.4), by differ-

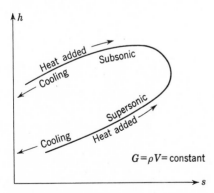

Fig. 6.7 Rayleigh line.

entiation,

$$\frac{dp}{d\rho} = \frac{G^2}{\rho^2} = V^2$$

using Eq. (6.7.1). Hence at the maximum s point of the Rayleigh line, $V = \sqrt{dp/d\rho}$ also and $\mathbf{M} = 1$, or sonic conditions prevail. The addition of heat to supersonic flow causes the Mach number of the flow to decrease toward $\mathbf{M} = 1$, and if just the proper amount of heat is added, $\mathbf{M}$ becomes 1. If more heat is added, choking results and conditions at the upstream end are altered to reduce the mass rate of flow. The addition of heat to subsonic flow causes an increase in the Mach number toward $\mathbf{M} = 1$, and again, too much heat transfer causes choking with an upstream adjustment of mass flow rate to a smaller value.

From Eq. (6.7.3) it is noted that the increase in isentropic stagnation pressure is a measure of the heat added. From $V^2 = \mathbf{M}^2 kRT$, $p = \rho RT$, and continuity,

$$pV = GRT \qquad \text{and} \qquad \rho V^2 = kp\mathbf{M}^2$$

Now, from the momentum equation,

$$p_1 + kp_1\mathbf{M}_1^2 = p_2 + kp_2\mathbf{M}_2^2$$

and

$$\frac{p_1}{p_2} = \frac{1 + k\mathbf{M}_2^2}{1 + k\mathbf{M}_1^2} \tag{6.7.5}$$

Writing this equation for the limiting case $p_2 = p^*$ when $\mathbf{M}_2 = 1$ gives

$$\frac{p}{p^*} = \frac{1 + k}{1 + k\mathbf{M}^2} \tag{6.7.6}$$

with p the pressure at any point in the duct where $\mathbf{M}$ is the corresponding Mach number. For the subsonic case, with $\mathbf{M}$ increasing to the right (Fig. 6.7), p must decrease, and for the supersonic case, as $\mathbf{M}$ decreases toward the right, p must increase.

To develop the other pertinent relations, the energy equation (6.7.3) is used,

$$c_p T_0 = \frac{kR}{k-1} T_0 = \frac{kR}{k-1} T + \frac{V^2}{2}$$

in which T_0 is the isentropic stagnation temperature and T the free-stream temperature at the same section. Applying this to section 1, after dividing through by $kRT_1/(k-1)$, yields

$$\frac{T_{01}}{T_1} = 1 + (k-1)\frac{\mathbf{M}_1^2}{2} \tag{6.7.7}$$

and for section 2

$$\frac{T_{02}}{T_2} = 1 + (k-1)\frac{\mathbf{M}_2^2}{2} \tag{6.7.8}$$

Dividing Eq. (6.7.7) by Eq. (6.7.8) gives

$$\frac{T_{01}}{T_{02}} = \frac{T_1}{T_2}\frac{2 + (k-1)\mathbf{M}_1^2}{2 + (k-1)\mathbf{M}_2^2} \tag{6.7.9}$$

The ratio T_1/T_2 is determined in terms of the Mach numbers as follows. From the perfect-gas law, $p_1 = \rho_1 RT_1$, $p_2 = \rho_2 RT_2$,

$$\frac{T_1}{T_2} = \frac{p_1}{p_2}\frac{\rho_2}{\rho_1} \tag{6.7.10}$$

From continuity $\rho_2/\rho_1 = V_1/V_2$, and by definition,

$$\mathbf{M}_1 = \frac{V_1}{\sqrt{kRT_1}} \qquad \mathbf{M}_2 = \frac{V_2}{\sqrt{kRT_2}}$$

so that

$$\frac{V_1}{V_2} = \frac{\mathbf{M}_1}{\mathbf{M}_2}\sqrt{\frac{T_1}{T_2}}$$

and

$$\frac{\rho_2}{\rho_1} = \frac{\mathbf{M}_1}{\mathbf{M}_2}\sqrt{\frac{T_1}{T_2}} \tag{6.7.11}$$

Now substituting Eqs. (6.7.5) and (6.7.11) into Eq. (6.7.10) and simplifying gives

$$\frac{T_1}{T_2} = \left(\frac{\mathbf{M}_1}{\mathbf{M}_2}\frac{1 + k\mathbf{M}_2^2}{1 + k\mathbf{M}_1^2}\right)^2 \tag{6.7.12}$$

This equation substituted into Eq. (6.7.9) leads to

$$\frac{T_{01}}{T_{02}} = \left(\frac{\mathbf{M}_1}{\mathbf{M}_2}\frac{1 + k\mathbf{M}_2^2}{1 + k\mathbf{M}_1^2}\right)^2 \frac{2 + (k-1)\mathbf{M}_1^2}{2 + (k-1)\mathbf{M}_2^2} \tag{6.7.13}$$

When this equation is applied to the downstream section where $T_{02} = T_0^*$ and $\mathbf{M}_2 = 1$ and the subscripts for the upstream section are dropped, the result is

$$\frac{T_0}{T_0^*} = \frac{\mathbf{M}^2(k+1)[2 + (k-1)\mathbf{M}^2]}{(1 + k\mathbf{M}^2)^2} \tag{6.7.14}$$

All the necessary equations for determination of frictionless flow with heat transfer in a constant-area duct are now available. Heat transfer per unit mass is given by $q_H = c_p(T_0^* - T_0)$ for $\mathbf{M} = 1$ at the exit. Use of the equations is illustrated in the following example.

EXAMPLE 6.15   Air at $V_1 = 300$ ft/s, $p = 40$ psia, $t = 60°F$ flows into a 4.0-in-diameter duct. How much heat transfer per unit mass is needed for sonic conditions at the exit? Determine pressure, temperature, and velocity at the exit and at the section where $\mathbf{M} = 0.70$.

$$\mathbf{M}_1 = \frac{V_1}{\sqrt{kRT_1}} = \frac{300}{\sqrt{1.4 \times 53.3 \times 32.17(460 + 60)}} = 0.268$$

The isentropic stagnation temperature at the entrance, from Eq. (6.7.7), is

$$T_{01} = T_1\left(1 + \frac{k-1}{2}\mathbf{M}_1^2\right) = 520(1 + 0.2 \times 0.268^2) = 527°R$$

The isentropic stagnation temperature at the exit, from Eq. (6.7.14), is

$$T_0^* = \frac{T_0(1 + k\mathbf{M}^2)^2}{(k+1)\mathbf{M}_2[2 + (k-1)\mathbf{M}^2]} = \frac{527(1 + 1.4 \times 0.268^2)^2}{2.4 \times 0.268^2(2 + 0.4 \times 0.268^2)}$$

$$= 1827°R$$

The heat transfer per slug of air flowing is

$$q_H = c_p(T_0^* - T_{01}) = 0.24 \times 32.17(1827 - 527) = 10{,}023 \text{ Btu/slug}$$

The pressure at the exit, Eq. (6.7.6), is

$$p^* = p \frac{1 + k\mathbf{M}^2}{k + 1} = \frac{40}{2.4} (1 + 1.4 \times 0.268^2) = 18.34 \text{ psia}$$

and the temperature, from Eq. (6.7.12),

$$T^* = T \left[\frac{1 + k\mathbf{M}^2}{(k + 1)\mathbf{M}} \right]^2 = 520 \left(\frac{1 + 1.4 \times 0.268^2}{2.4 \times 0.268} \right)^2 = 1522°\text{R}$$

At the exit,

$$V^* = c^* = \sqrt{kRT^*} = \sqrt{1.4 \times 53.3 \times 32.17 \times 1522} = 1911 \text{ ft/s}$$

At the section where $\mathbf{M} = 0.7$, from Eq. (6.7.6),

$$p = p^* \frac{k + 1}{1 + k\mathbf{M}^2} = \frac{18.35 \times 2.4}{1 + 1.4 \times 0.7^2} = 26.1 \text{ psia}$$

From Eq. (6.7.12)

$$T = T^* \left[\frac{(k + 1)\mathbf{M}}{1 + k\mathbf{M}^2} \right]^2 = 1520 \left(\frac{2.4 \times 0.7}{1 + 1.4 \times 0.7^2} \right)^2 = 1509°\text{R}$$

and

$$V = \mathbf{M}\sqrt{kRT} = 0.7 \sqrt{1.4 \times 53.3 \times 32.17 \times 1509} = 1332 \text{ ft/s}$$

The trends in flow properties are shown in Table 6.2.

For curves and tables tabulating the various equations, consult the books by Cambel and Jennings, Keenan and Kaye, and Shapiro, listed in the references at the end of this chapter.

6.8 STEADY ISOTHERMAL FLOW IN LONG PIPELINES

In the analysis of isothermal flow of a perfect gas through long ducts, neither the Fanno nor Rayleigh line is applicable, since the Fanno line applies to adiabatic flow and the Rayleigh line to frictionless flow. An analysis somewhat similar to those of the previous two sections is carried out to show the trend in properties with Mach number.

Table 6.2 *Trends in flow properties*

Property	Heating		Cooling	
	M > 1	**M** < 1	**M** > 1	**M** < 1
Pressure p	Increases	Decreases	Decreases	Increases
Velocity V	Decreases	Increases	Increases	Decreases
Isentropic stagnation temperature T_0	Increases	Increases	Decreases	Decreases
Density ρ	Increases	Decreases	Decreases	Increases
Temperature T	Increases	Increases for **M** < $1/\sqrt{k}$ Decreases for **M** > $1/\sqrt{k}$	Decreases	Decreases for **M** < $1/\sqrt{k}$ Increases for **M** > $1/\sqrt{k}$

The appropriate equations are

Momentum [Eq. (6.6.3)]: $$\frac{dp}{p} + \frac{f}{2D}\frac{\rho V^2}{p}\,dx + \frac{\rho V}{p}\,dV = 0 \qquad (6.8.1)$$

Equation of state: $$\frac{p}{\rho} = \text{const} \qquad \frac{dp}{p} = \frac{d\rho}{\rho} \qquad (6.8.2)$$

Continuity: $$\rho V = \text{const} \qquad \frac{d\rho}{\rho} = -\frac{dV}{V} \qquad (6.8.3)$$

Energy [Eq. (6.7.7)]: $$T_0 = T\left[1 + \frac{(k-1)}{2}\mathbf{M}^2\right] \qquad (6.8.4)$$

in which T_0 is the isentropic stagnation temperature at the section where the free-stream static temperature is T and the Mach number is **M**.

Stagnation pressure [Eq. (6.3.11)]: $$p_0 = p\left(1 + \frac{k-1}{2}\mathbf{M}^2\right)^{k/(k-1)} \qquad (6.8.5)$$

in which p_0 is the pressure (at the section of p and **M**) obtained by reducing the velocity to zero isentropically.

From definitions and the above equations

$$V = c\mathbf{M} = \sqrt{kRT}\,\mathbf{M} \qquad \frac{dV}{V} = \frac{d\mathbf{M}}{\mathbf{M}} = \frac{d\mathbf{M}^2}{2\mathbf{M}^2}$$

$$\frac{\rho V}{p} dV = \frac{V \, dV}{RT} = \frac{c^2}{RT} \mathbf{M} \, d\mathbf{M} = k\mathbf{M} \, d\mathbf{M}$$

$$\frac{\rho V^2}{p} = \frac{c^2 \mathbf{M}^2}{RT} = k\mathbf{M}^2$$

Substituting into the momentum equation gives

$$\frac{dp}{p} = \frac{d\rho}{\rho} = -\frac{dV}{V} = -\frac{1}{2}\frac{d\mathbf{M}^2}{\mathbf{M}^2} = -\frac{k\mathbf{M}^2}{1 - k\mathbf{M}^2}\frac{f \, dx}{2D} \tag{6.8.6}$$

The differential dx is positive in the downstream direction, and so one may conclude that the trends in properties vary according as $\mathbf{M}$ is less than or greater than $1/\sqrt{k}$. For $\mathbf{M} < 1/\sqrt{k}$, the pressure and density decrease and velocity and Mach number increase, with the opposite trends for $\mathbf{M} > 1/\sqrt{k}$; hence, the Mach number always approaches $1/\sqrt{k}$, in place of unity for adiabatic flow in pipelines.

To determine the direction of heat transfer differentiate Eq. (6.8.4) and then divide by it, remembering that T is constant:

$$\frac{dT_0}{T_0} = \frac{k - 1}{2 + (k - 1)\mathbf{M}^2} d\mathbf{M}^2 \tag{6.8.7}$$

Eliminating $d\mathbf{M}^2$ in this equation and Eq. (6.8.6) gives

$$\frac{dT_0}{T_0} = \frac{k(k - 1)\mathbf{M}^4}{(1 - k\mathbf{M}^2)[2 + (k - 1)\mathbf{M}^2]}\frac{f \, dx}{D} \tag{6.8.8}$$

which shows that the isentropic stagnation temperature increases for $\mathbf{M} < 1/\sqrt{k}$, indicating that heat is transferred to the fluid. For $M > 1/\sqrt{k}$ heat transfer is from the fluid.

From Eqs. (6.8.5) and (6.8.6)

$$\frac{dp_0}{p_0} = \frac{2 - (k + 1)\mathbf{M}^2}{2 + (k - 1)\mathbf{M}^2}\frac{k\mathbf{M}^2}{k\mathbf{M}^2 - 1}\frac{f \, dx}{2D} \tag{6.8.9}$$

Table 6.3 shows the trends of fluid properties.

By integration of the various Eqs. (6.8.6) in terms of $\mathbf{M}$, the change with Mach number is found. The last two terms yield

$$\frac{f}{D}\int_0^{L_{max}} dx = \frac{1}{k}\int_M^{1/\sqrt{k}}\frac{(1 - k\mathbf{M}^2)}{\mathbf{M}^4} d\mathbf{M}^2$$

Table 6.3 *Trends in fluid properties for isothermal flow*

Property	$\mathbf{M} < 1/\sqrt{k}$ subsonic	$\mathbf{M} > 1/\sqrt{k}$ subsonic or supersonic
Pressure p	Decreases	Increases
Density ρ	Decreases	Increases
Velocity V	Increases	Decreases
Mach number $\mathbf{M}$	Increases	Decreases
Stagnation temperature T_0	Increases	Decreases
Stagnation pressure p_0	Decreases	Increases for $\mathbf{M} < \sqrt{2}/(k+1)$
		Decreases for $\mathbf{M} > \sqrt{2}/(k+1)$

or

$$\frac{f}{D} L_{\max} = \frac{1 - k\mathbf{M}^2}{k\mathbf{M}^2} + \ln (k\mathbf{M}^2) \tag{6.8.10}$$

in which $L_{\max}$, as before, represents the maximum length of duct. For greater lengths choking occurs, and the mass rate is decreased. To find the pressure change,

$$\int_p^{p^{*t}} \frac{dp}{p} = -\frac{1}{2} \int_{\mathbf{M}}^{1/\sqrt{k}} \frac{d\mathbf{M}^2}{\mathbf{M}^2}$$

and

$$\frac{p^{*t}}{p} = \sqrt{k}\, \mathbf{M} \tag{6.8.11}$$

The superscript ^{*t} indicates conditions at $\mathbf{M} = 1/\sqrt{k}$, and $\mathbf{M}$ and p represent values at any upstream section.

EXAMPLE 6.16 Helium enters a 10-cm-ID pipe from a converging-diverging nozzle at $\mathbf{M} = 1.30$, $p = 14 \text{ kN/m}^2$ abs, $T = 225$ K. Determine for isothermal flow (*a*) the maximum length of pipe for no choking, (*b*) the downstream conditions, and (*c*) the length from the exit to the section where $\mathbf{M} = 1.0$. $f = 0.016$.

(*a*)   From Eq. (6.8.10) for $k = 1.66$

$$\frac{0.016 L_{\max}}{0.1 \text{ m}} = \frac{1 - 1.66 \times 1.3^2}{1.66 \times 1.3^2} + \ln (1.66 \times 1.3^2)$$

from which $L_{\max} = 2.425$ m.

(b)   From Eq. (6.8.11)

$$p^{*t} = p\sqrt{k}\,\mathbf{M} = 14\text{ kN/m}^2\,\sqrt{1.66}\ 1.3 = 23.45\text{ kN/m}^2\text{ abs}$$

The Mach number at the exit is $1/\sqrt{1.66} = 0.756$. From Eqs. (6.8.6)

$$\int_V^{V^{*t}} \frac{dV}{V} = \frac{1}{2}\int_\mathbf{M}^{1/\sqrt{k}} \frac{d\mathbf{M}^2}{\mathbf{M}^2} \qquad\text{or}\qquad \frac{V^{*t}}{V} = \frac{1}{\sqrt{k}\,\mathbf{M}}$$

At the upstream section

$$V = \mathbf{M}\,\sqrt{kRT} = 1.3\,\sqrt{1.66 \times 2077 \times 225} = 1145\text{ m/s}$$

and

$$V^{*t} = \frac{V}{\sqrt{k}\,\mathbf{M}} = \frac{1145\text{ m/s}}{\sqrt{1.66}\ 1.3} = 683.6\text{ m/s}$$

(c)   From Eq. (6.8.10) for $\mathbf{M} = 1$,

$$\frac{0.016}{0.1\text{ m}}L'_{\max} = \frac{1 - 1.66}{1.66} + \ln 1.66$$

or $L'_{\max} = 0.683$ m. $\mathbf{M} = 1$ occurs 0.683 m from the exit.

6.9   ANALOGY OF SHOCK WAVES TO OPEN-CHANNEL WAVES

Both the oblique and normal shock waves in a gas have their counterpart in open-channel flow. An elementary surface wave has a speed in still liquid of $\sqrt{gy}$, in which y is the depth in a wide, open channel. When flow in the channel is such that $V = V_c = \sqrt{gy}$, the Froude number is unity and flow is said to be *critical;* i.e., a small disturbance cannot be propagated upstream. This is analogous to sonic flow at the throat of a tube, with Mach number unity. For liquid velocities greater than $V_c = \sqrt{gy}$ the Froude number is greater than unity and the velocity is supercritical, analogous to supersonic gas flow. Changes in depth are analogous to changes in density in gas flow. The continuity equation in an open channel of constant width is

$$Vy = \text{const}$$

and the continuity equation for compressible flow in a tube of constant cross section is

$$V\rho = \text{const}$$

Compressible fluid density ρ and open-channel depth y are analogous.

The same analogy is also present in the energy equation. The energy equation for a horizontal open channel of constant width, neglecting friction, is

$$\frac{V^2}{2g} + y = \text{const}$$

After differentiating,

$$V\,dV + g\,dy = 0$$

By substitution from $V_c = \sqrt{gy}$ to eliminate g,

$$V\,dV + V_c^2\frac{dy}{y} = 0$$

which is to be compared with the energy equation for compressible flow [Eq. (6.3.4)]

$$V\,dV + c^2\frac{d\rho}{\rho} = 0$$

The two critical velocities V_c and c are analogous, and, hence, y and ρ are analogous.

By applying the momentum equation to a small depth change in horizontal open-channel flow, and to a sudden density change in compressible flow, the density and the open-channel depth can again be shown to be analogous. In effect, the analogy is between the Froude number and the Mach number.

Analogous to the normal shock wave is the hydraulic jump, which causes a sudden change in velocity and depth, and a change in Froude number from greater than unity to less than unity. Analogous to the oblique shock and rarefaction waves in gas flow are oblique liquid waves produced in a channel by changes in the direction of the channel walls or by changes in floor elevation.

A body placed in an open channel with flow at Froude number greater than unity causes waves on the surface that are analogous to shock and rarefaction waves on a similar (two-dimensional) body in a supersonic wind tunnel. Changes to greater depth are analogous to compression shock, and changes to lesser depth to rarefaction waves. Shallow water tanks, called *ripple tanks*, have been used to study supersonic flow situations.

PROBLEMS

6.1 3 kg of a perfect gas, molecular weight 36, had its temperature increased 2°C when 6.4 kJ of work was done on it in an insulated constant-volume chamber. Determine c_v and c_p.

6.2 A gas of molecular weight 48 has $c_p = 0.372$. What is c_v for this gas?

6.3 Calculate the specific heat ratio k for Probs. 6.1 and 6.2.

6.4 The enthalpy of a gas is increased by 0.4 Btu/lb$_m$·°R when heat is added at constant pressure, and the internal energy is increased by 0.3 Btu/lb$_m$·°R when the volume is maintained constant and heat is added. Calculate the molecular weight.

6.5 Calculate the enthalpy change of 2 kg carbon monoxide from $p_1 = 14$ kN/m² abs, $t_1 = 5°C$ to $p_2 = 30$ kN/m² abs, $t_2 = 170°C$.

6.6 Calculate the entropy change in Prob. 6.5.

6.7 From Eq. (6.1.13) and the perfect-gas law, derive the equation of state for isentropic flow.

6.8 Compute the enthalpy change per slug for helium from $t_1 = 0°F$, $p_1 = 15$ psia to $t_2 = 100°F$ in an isentropic process.

6.9 In an isentropic process 1 kg oxygen with a volume of 100 l at 15°C has its absolute pressure doubled. What is the final temperature?

6.10 Work out the expression for density change with temperature for a reversible polytropic process.

6.11 Hydrogen at 40 psia, 30°F, has its temperature increased to 120°F by a reversible polytropic process with $n = 1.20$. Calculate the final pressure.

6.12 A gas has a density decrease of 10 percent in a reversible polytropic process when the temperature decreases from 45 to 5°C. Compute the exponent n for the process.

6.13 A projectile moves through water at 80°F at 3000 ft/s. What is its Mach number?

6.14 If an airplane travels at 1350 km/h at sea level, $p = 101$ kPa abs, $t = 20°C$, and at the same speed in the stratosphere where $t = -55°C$, how much greater is the Mach number in the latter case?

6.15 What is the speed of sound through hydrogen at 80°F?

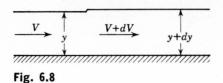

Fig. 6.8

6.16   Derive the equation for speed of a small liquid wave in an open channel by using the methods of Sec. 6.2 for determination of speed of sound (Fig. 6.8).

6.17   By using the energy equation

$$V \, dV + \frac{dp}{\rho} + d(\text{losses}) = 0$$

the continuity equation $\rho V = $ const, and $c = \sqrt{dp/d\rho}$, show that for subsonic flow in a pipe the velocity must increase in the downstream direction.

6.18   Isentropic flow of air occurs at a section of a pipe where $p = 40$ psia, $t = 90°F$, and $V = 537$ ft/s. An object is immersed in the flow which brings the velocity to zero. What are the temperature and pressure at the stagnation point?

6.19   What is the Mach number for the flow of Prob. 6.18?

6.20   How do the temperature and pressure at the stagnation point in isentropic flow compare with reservoir conditions?

6.21   Air flows from a reservoir at $70°C$, 7 atm. Assuming isentropic flow, calculate the velocity, temperature, pressure, and density at a section where $\mathbf{M} = 0.60$.

6.22   Oxygen flows from a reservoir where $p_0 = 100$ psia, $t_0 = 60°F$, to a 6-in-diameter section where the velocity is 600 ft/s. Calculate the mass rate of flow (isentropic) and the Mach number, pressure, and temperature at the 6-in section.

6.23   Helium discharges from a $\frac{1}{2}$-in-diameter converging nozzle at its maximum rate for reservoir conditions of $p = 4$ kg$_f$/cm^2 abs, $t = 25°C$. What restrictions are placed on the downstream pressure? Calculate the mass flow rate and velocity of the gas at the nozzle.

6.24   Air in a reservoir at 350 psia, $t = 290°F$, flows through a 2-in-diameter throat in a converging-diverging nozzle. For $\mathbf{M} = 1$ at the throat, calculate p, ρ, and T there.

6.25   What must be the velocity, pressure, density, temperature, and diameter at a cross section of the nozzle of Prob. 6.24 where $\mathbf{M} = 2.4$?

6.26   Nitrogen in sonic flow at a 25-mm-diameter throat section has a pressure of 50 kN/m^2 abs, $t = -20°C$. Determine the mass flow rate.

6.27   What is the Mach number for Prob. 6.26 at a 40-mm-diameter section in supersonic and in subsonic flow?

6.28 What diameter throat section is needed for critical flow of 0.5 lb_m/s carbon monoxide from a reservoir where $p = 300$ psia, $t = 100°F$?

6.29 A supersonic nozzle is to be designed for airflow with **M** $= 3$ at the exit section, which is 20 cm in diameter and has a pressure of 7 kN/m^2 abs and temperature of $-85°C$. Calculate the throat area and reservoir conditions.

6.30 In Prob. 6.29 calculate the diameter of cross section for **M** $= 1.5$, 2.0, and 2.5.

6.31 For reservoir conditions of $p_0 = 150$ psia, $t_0 = 120°F$, air flows through a converging-diverging tube with a 3.0 in-diameter throat with a maximum Mach number of 0.80. Determine the mass rate of flow and the diameter, pressure, velocity, and temperature at the exit where **M** $= 0.50$.

6.32 Calculate the exit velocity and the mass rate of flow of nitrogen from a reservoir where $p = 4$ atm, $t = 25°C$, through a converging nozzle of 5 cm diameter discharging to atmosphere.

6.33 Reduce Eq. (6.3.25) to its form for airflow. Plot p/p_0 vs. A^*/A for the range of p/p_0 from 0.98 to 0.02.

6.34 By utilizing the plot of Prob. 6.33, find the two pressure ratios for $A^*/A = 0.50$.

6.35 In a converging-diverging duct in supersonic flow of hydrogen, the throat diameter is 2.0 in. Determine the pressure ratios p/p_0 in the converging and diverging ducts where the diameter is 2.25 in.

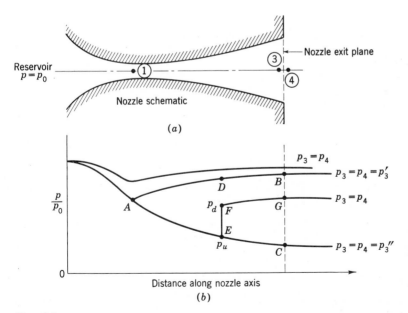

Fig. 6.9

6.36 A shock wave occurs in a duct carrying air where the upstream Mach number is 2.0 and upstream temperature and pressure are 15°C and 20 kN/m² abs. Calculate the Mach number, pressure, temperature, and velocity after the shock wave.

6.37 Show that entropy has increased across the shock wave of Prob. 6.36.

6.38 Conditions immediately before a normal shock wave in airflow are $p_u = 6$ psia, $t_u = 100°F$, $V_u = 1800$ ft/s. Find $\mathbf{M}_u$, $\mathbf{M}_d$, p_d, and t_d, where the subscript d refers to conditions just downstream from the shock wave.

6.39 For $A = 0.16$ ft² in Prob. 6.38, calculate the entropy increase across the shock wave in Btu per second per degree Rankine.

6.40 From Eqs. (6.3.1), (6.3.4), and (6.3.5) deduce that at the throat of a convergent-divergent (De Laval) nozzle, point 1 of Fig. 6.9a, $dp = 0$, $d\rho = 0$ for $\mathbf{M} \neq 1$ (cf. Fig. 6.9b). Are these differentials zero for $\mathbf{M} = 1$? Explain.

6.41 From Eqs. (6.3.1), (6.3.4), and (6.3.5) justify the slopes of the curves shown in Fig. 6.9b. Do not consider EFG.

6.42 For the nozzle described below, plot curves ADB and AEC (Fig. 6.9b). (*Suggestions:* Determine only one intermediate point. Use section VI.) The reservoir has air at 300 kPa abs and 40°C when sonic conditions are obtained at the throat.

	Section									
	I	II	III	IV	V	VI	VII	VIII	IX	X *exit*
Distance downstream from throat, cm	0.5	1.0	1.5	2.0	2.5	3.0	3.5	4.0	4.5	5.0
A/A^* ($A^* = 27$ cm²)	1.030	1.050	1.100	1.133	1.168	1.200	1.239	1.269	1.310	1.345

6.43 Using the data from Prob. 6.42 determine p_3/p_0 when a normal shock wave occurs at section VI.

6.44 Could a flow discontinuity occur at section VI of Prob. 6.42 so that the flow path would be described by $ADFG$ of Fig. 6.9b? (*Hint:* Determine the entropy changes.)

6.45 What is p_3/p_0 when a normal shock wave occurs just inside the nozzle exit? (*Hint:* $p_d = p_4$ and p_u is p_3 for isentropic flow up to section VI of Prob. 6.42.)

6.46 Suggest what might occur just outside the nozzle if there is a receiver pressure p_4 which is above that for which the gas flows isentropically throughout the nozzle into the receiver, point C of Fig. 6.9b, but below that for which a normal shock is possible at the nozzle exit (cf. Prob. 6.45).

6.47 Speculate on what occurs within and without the nozzle if the receiver pressure is below that corresponding to point C of Fig. 6.9b.

6.48 Show, from the equations of Sec. 6.6, that temperature, pressure, and density decrease in real, adiabatic duct flow for subsonic conditions and increase for supersonic conditions.

6.49 What length of 4-in-diameter insulated duct, $f = 0.018$, is needed when oxygen enters at $\mathbf{M} = 3.0$ and leaves at $\mathbf{M} = 2.0$?

6.50 Air enters an insulated pipe at $\mathbf{M} = 0.4$ and leaves at $\mathbf{M} = 0.6$. What portion of the duct length is required for the flow to occur at $\mathbf{M} = 0.5$?

6.51 Determine the maximum length, without choking, for the adiabatic flow of air in a 10-cm-diameter duct, $f = 0.025$, when upstream conditions are $t = 50°C$, $V = 200$ m/s, $p = 2$ kg$_f$/cm^2 abs. What are the pressure and temperature at the exit?

6.52 What minimum size insulated duct is required to transport 2 lb$_m$/s nitrogen 1000 ft? The upstream temperature is 80°F, and the velocity there is 200 ft/s. $f = 0.020$.

6.53 Find the upstream and downstream pressures in Prob. 6.52.

6.54 What is the maximum mass rate of flow of air from a reservoir, $t = 15°C$, through 6 m of insulated 25-mm-diameter pipe, $f = 0.020$, discharging to atmosphere? $p = 1$ kg$_f$/cm^2 abs.

6.55 In frictionless oxygen flow through a duct the following conditions prevail at inlet and outlet: $V_1 = 300$ ft/s; $t_1 = 80°F$; $\mathbf{M}_2 = 0.4$. Find the heat added per slug and the pressure ratio p_1/p_2.

6.56 In frictionless air the flow through a 10-cm-diameter duct 0.15 kg/s enters at $t = 0°C$, $p = 7$ kN/m^2 abs. How much heat, in kilocalories per kilogram, can be added without choking the flow?

6.57 Frictionless flow through a duct with heat transfer causes the Mach number to decrease from 2 to 1.75. $k = 1.4$. Determine the temperature, velocity, pressure, and density ratios.

6.58 In Prob. 6.57 the duct is 2 in square, $p_1 = 15$ psia, and $V_1 = 2000$ ft/s. Calculate the mass rate of flow for air flowing.

6.59 How much heat must be transferred per kilogram to cause the Mach number to increase from 2 to 2.8 in a frictionless duct carrying air? $V_1 = 500$ m/s.

6.60 Oxygen at $V_1 = 525$ m/s, $p = 80$ kN/m^2 abs, $t = -10°C$ flows in a 5-cm-diameter frictionless duct. How much heat transfer per kilogram is needed for sonic conditions at the exit?

6.61 Prove the density, pressure, and velocity trends given in Sec. 6.8 in the table of trends in flow properties.

6.62 Apply the first law of thermodynamics, Eq. (3.7.1), to isothermal flow of a perfect gas in a horizontal pipeline, and develop an expression for the heat added per slug flowing.

6.63 Air is flowing at constant temperature through a 3-in-diameter horizontal pipe, $f = 0.02$. At the entrance $V_1 = 300$ ft/s, $t = 120°F$, $p_1 = 30$ psia. What is the maximum pipe length for this flow, and how much heat is transferred to the air per pound mass?

6.64 Air at 15°C flows through a 25-mm-diameter pipe at constant temperature. At the entrance $V_1 = 60$ m/s, and at the exit $V_2 = 90$ m/s. $f = 0.016$. What is the length of the pipe?

6.65 If the pressure at the entrance of the pipe of Prob. 6.64 is 1.5 atm, what is the pressure at the exit and what is the heat transfer to the pipe per second?

6.66 Hydrogen enters a pipe from a converging nozzle at $\mathbf{M} = 1$, $p = 2$ psia, $t = 0°F$. Determine for isothermal flow the maximum length of pipe, in diameters, and the pressure change over this length. $f = 0.016$.

6.67 Oxygen flows at constant temperature of 20°C from a pressure tank, $p = 130$ atm, through 10 ft of 3-mm-ID tubing to another tank where $p = 110$ atm. $f = 0.016$. Determine the mass rate of flow.

6.68 In isothermal flow of nitrogen at 80°F, 2 lb$_m$/s is to be transferred 100 ft from a tank where $p = 200$ psia to a tank where $p = 160$ psia. What is the minimum size tubing, $f = 0.016$, that is needed?

6.69 Specific heat at constant volume is defined by

(a) kc_p (b) $\left(\dfrac{\partial u}{\partial T}\right)_p$ (c) $\left(\dfrac{\partial T}{\partial u}\right)_v$ (d) $\left(\dfrac{\partial u}{\partial T}\right)_v$ (e) none of these answers

6.70 Specific heat at constant pressure, for a perfect gas, is *not* given by

(a) kc_v (b) $\left(\dfrac{\partial h}{\partial T}\right)_p$ (c) $\dfrac{h_2 - h_1}{T_2 - T_1}$ (d) $\dfrac{\Delta u + \Delta(p/\rho)}{\Delta T}$

(e) any of these answers

6.71 For a perfect gas, the enthalpy

(a) always increases owing to losses
(b) depends upon the pressure only
(c) depends upon the temperature only
(d) may increase while the internal energy decreases
(e) satisfies none of these answers

6.72 The following classes of substances may be considered perfect gases:

(a) ideal fluids   (b) saturated steam, water vapor, and air
(c) fluids with a constant bulk modulus of elasticity
(d) water vapor, hydrogen, and nitrogen at low pressure
(e) none of these answers

6.73 c_p and c_v are related by

(a) $k = c_p/c_v$ (b) $k = c_p c_v$ (c) $k = c_v/c_p$ (d) $c_p = c_v{}^k$
(e) none of these answers

6.74 If $c_p = 0.30$ Btu/lb$_m \cdot °$R and $k = 1.66$, in foot-pounds per slug per degree Fahrenheit, c_v equals

(a) 0.582   (b) 1452   (c) 4524   (d) 7500   (e) none of these answers

6.75 If $c_p = 0.30$ kcal/kg$\cdot$K and $k = 1.33$, the gas constant in kilocalories per kilogram per kelvin is

(a) 0.075   (b) 0.099   (c) 0.399   (d) 0.699   (e) none of these answers

6.76 $R = 62$ ft$\cdot$lb/lb$_m \cdot °$R and $c_p = 0.279$ Btu/lb$_m \cdot °$F. The isentropic exponent k is

(a) 1.2   (b) 1.33   (c) 1.66   (d) 1.89   (e) none of these answers

6.77 The specific heat ratio is given by

(a) $\dfrac{1}{1 - R/c_p}$ (b) $1 + \dfrac{c_v}{R}$ (c) $\dfrac{c_p}{c_v} + R$ (d) $\dfrac{1}{1 - c_v/R}$

(e) none of these answers

6.78 The entropy change for a perfect gas is

(a) always positive   (b) a function of temperature only   (c) $(\Delta q_H/T)_{\text{rev}}$
(d) a thermodynamic property depending upon temperature and pressure
(e) a function of internal energy only

6.79 An isentropic process is always

(a) irreversible and adiabatic
(b) reversible and isothermal
(c) frictionless and adiabatic
(d) frictionless and irreversible
(e) none of these answers

6.80 The relation $p = $ const ρ^k holds only for those processes that are

(a) reversible polytropic
(b) isentropic
(c) frictionless isothermal
(d) adiabatic irreversible
(e) none of these answers

6.81 The reversible polytropic process is

(a) adiabatic frictionless
(b) given by $p/\rho = $ const
(c) given by $p\rho^k = $ const

(d) given by $p/\rho^n = $ const
(e) none of these answers

6.82 A reversible polytropic process could be given by

(a) $\dfrac{T_1}{T_2} = \left(\dfrac{\rho_1}{\rho_2}\right)^{n-1}$ (b) $\dfrac{p_1}{p_2} = \left(\dfrac{\rho_2}{\rho_1}\right)^{n}$ (c) $\dfrac{T_1}{T_2} = \left(\dfrac{p_1}{p_2}\right)^{n-1}$ (d) $\dfrac{T_1}{T_2} = \left(\dfrac{\rho_1}{\rho_2}\right)^{(n-1)/n}$

(e) none of these answers

6.83 In a reversible polytropic process

(a) some heat transfer occurs
(b) the entropy remains constant
(c) the enthalpy remains constant
(d) the internal energy remains constant
(e) the temperature remains constant

6.84 The differential equation for energy in isentropic flow may take the form

(a) $dp + d(\rho V^2) = 0$ (b) $\dfrac{dV}{V} + \dfrac{d\rho}{\rho} + \dfrac{dA}{A} = 0$

(c) $2V\,dV + \dfrac{dp}{\rho} = 0$ (d) $V\,dV + \dfrac{dp}{\rho} = 0$

(e) none of these answers

6.85 Select the expression that does *not* give the speed of a sound wave:

(a) $\sqrt{kRT}$ (b) $\sqrt{k\rho/p}$ (c) $\sqrt{dp/d\rho}$ (d) $\sqrt{kp/\rho}$ (e) $\sqrt{K/\rho}$

6.86 The speed of a sound wave in a gas is analogous to

(a) the speed of flow in an open channel
(b) the speed of an elementary wave in an open channel
(c) the change in depth in an open channel
(d) the speed of a disturbance traveling upstream in moving liquid
(e) none of these answers

6.87 The speed of sound in water, in feet per second, under ordinary conditions is about

(a) 460   (b) 1100   (c) 4600   (d) 11,000   (e) none of these answers

6.88 The speed of sound in an ideal gas varies directly as

(a) the density   (b) the absolute pressure   (c) the absolute temperature
(d) the bulk modulus of elasticity   (e) none of these answers

6.89 Select the correct statement regarding frictionless flow:

(*a*) In diverging conduits the velocity always decreases.
(*b*) The velocity is always sonic at the throat of a converging-diverging tube.
(*c*) In supersonic flow the area decreases for increasing velocity.
(*d*) Sonic velocity cannot be exceeded at the throat of a converging-diverging tube.
(*e*) At Mach zero the velocity is sonic.

6.90 In isentropic flow, the temperature

(*a*) cannot exceed the reservoir temperature
(*b*) cannot drop, then increase again downstream
(*c*) is independent of the Mach number
(*d*) is a function of Mach number only
(*e*) remains constant in duct flow

6.91 The critical pressure ratio for isentropic flow of carbon monoxide is

(*a*) 0.528   (*b*) 0.634   (*c*) 0.833   (*d*) 1.0   (*e*) none of these answers

6.92 Select the correct statement regarding flow through a converging-diverging tube.

(*a*) When the Mach number at exit is greater than unity no shock wave has developed in the tube.
(*b*) When the critical pressure ratio is exceeded, the Mach number at the throat is greater than unity.
(*c*) For sonic velocity at the throat, one and only one pressure or velocity can occur at a given downstream location.
(*d*) The Mach number at the throat is always unity.
(*e*) The density increases in the downstream direction throughout the converging portion of the tube.

6.93 In a normal shock wave in one-dimensional flow the

(*a*) velocity, pressure, and density increase
(*b*) pressure, density, and temperature increase
(*c*) velocity, temperature, and density increase
(*d*) pressure, density, and momentum per unit time increase
(*e*) entropy remains constant

6.94 A normal shock wave

(*a*) is reversible
(*b*) may occur in a converging tube
(*c*) is irreversible
(*d*) is isentropic
(*e*) is none of these answers

6.95 A normal shock wave is analogous to

(*a*) an elementary wave in still liquid
(*b*) the hydraulic jump

(*c*) open-channel conditions with $\mathbf{F} < 1$
(*d*) flow of liquid through an expanding nozzle
(*e*) none of these answers

6.96 Across a normal shock wave in a converging-diverging nozzle for adiabatic flow the following relationships are valid:

(*a*) continuity and energy equations, equation of state, isentropic relationship
(*b*) energy and momentum equations, equation of state, isentropic relationship
(*c*) continuity, energy, and momentum equations; equation of state
(*d*) equation of state, isentropic relationship, momentum equation, mass-conservation principle
(*e*) none of these answers

6.97 Across a normal shock wave there is an increase in

(*a*) p, $\mathbf{M}$, s (*b*) p, s; decrease in $\mathbf{M}$ (*c*) p; decrease in s, $\mathbf{M}$
(*d*) p, $\mathbf{M}$; no change in s (*e*) p, $\mathbf{M}$, T

6.98 A Fanno line is developed from the following equations:

(*a*) momentum and continuity
(*b*) energy and continuity
(*c*) momentum and energy
(*d*) momentum, continuity, and energy
(*e*) none of these answers

6.99 A Rayleigh line is developed from the following equations:

(*a*) momentum and continuity
(*b*) energy and continuity
(*c*) momentum and energy
(*d*) momentum, continuity, and energy
(*e*) none of these answers

6.100 Select the correct statement regarding a Fanno or Rayleigh line:

(*a*) Two points having the same value of entropy represent conditions before and after a shock wave.
(*b*) pV is held constant along the line.
(*c*) Mach number always increases with entropy.
(*d*) The subsonic portion of the curve is at higher enthalpy than the supersonic portion.
(*e*) Mach 1 is located at the maximum enthalpy point.

6.101 Choking in pipe flow means that

(*a*) a valve is closed in the line
(*b*) a restriction in flow area occurs

(*c*) the specified mass flow rate cannot occur
(*d*) shock waves always occur
(*e*) supersonic flow occurs somewhere in the line

6.102 In subsonic adiabatic flow with friction in a pipe

(*a*) V, **M**, s increase; p, T, ρ decrease   (*b*) p, V, **M** increase; T, ρ decrease
(*c*) p, **M**, s increase; V, T, ρ decrease   (*d*) ρ, **M**, s increase; V, T, p decrease
(*e*) T, V, s increase; **M**, p, ρ decrease

6.103 In supersonic adiabatic flow with friction in a pipe

(*a*) V, **M**, s increase; p, T, ρ decrease   (*b*) p, T, s increase; ρ, V, **M** decrease
(*c*) p, **M**, s increase; V, T, ρ decrease   (*d*) p, T, ρ, s increase; V, **M** decrease
(*e*) p, ρ, s increase; V, **M**, T decrease

6.104 Select the correct statement regarding frictionless duct flow with heat transfer:

(*a*) Adding heat to supersonic flow increases the Mach number.
(*b*) Adding heat to subsonic flow increases the Mach number.
(*c*) Cooling supersonic flow decreases the Mach number.
(*d*) The Fanno line is valuable in analyzing the flow.
(*e*) The isentropic stagnation temperature remains constant along the pipe.

6.105 Select the correct trends in flow properties for frictionless duct flow with heat transferred to the pipe **M** < 1:

(*a*) p, V increase; ρ, T, T_0 decrease   (*b*) V, T_0 increase; p, ρ decrease
(*c*) p, ρ, T increase; V, T_0 decrease   (*d*) V, T increase; p, ρ, T_0 decrease
(*e*) T_0, V, ρ increase; p, T decrease

6.106 Select the correct trends for cooling in frictionless duct flow **M** > 1:

(*a*) V increases; p, ρ, T, T_0 decrease   (*b*) p, V increase; ρ, T, T_0 decrease
(*c*) p, ρ, V increase; T, T_0 decrease   (*d*) p, ρ increase; V, T, T_0 decrease
(*e*) V, T, T_0 increase; p, ρ decrease

6.107 In steady, isothermal flow in long pipelines, the significant value of **M** for determining trends in flow properties is

(*a*) $1/k$ (*b*) $1/\sqrt{k}$ (*c*) 1 (*d*) $\sqrt{k}$ (*e*) k

6.108 Select the correct trends in fluid properties for isothermal flow in ducts for **M** < 0.5:

(*a*) V increases; **M**, T_0, p, p_0, ρ decrease   (*b*) V, **M** increase; T_0, p, p_0, ρ decrease
(*c*) V, **M**, T_0 increase; p, p_0, ρ decrease   (*d*) V, T_0 increase; **M**, p, p_0, ρ decrease
(*e*) V, **M**, p_0, T_0 increase; p, ρ decrease

REFERENCES

Cambel, A. B., and B. H. Jennings: "Gas Dynamics," McGraw-Hill, New York, 1958.

Keenan, J. H., and J. Kaye: "Gas Tables," Wiley, New York, 1948.

Liepmann, H. W., and A. Roshko: "Elements of Gas Dynamics," Wiley, New York, 1957.

Owczarek, J. A.: "Fundamentals of Gas Dynamics," International Textbook, Scranton, Pa., 1964.

Shapiro, A. H.: "The Dynamics and Thermodynamics of Compressible Fluid Flow," vol. 1, Ronald, New York, 1953.

7

IDEAL-FLUID FLOW

In the preceding chapters most of the relationships have been developed for one-dimensional flow, i.e., flow in which the average velocity at each cross section is used and variations across the section are neglected. Many design problems in fluid flow, however, require more exact knowledge of velocity and pressure distributions, such as in flow over curved boundaries along an airplane wing, through the passages of a pump or compressor, or over the crest of a dam. An understanding of two- and three-dimensional flow of a nonviscous, incompressible fluid provides the student with a much broader approach to many real fluid-flow situations. There are also analogies that permit the same methods to apply to flow through porous media.

In this chapter the principles of irrotational flow of an ideal fluid are developed and applied to elementary flow cases. After the flow requirements are established, the vector operator ∇ is introduced, Euler's equation is derived and the velocity potential is defined. Euler's equation is then integrated to obtain Bernoulli's equation, and stream functions and boundary conditions are developed. Flow cases are then studied in three and two dimensions.

7.1 REQUIREMENTS FOR IDEAL-FLUID FLOW

The Prandtl hypothesis, Sec. 5.6, states that, for fluids of low viscosity, the effects of viscosity are appreciable only in a narrow region surrounding the fluid boundaries. For incompressible-flow situations in which the boundary layer remains thin, ideal-fluid results may be applied to flow of a real fluid to a satisfactory degree of approximation. Converging or accelerating flow situations generally have thin boundary layers, but decelerating flow may have

389

separation of the boundary layer and development of a large wake that is difficult to predict analytically.

An ideal fluid must satisfy the following requirements:

1. The continuity equation, Sec. 3.4, div $\mathbf{q} = 0$, or

$$\frac{\partial u}{\partial x} + \frac{\partial v}{\partial y} + \frac{\partial w}{\partial z} = 0$$

2. Newton's second law of motion at every point at every instant.
3. Neither penetration of fluid into nor gaps between fluid and boundary at any solid boundary.

If, in addition to requirements 1, 2, and 3, the assumption of irrotational flow is made, the resulting fluid motion closely resembles real-fluid motion for fluids of low viscosity, outside boundary layers.

Using the above conditions, the application of Newton's second law to a fluid particle leads to the Euler equation, which, together with the assumption of irrotational flow, can be integrated to obtain the Bernoulli equation. The unknowns in a fluid-flow situation with given boundaries are velocity and pressure at every point. Unfortunately, in most cases it is impossible to proceed directly to equations for velocity and pressure distribution from the boundary conditions.

7.2 THE VECTOR OPERATOR $\triangledown$

The vector operator $\triangledown$, which may act on a vector as a scalar or vector product or may act on a scalar function, is most useful in developing ideal-fluid-flow theory.

Let U be the quantity acted upon by the operator. The operator $\triangledown$ is defined by

$$\triangledown U = \lim_{\mathcal{V} \to 0} \frac{1}{\mathcal{V}} \int_S \mathbf{n}_1 U \, dS \tag{7.2.1}$$

U may be interpreted as $\cdot \mathbf{a}$, $\times \mathbf{a}$, where $\mathbf{a}$ is any vector, or as a scalar, say ϕ. Consider a small volume $\mathcal{V}$ with surface S and surface element dS. $\mathbf{n}_1$ is a unit vector in the direction of the outwardly drawn normal n of the surface element dS (Fig. 7.1). This definition of the operator is now examined to develop the concepts of gradient, divergence, and curl.

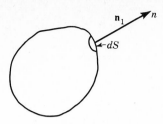

Fig. 7.1 Notation for unit vector $\mathbf{n}_1$ normal to area element dS.

When U is a scalar, say ϕ, the *gradient* of ϕ is

$$\text{grad } \phi = \boldsymbol{\nabla}\phi = \lim_{\upsilon \to 0} \frac{1}{\upsilon} \int_S \mathbf{n}_1 \phi \, dS \tag{7.2.2}$$

To interpret grad ϕ, the volume element is taken as a small prism of cross-sectional area dS, of height dn, with one end area in the surface $\phi(x,y,z) = c$ and the other end area in the surface

$$\phi + \left(\frac{\partial \phi}{\partial n}\right) dn = \text{const}$$

(Fig. 7.2). As there is no change in ϕ in surfaces parallel to the end faces, by symmetry, $\int \mathbf{n}_1 \phi \, dS$ over the curved surface of the element vanishes. Then

$$\int_S \mathbf{n}_1 \phi \, dS = \mathbf{n}_1 \left(\phi + \frac{\partial \phi}{\partial n} dn - \phi \right) dS$$

and the right-hand side of Eq. (7.2.2) becomes

$$\lim_{\upsilon \to 0} \frac{\mathbf{n}_1}{dS \, dn} \frac{\partial \phi}{\partial n} dn \, dS = \mathbf{n}_1 \frac{\partial \phi}{\partial n}$$

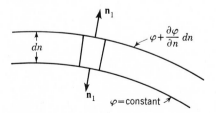

Fig. 7.2 Surfaces of constant scalar ϕ.

and

$$\text{grad } \phi = \nabla\phi = \mathbf{n_1}\frac{\partial\phi}{\partial n} \tag{7.2.3}$$

in which $\mathbf{n_1}$ is the unit vector, drawn normal to the surface over which ϕ is constant, positive in the direction of increasing ϕ; grad ϕ is a vector.

By interpreting U as the scalar (dot) product with ∇, the *divergence* is obtained. Let U be $\cdot\mathbf{q}$; then

$$\text{div }\mathbf{q} = \nabla \cdot \mathbf{q} = \lim_{\upsilon\to 0}\frac{1}{\upsilon}\int_S \mathbf{n_1} \cdot \mathbf{q}\, dS \tag{7.2.4}$$

This expression has been used (in somewhat different form) in deriving the general continuity equation in Sec. 3.4. It is the volume flux per unit volume at a point and is a scalar.

The *curl* $\nabla \times \mathbf{q}$ is a more difficult concept that deals with the *vorticity* or rotation of a fluid element:

$$\text{curl }\mathbf{q} = \nabla \times \mathbf{q} = \lim_{\upsilon\to 0}\frac{1}{\upsilon}\int_S \mathbf{n_1} \times \mathbf{q}\, dS \tag{7.2.5}$$

With reference to Fig. 7.3, $\mathbf{n_1} \times \mathbf{q}$ is the velocity component tangent to the surface element dS at a point, since the vector product is a vector at right angles to the plane of the two constituent vectors, with magnitude $q \sin\theta$, as $n_1 = 1$. Then $\mathbf{n_1} \times \mathbf{q}\, dS$ is an elemental vector that is the product of tangential velocity and surface area element. Summing up over the surface,

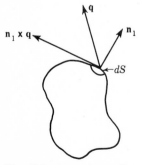

Fig. 7.3 Notation for curl of the velocity vector.

then dividing by the volume, with the limit taken as $\upsilon \rightarrow 0$, yields the curl **q** at a point.

A special type of fluid motion is examined to demonstrate the connection between curl and rotation. Let a small circular cylinder of fluid be rotating about its axis as if it were a solid (Fig. 7.4), with angular velocity ω, which is a vector parallel to the axis of rotation. The radius of the cylinder is r and the length l. $\mathbf{n}_1 \times \mathbf{q}$ at every point on the curved surface is a vector parallel to the axis having the magnitude $q = \omega r$. Over the end areas the vector $\mathbf{n}_1 \times \mathbf{q}$ is equal and opposite at corresponding points on each end and contributes nothing to the curl. Then, since $ds = lr\, d\alpha$,

$$\int_S \mathbf{n}_1 \times \mathbf{q}\, dS = \boldsymbol{\omega} \int_0^{2\pi} rlr\, d\alpha = 2\pi r^2 l \boldsymbol{\omega}$$

Equation (7.2.5) now yields

$$\text{curl } \mathbf{q} = \lim_{\upsilon \to 0} \frac{1}{\pi r^2 l}\, 2\pi r^2 l \boldsymbol{\omega} = 2\boldsymbol{\omega}$$

showing that for solid-body rotation the curl of the velocity at a point is twice the rotation vector. If one considers the pure translation of a small element moving as a solid, then the curl **q** is always zero. As any rigid-body motion is a combination of a translation and a rotation, the curl of the velocity vector is always twice the rotation vector.

A fluid, however, not only may translate and rotate but also may deform. The definition of curl **q** applies, and hence the *rotation of a fluid* at a point

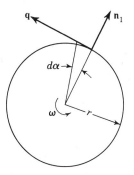

Fig. 7.4 Small fluid cylinder rotating as a solid.

is defined by

$$\boldsymbol{\omega} = \tfrac{1}{2}\operatorname{curl}\mathbf{q} = \tfrac{1}{2}(\boldsymbol{\nabla}\times\mathbf{q}) \tag{7.2.6}$$

When $\boldsymbol{\omega} = 0$ throughout certain portions of a fluid, the motion there is described as *irrotational*. The vorticity vector curl $\mathbf{q}$ has certain characteristics similar to the velocity vector $\mathbf{q}$. Vortex lines are everywhere tangent to the vorticity vector, and vortex tubes, comprising the vortex lines through a small closed curve, follow certain continuity principles; viz., the product of vorticity by area of the tube must remain constant along the vortex tube, or div (curl $\mathbf{q}$) $= \boldsymbol{\nabla}\cdot(\boldsymbol{\nabla}\times\mathbf{q}) = 0$.

The operator $\boldsymbol{\nabla}$ acts like a vector but must be applied to a scalar or a vector to have physical significance.

Scalar components of vector relationships

Any vector can be decomposed into three components along mutually perpendicular axes, say the x, y, z axes. The component is a scalar, as only magnitude and sign (sense) are needed to specify it; $f_x = -3$ indicates the x component of a vector $\mathbf{f}$ acting in the $-x$ direction.

The vector is expressed in terms of its scalar components by use of the fixed unit vectors, $\mathbf{i}$, $\mathbf{j}$, $\mathbf{k}$ parallel to the x, y, z axes, respectively:

$$\mathbf{a} = \mathbf{i}a_x + \mathbf{j}a_y + \mathbf{k}a_z$$

The unit vectors combine as follows:

$$\mathbf{i}\cdot\mathbf{i} = \mathbf{j}\cdot\mathbf{j} = \mathbf{k}\cdot\mathbf{k} = 1 \qquad \mathbf{i}\cdot\mathbf{j} = \mathbf{j}\cdot\mathbf{k} = \mathbf{k}\cdot\mathbf{i} = 0$$

$$\mathbf{i}\times\mathbf{j} = \mathbf{k} \qquad \mathbf{j}\times\mathbf{k} = \mathbf{i} \qquad \mathbf{k}\times\mathbf{i} = \mathbf{j} = -\mathbf{i}\times\mathbf{k} \qquad \text{etc.}$$

The scalar product of two vectors $\mathbf{a}\cdot\mathbf{b}$ is

$$\mathbf{a}\cdot\mathbf{b} = (\mathbf{i}a_x + \mathbf{j}a_y + \mathbf{k}a_z)\cdot(\mathbf{i}b_x + \mathbf{j}b_y + \mathbf{k}b_z)$$

$$= a_x b_x + a_y b_y + a_z b_z$$

The vector product of two vectors $\mathbf{a}\times\mathbf{b}$ is

$$\mathbf{a}\times\mathbf{b} = (\mathbf{i}a_x + \mathbf{j}a_y + \mathbf{k}a_z)\times(\mathbf{i}b_x + \mathbf{j}b_y + \mathbf{k}b_z)$$

$$= \mathbf{i}(a_y b_z - a_z b_y) + \mathbf{j}(a_z b_x - a_x b_z) + \mathbf{k}(a_x b_y - a_y b_x)$$

It is conveniently written in determinant form:

$$\mathbf{a} \times \mathbf{b} = \begin{vmatrix} \mathbf{i} & \mathbf{j} & \mathbf{k} \\ a_x & a_y & a_z \\ b_x & b_y & b_z \end{vmatrix}$$

To find the scalar components of $\nabla\phi$, first consider $\mathbf{a} \cdot \nabla\phi$ (Fig. 7.5), in which $\mathbf{a}$ is any vector. By Eq. (7.2.3)

$$\mathbf{a} \cdot \nabla\phi = \mathbf{a} \cdot \mathbf{n}_1 \frac{\partial\phi}{\partial n} = a \cos\theta \frac{\partial\phi}{\partial n}$$

as θ is the angle between $\mathbf{a}$ and $\mathbf{n}_1$ and $n_1 = 1$. A change da in magnitude of $\mathbf{a}$ corresponds to a change in $\mathbf{n}$, given by $da \cos\theta = dn$; hence

$$a \cos\theta \frac{\partial\phi}{\partial n} = a \frac{\partial\phi}{\partial a}$$

and

$$\mathbf{a} \cdot \nabla\phi = a \frac{\partial\phi}{\partial a} \tag{7.2.7}$$

The scalar components of $\nabla\phi$ are

$$\mathbf{i} \cdot \nabla\phi = \frac{\partial\phi}{\partial x} \qquad \mathbf{j} \cdot \nabla\phi = \frac{\partial\phi}{\partial y} \qquad \mathbf{k} \cdot \nabla\phi = \frac{\partial\phi}{\partial z}$$

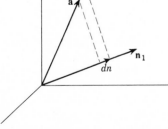

Fig. 7.5 Change of vector $\mathbf{a}$ corresponding to change in normal direction.

and

$$\nabla\phi = \mathbf{i}\frac{\partial\phi}{\partial x} + \mathbf{j}\frac{\partial\phi}{\partial y} + \mathbf{k}\frac{\partial\phi}{\partial z} \tag{7.2.8}$$

The operator ∇, in terms of its scalar components, is

$$\nabla = \mathbf{i}\frac{\partial}{\partial x} + \mathbf{j}\frac{\partial}{\partial y} + \mathbf{k}\frac{\partial}{\partial z} \tag{7.2.9}$$

The scalar product, say $\nabla \cdot \mathbf{q}$, becomes

$$\nabla \cdot \mathbf{q} = \left(\mathbf{i}\frac{\partial}{\partial x} + \mathbf{j}\frac{\partial}{\partial y} + \mathbf{k}\frac{\partial}{\partial z}\right) \cdot (\mathbf{i}u + \mathbf{j}v + \mathbf{k}w)$$

$$= \frac{\partial u}{\partial x} + \frac{\partial v}{\partial y} + \frac{\partial w}{\partial z} \tag{7.2.10}$$

as in Sec. 3.4.

The vector product $\nabla \times \mathbf{q}$, in scalar components, is

$$\nabla \times \mathbf{q} = \left(\mathbf{i}\frac{\partial}{\partial x} + \mathbf{j}\frac{\partial}{\partial y} + \mathbf{k}\frac{\partial}{\partial z}\right) \times (\mathbf{i}u + \mathbf{j}v + \mathbf{k}w)$$

$$= \mathbf{i}\left(\frac{\partial w}{\partial y} - \frac{\partial v}{\partial z}\right) + \mathbf{j}\left(\frac{\partial u}{\partial z} - \frac{\partial w}{\partial x}\right) + \mathbf{k}\left(\frac{\partial v}{\partial x} - \frac{\partial u}{\partial y}\right) \tag{7.2.11}$$

The quantities in parentheses are vorticity components, which are twice the value of rotation components ω_x, ω_y, ω_z, and so

$$\nabla \times \mathbf{q} = \mathbf{i}2\omega_x + \mathbf{j}2\omega_y + \mathbf{k}2\omega_z \tag{7.2.12}$$

7.3 EULER'S EQUATION OF MOTION

In Sec. 3.5, Euler's equation was derived for steady flow of a frictionless fluid along a streamline. The assumption is made here that the flow is frictionless, and a continuum is assumed. Newton's second law of motion is applied to a fluid particle of mass $\rho\,\delta\mathcal{V}$. Three terms enter: the body force, the surface force, and mass times acceleration. Let $\mathbf{F}$ be the body force (such as gravity) per unit mass acting on the particle. Then $\mathbf{F}\rho\,\delta\mathcal{V}$ is the body-force vector. The surface force, from the preceding section, is $-\int_s \mathbf{n}_1 p\,dS$ if the fluid is frictionless or nonviscous, so that only normal forces act. The mass-times-acceleration term is $\rho \, \delta\mathcal{V} \, d\mathbf{q}/dt$. Assembling these terms gives

$$\mathbf{F}\rho \, \delta\mathcal{V} - \int_S \mathbf{n}_1 p \, ds = \rho \, \delta\mathcal{V} \frac{d\mathbf{q}}{dt}$$

Now, dividing through by the mass of the element and taking the limit as $\delta\mathcal{V} \to 0$ yields

$$\mathbf{F} - \frac{1}{\rho} \lim_{\delta\mathcal{V} \to 0} \frac{1}{\delta\mathcal{V}} \int_S \mathbf{n}_1 p \, dS = \frac{d\mathbf{q}}{dt}$$

Use of the operator $\boldsymbol{\nabla}$ leads to

$$\mathbf{F} - \frac{1}{\rho} \boldsymbol{\nabla} p = \frac{d\mathbf{q}}{dt} \qquad (7.3.1)$$

This is Euler's equation of motion in vector notation. By forming the scalar product of each term with $\mathbf{i}$, then $\mathbf{j}$, then $\mathbf{k}$, the following scalar component equations are obtained:

$$X - \frac{1}{\rho} \frac{\partial p}{\partial x} = \frac{du}{dt} \qquad Y - \frac{1}{\rho} \frac{\partial p}{\partial y} = \frac{dv}{dt} \qquad Z - \frac{1}{\rho} \frac{\partial p}{\partial z} = \frac{dw}{dt} \qquad (7.3.2)$$

in which X, Y, Z are the body-force components per unit mass. The acceleration terms may be expanded. In general $u = u(x,y,z,t)$, and so (see Appendix B)

$$du = \frac{\partial u}{\partial x} \, dx + \frac{\partial u}{\partial y} \, dy + \frac{\partial u}{\partial z} \, dz + \frac{\partial u}{\partial t} \, dt$$

For du/dt to be the acceleration component of a particle in the x direction, the x, y, z coordinates of the moving particle become functions of time, and du may be divided by dt, yielding

$$a_x = \frac{du}{dt} = \frac{\partial u}{\partial x} \frac{dx}{dt} + \frac{\partial u}{\partial y} \frac{dy}{dt} + \frac{\partial u}{\partial z} \frac{dz}{dt} + \frac{\partial u}{\partial t}$$

But

$$u = \frac{dx}{dt} \qquad v = \frac{dy}{dt} \qquad w = \frac{dz}{dt}$$

and

$$\frac{du}{dt} = u\frac{\partial u}{\partial x} + v\frac{\partial u}{\partial y} + w\frac{\partial u}{\partial z} + \frac{\partial u}{\partial t} \tag{7.3.3}$$

Similarly

$$\frac{dv}{dt} = u\frac{\partial v}{\partial x} + v\frac{\partial v}{\partial y} + w\frac{\partial v}{\partial z} + \frac{\partial v}{\partial t} \tag{7.3.4}$$

$$\frac{dw}{dt} = u\frac{\partial w}{\partial x} + v\frac{\partial w}{\partial y} + w\frac{\partial w}{\partial z} + \frac{\partial w}{\partial t} \tag{7.3.5}$$

If the extraneous force is *conservative*, it may be derived from a potential ($\mathbf{F} = -\operatorname{grad}\Omega$):

$$X = -\frac{\partial\Omega}{\partial x} \qquad Y = -\frac{\partial\Omega}{\partial y} \qquad Z = -\frac{\partial\Omega}{\partial z} \tag{7.3.6}$$

In particular, if gravity is the only body force acting, $\Omega = gh$, with h a direction measured vertically upward; thus

$$X = -g\frac{\partial h}{\partial x} \qquad Y = -g\frac{\partial h}{\partial y} \qquad Z = -g\frac{\partial h}{\partial z} \tag{7.3.7}$$

Remembering that ρ is constant for an ideal fluid, substituting Eqs. (7.3.3) to (7.3.7) into Eqs. (7.3.2) gives

$$-\frac{1}{\rho}\frac{\partial}{\partial x}(p + \gamma h) = u\frac{\partial u}{\partial x} + v\frac{\partial u}{\partial y} + w\frac{\partial u}{\partial z} + \frac{\partial u}{\partial t} \tag{7.3.8}$$

$$-\frac{1}{\rho}\frac{\partial}{\partial y}(p + \gamma h) = u\frac{\partial v}{\partial x} + v\frac{\partial v}{\partial y} + w\frac{\partial v}{\partial z} + \frac{\partial v}{\partial t} \tag{7.3.9}$$

$$-\frac{1}{\rho}\frac{\partial}{\partial z}(p + \gamma h) = u\frac{\partial w}{\partial x} + v\frac{\partial w}{\partial y} + w\frac{\partial w}{\partial z} + \frac{\partial w}{\partial t} \tag{7.3.10}$$

The first three terms on the right-hand sides of the equations are *convective-acceleration* terms, depending upon changes of velocity with space. The last term is the *local acceleration*, depending upon velocity change with time at a point.

Natural coordinates in two-dimensional flow

Euler's equations in two dimensions are obtained from the general-component equations by setting $w = 0$ and $\partial/\partial z = 0$; thus

$$-\frac{1}{\rho}\frac{\partial}{\partial x}(p + \gamma h) = u\frac{\partial u}{\partial x} + v\frac{\partial u}{\partial y} + \frac{\partial u}{\partial t} \qquad (7.3.11)$$

$$-\frac{1}{\rho}\frac{\partial}{\partial y}(p + \gamma h) = u\frac{\partial v}{\partial x} + v\frac{\partial v}{\partial y} + \frac{\partial v}{\partial t} \qquad (7.3.12)$$

By taking particular directions for the x and y axes, they can be reduced to a form that makes them easier to understand. If the x axis, called the s axis, is taken parallel to the velocity vector at a point (Fig. 7.6), it is then tangent to the streamline through the point. The y axis, called the n axis, is drawn toward the center of curvature of the streamline. The velocity component u is v_s and the component v is v_n. As v_n is zero at the point, Eq. (7.3.11) becomes

$$-\frac{1}{\rho}\frac{\partial}{\partial s}(p + \gamma h) = v_s\frac{\partial v_s}{\partial s} + \frac{\partial v_s}{\partial t} \qquad (7.3.13)$$

Although v_n is zero at the point (s,n), its rates of change with respect to s and t are not necessarily zero. Equation (7.3.12) becomes

$$-\frac{1}{\rho}\frac{\partial}{\partial n}(p + \gamma h) = v_s\frac{\partial v_n}{\partial s} + \frac{\partial v_n}{\partial t} \qquad (7.3.14)$$

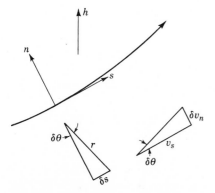

Fig. 7.6 Notation for natural coordinates.

When the velocity at s and at $s + \delta s$ along the streamline is considered, v_n changes from zero to δv_n. With r the radius of curvature of the streamline at s, from similar triangles (Fig. 7.6),

$$\frac{\delta s}{r} = \frac{\delta v_n}{v_s} \qquad \text{or} \qquad \frac{\partial v_n}{\partial s} = \frac{v_s}{r}$$

Substituting into Eq. (7.3.14) gives

$$-\frac{1}{\rho}\frac{\partial}{\partial n}(p + \gamma h) = \frac{v_s^2}{r} + \frac{\partial v_n}{\partial t} \tag{7.3.15}$$

For steady flow of an incompressible fluid Eqs. (7.3.11) and (7.3.15) may be written

$$-\frac{1}{\rho}\frac{\partial}{\partial s}(p + \gamma h) = \frac{\partial}{\partial s}\left(\frac{v_s^2}{2}\right) \tag{7.3.16}$$

and

$$-\frac{1}{\rho}\frac{\partial}{\partial n}(p + \gamma h) = \frac{v_s^2}{r} \tag{7.3.17}$$

Equation (7.3.16) can be integrated with respect to s to produce Eq. (3.9.1), with the constant of integration varying with n, that is, from one streamline to another. Equation (7.3.17) shows how pressure head varies across streamlines. With v_s and r known functions of n, Eq. (7.3.17) can be integrated.

EXAMPLE 7.1 A container of liquid is rotated with angular velocity ω about a vertical axis as a solid. Determine the variation of pressure in the liquid.

n is the radial distance, measured inwardly, $n = -r$, $dn = -dr$, and $v_s = \omega r$. By integrating Eq. (7.3.17),

$$-\frac{1}{\rho}(p + \gamma h) = -\int \frac{\omega^2 r^2 \, dr}{r} + \text{const}$$

or

$$\frac{1}{\rho}(p + \gamma h) = \frac{\omega^2 r^2}{2} + \text{const}$$

To evaluate the constant, if $p = p_0$ when $r = 0$ and $h = 0$, then

$$p = p_0 - \gamma h + \rho \frac{\omega^2 r^2}{2}$$

which shows that the pressure is hydrostatic along a vertical line and increases as the square of the radius. Integration of Eq. (7.3.16) shows that the pressure is constant for a given h and v_s, that is, along a streamline.

7.4   IRROTATIONAL FLOW; VELOCITY POTENTIAL

In this section it is shown that the assumption of irrotational flow leads to the existence of a velocity potential. By use of these relations and the assumption of a conservative body force, the Euler equations can be integrated.

The individual particles of a frictionless incompressible fluid initially at rest cannot be caused to rotate. This can be visualized by considering a small free body of fluid in the shape of a sphere. Surface forces act normal to its surface, since the fluid is frictionless, and therefore act through the center of the sphere. Similarly the body force acts at the mass center. Hence no torque can be exerted on the sphere, and it remains without rotation. Likewise, once an ideal fluid has rotation, there is no way of altering it, as no torque can be exerted on an elementary sphere of the fluid.

By assuming that the fluid has no rotation, i.e., it is irrotational, curl $\mathbf{q} = 0$, or from Eq. (7.2.11)

$$\frac{\partial v}{\partial x} = \frac{\partial u}{\partial y} \qquad \frac{\partial w}{\partial y} = \frac{\partial v}{\partial z} \qquad \frac{\partial u}{\partial z} = \frac{\partial w}{\partial x} \qquad (7.4.1)$$

These restrictions on the velocity must hold at every point (except special singular points or lines). The first equation is the irrotational condition for two-dimensional flow. It is the condition that the differential expression

$$u \, dx + v \, dy$$

is exact, say

$$u \, dx + v \, dy = -d\phi = -\frac{\partial \phi}{\partial x} dx - \frac{\partial \phi}{\partial y} dy \qquad (7.4.2)$$

The minus sign is arbitrary; it is a convention that causes the value of ϕ to decrease in the direction of the velocity. By comparing terms in Eq. (7.4.2),

$u = -\partial\phi/\partial x$, $v = -\partial\phi/\partial y$. This proves the existence, in two-dimensional flow, of a function ϕ such that its negative derivative with respect to any direction is the velocity component in that direction. It can also be demonstrated for three-dimensional flow. In vector form,

$$\mathbf{q} = -\text{grad } \phi = -\nabla\phi \tag{7.4.3}$$

is equivalent to

$$u = -\frac{\partial\phi}{\partial x} \qquad v = -\frac{\partial\phi}{\partial y} \qquad w = -\frac{\partial\phi}{\partial z} \tag{7.4.4}$$

The assumption of a velocity potential is equivalent to the assumption of irrotational flow, as

$$\text{curl } (-\text{grad } \phi) = -\nabla \times \nabla\phi = 0 \tag{7.4.5}$$

because $\nabla \times \nabla = 0$. This is shown from Eq. (7.4.4) by cross-differentiation:

$$\frac{\partial u}{\partial y} = -\frac{\partial^2\phi}{\partial x\,\partial y} \qquad \frac{\partial v}{\partial x} = -\frac{\partial^2\phi}{\partial y\,\partial x}$$

proving $\partial v/\partial x = \partial u/\partial y$, etc.

Substitution of Eqs. (7.4.4) into the continuity equation

$$\frac{\partial u}{\partial x} + \frac{\partial v}{\partial y} + \frac{\partial w}{\partial z} = 0$$

yields

$$\frac{\partial^2\phi}{\partial x^2} + \frac{\partial^2\phi}{\partial y^2} + \frac{\partial^2\phi}{\partial z^2} = 0 \tag{7.4.6}$$

In vector form this is

$$\nabla \cdot \mathbf{q} = -\nabla \cdot \nabla\phi = -\nabla^2\phi = 0 \tag{7.4.7}$$

and is written $\nabla^2\phi = 0$. Equation (7.4.6) or (7.4.7) is the *Laplace* equation. Any function ϕ that satisfies the Laplace equation is a possible irrotational fluid-flow case. As there are an infinite number of solutions to the Laplace equation, each of which satisfies certain flow boundaries, the main problem is the selection of the proper function for the particular flow case.

Because ϕ appears to the first power in each term of Eq. (7.4.6), it is a linear equation, and the sum of two solutions is also a solution; e.g., if ϕ_1 and ϕ_2 are solutions of Eq. (7.4.6) then $\phi_1 + \phi_2$ is a solution; thus

$$\nabla^2\phi_1 = 0 \qquad \nabla^2\phi_2 = 0$$

then

$$\nabla^2(\phi_1 + \phi_2) = \nabla^2\phi_1 + \nabla^2\phi_2 = 0$$

Similarly if ϕ_1 is a solution $C\phi_1$ is a solution if C is constant.

7.5 INTEGRATION OF EULER'S EQUATIONS; BERNOULLI EQUATION

Equation (7.3.8) can be rearranged so that every term contains a partial derivative with respect to x. From Eq. (7.4.1)

$$v\,\frac{\partial u}{\partial y} = v\,\frac{\partial v}{\partial x} = \frac{\partial}{\partial x}\frac{v^2}{2} \qquad w\,\frac{\partial u}{\partial z} = w\,\frac{\partial w}{\partial x} = \frac{\partial}{\partial x}\frac{w^2}{2}$$

and from Eq. (7.4.4)

$$\frac{\partial u}{\partial t} = -\frac{\partial}{\partial x}\frac{\partial \phi}{\partial t}$$

Making these substitutions into Eq. (7.3.8) and rearranging give

$$\frac{\partial}{\partial x}\left(\frac{p}{\rho} + gh + \frac{u^2}{2} + \frac{v^2}{2} + \frac{w^2}{2} - \frac{\partial \phi}{\partial t}\right) = 0$$

As $u^2 + v^2 + w^2 = q^2$, the square of the speed,

$$\frac{\partial}{\partial x}\left(\frac{p}{\rho} + gh + \frac{q^2}{2} - \frac{\partial \phi}{\partial t}\right) = 0 \tag{7.5.1}$$

Similarly for the y and z directions,

$$\frac{\partial}{\partial y}\left(\frac{p}{\rho} + gh + \frac{q^2}{2} - \frac{\partial \phi}{\partial t}\right) = 0 \tag{7.5.2}$$

$$\frac{\partial}{\partial z}\left(\frac{p}{\rho} + gh + \frac{q^2}{2} - \frac{\partial \phi}{\partial t}\right) = 0 \tag{7.5.3}$$

The quantities within the parentheses are the same in Eqs. (7.5.1) to (7.5.3). Equation (7.5.1) states that the quantity is not a function of x, since the derivative with respect to x is zero. Similarly the other equations show that the quantity is not a function of y or z. Therefore it can be a function of t only, say $F(t)$:

$$\frac{p}{\rho} + gh + \frac{q^2}{2} - \frac{\partial \phi}{\partial t} = F(t) \tag{7.5.4}$$

In steady flow $\partial \phi / \partial t = 0$ and $F(t)$ becomes a constant E:

$$\frac{p}{\rho} + gh + \frac{q^2}{2} = E \tag{7.5.5}$$

The available energy is everywhere constant throughout the fluid. This is Bernoulli's equation for an irrotational fluid.

The pressure term can be separated into two parts, the hydrostatic pressure p_s and the dynamic pressure p_d, so that $p = p_s + p_d$. Inserting in Eq. (7.5.5) gives

$$gh + \frac{p_s}{\rho} + \frac{p_d}{\rho} + \frac{q^2}{2} = E$$

The first two terms may be written

$$gh + \frac{p_s}{\rho} = \frac{1}{\rho}(p_s + \gamma h)$$

with h measured vertically upward. The expression is a constant, since it expresses the hydrostatic law of variation of pressure. These two terms may be included in the constant E. After dropping the subscript on the dynamic pressure, there remains

$$\frac{p}{\rho} + \frac{q^2}{2} = E \tag{7.5.6}$$

This simple equation permits the variation in pressure to be determined if the speed is known or vice versa. Assuming both the speed q_0 and the dynamic pressure p_0 to be known at one point,

$$\frac{p_0}{\rho} + \frac{q_0^2}{2} = \frac{p}{\rho} + \frac{q^2}{2}$$

or

$$p = p_0 + \frac{\rho q_0^2}{2}\left[1 - \left(\frac{q}{q_0}\right)^2\right]$$ (7.5.7)

EXAMPLE 7.2 A submarine moves through water at a speed of 30 ft/s. At a point A on the submarine 5 ft above the nose, the velocity of the submarine relative to the water is 50 ft/s. Determine the dynamic pressure difference between this point and the nose, and determine the difference in total pressure between the two points.

If the submarine is stationary and the water is moving past it, the velocity at the nose is zero, and the velocity at A is 50 ft/s. By selecting the dynamic pressure at infinity as zero, from Eq. (7.5.6)

$$E = 0 + \frac{q_0^2}{2} = \frac{30^2}{2} = 450 \text{ ft} \cdot \text{lb/slug}$$

For the nose

$$\frac{p}{\rho} = E = 450 \qquad p = 450 \times 1.935 = 870 \text{ lb/ft}^2$$

for point A

$$\frac{p}{\rho} = E - \frac{q^2}{2} = 450 - \frac{50^2}{2} \qquad \text{and} \qquad p = 1.935\left(\frac{30^2}{2} - \frac{50^2}{2}\right) = -1548 \text{ lb/ft}^2$$

Therefore the difference in dynamic pressure is

$$-1548 - 870 = -2418 \text{ lb/ft}^2$$

The difference in total pressure can be obtained by applying Eq. (7.5.5) to point A and to the nose n,

$$gh_A + \frac{p_A}{\rho} + \frac{q_A^2}{2} = gh_n + \frac{p_n}{\rho} + \frac{q_n^2}{2}$$

Hence

$$p_A - p_n = \rho\left(gh_n - gh_A + \frac{q_n^2 - q_A^2}{2}\right) = 1.935\left(-5g - \frac{50^2}{2}\right)$$

$$= -2740 \text{ lb/ft}^2$$

It may also be reasoned that the actual pressure difference varies by 5γ from the dynamic pressure difference since A is 5 ft above the nose, or $-2418 - 5 \times 62.4 = -2740 \, \text{lb/ft}^2$.

7.6 STREAM FUNCTIONS; BOUNDARY CONDITIONS

Two stream functions are defined: one for two-dimensional flow, where all lines of motion are parallel to a fixed plane, say the xy plane, and the flow is identical in each of these planes, and the other for three-dimensional flow with axial symmetry, i.e., all flow lines are in planes intersecting the same line or axis, and the flow is identical in each of these planes.

Two-dimensional stream function

If A, P represent two points in one of the flow planes, e.g., the xy plane (Fig. 7.7), and if the plane has unit thickness, the rate of flow across any two lines ACP, ABP must be the same if the density is constant and no fluid is created or destroyed within the region, as a consequence of continuity. Now, if A is a fixed point and P a movable point, the flow rate across any line connecting the two points is a function of the position of P. If this function is ψ, and if it is taken as a sign convention that it denotes the flow rate from right to left as the observer views the line from A looking toward P, then

$$\psi = \psi(x,y)$$

is defined as the stream function.

If ψ_1, ψ_2 represent the values of stream function at points P_1, P_2 (Fig. 7.8), respectively, then $\psi_2 - \psi_1$ is the flow across P_1P_2 and is independent of

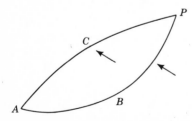

Fig. 7.7 Fluid region showing the positive flow direction used in the definition of a stream function.

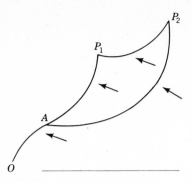

Fig. 7.8 Flow between two points in a fluid region.

the location of A. Taking another point O in the place of A changes the values of ψ_1, ψ_2 by the same amount, viz., the flow across OA. Then ψ is indeterminate to the extent of an arbitrary constant.

The velocity components u, v in the x, y directions can be obtained from the stream function. In Fig. 7.9a, the flow $\delta\psi$ across $\overline{AP} = \delta y$, from right to left, is $-u\,\delta y$, or

$$u = -\frac{\delta\psi}{\delta y} = -\frac{\partial\psi}{\partial y} \tag{7.6.1}$$

and similarly

$$v = \frac{\delta\psi}{\delta x} = \frac{\partial\psi}{\partial x} \tag{7.6.2}$$

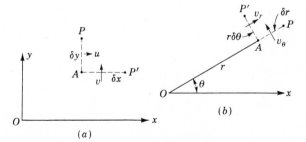

Fig. 7.9 Selection of path to show relation of velocity components to stream function.

In words, the partial derivative of the stream function with respect to any direction gives the velocity component $+90°$ (counterclockwise) to that direction. In plane polar coordinates

$$v_r = -\frac{1}{r}\frac{\partial \psi}{\partial \theta} \qquad v_\theta = \frac{\partial \psi}{\partial r}$$

from Fig. 7.9b.

When the two points P_1, P_2 of Fig. 7.8 lie on the same streamline, $\psi_1 - \psi_2 = 0$ as there is no flow across a streamline. Hence, a streamline is given by $\psi = $ const. By comparing Eqs. (7.4.4) with Eqs. (7.6.1) and (7.6.2),

$$\frac{\partial \phi}{\partial x} = \frac{\partial \psi}{\partial y} \qquad \frac{\partial \phi}{\partial y} = -\frac{\partial \psi}{\partial x} \tag{7.6.3}$$

These are the Cauchy-Riemann equations.

By Eqs. (7.6.3) a stream function can be found for each velocity potential. If the velocity potential satisfies the Laplace equation the stream function also satisfies it. Hence, the stream function may be considered as velocity potential for another flow case.

Stokes' stream function for axially symmetric flow

In any one of the planes through the axis of symmetry select two points A, P such that A is fixed and P is variable. Draw a line connecting AP. The flow through the surface generated by rotating AP about the axis of symmetry is a function of the position of P. Let this function be $2\pi\psi$, and let the axis of symmetry be the x axis of a cartesian system of reference. Then ψ is a function of x and $\hat\omega$, where

$$\hat\omega = \sqrt{y^2 + z^2}$$

is the distance from P to the x axis. The surfaces $\psi = $ const are stream surfaces.

To find the relation between ψ and the velocity components u, v' parallel to the x axis and the $\hat\omega$ axis (perpendicular to the x axis), respectively, a procedure is employed similar to that for two-dimensional flow. Let PP' be an infinitesimal step first parallel to $\hat\omega$ and then to x; that is, $PP' = \delta\hat\omega$ and then $PP' = \delta x$. The resulting relations between stream function and velocity are given by

$$-2\pi\hat\omega\,\delta\hat\omega\,u = 2\pi\,\delta\psi \qquad \text{and} \qquad 2\pi\hat\omega\,\delta x\,v' = 2\pi\,\delta\psi$$

Solving for u, v' gives

$$u = -\frac{1}{\tilde{\omega}}\frac{\partial \psi}{\partial \tilde{\omega}} \qquad v' = \frac{1}{\tilde{\omega}}\frac{\partial \psi}{\partial x} \tag{7.6.4}$$

The same sign convention is used as in the two-dimensional case.

The relations between stream function and potential function are

$$\frac{\partial \phi}{\partial x} = \frac{1}{\tilde{\omega}}\frac{\partial \psi}{\partial \tilde{\omega}} \qquad \frac{\partial \phi}{\partial \tilde{\omega}} = -\frac{1}{\tilde{\omega}}\frac{\partial \psi}{\partial x} \tag{7.6.5}$$

In three-dimensional flow with axial symmetry, ψ has the dimensions $L^3\,T^{-1}$, or volume per unit time.

The stream function is used for flow about bodies of revolution that are frequently expressed most readily in spherical polar coordinates. Let r be the distance from the origin and θ be the polar angle; the meridian angle is not needed because of axial symmetry. Referring to Fig. 7.10a and b,

$$2\pi r \sin \theta\, \delta r\, v_\theta = 2\pi\, \delta \psi \qquad -2\pi r \sin \theta\, r\, \delta\theta\, v_r = 2\,\pi\delta\psi$$

from which

$$v_\theta = \frac{1}{r \sin \theta}\frac{\partial \psi}{\partial r} \qquad v_r = -\frac{1}{r^2 \sin \theta}\frac{\partial \psi}{\partial \theta} \tag{7.6.6}$$

and

$$\frac{1}{\sin \theta}\frac{\partial \psi}{\partial \theta} = r^2 \frac{\partial \phi}{\partial r} \qquad \frac{\partial \psi}{\partial r} = -\sin \theta\, \frac{\partial \phi}{\partial \theta} \tag{7.6.7}$$

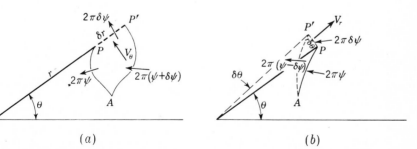

$$(a) \qquad\qquad\qquad (b)$$

Fig. 7.10 Displacement of P to show the relation between velocity components and Stokes' stream function.

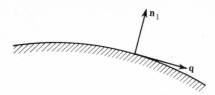

Fig. 7.11 Notation for boundary condition at a fixed boundary.

These expressions are useful in dealing with flow about spheres, ellipsoids, and disks and through apertures.

Boundary conditions

At a fixed boundary the velocity component normal to the boundary must be zero at every point on the boundary (Fig. 7.11):

$$\mathbf{q} \cdot \mathbf{n}_1 = 0 \tag{7.6.8}$$

$\mathbf{n}_1$ is a unit vector normal to the boundary. In scalar notation this is easily expressed in terms of the velocity potential

$$\frac{\partial \phi}{\partial n} = 0 \tag{7.6.9}$$

at all points on the boundary. For a moving boundary (Fig. 7.12), where the boundary point has the velocity **V**, the fluid-velocity component normal to the boundary must equal the velocity of the boundary normal to the boundary; thus

$$\mathbf{q} \cdot \mathbf{n}_1 = \mathbf{V} \cdot \mathbf{n}_1 \tag{7.6.10}$$

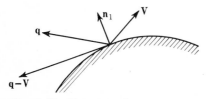

Fig. 7.12 Notation for boundary condition at a moving boundary.

or

$$(\mathbf{q} - \mathbf{V}) \cdot \mathbf{n}_1 = 0 \qquad (7.6.11)$$

For two fluids in contact, a dynamical boundary condition is required; viz., the pressure must be continuous across the interface.

A stream surface in steady flow (fixed boundaries) satisfies the condition for a boundary and may be taken as a solid boundary.

7.7 THE FLOW NET

In two-dimensional flow the flow net is of great benefit; it is taken up in this section.

The line given by $\phi(x,y) = $ const is called an *equipotential* line. It is a line along which the value of ϕ (the velocity potential) does not change. Since velocity v_s in any direction s is given by

$$v_s = -\frac{\partial \phi}{\partial s} = -\lim_{\Delta s \to 0} \frac{\Delta \phi}{\Delta s}$$

and $\Delta \phi$ is zero for two closely spaced points on an equipotential line, the velocity vector has no component in the direction defined by the line through the two points. In the limit as $\Delta s \to 0$ this proves that there is no velocity component tangent to an equipotential line and, therefore, the velocity vector must be everywhere *normal* to an equipotential line (except at singular points where the velocity is zero or infinite).

The line $\psi(x,y) = $ const is a streamline and is everywhere tangent to the velocity vector. Streamlines and equipotential lines are therefore *orthogonal*; i.e., they intersect at right angles, except at singular points. A *flow net* is composed of a family of equipotential lines and a corresponding family of streamlines with the constants varying in arithmetical progression. It is customary to let the change in constant between adjacent equipotential lines and adjacent streamlines be the same, for example, Δc. In Fig. 7.13, if the distance between streamlines is Δn and the distance between equipotential lines is Δs at some small region in the flow net, the approximate velocity v_s is then given in terms of the spacing of the equipotential lines [Eq. (7.4.4)],

$$v_s \approx -\frac{\Delta \phi}{\Delta s} = -\frac{-\Delta c}{\Delta s} = \frac{\Delta c}{\Delta s}$$

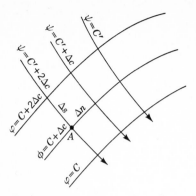

Fig. 7.13 Elements of a flow net.

or in terms of the spacing of streamlines [Eqs. (7.6.1) and (7.6.2)],

$$v_s \approx \frac{\Delta\psi}{\Delta n} = \frac{\Delta c}{\Delta n}$$

These expressions are approximate when Δc is finite, but when Δc becomes very small, the expressions become exact and yield velocity at a point. As both velocities referred to are the same, the equations show that $\Delta s = \Delta n$, or that the flow net consists of an orthogonal grid that reduces to perfect squares in the limit as the grid size approaches zero.

Once a flow net has been found by any means to satisfy the boundary conditions and to form an orthogonal net reducing to perfect squares in the limit as the number of lines is increased, the flow net is the only solution for the particular boundaries, as uniqueness theorems in hydrodynamics prove. In steady flow when the boundaries are stationary, the boundaries themselves become part of the flow net, as they are streamlines. The problem of finding the flow net to satisfy given fixed boundaries may be considered purely as a *graphical* exercise, i.e., the construction of an orthogonal system of lines that compose the boundaries and reduce to perfect squares in the limit as the number of lines increases. This is one of the practical methods employed in two-dimensional-flow analysis, although it usually requires many attempts and much erasing.

Another practical method of obtaining a flow net for a particular set of fixed boundaries is the *electrical analogy*. The boundaries in a model are formed out of strips of nonconducting material mounted on a flat nonconducting surface, and the end equipotential lines are formed out of a conducting strip, e.g., brass or copper. An electrolyte (conducting liquid) is placed at uniform depth in the flow space and a voltage potential applied to the two end conducting strips. By means of a probe and a voltmeter, lines with constant drop in voltage from one end are mapped out and plotted. These are equipotential lines. By reversing the process and making the flow boundaries out of conducting material and the end equipotential lines from nonconducting material, the streamlines are mapped.

A special conducting paper, called *Teledeltos* paper, may be used in place of a tank with an electrolyte. Silver ink is used to form a conducting strip or line having constant voltage. One cuts the paper to the size and shape needed, places the constant-voltage lines on the paper with a heavy line of silver ink, then marks the intermediate points of constant voltage directly on the paper, using the same circuits as with an electrolyte.

The relaxation method[1] numerically determines the value of potential function at points throughout the flow, usually located at the intersections of a square grid. The Laplace equation is written as a difference equation, and it is shown that the value of potential function at a grid point is the average of the four values at the neighboring grid points. Near the boundaries special formulas are required. With values known at the boundaries, each grid point is computed based on the assumed values at the neighboring grid points; then these values are improved by repeating the process until the changes are within the desired accuracy. This method is particularly convenient for solution with high-speed digital computers.

Use of the flow net

After a flow net for a given boundary configuration has been obtained, it may be used for all irrotational flows with geometrically similar boundaries. It is necessary to know the velocity at a single point and the pressure at one point. Then, by use of the flow net, the velocity can be determined at every other point. Application of the Bernoulli equation [Eq. (7.5.7)] produces the dynamic pressure. If the velocity is known, e.g., at A (Fig. 7.13), Δn or Δs can be scaled from the adjacent lines. Then $\Delta c \approx \Delta n\, v_s \approx \Delta s\, v_s$. With the constant Δc determined for the whole grid in this manner, measurement of Δs or Δn at any other point permits the velocity to be computed there,

$$v_s \approx \frac{\Delta c}{\Delta s} = \frac{\Delta c}{\Delta n}$$

[1] C.-S. Yih, Ideal-Fluid Flow, p. **4**–67 in V. L. Streeter (ed.), "Handbook of Fluid Dynamics," McGraw-Hill, New York, 1961.

The concepts underlying the flow net have been developed for irrotational flow of an ideal fluid.   Because of the similarity of differential equations describing groundwater flow and irrotational flow, the flow net can also be used to determine streamlines and lines of constant piezometric head $(h + p/\gamma)$ for percolation through homogeneous porous media. The flow cases of Sec. 7.9 may then be interpreted in terms of the very rotational, viscous flow at extremely small velocities through a porous medium.

7.8 THREE-DIMENSIONAL FLOW

Because of space limitations only a few three-dimensional cases are considered.   They are sources and sinks, the doublet, and uniform flow singly or combined.

Three-dimensional sources and sinks

A source in three-dimensional flow is a point from which fluid issues at a uniform rate in all directions.   It is entirely fictitious, as there is nothing resembling it in nature, but that does not reduce its usefulness in obtaining flow patterns.   The *strength* of the source m is the rate of flow passing through any surface enclosing the source.

As the flow is outward and is uniform in all directions, the velocity, a distance r from the source, is the strength divided by the area of the sphere through the point with center at the source, or

$$v_r = \frac{m}{4\pi r^2}$$

Since $v_r = -\partial\phi/\partial r$ and $v_\theta = 0$, it follows that $\partial\phi/\partial\theta = 0$, and the velocity potential can be found.

$$-\frac{\partial\phi}{\partial r} = \frac{m}{4\pi r^2}$$

and

$$\phi = \frac{m}{4\pi r} \tag{7.8.1}$$

A *negative source* is a *sink*. Fluid is assumed to flow uniformly into a sink and there disappear.

Three-dimensional doublets

A *doublet*, or *double source*, is a combination of a source and a sink of equal strength, which are allowed to approach each other in such a manner that the product of their strength and the distance between them remains a constant in the limit.

In Fig. 7.14, a source of strength m is located at $(a,0)$ and a sink of the same strength at $(-a,0)$. Since each satisfies the Laplace equation, their sum also satisfies it:

$$\phi = \frac{m}{4\pi} \left(\frac{1}{r_1} - \frac{1}{r_2} \right) \tag{7.8.2}$$

By the law of sines and Fig. 7.14,

$$\frac{r_1}{\sin \theta_2} = \frac{r_2}{\sin \theta_1} = \frac{2a}{\sin (\theta_1 - \theta_2)} = \frac{2a}{2 \sin \tfrac{1}{2} (\theta_1 - \theta_2) \cos \tfrac{1}{2} (\theta_1 - \theta_2)}$$

as the angle between r_2 and r_1 at P is $\theta_1 - \theta_2$. Solving for $r_2 - r_1$ gives

$$r_2 - r_1 = \frac{a(\sin \theta_1 - \sin \theta_2)}{\sin \tfrac{1}{2} (\theta_1 - \theta_2) \cos \tfrac{1}{2} (\theta_1 - \theta_2)} = \frac{2a \cos \tfrac{1}{2} (\theta_1 + \theta_2)}{\cos \tfrac{1}{2} (\theta_1 - \theta_2)}$$

From Eq. (7.8.2)

$$\phi = \frac{m}{4\pi} \frac{r_2 - r_1}{r_1 r_2} = \frac{2am \cos \tfrac{1}{2} (\theta_1 + \theta_2)}{4\pi r_1 r_2 \cos \tfrac{1}{2} (\theta_1 - \theta_2)} = \frac{\mu}{r_1 r_2} \frac{\cos \tfrac{1}{2} (\theta_1 + \theta_2)}{\cos \tfrac{1}{2} (\theta_1 - \theta_2)}$$

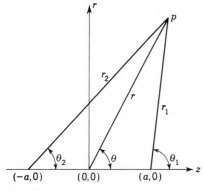

Fig. 7.14   Auxiliary coordinate systems used for Rankine body.

Fig. 7.15 Streamlines and equipotential lines for a three-dimensional doublet.

In the limit as a approaches zero, $\theta_2 = \theta_1 = \theta$, $r_2 = r_1 = r$, and

$$\phi = \frac{\mu}{r^2} \cos \theta \tag{7.8.3}$$

which is the velocity potential for a doublet[1] at the origin with axis in the positive x direction. Equation (7.8.3) can be converted into the stream function by Eqs. (7.6.7). The stream function is

$$\psi = -\frac{\mu \hat{\omega}^2}{r^3} = -\frac{\mu \sin^2 \theta}{r} \tag{7.8.4}$$

Streamlines and equipotential lines for the doublet are drawn in Fig. 7.15.

[1] L. M. Milne-Thompson, "Theoretical Hydrodynamics," p. 414, Macmillan, London, 1938.

Source in a uniform stream

The radial velocity v_r due to a source at the origin

$$\phi = \frac{m}{4\pi r} \tag{7.8.1}$$

is

$$v_r = -\frac{\partial \phi}{\partial r} = \frac{m}{4\pi r^2}$$

which, when multiplied by the surface area of the sphere concentric with it, gives the strength m. Since the flow from the source has axial symmetry, Stokes' stream function is defined. For spherical polar coordinates, from Eqs. (7.6.7),

$$\frac{\partial \psi}{\partial r} = -\sin\theta \frac{\partial \phi}{\partial \theta} \qquad \frac{\partial \psi}{\partial \theta} = r^2 \sin\theta \frac{\partial \phi}{\partial r}$$

With Eq. (7.8.1),

$$\frac{\partial \psi}{\partial r} = 0 \qquad \frac{\partial \psi}{\partial \theta} = -\frac{m}{4\pi}\sin\theta$$

Integrating gives

$$\psi = \frac{m}{4\pi}\cos\theta \tag{7.8.5}$$

the stream function for a source at the origin. Equipotential lines and streamlines are shown in Fig. 7.16 for constant increments of ϕ and ψ.

A uniform stream of fluid having a velocity U in the negative x direction throughout space is given by

$$-\frac{\partial \phi}{\partial x} = -U \qquad \frac{\partial \phi}{\partial \hat{\omega}} = 0$$

Integrating gives

$$\phi = Ux = Ur\cos\theta \tag{7.8.6}$$

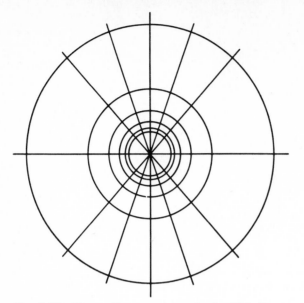

Fig. 7.16 Streamlines and equipotential lines for a source.

The stream function is found in the same manner as above to be

$$\psi = \frac{U}{2}\,\omega^2 = \frac{Ur^2}{2}\,\sin^2\theta \qquad\qquad (7.8.7)$$

The flow network is shown in Fig. 7.17.

Combining the uniform flow and the source flow, which may be accomplished by adding the two velocity potentials and the two stream functions, gives

$$\phi = \frac{m}{4\pi r} + Ur\cos\theta \qquad \psi = \frac{m}{4\pi}\cos\theta + \frac{Ur^2}{2}\,\sin^2\theta \qquad\qquad (7.8.8)$$

The resulting flow is everywhere the same as if the separate velocity vectors were added for each point in space.

A stagnation point is a point in the fluid where the velocity is zero. The conditions for stagnation point, where spherical polar coordinates are used and when the flow has axial symmetry, are

$$v_r = -\frac{\partial\phi}{\partial r} = 0 \qquad v_\theta = -\frac{1}{r}\frac{\partial\phi}{\partial\theta} = 0$$

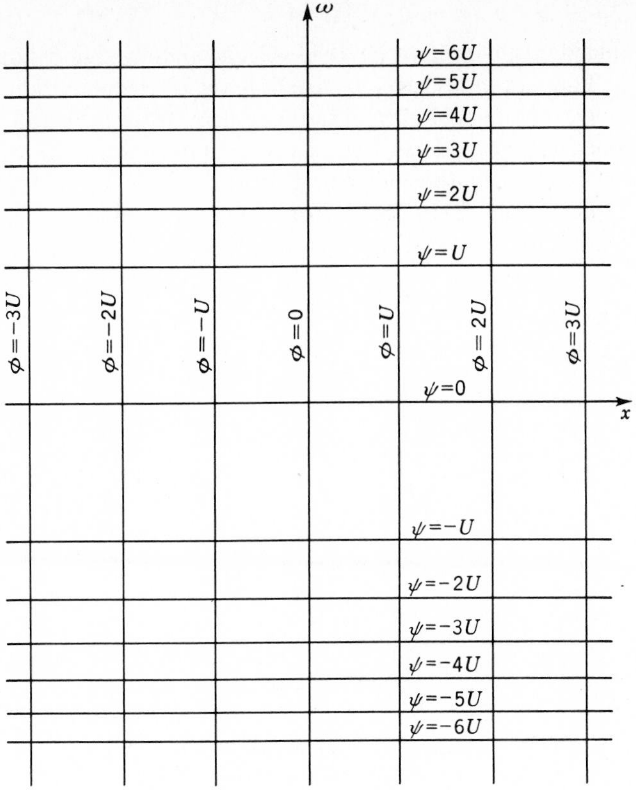

Fig. 7.17 Streamlines and equipotential lines for uniform flow in negative x direction.

Use of these expressions with Eqs. (7.8.8) gives

$$\frac{m}{4\pi r^2} - U\cos\theta = 0 \qquad U\sin\theta = 0$$

which are satisfied by only one point in space, viz.,

$$\theta = 0 \qquad r = \sqrt{\frac{m}{4\pi U}}$$

Substituting this point back into the stream function gives $\psi = m/4\pi$, which is the stream surface through the stagnation point. The equation of this surface is found from Eqs. (7.8.8):

$$\cos\theta + \frac{2\pi U}{m} r^2 \sin^2\theta = 1 \tag{7.8.9}$$

The flow under consideration is steady, as the velocity potential does not change with the time. Therefore, any stream surface satisfies the conditions for a boundary: the velocity component normal to the stream surface in steady flow is always zero. Since stream surfaces through stagnation points usually split the flow, they are frequently the most interesting possible boundary. This stream surface is plotted in Fig. 7.18. Substituting $\hat{\omega} = r \sin \theta$ in Eq. (7.8.9), the distance of a point (r,θ) from the x axis is given by

$$\hat{\omega}^2 = \frac{m}{2\pi U} (1 - \cos \theta)$$

which shows that $\hat{\omega}$ has a maximum value $\sqrt{m/\pi U}$ as θ approaches π, that is, as r approaches infinity. Hence, $\hat{\omega} = \sqrt{m/\pi U}$ is an asymptotic surface to the dividing stream surface. Equation (7.8.9) may be expressed in the form

$$r = \frac{1}{2} \sqrt{\frac{m}{\pi U}} \sec \frac{\theta}{2} \qquad\qquad (7.8.10)$$

from which the surface is easily plotted. Such a figure of revolution is called a *half-body*, as it extends to negative infinity, surrounding the negative x axis.

The pressure at any point, i.e., the dynamic pressure from Eq. (7.5.7), is

$$p = \frac{\rho}{2} (U^2 - q^2)$$

in which the dynamic pressure at infinity is taken as zero. q is the speed at

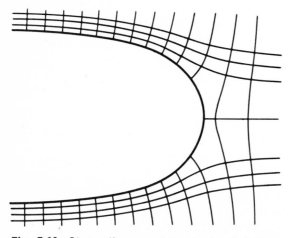

Fig. 7.18 Streamlines and equipotential lines for a half-body.

any point. Evaluating q from Eqs. (7.8.8) gives

$$q^2 = \left(\frac{\partial\phi}{\partial r}\right)^2 + \frac{1}{r^2}\left(\frac{\partial\phi}{\partial\theta}\right)^2 = U^2 + \frac{m^2}{16\pi^2 r^2} - \frac{mU\cos\theta}{2\pi r^2}$$

and

$$p = \frac{\rho}{2}U^2\left(\frac{m\cos\theta}{2\pi r^2 U} - \frac{m^2}{16\pi^2 r^4 U^2}\right) \tag{7.8.11}$$

from which the pressure can be found for any point except the origin, which is a singular point. When Eq. (7.8.10) is substituted into Eq. (7.8.11), the pressure is given in terms of r for any point on the half body; thus

$$p = \frac{\rho}{2}U^2\left(\frac{3m^2}{16\pi^2 r^4 U^2} - \frac{m}{2\pi r^2 U}\right) \tag{7.8.12}$$

This shows that the dynamic pressure approaches zero as r increases downstream along the body.

Source and sink of equal strength in a uniform stream; Rankine bodies

A source of strength m, located at $(a,0)$, has the velocity potential at any point P given by

$$\phi_1 = \frac{m}{4\pi r_1}$$

in which r_1 is the distance from $(a,0)$ to P, as shown in Fig. 7.14. Similarly, the potential function for a sink of strength m at $(-a,0)$ is

$$\phi_2 = -\frac{m}{4\pi r_2}$$

Since both ϕ_1 and ϕ_2 satisfy the Laplace equation, their sum will also be a solution,

$$\phi = \frac{m}{4\pi}\left(\frac{1}{r_1} - \frac{1}{r_2}\right) \tag{7.8.13}$$

Because r_1, r_2 are measured from different points, this expression must be handled differently from the usual algebraic equation.

The stream functions for the source and sink may also be added to give the stream function for the combined flow,

$$\psi = \frac{m}{4\pi} (\cos \theta_1 - \cos \theta_2) \qquad (7.8.14)$$

The stream surfaces and equipotential surfaces take the form shown in Fig. 7.19, which is plotted from Eqs. (7.8.13) and (7.8.14) by taking constant values of ϕ and ψ.

Superposing a uniform flow of velocity U in the negative x direction, $\phi = Ux$, $\psi = \frac{1}{2}U\hat{\omega}^2$, the potential and stream functions for source and sink of equal strength in a uniform flow (in direction of source to sink) are

$$\phi = Ux + \frac{m}{4\pi} \left(\frac{1}{r_1} - \frac{1}{r_2} \right)$$

$$= Ux + \frac{m}{4\pi} \left[\frac{1}{\sqrt{(x-a)^2 + \hat{\omega}^2}} - \frac{1}{\sqrt{(x+a)^2 + \hat{\omega}^2}} \right] \qquad (7.8.15)$$

$$\psi = \frac{1}{2}Ur^2 \sin^2 \theta + \frac{m}{4\pi} (\cos \theta_1 - \cos \theta_2) \qquad (7.8.16)$$

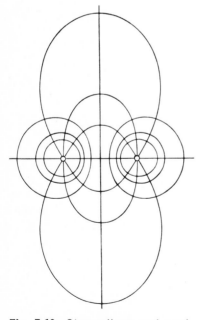

Fig. 7.19 Streamlines and equipotential lines for a source and sink of equal strength.

As any stream surface may be taken as a solid boundary in steady flow, the location of a closed surface for this flow case will represent flow of a uniform stream around a body. Examining the stream function, for $x > a$ and $\theta_1 = \theta_2 = 0$, $\psi = 0$. For $x < -a$ and $\theta_1 = \theta_2 = \theta = \pi$, $\psi = 0$. Therefore, $\psi = 0$ must be the dividing streamline, since the x axis is the axis of symmetry. The equation of the dividing streamline is, from Eq. (7.8.16),

$$\hat{\omega}^2 + \frac{m}{2\pi U}(\cos \theta_1 - \cos \theta_2) = 0 \qquad (7.8.17)$$

in which $\hat{\omega} = r \sin \theta$ is the distance of a point on the dividing stream surface from the x axis. Since $\cos \theta_1$ and $\cos \theta_2$ are never greater than unity, $\hat{\omega}$ cannot exceed $\sqrt{m/\pi U}$, which shows that the surface is closed and hence can be replaced by a solid body of exactly the same shape. By changing the signs of m and U the flow is reversed and the body should change end for end. From Eq. (7.8.17) it is seen that the equation is unaltered; hence, the body has symmetry with respect to the plane $x = 0$. It is necessarily a body of revolution because of axial symmetry of the equations.

To locate the stagnation points C, D (Fig. 7.20), which must be on the x axis, it is known that the velocity is along the x axis (it is a streamline). From Eq. (7.8.15) the velocity potential ϕ_x for points on the x axis is given by

$$\phi_x = \frac{ma}{2\pi}\frac{1}{x^2 - a^2} + Ux$$

since

$$r_1 = x - a \qquad r_2 = x + a$$

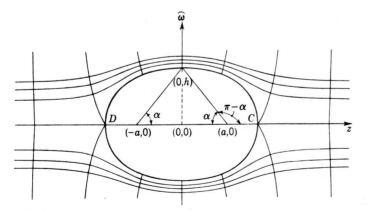

Fig. 7.20 Rankine body.

Differentiating with respect to x and setting the result equal to zero yield

$$\frac{\partial \phi_x}{\partial x} = U - \frac{max_0}{\pi (x_0{}^2 - a^2)^2} = 0 \tag{7.8.18}$$

where x_0 is the x coordinate of the stagnation point. This gives the point $C(x_0, 0)$ (a trial solution). The half-breadth h is determined as follows. From Fig. 7.20.,

$$\theta_1 = \pi - \alpha \qquad \theta_2 = \alpha$$

in which

$$\cos \alpha = \frac{a}{\sqrt{h^2 + a^2}}$$

Substituting into Eq. (7.8.17) gives

$$h^2 = \frac{m}{\pi U} \frac{a}{\sqrt{h^2 + a^2}} \tag{7.8.19}$$

from which h can be determined (also by trial solution).

Eliminating m/U between Eqs. (7.8.18) and (7.8.19) leads to

$$\frac{m}{U} = \frac{(x_0{}^2 - a^2)^2 \, \pi}{x_0 \, a} = \frac{\pi}{a} h^2 \sqrt{h^2 + a^2}$$

The value of a can be obtained for a predetermined body (x_0, h specified). Hence, U can be given any positive value, and the pressure and velocity distribution can be determined.

In determining the velocity at points throughout the region it is convenient to find the velocity at each point due to each component of the flow, i.e., due to the source, the sink, and the uniform flow, separately, and add the components graphically or by $\hat{\omega}$ and x components.

Bodies obtained from source-sink combinations with uniform flow are called *Rankine bodies*.

Translation of a sphere in an infinite fluid

The velocity potential for a solid moving through an infinite fluid otherwise at rest must satisfy the following conditions:[1]

1.   The Laplace equation $\nabla^2 \phi = 0$ everywhere except at singular points.

[1] G. G. Stokes, "Mathematical and Physical Papers," vol. 1, pp. 38–43, Cambridge University Press, London, 1880.

2. The fluid must remain at rest at infinity; hence, the space derivatives of ϕ must vanish at infinity.
3. The boundary conditions at the surface of the solid must be satisfied.

For a sphere of radius a with center at the origin moving with velocity U in the positive x direction, the velocity of the surface normal to itself is $U \cos \theta$, from Fig. 7.21. The fluid velocity normal to the surface is $-\partial \phi / \partial r$; hence the boundary condition is

$$-\frac{\partial \phi}{\partial r} = U \cos \theta$$

The velocity potential for the doublet [Eq. (7.8.3)]

$$\phi = \frac{\mu \cos \theta}{r^2}$$

satisfies $\nabla^2 \phi = 0$ for any constant value of μ. Substituting it into the boundary condition gives

$$-\frac{\partial \phi}{\partial r} = \frac{2\mu}{r^3} \cos \theta = U \cos \theta$$

which is satisfied for $r = a$ if $\mu = Ua^3/2$. It may also be noted that the velocity components,

$$-\frac{\partial \phi}{\partial r} \qquad \text{and} \qquad -\frac{1}{r}\frac{\partial \phi}{\partial \theta}$$

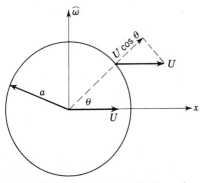

Fig. 7.21 Sphere translating in the positive x direction.

are zero at infinity. Therefore,

$$\phi = \frac{Ua^3}{2r^2} \cos \theta \qquad (7.8.20)$$

satisfies all the conditions for translation of a sphere in an infinite fluid. This case is one of unsteady flow, solved for the instant when the center of the sphere is at the origin. Because this equation has been specialized for a particular instant, the pressure distribution cannot be found from it by use of Eq. (7.5.7). Streamlines and equipotential lines for the sphere are shown in Fig. 7.22.

The stream function for this flow case is

$$\psi = -\frac{Ua^3}{2r} \sin^2 \theta \qquad (7.8.21)$$

Steady flow of an infinite fluid around a sphere

The unsteady-flow case in the preceding section can be converted into a steady-flow case by superposing upon the flow a uniform stream of magnitude U in the negative x direction. To prove this, add $\phi = Ux = Ur \cos \theta$ to the potential function [Eq. (7.8.20)]; thus

$$\phi = \frac{Ua^3}{2r^2} \cos \theta + Ur \cos \theta \qquad (7.8.22)$$

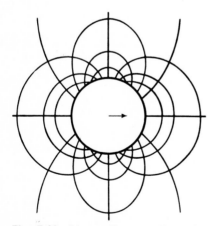

Fig. 7.22 Streamlines and equipotential lines for a sphere moving through fluid.

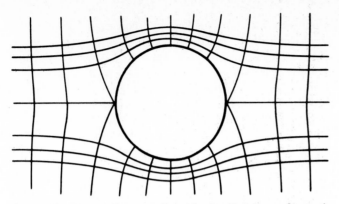

Fig. 7.23 Streamlines and equipotential lines for uniform flow about a sphere at rest.

The stream function corresponding to this is

$$\psi = -\frac{Ua^3}{2r} \sin^2 \theta + \frac{Ur^2}{2} \sin^2 \theta \qquad (7.8.23)$$

Then from Eq. (7.8.23), $\psi = 0$ when $\theta = 0$ and when $r = a$. Hence, the stream surface $\psi = 0$ is the sphere $r = a$, which may be taken as a solid, fixed boundary. Streamlines and equipotential lines are shown in Fig. 7.23. Perhaps it should be mentioned that the equations give a flow pattern for the interior portion of the sphere as well. No fluid passes through the surface of the sphere, however.

The velocity at any point on the surface of the sphere is

$$-\frac{1}{r}\frac{\partial \phi}{\partial \theta}\bigg]_{r=a} = q = \tfrac{3}{2}U \sin \theta$$

The stagnation points are at $\theta = 0$, $\theta = \pi$. The maximum velocity $\frac{3}{2}U$ occurs at $\theta = \pi/2$. The dynamic pressure distribution over the surface of the sphere is

$$p = \frac{\rho U^2}{2}(1 - \tfrac{9}{4} \sin^2 \theta) \qquad (7.8.24)$$

for dynamic pressure of zero at infinity.

In Secs. 5.6 and 5.7 the flow around a sphere in a real fluid is discussed. Figure 5.20 shows the actual wake formation due to laminar and turbulent boundary-layer separation. The drag on a sphere in real fluids is given by the drag-coefficient data in Fig. 5.21. In the ideal-fluid-flow case the boundary condition does not permit separation. If the x component of force exerted on the sphere by dynamic pressure is obtained from Eq. (7.8.24) by integration, it will be found to be zero. There is no drag on a body in ideal-fluid flow because the energy is everywhere constant and the high velocity of fluid obtained by flow over the front portion to the largest cross section is completely converted back into pressure in flowing over the downstream portion.

7.9 TWO-DIMENSIONAL FLOW

Two simple flow cases that may be interpreted for flow along straight boundaries are first examined; then the source, vortex, doublet, uniform flow, and flow around a cylinder, with and without circulation, are discussed.

Flow around a corner

The potential function

$$\phi = A\,(x^2 - y^2)$$

has as its stream function

$$\psi = 2Axy = Ar^2 \sin 2\theta$$

Fig. 7.24 Flow net for flow around 90° bend.

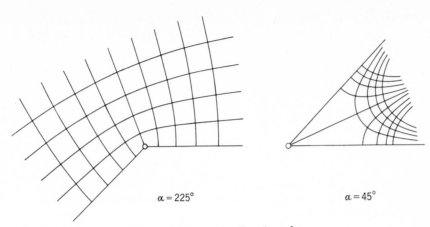

$\alpha = 225°$ $\alpha = 45°$

Fig. 7.25   Flow net for flow along two inclined surfaces.

in which r and θ are polar coordinates.   It is plotted for equal-increment changes in ϕ and ψ in Fig. 7.24.   Conditions at the origin are not defined, as it is a stagnation point.   As any of the streamlines may be taken as fixed boundaries, the plus axes may be taken as walls, yielding flow into a 90° corner.   The equipotential lines are hyperbolas having axes coincident with the coordinate axes and asymptotes given by $y = \pm x$.   The streamlines are rectangular hyperbolas, having $y = \pm x$ as axes and the coordinate axes as asymptotes.   From the polar form of the stream function it is noted that the two lines $\theta = 0$ and $\theta = \pi/2$ are the streamline $\psi = 0$.

This case may be generalized to yield flow around a corner with angle α. By examining

$$\phi = A r^{\pi/\alpha} \cos \frac{\pi\theta}{\alpha} \qquad \psi = A r^{\pi/\alpha} \sin \frac{\pi\theta}{\alpha}$$

it is noted the streamline $\psi = 0$ is now given by $\theta = 0$ and $\theta = \alpha$.   Two flow nets are shown in Fig. 7.25, for the cases $\alpha = 225°$ and $\alpha = 45°$.

Source

A line normal to the xy plane, from which fluid is imagined to flow uniformly in all directions *at right angles* to it, is a source.   It appears as a point in the customary two-dimensional flow diagram.   The total flow per unit time per unit length of line is called the *strength* of the source.   As the flow is in radial lines from the source, the velocity a distance r from the source is determined by the strength divided by the flow area of the cylinder, or $2\pi\mu/2\pi r$, in which the strength is $2\pi\mu$. Then, since by Eq. (7.4.4) the velocity in any direction is given by the negative derivative of the velocity potential with respect to the direction,

$$-\frac{\partial\phi}{\partial r} = \frac{\mu}{r} \qquad \frac{\partial\phi}{\partial\theta} = 0$$

and

$$\phi = -\mu \ln r$$

is the velocity potential, in which ln indicates the natural logarithm and r is the distance from the source. This value of ϕ satisfies the Laplace equation in two dimensions.

The streamlines are radial lines from the source, i.e.,

$$\frac{\partial\psi}{\partial r} = 0 \qquad -\frac{1}{r}\frac{\partial\psi}{\partial\theta} = \frac{\mu}{r}$$

From the second equation

$$\psi = -\mu\theta$$

Lines of constant ϕ (equipotential lines) and constant ψ are shown in Fig. 7.26. A *sink* is a negative source, a line into which fluid is flowing.

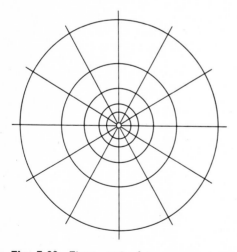

Fig. 7.26 Flow net for source or vortex.

Vortex

In examining the flow case given by selecting the stream function for the source as a velocity potential,

$$\phi = -\mu\theta \qquad \psi = \mu \ln r$$

which also satisfies the Laplace equation, it is seen that the equipotential lines are radial lines and the streamlines are circles. The velocity is in a tangential direction only, since $\partial\phi/\partial r = 0$. It is

$$q = -\frac{1}{r}\frac{\partial\phi}{\partial\theta} = \frac{\mu}{r}$$

since $r\,\delta\theta$ is the length element in the tangential direction.

In referring to Fig. 7.27, the *flow along a closed curve* is called the *circulation*. The flow along an element of the curve is defined as the product of the length element δs of the curve and the component of the velocity tangent to the curve, $q\cos\alpha$. Hence the circulation Γ around a closed path C is

$$\Gamma = \int_C q\cos\alpha\,ds = \int_C \mathbf{q}\cdot d\mathbf{s}$$

The velocity distribution given by the equation $\phi = -\mu\theta$ is for the *vortex* and is such that the circulation around any closed path that contains the vortex is constant. The value of the circulation is the strength of the vortex. By selecting any circular path of radius r to determine the circulation, $\alpha = 0°$, $q = \mu/r$, and $ds = r\,d\theta$; hence,

$$\Gamma = \int_C q\cos\alpha\,ds = \int_0^{2\pi} \frac{\mu}{r} r\,d\theta = 2\pi\mu$$

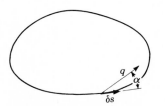

Fig. 7.27 Notation for definition of circulation.

At the point $r = 0$, $q = \mu/r$ goes to infinity; hence, this point is called a singular point. Figure 7.26 shows the equipotential lines and streamlines for the vortex.

Doublet

The two-dimensional doublet is defined as the limiting case as a source and sink of equal strength approach each other so that the product of their strength and the distance between them remains a constant $2\pi\mu$. μ is called the *strength* of the doublet. The axis of the doublet is from the sink toward the source, i.e., the line along which they approach each other.

In Fig. 7.28 a source is located at $(a,0)$ and a sink of equal strength at $(-a,0)$. The velocity potential for both, at some point P, is

$$\phi = -m \ln r_1 + m \ln r_2$$

with r_1, r_2 measured from source and sink, respectively, to the point P. Thus $2\pi m$ is the strength of source and sink. To take the limit as a approaches zero for $2am = \mu$, the form of the expression for ϕ must be altered. The terms r_1 and r_2 may be expressed in terms of the polar coordinates r, θ by the cosine law, as follows:

$$r_1^2 = r^2 + a^2 - 2ar \cos\theta = r^2 \left[1 + \left(\frac{a}{r}\right)^2 - 2\frac{a}{r} \cos\theta \right]$$

$$r_2^2 = r^2 + a^2 + 2ar \cos\theta = r^2 \left[1 + \left(\frac{a}{r}\right)^2 + 2\frac{a}{r} \cos\theta \right]$$

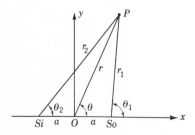

Fig. 7.28 Notation for derivation of a two-dimensional doublet.

Rewriting the expression for ϕ with these relations gives

$$\phi = -\frac{m}{2}(\ln r_1^2 - \ln r_2^2) = -\frac{m}{2}\left\{\ln r^2 + \ln\left[1 + \left(\frac{a}{r}\right)^2 - 2\frac{a}{r}\cos\theta\right]\right.$$

$$\left. -\ln r^2 - \ln\left[1 + \left(\frac{a}{r}\right)^2 + 2\frac{a}{r}\cos\theta\right]\right\}$$

By the series expression,

$$\ln(1+x) = x - \frac{x^2}{2} + \frac{x^3}{3} - \frac{x^4}{4} + \cdots$$

$$\phi = -\frac{m}{2}\left\{\left(\frac{a}{r}\right)^2 - 2\frac{a}{r}\cos\theta - \frac{1}{2}\left[\left(\frac{a}{r}\right)^2 - 2\frac{a}{r}\cos\theta\right]^2\right.$$

$$+\frac{1}{3}\left[\left(\frac{a}{r}\right)^2 - 2\frac{a}{r}\cos\theta\right]^3 - \cdots - \left[\left(\frac{a}{r}\right)^2 + 2\frac{a}{r}\cos\theta\right]$$

$$\left. +\frac{1}{2}\left[\left(\frac{a}{r}\right)^2 + 2\frac{a}{r}\cos\theta\right]^2 - \frac{1}{3}\left[\left(\frac{a}{r}\right)^2 + 2\frac{a}{r}\cos\theta\right]^3 + \cdots\right\}$$

After simplifying,

$$\phi = 2am\left[\frac{\cos\theta}{r} + \left(\frac{a}{r}\right)^2\frac{\cos\theta}{r} - \left(\frac{a}{r}\right)^4\frac{\cos\theta}{r} - \frac{4}{3}\left(\frac{a}{r}\right)^2\frac{\cos^3\theta}{r} + \cdots\right]$$

Now, if $2am = \mu$ and if the limit is taken as a approaches zero,

$$\phi = \frac{\mu\cos\theta}{r}$$

which is the velocity potential for a two-dimensional doublet at the origin, with axis in the $+x$ direction.

Using the relations

$$v_r = -\frac{\partial\phi}{\partial r} = -\frac{1}{r}\frac{\partial\psi}{\partial\theta} \qquad v_\theta = -\frac{1}{r}\frac{\partial\phi}{\partial\theta} = \frac{\partial\psi}{\partial r}$$

gives for the doublet

$$\frac{\partial \psi}{\partial \theta} = - \frac{\mu \cos \theta}{r} \qquad \frac{\partial \psi}{\partial r} = \frac{\mu}{r^2} \sin \theta$$

After integrating,

$$\psi = - \frac{\mu \sin \theta}{r}$$

is the stream function for the doublet. The equations in cartesian coordinates are

$$\phi = \frac{\mu x}{x^2 + y^2} \qquad \psi = - \frac{\mu y}{x^2 + y^2}$$

Rearranging gives

$$\left(x - \frac{\mu}{2\phi}\right)^2 + y^2 = \frac{\mu^2}{4\phi^2} \qquad x^2 + \left(y + \frac{\mu}{2\psi}\right)^2 = \frac{\mu^2}{4\psi^2}$$

Fig. 7.29 Equipotential lines and streamlines for the two-dimensional doublet.

The lines of constant ϕ are circles through the origin with centers on the x axis, and the streamlines are circles through the origin with centers on the y axis, as shown in Fig. 7.29. The origin is a singular point where the velocity goes to infinity.

Uniform flow

Uniform flow in the $-x$ direction, $u = -U$, is expressed by

$$\phi = Ux \qquad \psi = Uy$$

In polar coordinates,

$$\phi = Ur\cos\theta \qquad \psi = Ur\sin\theta$$

Flow around a circular cylinder

The addition of the flow due to a doublet and a uniform flow results in flow around a circular cylinder; thus

$$\phi = Ur\cos\theta + \frac{\mu\cos\theta}{r} \qquad \psi = Ur\sin\theta - \frac{\mu\sin\theta}{r}$$

As a streamline in steady flow is a possible boundary, the streamline $\psi = 0$ is given by

$$0 = \left(Ur - \frac{\mu}{r}\right)\sin\theta$$

which is satisfied by $\theta = 0$, π, or by the value of r that makes

$$Ur - \frac{\mu}{r} = 0$$

If this value is $r = a$, which is a circular cylinder, then

$$\mu = Ua^2$$

and the streamline $\psi = 0$ is the x axis and the circle $r = a$. The potential and stream functions for uniform flow around a circular cylinder of radius a are, by substitution of the value of μ,

$$\phi = U\left(r + \frac{a^2}{r}\right)\cos\theta \qquad \psi = U\left(r - \frac{a^2}{r}\right)\sin\theta$$

for the uniform flow in the $-x$ direction. The equipotential lines and stream-lines for this case are shown in Fig. 7.30.

The velocity at any point in the flow can be obtained from either the velocity potential or the stream function. On the surface of the cylinder the velocity is necessarily tangential and is expressed by $\partial\psi/\partial r$ for $r = a$; thus

$$q\Big|_{r=a} = U\left(1 + \frac{a^2}{r^2}\right)\sin\theta\Big|_{r=a} = 2U\sin\theta$$

The velocity is zero (stagnation point) at $\theta = 0$, π and has maximum values of $2U$ at $\theta = \pi/2$, $3\pi/2$. For the dynamic pressure zero at infinity, with Eq. (7.5.7) for $p_0 = 0$, $q_0 = U$,

$$p = \frac{\rho}{2}U^2\left[1 - \left(\frac{q}{U}\right)^2\right]$$

which holds for any point in the plane except the origin. For points on the cylinder,

$$p = \frac{\rho}{2}U^2(1 - 4\sin^2\theta)$$

The maximum pressure, which occurs at the stagnation points, is $\rho U^2/2$;

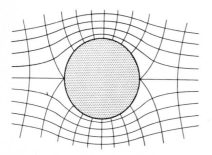

Fig. 7.30 Equipotential lines and streamlines for flow around a circular cylinder.

and the minimum pressure, at $\theta = \pi/2,\ 3\pi/2$, is $-3\rho U^2/2$. The points of zero dynamic pressure are given by $\sin\theta = \pm\frac{1}{2}$, or $\theta = \pm\pi/6,\ \pm 5\pi/6$. A cylindrical pitot-static tube is made by providing three openings in a cylinder, at 0 and $\pm 30°$, as the difference in pressure between 0 and $\pm 30°$ is the dynamic pressure $\rho U^2/2$.

The drag on the cylinder is shown to be zero by integration of the x component of the pressure force over the cylinder; thus

$$\text{Drag} = \int_0^{2\pi} pa\cos\theta\ d\theta = \frac{\rho a U^2}{2} \int_0^{2\pi} (1 - 4\sin^2\theta)\cos\theta\ d\theta = 0$$

Similarly, the lift force on the cylinder is zero.

Flow around a circular cylinder with circulation

The addition of a vortex to the doublet and the uniform flow results in flow around a circular cylinder with circulation,

$$\phi = U\left(r + \frac{a^2}{r}\right)\cos\theta - \frac{\Gamma}{2\pi}\theta \qquad \psi = U\left(r - \frac{a^2}{r}\right)\sin\theta + \frac{\Gamma}{2\pi}\ln r$$

The streamline $\psi = (\Gamma/2\pi)\ln a$ is the circular cylinder $r = a$, and, at great distances from the origin, the velocity remains $u = -U$, showing that flow around a circular cylinder is maintained with addition of the vortex. Some of the streamlines are shown in Fig. 7.31.

The velocity at the surface of the cylinder, necessarily tangent to the

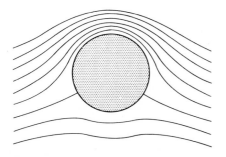

Fig. 7.31 Streamlines for flow around a circular cylinder with circulation.

cylinder, is

$$q = \frac{\partial \psi}{\partial r}\bigg|_{r=a} = 2U \sin \theta + \frac{\Gamma}{2\pi a}$$

Stagnation points occur when $q = 0$; that is,

$$\sin \theta = -\frac{\Gamma}{4\pi U a}$$

When the circulation is $4\pi U a$, the two stagnation points coincide at $r = a$, $\theta = -\pi/2$. For larger circulation, the stagnation point moves away from the cylinder.

The pressure at the surface of the cylinder is

$$p = \frac{\rho U^2}{2}\left[1 - \left(2 \sin \theta + \frac{\Gamma}{2\pi a U}\right)^2\right]$$

The drag again is zero. The lift, however, becomes

$$\text{Lift} = -\int_0^{2\pi} pa \sin \theta \, d\theta$$

$$= -\frac{\rho a U^2}{2}\int_0^{2\pi}\left[1 - \left(2 \sin \theta + \frac{\Gamma}{2\pi a U}\right)^2\right]\sin \theta \, d\theta = \rho U \Gamma$$

showing that the lift is directly proportional to the density of fluid, the approach velocity U, and the circulation Γ. This thrust, which acts at right angles to the approach velocity, is referred to as the *Magnus effect*. The Flettner rotor ship was designed to utilize this principle by mounting circular cylinders with axes vertical on the ship and then mechanically rotating the cylinders to provide circulation. Airflow around the rotors produces the thrust at right angles to the relative wind direction. The close spacing of streamlines along the upper side of Fig. 7.31 indicates that the velocity is high there and that the pressure must then be correspondingly low.

The theoretical flow around a circular cylinder with circulation can be transformed[1] into flow around an airfoil with the same circulation and the same lift.

The airfoil develops its lift by producing a circulation around it due to its shape. It can be shown[1] that the lift is $\rho U \Gamma$ for any cylinder in two-dimensional

[1] V. L. Streeter, "Fluid Dynamics," pp. 137–155, McGraw-Hill, New York, 1948.

flow. The angle of inclination of the airfoil relative to the approach velocity (angle of attack) greatly affects the circulation. For large angles of attack, the flow does not follow the wing profile, and the theory breaks down.

It should be mentioned that all two-dimensional ideal-fluid-flow cases may be conveniently handled by complex-variable theory and by a system of *conformal mapping*, which transforms a flow net from one configuration to another by a suitable complex-variable mapping function.

EXAMPLE 7.3   A source with strength 0.2 m³/s·m and a vortex with strength 1 m²/s are located at the origin. Determine the equation for velocity potential and stream function. What are the velocity components at $x = 1$ m, $y = 0.5$ m?

The velocity potential for the source is

$$\phi = -\frac{0.2}{2\pi} \ln r \qquad \text{m}^2/\text{s}$$

and the corresponding stream function is

$$\psi = -\frac{0.2}{2\pi} \theta \qquad \text{m}^2/\text{s}$$

The velocity potential for the vortex is

$$\phi = -\frac{1}{2\pi} \theta \qquad \text{m}^2/\text{s}$$

and the corresponding stream function is

$$\psi = \frac{1}{2\pi} \ln r \qquad \text{m}^2/\text{s}$$

Adding the respective functions gives

$$\phi = -\frac{1}{\pi}\left(0.1 \ln r + \frac{\theta}{2}\right) \qquad \text{and} \qquad \psi = -\frac{1}{\pi}(0.1\theta - \tfrac{1}{2}\ln r)$$

The radial and tangential velocity components are

$$v_r = -\frac{\partial \phi}{\partial r} = \frac{1}{10\pi r} \qquad v_\theta = -\frac{1}{r}\frac{\partial \phi}{\partial \theta} = \frac{1}{2\pi r}$$

At $(1,0.5)$, $r = \sqrt{1^2 + 0.5^2} = 1.117$ m, $v_r = 0.0285$ m/s, $v_\theta = 0.143$ m/s.

PROBLEMS

7.1 Compute the gradient of the following two-dimensional scalar functions:

(a) $\phi = -2 \ln (x^2 + y^2)$ (b) $\phi = Ux + Vy$ (c) $\phi = 2xy$

7.2 Compute the divergence of the gradients of ϕ found in Prob. 7.1.

7.3 Compute the curl of the gradients of ϕ found in Prob. 7.1.

7.4 For $\mathbf{q} = \mathbf{i}(x + y) + \mathbf{j}(y + z) + \mathbf{k}(x^2 + y^2 + z^2)$ find the components of rotation at (2,2,2).

7.5 Derive the equation of continuity for two-dimensional flow in polar coordinates by equating the net efflux from a small polar element to zero (Fig. 7.32). It is

$$\frac{\partial v_r}{\partial r} + \frac{v_r}{r} + \frac{1}{r}\frac{\partial v_\theta}{\partial \theta} = 0$$

7.6 The x component of velocity is $u = x^2 + z^2 + 5$, and the y component is $v = y^2 + z^2$. Find the simplest z component of velocity that satisfies continuity.

7.7 A velocity potential in two-dimensional flow is $\phi = y + x^2 - y^2$. Find the stream function for this flow.

7.8 The two-dimensional stream function for a flow is $\psi = 9 + 6x - 4y + 7xy$. Find the velocity potential.

7.9 Derive the partial differential equations relating ϕ and ψ for two-dimensional flow in plane polar coordinates.

7.10 From the continuity equation in polar coordinates in Prob. 7.5, derive the Laplace equation in the same coordinate system.

7.11 Does the function $\phi = 1/r$ satisfy the Laplace equation in two dimensions? In three-dimensional flow is it satisfied?

7.12 By use of the equations developed in Prob. 7.9 find the two-dimensional stream function for $\phi = \ln r$.

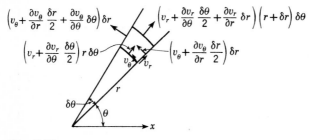

Fig. 7.32

7.13   Find the Stokes stream function for $\phi = 1/r$.

7.14   For the Stokes stream function $\psi = 9r^2 \sin^2 \theta$, find ϕ in cartesian coordinates.

7.15   In Prob. 7.14 what is the discharge between stream surfaces through the points $r = 1, \theta = 0$ and $r = 1, \theta = \pi/4$?

7.16   Write the boundary conditions for steady flow around a sphere, of radius a, at its surface and at infinity.

7.17   A circular cylinder of radius a has its center at the origin and is translating with velocity V in the y direction. Write the boundary condition in terms of ϕ that is to be satisfied at its surface and at infinity.

7.18   A source of strength 30 cfs is located at the origin, and another source of strength 20 cfs is located at $(1,0,0)$. Find the velocity components u, v, w at $(-1,0,0)$ and $(1,1,1)$.

7.19   If the dynamic pressure is zero at infinity in Prob. 7.18, for $\rho = 2.00$ slugs/ft³ calculate the dynamic pressure at $(-1,0,0)$ and $(1,1,1)$.

7.20   A source of strength m at the origin and a uniform flow of 5 m/s are combined in three-dimensional flow so that a stagnation point occurs at $(1,0,0)$. Obtain the velocity potential and stream function for this flow case.

7.21   By use of symmetry obtain the velocity potential for a three-dimensional sink of strength 1 m³/s located 1 m from a plane barrier.

7.22   Equations are wanted for flow of a uniform stream of 12 ft/s around a Rankine body 4 ft long and 2 ft thick in a transverse direction.

7.23   A source of strength 10 cfs at $(1,0,0)$ and a sink of the same strength at $(-1,0,0)$ are combined with a uniform flow of 25 ft/s in the $-x$ direction. Determine the size of Rankine body formed by this flow.

7.24   A sphere of radius 1 m, with center at the origin, has a uniform flow of 6 m/s in the $-x$ direction flowing around it. At $(1.25,0,0)$ the dynamic pressure is 500 N/m² and $\rho = 1000$ kg/m³. Find the equation for pressure distribution over the surface of the sphere.

7.25   By integration over the surface of the sphere of Prob. 7.24 show that the drag on the sphere is zero.

7.26   Show that if two stream functions ψ_1 and ψ_2 both satisfy the Laplace equation, $\nabla^2 \psi = 0$ for $\psi = \psi_1 + \psi_2$.

7.27   Show that if u_1, v_1 and u_2, v_2 are the velocity components of two velocity potentials ϕ_1 and ϕ_2 which satisfy the Laplace equation, then for $\phi = \phi_1 + \phi_2$ the velocity components are $u = u_1 + u_2$ and $v = v_1 + v_2$.

7.28   A two-dimensional source is located at $(1,0)$ and another one of the same strength at $(-1,0)$. Construct the velocity vector at $(0,0)$, $(0,1)$, $(0,-1)$, $(0,-2)$, and $(1,1)$. (*Hint:* Using the results of Prob. 7.27, draw the velocity components by adding the individual velocity components induced at the point in question by each source, without regard to the other, due to its strength and location.)

7.29 Determine the velocity potential for a source located at (1,0). Write the equation for the velocity potential for the source system described in Prob. 7.28.

7.30 Draw a set of streamlines for each of the sources described in Prob. 7.28 and from this diagram construct the streamlines for the combined flow. (*Hint:* For each of the sources draw streamlines separated by an angle of $\pi/6$. Finally combine the intersection points of those rays for which $\psi_1 + \psi_2$ is constant.)

7.31 Does the line $x = 0$ form a line in the flow field described in Prob. 7.28 for which there is no velocity component normal to it? Is this line a streamline? Could this line be the trace of a solid plane lamina that was submerged in the flow? Does the velocity potential determined in Prob. 7.29 describe the flow in the region $x > 0$ for a source located at a distance of unity from a plane wall? Justify your answers.

7.32 Determine the equation for the velocity on the line $x = 0$ for the flow described in Prob. 7.28. Find an equation for the pressure on the surface whose trace is $x = 0$. What is the force on one side of this plane due to the source located at distance of unity from it? Water is the fluid.

7.33 In two-dimensional flow what is the nature of the flow given by $\phi = 7x + 2 \ln r$?
7.34 By using a method similar to that suggested in Prob. 7.30 draw the potential lines for the flow given in Prob. 7.33.

7.35 By using the suggestion in Prob. 7.30 draw a flow net for a flow consisting of a source and a vortex which are located at the origin. Use the same value for μ in both the source and the vortex.

7.36 A source discharging 20 cfs/ft is located at $(-1,0)$, and a sink of twice the strength is located at (2,0). For dynamic pressure at the origin of 100 lb/ft², $\rho = 1.8$ slugs/ft³, find the velocity and dynamic pressure at (0,1) and (1,1).

7.37 Select the strength of doublet needed to portray a uniform flow of 20 m/s around a cylinder of radius 2 m.

7.38 Develop the equations for flow around a Rankine cylinder formed by a source, an equal sink, and a uniform flow. If $2a$ is the distance between source and sink, their strength is $2\pi\mu$, and U is the uniform velocity, develop an equation for length of the body. ·

7.39 A circular cylinder 8 ft in diameter rotates at 500 rpm. When in an airstream, $\rho = 0.002$ slug/ft³, moving at 400 ft/s, what is the lift force per foot of cylinder, assuming 90 percent efficiency in developing circulation from the rotation?

7.40 An unsteady-flow case may be transformed into a steady-flow case

(*a*) regardless of the nature of the problem
(*b*) when two bodies are moving toward each other in an infinite fluid
(*c*) when an unsymmetrical body is rotating in an infinite fluid

(d) when a single body translates in an infinite fluid

(e) under no circumstances

7.41 Select the value of ϕ that satisfies continuity.

(a) $x^2 + y^2$ (b) $\sin x$ (c) $\ln(x + y)$ (d) $x + y$

(e) none of these answers

7.42 The units for Euler's equations of motion are given by

(a) force per unit mass

(b) velocity

(c) energy per unit weight

(d) force per unit weight

(e) none of these answers

7.43 Euler's equations of motion can be integrated when it is assumed that

(a) the continuity equation is satisfied

(b) the fluid is incompressible

(c) a velocity potential exists and the density is constant

(d) the flow is rotational and incompressible

(e) the fluid is nonviscous

7.44 Euler's equations of motion are a mathematical statement that at every point

(a) rate of mass inflow equals rate of mass outflow

(b) force per unit mass equals acceleration

(c) the energy does not change with the time

(d) Newton's third law of motion holds

(e) the fluid momentum is constant

7.45 In irrotational flow of an ideal fluid

(a) a velocity potential exists

(b) all particles must move in straight lines

(c) the motion must be uniform

(d) the flow is always steady

(e) the velocity must be zero at a boundary

7.46 A function ϕ that satisfies the Laplace equation

(a) must be linear in x and y

(b) is a possible case of rotational fluid flow

(c) does not necessarily satisfy the continuity equation

(d) is a possible fluid-flow case

(e) is none of these answers

7.47 If ϕ_1 and ϕ_2 are each solutions of the Laplace equation, which of the following is also a solution?

(a) $\phi_1 - 2\phi_2$ (b) $\phi_1\phi_2$ (c) ϕ_1/ϕ_2 (d) $\phi_1{}^2$ (e) none of these answers

7.48   Select the relation that must hold if the flow is irrotational.

(*a*) $\partial u/\partial y + \partial v/\partial x = 0$ (*b*) $\partial u/\partial x = \partial v/\partial y$ (*c*) $\partial^2 u/\partial x^2 + \partial^2 v/\partial y^2 = 0$
(*d*) $\partial u/\partial y = \partial v/\partial x$ (*e*) none of these answers

7.49   The Bernoulli equation in steady ideal-fluid flow states that

(*a*) the velocity is constant along a streamline
(*b*) the energy is constant along a streamline but may vary across streamlines
(*c*) when the speed increases, the pressure increases
(*d*) the energy is constant throughout the fluid
(*e*) the net flow rate into any small region must be zero

7.50   The Stokes stream function applies to

(*a*) all three-dimensional ideal-fluid-flow cases
(*b*) ideal (nonviscous) fluids only
(*c*) irrotational flow only
(*d*) cases of axial symmetry
(*e*) none of these cases

7.51   The Stokes stream function has the value $\psi = 1$ at the origin and the value $\psi = 2$ at $(1,1,1)$. The discharge through the surface between these points is

(*a*) 1   (*b*) π (*c*) 2π (*d*) 4   (*e*) none of these answers

7.52   Select the relation that must hold in two-dimensional, irrotational flow:

(*a*) $\partial \phi/\partial x = \partial \psi/\partial y$ (*b*) $\partial \phi/\partial x = -\partial \psi/\partial y$ (*c*) $\partial \phi/\partial y = \partial \psi/\partial x$
(*d*) $\partial \phi/\partial x = \partial \phi/\partial y$ (*e*) none of these answers

7.53   The two-dimensional stream function
(*a*) is constant along an equipotential surface
(*b*) is constant along a streamline
(*c*) is defined for irrotational flow only
(*d*) relates velocity and pressure
(*e*) is none of these answers

7.54   In two-dimensional flow $\psi = 4$ ft²/s at $(0,2)$ and $\psi = 2$ ft²/s at $(0,1)$. The discharge between the two points is

(*a*) from left to right   (*b*) 4π cfs/ft   (*c*) 2 cfs/ft   (*d*) $1/\pi$ cfs/ft
(*e*) none of these answers

7.55   The boundary condition for steady flow of an ideal fluid is that the

(*a*) velocity is zero at the boundary
(*b*) velocity component normal to the boundary is zero
(*c*) velocity component tangent to the boundary is zero
(*d*) boundary surface must be stationary
(*e*) continuity equation must be satisfied

7.56 An equipotential surface

(a) has no velocity component tangent to it
(b) is composed of streamlines
(c) is a stream surface
(d) is a surface of constant dynamic pressure
(e) is none of these answers

7.57 A source in two-dimensional flow
(a) is a point from which fluid is imagined to flow outward uniformly in all directions
(b) is a line from which fluid is imagined to flow uniformly in all directions at right angles to it
(c) has a strength defined as the speed at unit radius
(d) has streamlines that are concentric circles
(e) has a velocity potential independent of the radius

7.58 The two-dimensional vortex

(a) has a strength given by the circulation around a path enclosing the vortex
(b) has radial streamlines
(c) has a zero circulation around it
(d) has a velocity distribution that varies directly as the radial distance from the vortex
(e) creates a velocity distribution that has rotation throughout the fluid

2

APPLICATIONS OF FLUID MECHANICS

In Part 1 the fundamental concepts and equations have been developed and illustrated by many examples and simple applications. Fluid resistance, dimensional analysis, compressible flow, and ideal-fluid flow have been presented. In Part 2 several of the important fields of application of fluid mechanics are explored: measurement of flow, turbomachinery, closed-conduit and open-channel flow, and unsteady flow.

8

FLUID MEASUREMENT

Fluid measurements include the determination of pressure, velocity, dis charge, shock waves, density gradients, turbulence, and viscosity. There are many ways these measurements may be taken, e.g., direct, indirect, gravimetric, volumetric, electronic, electromagnetic, and optical. Direct measurements for discharge consist in the determination of the volume or weight of fluid that passes a section in a given time interval. Indirect methods of discharge measurement require the determination of head, difference in pressure, or velocity at several points in a cross section, and with these computing the discharge. The most precise methods are the gravimetric or volumetric determinations, in which the weight or volume is measured by weigh scales or by a calibrated tank for a time interval that is measured by a stop watch.

Pressure and velocity measurements are first undertaken in this chapter, followed by positive-displacement meters, rate meters, river-flow measurement, and turbulence and viscosity measurement.

8.1 PRESSURE MEASUREMENT

The measurement of pressure is required in many devices that determine the velocity of a fluid stream or its rate of flow, because of the relationship between velocity and pressure given by the energy equation. The static pressure of a fluid in motion is its pressure when the velocity is undisturbed by the measurement. Figure 8.1 indicates one method of measuring static pressure, the *piezometer* opening. When the flow is parallel, as indicated, the pressure variation is hydrostatic normal to the streamlines; hence, by measuring the

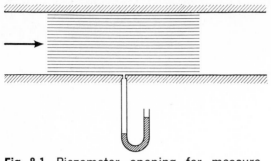

Fig. 8.1 Piezometer opening for measurement of static pressure.

pressure at the wall, the pressure at any other point in the cross section can be determined. The piezometer opening should be small, with length of opening at least twice its diameter, and should be normal to the surface, with no burrs at its edges because small eddies form and distort the measurement. A small amount of rounding of the opening is permissible. Any slight misalignment or roughness at the opening may cause errors in measurement; therefore, it is advisable to use several piezometer openings connected together into a *piezometer ring*. When the surface is rough in the vicinity of the opening, the reading is unreliable. For small irregularities it may be possible to smooth the surface around the opening.

For rough surfaces, the *static tube* (Fig. 8.2) may be used. It consists of a tube that is directed upstream with the end closed. It has radial holes in the cylindrical portion downstream from the nose. The flow is presumed to be moving by the openings as if it were undisturbed. There are disturbances, however, due to both the nose and the right-angled leg that is normal to the

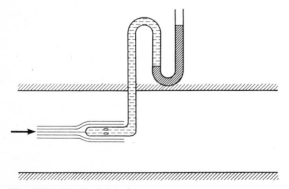

Fig. 8.2 Static tube.

flow. The static tube should be calibrated, as it may read too high or too low. If it does not read true static pressure, the discrepancy Δh normally varies as the square of the velocity of flow by the tube; i.e.,

$$\Delta h = C\,\frac{v^2}{2g}$$

in which C is determined by towing the tube in still fluid where pressure and velocity are known or by inserting it into a smooth pipe that contains a piezometer ring.

Such tubes are relatively insensitive to the Reynolds number and to Mach numbers below unity. Their alignment with the flow is not critical, so that an error of but a few percent is to be expected for a yaw misalignment of 15°.

The piezometric opening may lead to a bourdon gage, a manometer, a micromanometer, or an electronic transducer. The transducers depend upon very small deformations of a diaphragm due to pressure change to create an electronic signal. The principle may be that of a strain gage and a Wheatstone bridge circuit, or it may rely on motion in a differential transformer, a capacitance chamber, or the piezoelectric behavior of a crystal under stress.

8.2 VELOCITY MEASUREMENT

Since determining velocity at a number of points in a cross section permits evaluating the discharge, velocity measurement is an important phase of measuring flow. Velocity can be found by measuring the time an identifiable particle takes to move a known distance. This is done whenever it is convenient or necessary. This technique has been developed to study flow in regions which are so small that the normal flow would be greatly disturbed and perhaps disappear if an instrument were introduced to measure the velocity. A transparent viewing region must be made available, and by means of a strong light and a powerful microscope the very minute impurities in the fluid can be photographed with a high-speed motion-picture camera. From such motion pictures the velocity of the particles, and therefore the velocity of the fluid in a small region, can be determined.

Normally, however, a device is used which does not measure velocity directly but yields a measurable quantity that can be related to velocity. The *pitot tube* operates on such a principle and is one of the most accurate methods of measuring velocity. In Fig. 8.3 a glass tube or hypodermic needle with a right-angled bend is used to measure the velocity v in an open channel. The tube opening is directed upstream so that the fluid flows into the opening

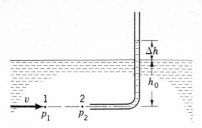

Fig. 8.3 Simple pitot tube.

until the pressure builds up in the tube sufficiently to withstand the impact of velocity against it. Directly in front of the opening the fluid is at rest. The streamline through 1 leads to the point 2, called the *stagnation point*, where the fluid is at rest, and there divides and passes around the tube. The pressure at 2 is known from the liquid column within the tube. Bernoulli's equation, applied between points 1 and 2, produces

$$\frac{v^2}{2g} + \frac{p_1}{\gamma} = \frac{p_2}{\gamma} = h_0 + \Delta h$$

since both points are at the same elevation. As $p_1/\gamma = h_0$, the equation reduces to

$$\frac{v^2}{2g} = \Delta h \qquad\qquad (8.2.1)$$

or

$$v = \sqrt{2g\,\Delta h} \qquad\qquad (8.2.2)$$

Practically, it is very difficult to read the height Δh from a free surface.

The pitot tube measures the stagnation pressure, which is also referred to as the *total* pressure. The total pressure is composed of two parts, the static pressure h_0 and the dynamic pressure Δh, expressed in length of a column of the flowing fluid (Fig. 8.3). The dynamic pressure is related to velocity head by Eq. (8.2.1).

By combining the static-pressure measurement and the total-pressure measurement, i.e., measuring each and connecting to opposite ends of a differential manometer, the dynamic pressure head is obtained. Figure 8.4 illustrates one arrangement. Bernoulli's equation applied from 1 to 2 is

$$\frac{v^2}{2g} + \frac{p_1}{\gamma} = \frac{p_2}{\gamma} \qquad\qquad (8.2.3)$$

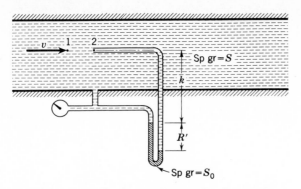

Fig. 8.4 Use of pitot tube and piezometer opening for measuring velocity.

The equation for the pressure through the manometer, in feet of water, is

$$\frac{p_1}{\gamma} S + kS + R'S_0 - (k + R')S = \frac{p_2}{\gamma} S$$

Simplifying gives

$$\frac{p_2 - p_1}{\gamma} = R'\left(\frac{S_0}{S} - 1\right) \tag{8.2.4}$$

Substituting for $(p_2 - p_1)/\gamma$ in Eq. (8.2.3) and solving for v results in

$$v = \sqrt{2gR'\left(\frac{S_0}{S} - 1\right)} \tag{8.2.5}$$

The pitot tube is also insensitive to flow alignment, and an error of only a few percent occurs if the tube has a yaw misalignment of less than 15°.

The static tube and pitot tube may be combined into one instrument, called a *pitot-static tube* (Fig. 8.5). Analyzing this system in a manner similar to that in Fig. 8.4 shows that the same relations hold; Eq. (8.2.5) expresses the velocity, but the uncertainty in the measurement of static pressure requires a corrective coefficient C to be applied:

$$v = C\sqrt{2gR'\left(\frac{S_0}{S} - 1\right)} \tag{8.2.6}$$

A particular form of pitot-static tube with a blunt nose, the *Prandtl tube,*

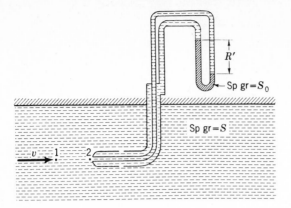

Fig. 8.5 Pitot-static tube.

has been designed so that the disturbances due to nose and leg cancel, leaving $C = 1$ in the equation. For other pitot-static tubes the constant C must be determined by calibration.

The *current meter* (Fig. 8.6) is used to measure the velocity of liquid flow in open channels. The cups are shaped so that the drag varies with orientation, causing a relatively slow rotation. With an electric circuit and headphones, an audible signal is detected for a fixed number of revolutions. The number of signals in a given time is a function of the velocity. The meters are calibrated by towing them through liquid at known speeds. For measuring high-velocity flow a current meter with a propeller as rotating element is used, as it offers less resistance to the flow.

Air velocities are measured with cup-type or vane (propeller) anemometers (Fig. 8.7) which drive generators indicating air velocity directly or drive counters indicating the number of revolutions.

By designing the vanes so that they have very low inertia, employing precision bearings and optical tachometers which effectively take no power to drive them, anemometers can be made to read very low air velocities. They can be sensitive enough to measure the convection air currents which the human body causes by its heat emission to the atmosphere.

Velocity measurement in compressible flow

The pitot-static tube may be used for velocity determinations in compressible flow. In Fig. 8.5 the velocity reduction from free-stream velocity at 1 to zero at 2 takes place very rapidly without significant heat transfer, and friction plays a very small part, so that the compression may be assumed to be isentropic. Applying Eq. (6.3.7) to points 1 and 2 of Fig. 8.5 with $V_2 = 0$

gives

$$\frac{V_1^2}{2} = \frac{k}{k-1}\left(\frac{p_2}{\rho_2} - \frac{p_1}{\rho_1}\right) = \frac{kR}{k-1}\,(T_2 - T_1) = c_p T_1 \left(\frac{T_2}{T_1} - 1\right) \qquad (8.2.7)$$

The substitution of c_p is from Eq. (6.1.8). Equation (6.1.17) then gives

$$\frac{V_1^2}{2} = c_p T_1 \left[\left(\frac{p_2}{p_1}\right)^{(k-1)/k} - 1\right] = c_p T_2 \left[1 - \left(\frac{p_1}{p_2}\right)^{(k-1)/k}\right] \qquad (8.2.8)$$

The static pressure p_1 may be obtained from the side openings of the pitot tube, and the stagnation pressure may be obtained from the impact opening leading to a simple manometer, or $p_2 - p_1$ may be found from the differential manometer. If the tube is not designed so that true static pressure is measured, it must be calibrated and the true static pressure computed.

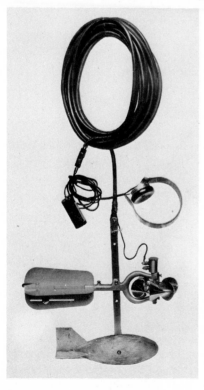

Fig. 8.6 Price current meter.
(*W. and L. E. Gurley.*)

Fig. 8.7 Air anemometer. (*Taylor Instrument Co.*)

Gas velocities may be measured with a *hot-wire anemometer*, which works on the principle that the resistance to flow through a fine platinum wire is a function of cooling due to gas flow around it. Cooled film sensors are also used for gas flow and have been adapted to liquid flow.

8.3 POSITIVE-DISPLACEMENT METERS

One volumetric meter, the positive-displacement meter, has pistons or partitions which are displaced by the flowing fluid and a counting mechanism that records the number of displacements in any convenient unit, such as gallons or cubic feet.

A common meter is the *disk* meter, or *wobble* meter (Fig. 8.8), used on many domestic water-distribution systems. The disk oscillates in a passageway so that a known volume of fluid moves through the meter for each oscillation. A stem normal to the disk operates a gear train, which in turn operates a counter. In good condition, these meters are accurate to within 1 percent. When they are worn, the error may be very large for small flows, such as those caused by a leaky faucet.

Fig. 8.8 Disk meter. (*Neptune Meter Co.*)

The flow of domestic gas at low pressure is usually measured by a volumetric meter with a traveling partition. The partition is displaced by gas inflow to one end of the chamber in which it operates, and then, by a change in valving, it is displaced to the opposite end of the chamber. The oscillations operate a counting mechanism.

Oil flow or high-pressure gas flow in a pipeline is frequently measured by a rotary meter in which cups or vanes move about an annular opening and displace a fixed volume of fluid for each revolution. Radial or axial pistons may be arranged so that the volume of a continuous flow through them is determined by rotations of a shaft.

Positive-displacement meters normally have no timing equipment that measures the rate of flow. The rate of steady flow may be determined with a stop watch to record the time for displacement of a given volume of fluid.

8.4 RATE METERS

A *rate meter* is a device that determines, generally by a single measurement, the quantity (weight or volume) per unit time that passes a given cross section. Included among rate meters are the *orifice, nozzle, venturi meter, rotameter* and *weir*, discussed in this section.

Orifice in a reservoir

An orifice may be used for measuring the rate of flow out of a reservoir or through a pipe. An orifice in a reservoir or tank may be in the wall or in the bottom. It is an opening, usually round, through which the fluid flows, as in Fig. 8.9. It may be square-edged, as shown, or rounded, as in Fig. 3.14. The area of the orifice is the area of the opening. With the square-edged orifice, the fluid jet contracts during a short distance of about one-half diameter downstream from the opening. The portion of the flow that approaches along the wall cannot make a right-angled turn at the opening and therefore maintains a radial velocity component that reduces the jet area. The cross section where the contraction is greatest is called the *vena contracta*. The streamlines are parallel throughout the jet at this section, and the pressure is atmospheric.

The head H on the orifice is measured from the center of the orifice to the free surface. The head is assumed to be held constant. Bernoulli's equation applied from a point 1 on the free surface to the center of the *vena contracta*, point 2, with local atmospheric pressure as datum and point 2 as elevation datum, neglecting losses, is written

$$\frac{V_1^2}{2g} + \frac{p_1}{\gamma} + z_1 = \frac{V_2^2}{2g} + \frac{p_2}{\gamma} + z_2$$

Inserting the values gives

$$0 + 0 + H = \frac{V_2^2}{2g} + 0 + 0$$

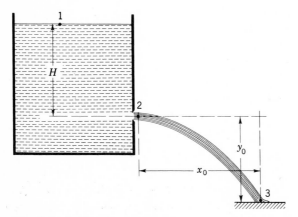

Fig. 8.9 Orifice in a reservoir.

or

$$V_2 = \sqrt{2gH} \qquad (8.4.1)$$

This is only the *theoretical* velocity because the losses between the two points were neglected. The ratio of the *actual* velocity V_a to the theoretical velocity V_t is called the *velocity coefficient* C_v; that is,

$$C_v = \frac{V_a}{V_t} \qquad (8.4.2)$$

Hence

$$V_{2a} = C_v \sqrt{2gH} \qquad (8.4.3)$$

The actual discharge Q_a from the orifice is the product of the actual velocity at the *vena contracta* and the area of the jet. The ratio of jet area A_2 at *vena contracta* to area of orifice A_0 is symbolized by another coefficient, called the *coefficient of contraction* C_c:

$$C_c = \frac{A_2}{A_0} \qquad (8.4.4)$$

The area at the *vena contracta* is $C_c A_0$. The actual discharge is thus

$$Q_a = C_v C_c A_0 \sqrt{2gH} \qquad (8.4.5)$$

It is customary to combine the two coefficients into a *discharge coefficient* C_d,

$$C_d = C_v C_c \qquad (8.4.6)$$

from which

$$Q_a = C_d A_0 \sqrt{2gH} \qquad (8.4.7)$$

There is no way to compute the losses between points 1 and 2; hence, C_v must be determined experimentally. It varies from 0.95 to 0.99 for the square-edged or rounded orifice. For most orifices, such as the square-edged one, the amount of contraction cannot be computed, and test results must be used. There are several methods for obtaining one or more of the coefficients. By measuring area A_0, the head H, and the discharge Q_a (by gravimetric or volumetric means), C_d is obtained from Eq. (8.4.7). Determination of either

C_v or C_c then permits determination of the other by Eq. (8.4.6). Several methods follow.

1. Trajectory method. By measuring the position of a point on the trajectory of the free jet downstream from the *vena contracta* (Fig. 8.9) the actual velocity V_a can be determined if air resistance is neglected. The x component of velocity does not change; therefore, $V_a t = x_0$, in which t is the time for a fluid particle to travel from the *vena contracta* to point 3. The time for a particle to drop a distance y_0 under the action of gravity when it has no initial velocity in that direction is expressed by $y_0 = gt^2/2$. After t is eliminated in the two relations,

$$V_a = \frac{x_0}{\sqrt{2y_0/g}}$$

With V_{2t} determined by Eq. (8.4.1) the ratio $V_a/V_t = C_v$ is known.

2. Direct measuring of V_a. With a pitot tube placed at the *vena contracta*, the actual velocity V_a is determined.

3. Direct measuring of jet diameter. With outside calipers, the diameter of jet at the *vena contracta* may be approximated. This is not a precise measurement and in general is less satisfactory than the other methods.

4. Use of momentum equation. When the reservoir is small enough to be suspended on knife-edges, as in Fig. 8.10, it is possible to determine the force F that creates the momentum in the jet. With the orifice opening closed, the tank is leveled by adding or subtracting weights. With the orifice discharging, a force creates the momentum in the jet and an equal and opposite force F' acts against the tank. By addition of sufficient weights W the tank is again leveled. From the figure, $F' = W x_0/y_0$. With

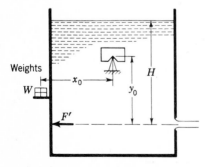

Fig. 8.10 Momentum method for determination of C_v and C_c.

the momentum equation,

$$\Sigma F_x = \frac{Q\gamma}{g}\left(V_{x\text{out}} - V_{x\text{in}}\right) \quad \text{or} \quad \frac{Wx_0}{y_0} = \frac{Q_a\gamma V_a}{g}$$

as $V_{x\text{in}}$ is zero and V_a is the final velocity. Since the actual discharge is measured, V_a is the only unknown in the equation.

Losses in orifice flow

The head loss in flow through an orifice is determined by applying the energy equation with a loss term for the distance between points 1 and 2 (Fig. 8.9),

$$\frac{V_{1a}^2}{2g} + \frac{p_1}{\gamma} + z_1 = \frac{V_{2a}^2}{2g} + \frac{p_2}{\gamma} + z_2 + \text{losses}$$

Substituting the values for this case gives

$$\text{Losses} = H - \frac{V_{2a}^2}{2g} = H(1 - C_v^2) = \frac{V_{2a}^2}{2g}\left(\frac{1}{C_v^2} - 1\right) \tag{8.4.8}$$

in which Eq. (8.4.3) has been used to obtain the losses in terms of H and C_v or V_{2a} and C_v.

EXAMPLE 8.1 A 3-in diameter orifice under a head of 16.0 ft discharges 2000 lb water in 32.6 s. The trajectory was determined by measuring $x_0 = 15.62$ ft for a drop of 4.0 ft. Determine C_v, C_c, C_d, the head loss per unit weight, and the horsepower loss.

The theoretical velocity V_{2t} is

$$V_{2t} = \sqrt{2gH} = \sqrt{64.4 \times 16} = 32.08 \text{ ft/s}$$

The actual velocity is determined from the trajectory. The time to drop 4 ft is

$$t = \sqrt{\frac{2y_0}{g}} = \sqrt{\frac{2 \times 4}{32.2}} = 0.498 \text{ s}$$

and the velocity is expressed by

$$x_0 = V_{2a}t \qquad V_{2a} = \frac{15.62}{0.498} = 31.4 \text{ ft/s}$$

Then

$$C_v = \frac{V_{2a}}{V_{2t}} = \frac{31.4}{32.08} = 0.98$$

The actual discharge Q_a is

$$Q_a = \frac{2000}{62.4 \times 32.6} = 0.984 \text{ cfs}$$

With Eq. (8.6.7)

$$C_d = \frac{Q_a}{A_0 \sqrt{2gH}} = \frac{0.984}{(\pi/64) \sqrt{64.4 \times 16}} = 0.625$$

Hence, from Eq. (8.4.6),

$$C_c = \frac{C_d}{C_v} = \frac{0.625}{0.98} = 0.638$$

The head loss, from Eq. (8.4.8), is

$$\text{Loss} = H(1 - C_v{}^2) = 16(1 - 0.98^2) = 0.63 \text{ ft·lb/lb}$$

The horsepower loss is

$$\frac{0.63 \times 2000}{550 \times 32.6} = 0.070 \text{ hp}$$

The Borda mouthpiece (Fig. 8.11), a short, thin-walled tube about one diameter long that projects into the reservoir (re-entrant), permits appli-

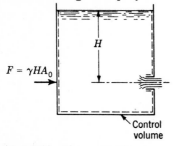

$F = \gamma H A_0$

H

Control volume

Fig. 8.11 The Borda mouthpiece.

cation of the momentum equation, which yields one relation between C_v and C_d. The velocity along the wall of the tank is almost zero at all points; hence, the pressure distribution is hydrostatic. With the component of force exerted on the liquid by the tank parallel to the axis of the tube, there is an unbalanced force due to the opening, which is $\gamma H A_0$. The final velocity is V_{2a}, the initial velocity is zero, and Q_a is the actual discharge. Then

$$\gamma H A_0 = Q_a \frac{\gamma}{g} V_{2a}$$

and

$$Q_a = C_d A_0 \sqrt{2gH} \qquad V_{2a} = C_v \sqrt{2gH}$$

Substituting for Q_a and V_{2a} and simplifying lead to

$$1 = 2C_d C_v = 2C_v{}^2 C_c$$

In the orifice situations considered, the liquid surface in the reservoir has been assumed to be held constant. An unsteady-flow case of some practical interest is that of determining the time to lower the reservoir surface a given distance. Theoretically, Bernoulli's equation applies only to steady flow, but if the reservoir surface drops slowly enough, the error from using Bernoulli's equation is negligible. The volume discharged from the orifice in time δt is $Q \, \delta t$, which must just equal the reduction in volume in the reservoir in the same time increment (Fig. 8.12), $A_R(-\delta y)$, in which A_R is the area of liquid surface at height y above the orifice. Equating the two expres-

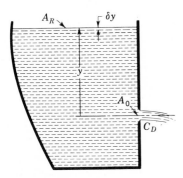

Fig. 8.12 Notation for falling head.

sions gives

$$Q \, \delta t = -A_R \, \delta y$$

Solving for δt and integrating between the limits $y = y_1$, $t = 0$ and $y = y_2$, $t = t$ yield

$$t = \int_0^t dt = -\int_{y_1}^{y_2} \frac{A_R \, dy}{Q}$$

The orifice discharge Q is $C_d A_0 \sqrt{2gy}$. After substitution for Q,

$$t = -\frac{1}{C_d A_0 \sqrt{2g}} \int_{y_1}^{y_2} A_R y^{-1/2} \, dy$$

When A_R is known as a function of y, the integral can be evaluated. Consistent with other SI or English units, t is in seconds. For the special case of a tank with constant cross section,

$$t = -\frac{A_R}{C_d A_0 \sqrt{2g}} \int_{y_1}^{y_2} y^{-1/2} \, dy = \frac{2A_R}{C_d A_0 \sqrt{2g}} \left(\sqrt{y_1} - \sqrt{y_2} \right)$$

EXAMPLE 8.2   A tank has a horizontal cross-sectional area of $2 \, m^2$ at the elevation of the orifice, and the area varies linearly with elevation so that it is $1 \, m^2$ at a horizontal cross section 3 m above the orifice. For a 10-cm-diameter orifice, $C_d = 0.65$, compute the time in seconds to lower the surface from 2.5 to 1 m above the orifice.

$$A_R = 2 - \frac{y}{3} \quad m^2$$

and

$$t = -\frac{1}{0.65\pi(0.05^2) \sqrt{2 \times 9.806}} \int_{2.5}^{1} \left(2 - \frac{y}{3} \right) y^{-1/2} \, dy = 73.7 \text{ s}$$

Venturi meter

The venturi meter is used to measure the rate of flow in a pipe. It is generally a casting (Fig. 8.13) consisting of an upstream section which is the same size as the pipe, has a bronze liner, and contains a piezometer ring for measuring

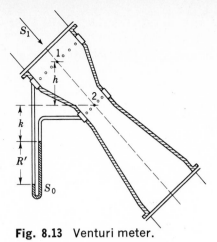

Fig. 8.13 Venturi meter.

static pressure; a converging conical section; a cylindrical throat with a bronze liner containing a piezometer ring; and a gradually diverging conical section leading to a cylindrical section the size of the pipe. A differential manometer is attached to the two piezometer rings. The size of a venturi meter is specified by the pipe and throat diameter; e.g., a 6 by 4 in venturi meter fits a 6-in-diameter pipe and has a 4-in-diameter throat. For accurate results the venturi meter should be preceded by at least 10 diameters of straight pipe. In the flow from the pipe to the throat, the velocity is greatly increased and the pressure correspondingly decreased. The amount of discharge in incompressible flow is shown to be a function of the manometer reading.

The pressures at the upstream section and throat are *actual pressures*, and the velocities from Bernoulli's equation without a loss term are *theoretical velocities*. When losses are considered in the energy equation, the velocities are *actual velocities*. First, with the Bernoulli equation (i.e., without a head-loss term) the theoretical velocity at the throat is obtained. Then by multiplying this by the velocity coefficient C_v, the actual velocity is obtained. The actual velocity times the actual area of the throat determines the actual discharge. From Fig. 8.13

$$\frac{V_{1t}^2}{2g} + \frac{p_1}{\gamma} + h = \frac{V_{2t}^2}{2g} + \frac{p_2}{\gamma} \tag{8.4.9}$$

in which elevation datum is taken through point 2. V_1 and V_2 are average velocities at sections 1 and 2, respectively; hence, α_1, α_2 are assumed to be unity. With the continuity equation $V_1 D_1^2 = V_2 D_2^2$,

$$\frac{V_1^2}{2g} = \frac{V_2^2}{2g} \left(\frac{D_2}{D_1}\right)^4 \tag{8.4.10}$$

which holds for either the actual velocities or the theoretical velocities. Equation (8.4.9) may be solved for V_{2t},

$$\frac{V_{2t}^2}{2g}\left[1 - \left(\frac{D_2}{D_1}\right)^4\right] = \frac{p_1 - p_2}{\gamma} + h$$

and

$$V_{2t} = \sqrt{\frac{2g[h + (p_1 - p_2)/\gamma]}{1 - (D_2/D_1)^4}} \tag{8.4.11}$$

Introducing the velocity coefficient $V_{2a} = C_v V_{2t}$ gives

$$V_{2a} = C_v \sqrt{\frac{2g[h + (p_1 - p_2)/\gamma]}{1 - (D_2/D_1)^4}} \tag{8.4.12}$$

After multiplying by A_2, the actual discharge Q is determined to be

$$Q = C_v A_2 \sqrt{\frac{2g[h + (p_1 - p_2)/\gamma]}{1 - (D_2/D_1)^4}} \tag{8.4.13}$$

The gage difference R' may now be related to the pressure difference by writing the equation for the manometer. In units of length of water (S_1 is the specific gravity of flowing fluid and S_0 the specific gravity of manometer liquid),

$$\frac{p_1}{\gamma} S_1 + (h + k + R') S_1 - R' S_0 - k S_1 = \frac{p_2}{\gamma} S_1$$

Simplifying gives

$$h + \frac{p_1 - p_2}{\gamma} = R'\left(\frac{S_0}{S_1} - 1\right) \tag{8.4.14}$$

By substituting into Eq. (8.4.13),

$$Q = C_v A_2 \sqrt{\frac{2gR'(S_0/S_1 - 1)}{1 - (D_2/D_1)^4}} \tag{8.4.15}$$

which is the venturi-meter equation for incompressible flow. The contraction coefficient is unity; hence, $C_v = C_d$. It should be noted that h has dropped out of the equation. The discharge depends upon the gage difference R' regardless

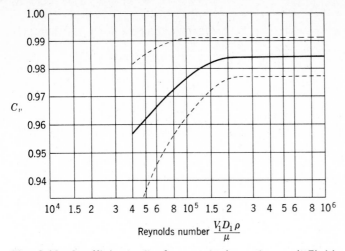

Fig. 8.14 Coefficient C_v for venturi meters. ("*Fluid Meters: Their Theory and Application,*" *5th ed., American Society of Mechanical Engineers, 1956.*)

of the orientation of the venturi meter; whether it is horizontal, vertical, or inclined, exactly the same equation holds.

C_v is determined by calibration, i.e., by measuring the discharge and the gage difference and solving for C_v, which is usually plotted against the Reynolds number. Experimental results for venturi meters are given in Fig. 8.14. They are applicable to diameter ratios D_2/D_1 from 0.25 to 0.75 within the tolerances shown by the dotted lines. Where feasible, a venturi meter should be selected so that its coefficient is constant over the range of Reynolds numbers for which it is to be used.

The coefficient may be slightly greater than unity for venturi meters that are unusually smooth inside. This does not mean that there are no losses but results from neglecting the kinetic-energy correction factors α_1, α_2 in the Bernoulli equation. Generally α_1 is greater than α_2 since the reducing section acts to make the velocity distribution uniform across section 2.

The venturi meter has a low overall loss, due to the gradually expanding conical section, which aids in reconverting the high kinetic energy at the throat into pressure energy. The loss is about 10 to 15 percent of the head change between sections 1 and 2.

Venturi meter for compressible flow

The theoretical flow of a compressible fluid through a venturi meter is substantially isentropic and is obtained from Eqs. (6.3.2), (6.3.6), and (6.3.7).

When multiplied by C_v, the velocity coefficient, it yields for mass flow rate

$$\dot{m} = C_v A_2 \sqrt{\frac{[2k/(k-1)]p_1\rho_1(p_2/p_1)^{2/k}[1-(p_2/p_1)^{(k-1)/k}]}{1-(p_2/p_1)^{2/k}(A_2/A_1)^2}} \qquad (8.4.16)$$

The velocity coefficient is the same as for liquid flow. Equation (8.4.13) when reduced to horizontal flow and modified by insertion of an expansion factor can be applied to compressible flow

$$\dot{m} = C_v Y A_2 \sqrt{\frac{2\rho_1\,\Delta p}{1-(D_2/D_1)^4}} \qquad (8.4.17)$$

Y can be found by solving Eqs. (8.4.17) and (8.4.16) and is shown to be a function of k, p_2/p_1, and A_2/A_1. Values of Y are plotted in Fig. 8.15 for $k = 1.40$; hence, by the use of Eq. (8.4.17) and Fig. 8.15 compressible flow can be computed for a venturi meter.

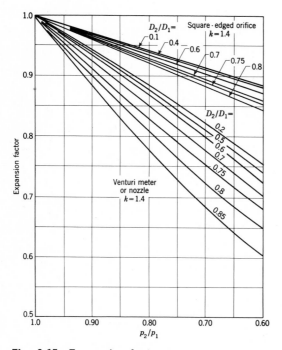

Fig. 8.15 Expansion factors.

Flow nozzle

The ISA (Instrument Society of America) flow nozzle (originally the VDI flow nozzle) is shown in Fig. 8.16. It has no contraction of the jet other than that of the nozzle opening; therefore the contraction coefficient is unity.

Equations (8.4.13) and (8.4.15) hold equally well for the flow nozzle. For a horizontal pipe ($h = 0$), Eq. (8.4.13) may be written

$$Q = CA_2 \sqrt{\frac{2\Delta p}{\rho}} \tag{8.4.18}$$

in which

$$C = \frac{C_v}{\sqrt{1 - (D_2/D_1)^4}} \tag{8.4.19}$$

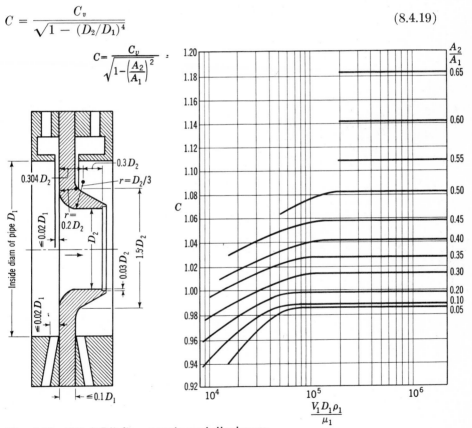

Fig. 8.16 ISA (VDI) flow nozzle and discharge coefficients. (*Ref. 11 in NACA Tech. Mem. 952.*)

and $\Delta p = p_1 - p_2$. The value of coefficient C in Fig. 8.16 is for use in Eq. (8.4.18). When the coefficient given in the figure is to be used, it is important that the dimensions shown be closely adhered to, particularly in the location of the piezometer openings (two methods shown) for measuring pressure drop. At least 10 diameters of straight pipe should precede the nozzle.

The flow nozzle is less costly than the venturi meter. It has the disadvantage that the overall losses are much higher because of the lack of guidance of jet downstream from the nozzle opening.

Compressible flow through a nozzle is found by Eq. (8.4.17) and Fig. 8.15 if $k = 1.4$. For other values of specific-heat ratio k, Eq. (8.4.16) may be used.

EXAMPLE 8.3 Determine the flow through a 6-in-diameter water line that contains a 4-in diameter flow nozzle. The mercury-water differential manometer has a gage difference of 10 in. Water temperature is 60°F.

From the data given, $S_0 = 13.6$, $S_1 = 1.0$, $R' = \frac{10}{12} = 0.833$ ft, $A_2 = \pi/36 = 0.0873$ ft², $\rho = 1.938$ slugs/ft³, $\mu = 2.359 \times 10^{-5}$ lb·s/ft². Substituting Eq. (8.4.19) into Eq. (8.4.15) gives

$$Q = CA_2 \sqrt{2gR'\left(\frac{S_0}{S_1} - 1\right)}$$

From Fig. 8.16, for $A_2/A_1 = (\frac{4}{6})^2 = 0.444$, assume that the horizontal region of the curves applies; hence, $C = 1.056$; then compute the flow and the Reynolds number.

$$Q = 1.056 \times 0.0873 \sqrt{64.4 \times 0.833 \left(\frac{13.6}{1.0} - 1.0\right)} = 2.40 \text{ cfs}$$

Then

$$V_1 = \frac{Q}{A_1} = \frac{2.40}{\pi/16} = 12.21 \text{ ft/s}$$

and

$$\mathbf{R} = \frac{V_1 D_1 \rho}{\mu} = \frac{12.21 \times 1.938}{2 \times 2.359 \times 10^{-5}} = 502,000$$

The chart shows the value of C to be correct; therefore, the discharge is 2.40 cfs.

Orifice in a pipe

The square-edged orifice in a pipe (Fig. 8.17) causes a contraction of the jet downstream from the orifice opening. For incompressible flow Bernoulli's equation applied from section 1 to the jet at its *vena contracta*, section 2, is

$$\frac{V_{1t}^2}{2g} + \frac{p_1}{\gamma} = \frac{V_{2t}^2}{2g} + \frac{p_2}{\gamma}$$

The continuity equation relates V_{1t} and V_{2t} with the contraction coefficient $C_c = A_2/A_0$,

$$V_1 \frac{\pi D_1^2}{4} = V_2 C_c \frac{\pi D_0^2}{4} \tag{8.4.20}$$

After eliminating V_{1t},

$$\frac{V_{2t}^2}{2g}\left[1 - C_c^2\left(\frac{D_0}{D_1}\right)^4\right] = \frac{p_1 - p_2}{\gamma}$$

and by solving for V_{2t} the result is

$$V_{2t} = \sqrt{\frac{2g(p_1 - p_2)/\gamma}{1 - C_c^2(D_0/D_1)^4}}$$

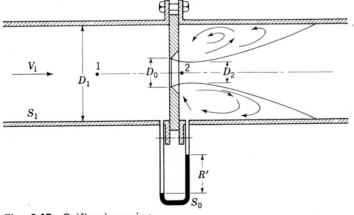

Fig. 8.17 Orifice in a pipe.

Multiplying by C_v to obtain the actual velocity at the *vena contracta* gives

$$V_{2a} = C_v \sqrt{\frac{2(p_1 - p_2)/\rho}{1 - C_c^2(D_0/D_1)^4}}$$

and, finally multiplying by the area of the jet, $C_c A_0$, produces the actual discharge Q,

$$Q = C_d A_0 \sqrt{\frac{2(p_1 - p_2)/\rho}{1 - C_c^2(D_0/D_1)^4}} \tag{8.4.21}$$

in which $C_d = C_v C_c$. In terms of the gage difference R', Eq. (8.4.21) becomes

$$Q = C_d A_0 \sqrt{\frac{2gR'(S_0/S_1 - 1)}{1 - C_c^2(D_0/D_1)^4}} \tag{8.4.22}$$

Because of the difficulty in determining the two coefficients separately, a simplified formula is generally used, Eq. (8.4.18),

$$Q = CA_0 \sqrt{\frac{2\Delta p}{\rho}} \tag{8.4.23}$$

or its equivalent,

$$Q = CA_0 \sqrt{2gR'\left(\frac{S_0}{S_1} - 1\right)} \tag{8.4.24}$$

Values of C are given in Fig. 8.18 for the VDI orifice.

Experimental values of expansion factor, for $k = 1.4$, are given in Fig. 8.15. Equation (8.4.23) for actual mass flow rate in compressible flow becomes

$$\dot{m} = CYA_0 \sqrt{2\rho_1 \Delta p} \tag{8.4.25}$$

The location of the pressure taps is usually specified so that an orifice can be installed in a conduit and used with sufficient accuracy without performing a calibration at the site.

Elbow meter

The elbow meter for incompressible flow is one of the simplest flow-rate measuring devices. Piezometer openings on the inside and on the outside of the elbow are connected to a differential manometer. Because of centrifugal force

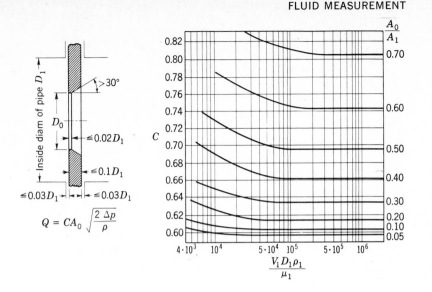

$$Q = CA_0 \sqrt{\frac{2\,\Delta p}{\rho}}$$

Fig. 8.18 VDI orifice and discharge coefficients. (*Ref. 11 in NACA Tech. Mem. 952.*)

at the bend, the difference in pressures is related to the discharge. A straight calming length should precede the elbow, and, for accurate results, the meter should be calibrated in place.[1] As most pipelines have an elbow, it may be used as the meter. After calibration the results are as reliable as with a venturi meter or a flow nozzle.

Rotameter

The rotameter (Fig. 8.19) is a variable-area meter that consists of an enlarging transparent tube and a metering "float" (actually heavier than the liquid) that is displaced upward by the upward flow of fluid through the tube. The tube is graduated to read the flow directly. Notches in the float cause it to rotate and thus maintain a central position in the tube. The greater the flow, the higher the position the float assumes.

Weirs

Open-channel flow may be measured by a *weir*, which is an obstruction in the channel that causes the liquid to back up behind it and to flow over it or

[1] W. M. Lansford, The Use of an Elbow in a Pipe Line for Determining the Rate of Flow in a Pipe, *Univ. Ill. Eng. Exp. Stn. Bull.* 289, December 1936.

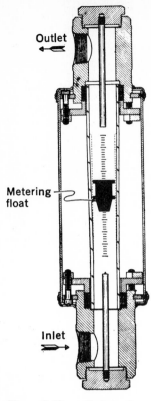

Fig.   8.19   Rotameter.
(*Fischer & Porter Co.*)

through it. By measuring the height of upstream liquid surface, the rate of
flow is determined. Weirs constructed from a sheet of metal or other material
so that the jet, or *nappe*, springs free as it leaves the upstream face are called
sharp-crested weirs. Other weirs such as the *broad-crested* weir support the flow
in a longitudinal direction.

The sharp-crested rectangular weir (Fig. 8.20) has a horizontal crest.
The nappe is contracted at top and bottom as shown. An equation for discharge can be derived if the contractions are neglected. Without contractions
the flow appears as in Fig. 8.21. The nappe has parallel streamlines with
atmospheric pressure throughout.

Bernoulli's equation applied between 1 and 2 is

$$H + 0 + 0 = \frac{v^2}{2g} + H - y + 0$$

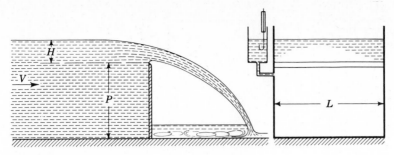

Fig. 8.20 Sharp-crested rectangular weir.

in which the velocity head at section 1 is neglected. By solving for v,

$$v = \sqrt{2gy}$$

The theoretical discharge Q_t is

$$Q_t = \int v \, dA = \int_0^H vL \, dy = \sqrt{2g} \, L \int_0^H y^{1/2} \, dy = \tfrac{2}{3} \sqrt{2g} \, LH^{3/2}$$

in which L is the width of weir. Experiment shows that the exponent of H is correct but that the coefficient is too great. The contractions and losses reduce the actual discharge to about 62 percent of the theoretical, or

$$Q = \begin{cases} 3.33LH^{3/2} & \text{English units} \\ 1.84LH^{3/2} & \text{SI units} \end{cases} \qquad (8.4.26)$$

When the weir does not extend completely across the width of the

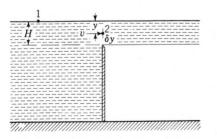

Fig. 8.21 Weir nappe without contractions.

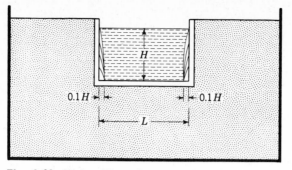

Fig. 8.22 Weir with end contractions.

channel, it has *end contractions*, illustrated in Fig. 8.22. An empirical correction for the reduction of flow is accomplished by subtracting $0.1H$ from L for each end contraction. The weir in Fig. 8.20 is said to have its end contractions *suppressed*.

The head H is measured upstream from the weir a sufficient distance to avoid the surface contraction. A hook gage mounted in a stilling pot connected to a piezometer opening determines the water-surface elevation from which the head is determined.

When the height P of weir (Fig. 8.20) is small, the velocity head at 1 cannot be neglected. A correction may be added to the head,

$$Q = CL\left(H + \alpha\frac{V^2}{2g}\right)^{3/2} \tag{8.4.27}$$

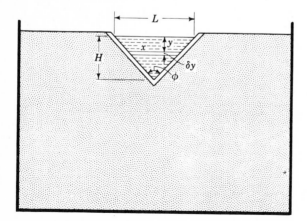

Fig. 8.23 V-notch weir.

in which V is velocity and α is greater than unity, usually taken around 1.4, which accounts for the nonuniform velocity distribution. Equation (8.4.27) must be solved for Q by trial since Q and V are both unknown. As a first trial, the term $\alpha V^2/2g$ may be neglected to approximate Q; then with this trial discharge a value of V is computed, since

$$V = \frac{Q}{L(P + H)}$$

For small discharges the V-notch weir is particularly convenient. The contraction of the nappe is neglected, and the theoretical discharge is computed (Fig. 8.23) as follows.

The velocity at depth y is $v = \sqrt{2gy}$, and the theoretical discharge is

$$Q_t = \int v \, dA = \int_0^H vx \, dy$$

By similar triangles, x may be related to y,

$$\frac{x}{H - y} = \frac{L}{H}$$

After substituting for v and x,

$$Q_t = \sqrt{2g} \frac{L}{H} \int_0^H y^{1/2}(H - y) \, dy = \tfrac{4}{15} \sqrt{2g} \frac{L}{H} H^{5/2}$$

Expressing L/H in terms of the angle ϕ of the V notch gives

$$\frac{L}{2H} = \tan \frac{\phi}{2}$$

Hence,

$$Q_t = \tfrac{8}{15} \sqrt{2g} \tan \frac{\phi}{2} H^{5/2}$$

The exponent in the equation is approximately correct, but the coefficient must be reduced by about 42 percent because of the neglected contractions. An approximate equation for a 90° V-notch weir is

$$Q = \begin{cases} 2.50 H^{2.50} & \text{English units} \\ 1.38 H^{2.50} & \text{SI units} \end{cases} \qquad (8.4.28)$$

Experiments show that the coefficient is increased by roughening the upstream side of the weir plate, which causes the boundary layer to grow thicker. The greater amount of slow-moving liquid near the wall is more easily turned, and hence there is less contraction of the nappe.

The broad-crested weir (Fig. 8.24a) supports the nappe so that the pressure variation is hydrostatic at section 2. Bernoulli's equation applied between points 1 and 2 can be used to find the velocity v_2 at height z, neglecting the velocity of approach,

$$H + 0 + 0 = \frac{v_2{}^2}{2g} + z + (y - z)$$

In solving for v_2,

$$v_2 = \sqrt{2g(H - y)}$$

z drops out; hence, v_2 is constant at section 2. For a weir of width L normal to the plane of the figure, the theoretical discharge is

$$Q = v_2 L y = L y \sqrt{2g(H - y)} \tag{8.4.29}$$

A plot of Q as abscissa against the depth y as ordinate, for constant H, is given in Fig. 8.24b. The depth is shown to be that which yields the maximum discharge, by the following reasoning.

A gate or other obstruction placed at section 3 of Fig. 8.24a can completely stop the flow by making $y = H$. Now, if a small flow is permitted to pass section 3 (holding H constant), the depth y becomes a little less than H and the discharge is, for example, as shown by point a on the depth-discharge curve. By further lifting of the gate or obstruction at section 3 the discharge-depth relationship follows the upper portion of the curve until the maximum discharge is reached. Any additional removal of downstream obstructions, however, has no effect upon the discharge, because the velocity of flow at

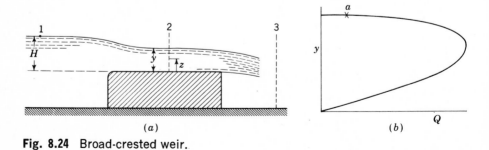

Fig. 8.24 Broad-crested weir.

section 2 is $\sqrt{gy}$, which is exactly the speed an elementary wave can travel in still liquid of depth y. Hence, the effect of any additional lowering of the downstream surface elevation cannot travel upstream to affect further the value of y, and the discharge occurs at the maximum value. This depth y, called the *critical depth*, is discussed in Sec. 11.4. The speed of an elementary wave is derived in Sec. 12.10.

By taking dQ/dy and with the result set equal to zero, for constant H,

$$\frac{dQ}{dy} = 0 = L\sqrt{2g(H-y)} + Ly\,\frac{1}{2}\,\frac{-2g}{\sqrt{2g(H-y)}}$$

and solving for y gives

$$y = \tfrac{2}{3}H$$

Inserting the value of H, that is, $3y/2$, into the equation for velocity v_2 gives

$$v_2 = \sqrt{gy}$$

and substituting the value of y into Eq. (8.4.29) leads to

$$Q_t = \begin{cases} 3.09LH^{3/2} & \text{English units} \\ 1.705LH^{3/2} & \text{SI units} \end{cases} \tag{8.4.30}$$

Experiments show that for a well-rounded upstream edge the discharge is

$$Q = \begin{cases} 3.03LH^{3/2} & \text{English units} \\ 1.67LH^{3/2} & \text{SI units} \end{cases} \tag{8.4.31}$$

which is within 2 percent of the theoretical value. The flow, therefore, adjusts itself to discharge at the maximum rate.

Since viscosity and surface tension have a minor effect on the discharge coefficients of weirs, a weir should be calibrated with the liquid that it will measure.

EXAMPLE 8.4 Tests on a 60° V-notch weir yield the following values of head H on the weir and discharge Q:

H, ft	0.345	0.356	0.456	0.537	0.568	0.594	0.619	0.635	0.654	0.665
Q, cfs	0.107	0.110	0.205	0.303	0.350	0.400	0.435	0.460	0.490	0.520

By means of the theory of least squares, determine the constants in $Q = CH^m$ for this weir.

By taking the logarithm of each side of the equation

$$\ln Q = \ln C + m \ln H \qquad \text{or} \qquad Y = B + mX$$

it is noted that the best values of B and m are needed for a straight line through the data when plotted on log-log paper.

By the theory of least squares, the best straight line through the data points is the one yielding a minimum value of the sums of the squares of vertical displacements of each point from the line; or, from Fig. 8.25,

$$F = \sum_{i=1}^{i=n} s_i^2 = \Sigma[y_i - (B + mx_i)]^2$$

n is the number of experimental points.

To minimize F, $\partial F/\partial B$ and $\partial F/\partial m$ are taken and set equal to zero, yielding two equations in the two unknowns B and m, as follows:

$$\frac{\partial F}{\partial B} = 0 = 2\Sigma[y_i - (B + mx_i)](-1)$$

from which

$$\Sigma y_i - nB - m\Sigma x_i = 0 \tag{1}$$

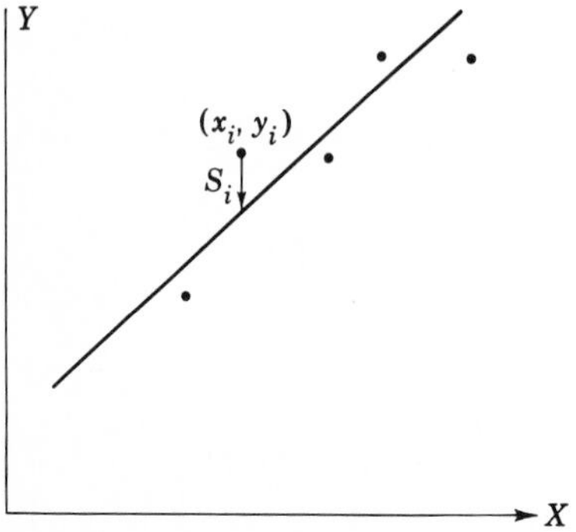

Fig. 8.25 Log-log plot of Q vs. H for V-notch weir.

and

$$\frac{\partial F}{\partial m} = 0 = 2\Sigma[y_i - (B + mx_i)](-x_i)$$

or

$$\Sigma x_i y_i - B\Sigma x_i - m\Sigma x_i^2 = 0 \qquad (2)$$

Solving Eqs. (1) and (2) for m gives

$$m = \frac{\Sigma x_i y_i/\Sigma x_i - \Sigma y_i/n}{\Sigma x_i^2/\Sigma x_i - \Sigma x_i/n} \qquad B = \frac{\Sigma y_i - m\Sigma x_i}{n}$$

The logarithms of Q_i and H_i may be looked up, and the summations taken by desk calculator. This problem, however, is very easily handled by digital computer with the following program:

```
   READ M,LH,LQ
   DIMENSION Q(10),H(10)
   NAMELIST/DIN/Q,H,N,M,C
   READ(5,DIN,END=99)
   X=.0
   Y=.0
   XX=.0
   XY=.0
   DO 1 I=1,N
   LH=ALOG(H(I))
   LQ=ALOG(Q(I))
   X=X+LH
   Y=Y+LQ
   XX=XX+LH**2
 1 XY=XY+LH*LQ
   M=(XY/X-Y/N)/(XX/X-X/N)
   B= (Y-M*X)/N
   C=EXP(B)
   WRITE(6,DIN)
99 CALL SYSTEM
   END
N = 10,Q(1) = .107, .110, .205, .303, .35, .4, .435, .46, .49, .52
      H(1) = .345, .356, .456, .537, .568, .594, .619, .635, .654, .665
```
For the data of this problem, $m = 2.437$, $C = 1.395$.

8.5 ELECTROMAGNETIC FLOW DEVICES

If a magnetic field is set up across a nonconducting tube and a conducting fluid flows through the tube, an induced voltage is produced across the flow which can be measured if electrodes are embedded in the tube walls.[1] The voltage is a linear function of the volume rate passing through the tube. Either an ac or a dc field may be used, with a corresponding signal generated at the electrodes. A disadvantage of the method is the small signal received and the large amount of amplification needed. The device has been used to measure the flow in blood vessels.

8.6 MEASUREMENT OF RIVER FLOW

Daily records of the discharge of rivers over long periods of time are essential to economic planning for utilization of their water resources or protection against floods. The daily measurement of discharge by determining velocity distribution over a cross section of the river is costly. To avoid this and still obtain daily records, *control sections* are established where the river channel is stable, i.e., with little change in bottom or sides of the stream bed. The control section is frequently at a break in slope of the river bottom where it becomes steeper downstream.

A gage rod is mounted at the control section so that the elevation of water surface is determined by reading the waterline on the rod; in some installations float-controlled recording gages keep a continuous record of river elevation. A *gage height-discharge* curve is established by taking current-meter measurements from time to time as the river discharge changes and plotting the resulting discharge against the gage height.

With a stable control section the gage height-discharge curve changes very little, and current-meter measurements are infrequent. For unstable control sections the curve changes continuously, and discharge measurements must be made every few days to maintain an accurate curve.

Daily readings of gage height produce a daily record of the river discharge.

8.7 MEASUREMENT OF TURBULENCE

Turbulence is a characteristic of the flow. It affects the calibration of measuring instruments and has an important effect upon heat transfer, evaporation, diffusion, and many other phenomena connected with fluid movement.

---

[1] H. G. Elrod, Jr., and R. R. Fouse, An Investigation of Electromagnetic Flowmeters, *Trans. ASME*, vol. 74, p. 589, May 1952.

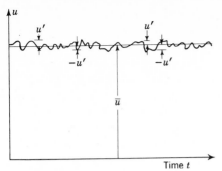

Fig. 8.26 Turbulent fluctuations in direction of flow.

Turbulence is generally specified by two quantities, the *size* and the *intensity* of the fluctuations. In steady flow the temporal mean velocity components at a point are constant. If these mean values are $\bar{u}$, $\bar{v}$, $\bar{w}$ and the velocity components at an instant are u, v, w, the fluctuations are given by u', v', w', in

$$u = \bar{u} + u' \qquad v = \bar{v} + v' \qquad w = \bar{w} + w'$$

The root mean square of measured values of the fluctuations (Fig. 8.26) is a measure of the intensity of the turbulence. These are $\sqrt{(\bar{u'})^2}$, $\sqrt{(\bar{v'})^2}$, $\sqrt{(\bar{w'})^2}$.

The size of the fluctuation is an average measure of the size of eddy, or vortex, in the flow. When two velocity measuring instruments (hot-wire anemometers) are placed adjacent to each other in a flow, the velocity fluctuations are correlated, i.e., they tend to change in unison. Separating these instruments reduces this correlation. The distance between instruments for zero correlation is a measure of the size of the fluctuation. Another method for determining turbulence is discussed in Sec. 5.6.

8.8 MEASUREMENT OF VISCOSITY

The treatment of fluid measurement is concluded with a discussion of methods for determining viscosity. Viscosity may be measured in a number of ways: (1) by use of Newton's law of viscosity; (2) by use of the Hagen-Poiseuille equation; (3) by methods that require calibration with fluids of known viscosity.

By measurement of the velocity gradient du/dy and the shear stress τ, in Newton's law of viscosity [Eq. (1.1.1)],

$$\tau = \mu \frac{du}{dy} \tag{8.8.1}$$

the dynamic or absolute viscosity can be computed. This is the most basic method as it determines all other quantities in the defining equation for viscosity. By means of a cylinder that rotates at a known speed with respect to an inner concentric stationary cylinder, du/dy is determined. By measurement of torque on the stationary cylinder, the shear stress can be computed. The ratio of shear stress to rate of change of velocity expresses the viscosity.

A schematic view of a concentric-cylinder viscometer is shown in Fig. 8.27. When the speed of rotation is N rpm and the radius is r_2 ft, the fluid velocity at the surface of the outer cylinder is $2\pi r_2 N/60$. With clearance b ft

$$\frac{du}{dy} = \frac{2\pi r_2 N}{60b}$$

The equation is based on $b \ll r_2$. The torque T_c on the inner cylinder is measured by a torsion wire from which it is suspended. By attaching a disk to the wire, its rotation can be determined by a fixed pointer. If the torque due to

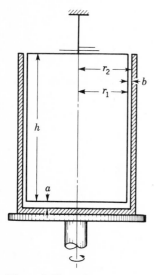

Fig. 8.27 Concentric-cylinder viscometer.

fluid below the bottom of the inner cylinder is neglected, the shear stress is

$$\tau = \frac{T_c}{2\pi r_1^2 h}$$

Substituting into Eq. (8.8.1) and solving for the viscosity give

$$\mu = \frac{15T_c b}{\pi^2 r_1^2 r_2 h N} \tag{8.8.2}$$

When the clearance a is so small that the torque contribution from the bottom is appreciable, it can be calculated in terms of the viscosity.
   Referring to Fig. 8.28,

$$\delta T = r\tau\, \delta A = r\mu\, \frac{\omega r}{a}\, r\, \delta r\, \delta\theta$$

in which the velocity change is ωr in the distance a ft. Integrating over the circular area of the disk and letting $\omega = 2\pi N/60$ lead to

$$T_d = \frac{\mu}{a}\frac{\pi}{30}N\int_0^{r_1}\int_0^{2\pi} r^3\, dr\, d\theta = \frac{\mu\pi^2}{a60}Nr_1^4 \tag{8.8.3}$$

The torque due to disk and cylinder must equal the torque T in the torsion wire, so that

$$T = \frac{\mu\pi^2 N r_1^4}{a60} + \frac{\mu\pi^2 r_1^2 r_2 hN}{15b} = \frac{\mu\pi^2 N r_1^2}{15}\left(\frac{r_1^2}{4a} + \frac{r_2 h}{b}\right) \tag{8.8.4}$$

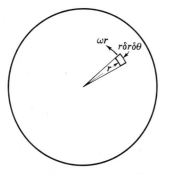

Fig. 8.28 Notation for determination of torque on a disk.

in which all quantities are known except μ. The flow between the surfaces must be laminar for Eqs. (8.8.2) to (8.8.4) to be valid.

Often the geometry of the inner cylinder is altered to eliminate the torque which acts on the lower surface. If the bottom surface of the inner cylinder is made concave, a pocket of air will be trapped between the bottom surface of the inner cylinder and the fluid in the rotating outer cup. A well-designed cup and a careful filling procedure will ensure the condition whereby the torque measured will consist of that produced in the annulus between the two cylinders and a minute amount resulting from the action of the air on the bottom surface. Naturally the viscometer must be provided with a temperature-controlled bath and a variable-speed drive which can be carefully regulated. Such design refinements are needed in order to obtain the rheological diagrams (cf. Fig. 1.2) for the fluid under test.

The measurement of all quantities in the Hagen-Poiseuille equation, except μ, by a suitable experimental arrangement, is another basic method for determination of viscosity. A setup as in Fig. 8.29 may be used. Some distance is required for the fluid to develop its characteristic velocity distribution after it enters the tube; therefore, the head or pressure must be measured by some means at a point along the tube. The volume υ of flow can be measured over a time t where the reservoir surface is held at a constant level. This yields Q, and, by determining γ, Δp can be computed. Then with L and D known, from Eq. (5.2.10a),

$$\mu = \frac{\Delta p \pi D^4}{128 Q L}$$

An adaptation of the capillary tube for industrial purposes is the *Saybolt viscometer* (Fig. 8.30). A short capillary tube is utilized, and the time is

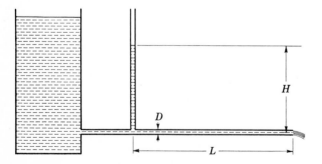

Fig. 8.29 Determination of viscosity by flow through a capillary tube.

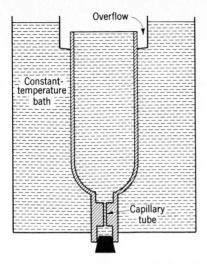

Fig. 8.30 Schematic view of Saybolt viscometer.

measured for 60 cm³ of fluid to flow through the tube under a falling head. The time in seconds is the Saybolt reading. This device measures kinematic viscosity, evident from a rearrangement of Eq. (5.2.10a). When $\Delta p = \rho g h$, $Q = \mathcal{V}/t$, and when the terms are separated that are the same regardless of the fluid,

$$\frac{\mu}{\rho t} = \frac{g h \pi D^4}{128 \, \mathcal{V} L} = C_1$$

Although the head h varies during the test, it varies over the same range for all liquids; and the terms on the right-hand side may be considered as a constant of the particular instrument. Since $\mu/\rho = \nu$, the kinematic viscosity is

$$\nu = C_1 t$$

which shows that the kinematic viscosity varies directly as the time t. The capillary tube is quite short, and so the velocity distribution is not established. The flow tends to enter uniformly, and then, owing to viscous drag at the walls, to slow down there and speed up in the center region. A correction in the above equation is needed, which is of the form C/t; hence

$$\nu = C_1 t + \frac{C_2}{t}$$

The approximate relationship between viscosity and Saybolt seconds is expressed by

$$\nu = 0.0022t - \frac{1.80}{t}$$

in which ν is in stokes and t in seconds.

For measuring viscosity there are many other industrial methods that generally have to be calibrated for each special case to convert to the absolute units. One consists of several tubes containing "standard" liquids of known graduated viscosities with a steel ball in each of the tubes. The time for the ball to fall the length of the tube depends upon the viscosity of the liquid. By placing the test sample in a similar tube, its viscosity can be approximated by comparison with the other tubes.

The flow of a fluid in a capillary tube is the basis for viscometers of the Oswald-Cannon-Fenske or Ubbelohde type. In essence the viscometer is a U tube, one leg of which is a fine capillary tube connected to a reservoir above. The tube is held vertically, and a known quantity of fluid is placed in the reservoir and allowed to flow by gravity through the capillary. The time is recorded for the free surface in the reservoir to fall between two scribed marks. A calibration constant for each instrument takes into account the variation of the capillary's bore from the standard, the bore's uniformity, entrance conditions, and the slight unsteadiness due to the falling head during the 1- to 2-min test. Various bore sizes can be obtained to cover a wide range of viscosities. Exact procedures for carrying out the tests are contained in the standards of the American Society for Testing and Materials.

PROBLEMS

8.1 A static tube (Fig. 8.2) indicates a static pressure that is 0.12 psi too low when liquid is flowing at 8 ft/s. Calculate the correction to be applied to the indicated pressure for the liquid flowing at 18 ft/s.

8.2 Four piezometer openings in the same cross section of a cast-iron pipe indicate the following pressures for simultaneous readings: 4.30, 4.26, 4.24, 3.7 cm Hg. What value should be taken for the pressure?

8.3 A simple pitot tube (Fig. 8.3) is inserted into a small stream of flowing oil, $\gamma = 55$ lb/ft³, $\mu = 0.65$ P, $\Delta h = 1.5$ in, $h_0 = 5$ in. What is the velocity at point 1?

8.4 A stationary body immersed in a river has a maximum pressure of 69 kPa exerted on it at a distance of 5.4 m below the free surface. Calculate the river velocity at this depth.

8.5 From Fig. 8.4 derive the equation for velocity at 1.

8.6 In Fig. 8.4 air is flowing ($p = 16$ psia, $t = 40°F$) and water is in the manometer. For $R' = 1.2$ in, calculate the velocity of air.

8.7 In Fig. 8.4 air is flowing ($p = 101$ kPa abs, $t_1 = 5°C$) and mercury is in the manometer. For $R' = 20$ cm, calculate the velocity at 1 (a) for isentropic compression of air between 1 and 2 and (b) for air considered incompressible.

8.8 A pitot-static tube directed into a 12 ft/s water stream has a gage difference of 1.47 in on a water-mercury differential manometer. Determine the coefficient for the tube.

8.9 A pitot-static tube, $C = 1.12$, has a gage difference of 1 cm on a water-mercury manometer when directed into a water stream. Calculate the velocity.

8.10 A pitot-static tube of the Prandtl type has the following value of gage difference R' for the radial distance from center of a 3-ft-diameter pipe:

r, ft	0.0	0.3	0.6	0.9	1.2	1.48
R', in	4.00	3.91	3.76	3.46	3.02	2.40

Water is flowing, and the manometer fluid has a specific gravity of 2.93. Calculate the discharge.

8.11 What would be the gage difference on a water-nitrogen manometer for flow of nitrogen at 200 m/s, using a pitot-static tube? The static pressure is 1.75 kg$_f$/cm^2 abs and corresponding temperature 25°C. True static pressure is measured by the tube.

8.12 Measurements in an airstream indicate that the stagnation pressure is 15 psia, the static pressure is 10 psia, and the stagnation temperature is 102°F. Determine the temperature and velocity of the airstream.

8.13 0.5 kg/s nitrogen flows through a 5-cm-diameter tube with stagnation temperature of 38°C and undisturbed temperature of 20°C. Find the velocity and static and stagnation pressures.

8.14 A disk meter has a volumetric displacement of 1.73 in^3 for one complete oscillation. Calculate the flow in gallons per minute for 86.5 oscillations per minute.

8.15 A disk water meter with volumetric displacement of 40 cm^3 per oscillation requires 470 oscillations per minute to pass 0.32 l/s and 3840 oscillations per minute to pass 2.57 l/s. Calculate the percent error, or slip, in the meter.

8.16 A volumetric tank 4 ft in diameter and 5 ft high was filled with oil in 16 min 32.4 s. What is the average discharge in gallons per minute?

8.17 A weigh tank receives 75 N liquid, sp gr 0.86, in 14.9 s. What is the flow rate in liters per minute?

8.18 Determine the equation for trajectory of a jet discharging horizontally from a small orifice with head of 15 ft and velocity coefficient of 0.96. Neglect air resistance.

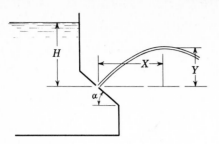

Fig. 8.31

8.19 An orifice of area 30 cm² in a vertical plate has a head of 1.1 m of oil, sp gr 0.91. It discharges 6790 N of oil in 79.3 s. Trajectory measurements yield $x_0 = 2.25$ m, $y_0 = 1.23$ m. Determine C_v, C_c, C_d.

8.20 Calculate Y, the maximum rise of a jet from an inclined plate (Fig. 8.31), in terms of H and α. Neglect losses.

8.21 In Fig. 8.31, for $\alpha = 45°$, $Y = 0.48H$. Neglecting air resistance of the jet, find C_v for the orifice.

8.22 Show that the locus of maximum points of the jet of Fig. 8.31 is given by $X^2 = 4Y(H - Y)$ when losses are neglected.

8.23 A 3-in-diameter orifice discharges 64 ft³ liquid, sp gr 1.07, in 82.2 s under a 9-ft head. The velocity at the *vena contracta* is determined by a pitot-static tube with coefficient 1.0. The manometer liquid is acetylene tetrabromide, sp gr 2.96, and the gage difference is $R' = 3.35$ ft. Determine C_v, C_c, and C_d.

8.24 A 10-cm-diameter orifice discharges 44.6 l/s water under a head of 2.75 m. A flat plate held normal to the jet just downstream from the *vena contracta* requires a force of 320 N to resist impact of the jet. Find C_d, C_v, and C_c.

8.25 Compute the discharge from the tank shown in Fig. 8.32.

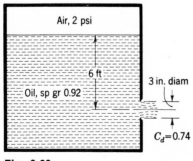

Fig. 8.32

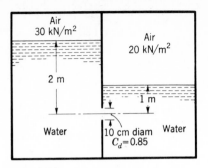

Fig. 8.33

8.26 For $C_v = 0.96$ in Fig. 8.32, calculate the losses in foot-pounds per pound and in foot-pounds per second.

8.27 Calculate the discharge through the orifice of Fig. 8.33.

8.28 For $C_v = 0.93$ in Fig. 8.33, determine the losses in joules per newton and in watts.

8.29 A 4-in-diameter orifice discharges 1.60 cfs liquid under a head of 11.8 ft. The diameter of jet at the *vena contracta* is found by calipering to be 3.47 in. Calculate C_v, C_d, and C_c.

8.30 A Borda mouthpiece 5 cm in diameter has a discharge coefficient of 0.51. What is the diameter of the issuing jet?

8.31 A 3-in-diameter orifice, $C_d = 0.82$, is placed in the bottom of a vertical tank that has a diameter of 4 ft. How long does it take to draw the surface down from 8 to 6 ft?

8.32 Select the size of orifice that permits a tank of horizontal cross section 1.5 m² to have the liquid surface drawn down at the rate of 18 cm/s for 3.35 m head on the orifice. $C_d = 0.63$.

8.33 A 4-in-diameter orifice in the side of a 6-ft-diameter tank draws the surface down from 8 to 4 ft above the orifice in 83.7 s. Calculate the discharge coefficient.

8.34 Select a reservoir of such size and shape that the liquid surface drops 1 m/min over a 3-m distance for flow through a 10-cm-diameter orifice. $C_d = 0.74$.

8.35 In Fig. 8.34 the truncated cone has an angle $\theta = 60°$. How long does it take to draw the liquid surface down from $y = 12$ ft to $y = 4$ ft?

8.36 Calculate the dimensions of a tank such that the surface velocity varies inversely as the distance from the centerline of an orifice draining the tank. When the head is 30 cm, the velocity of fall of the surface is 3 cm/s; orifice diameter is 1.25 cm, $C_d = 0.66$.

8.37 Determine the time required to raise the right-hand surface of Fig. 8.35 by 2 ft.

8.38 How long does it take to raise the water surface of Fig. 8.36 2 m? The left-hand surface is a large reservoir of constant water-surface elevation.

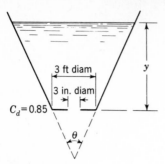

3 ft diam

3 in. diam

y

$C_d = 0.85$

θ

Fig. 8.34

8.39 Show that for incompressible flow the losses per unit weight of fluid between the upstream section and throat of a venturi meter are $K V_2^2/2g$ if

$$K = [(1/C_v)^2 - 1][1 - (D_2/D_1)^4]$$

8.40 A 200 by 100 in venturi meter carries water at 80°F. A water-air differential manometer has a gage difference of 2.4 in. What is the discharge?

8.41 What is the pressure difference between the upstream section and throat of a 16 by 8 cm horizontal venturi meter carrying 50 l/s water at 48°C?

8.42 A 12 by 6 in venturi meter is mounted in a vertical pipe with the flow upward. 2000 gpm oil, sp gr 0.80, $\mu = 0.1$ P, flows through the pipe. The throat section is 4 in above the upstream section. What is $p_1 - p_2$?

8.43 Air flows through a venturi meter in a 5-cm-diameter pipe having a throat diameter of 3 cm, $C_v = 0.97$. For $p_1 = 830$ kPa abs, $t_1 = 15°C$, $p_2 = 550$ kPa abs, calculate the mass per second flowing.

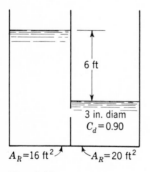

6 ft

3 in. diam
$C_d = 0.90$

$A_R = 16$ ft^2

$A_R = 20$ ft^2

Fig. 8.35

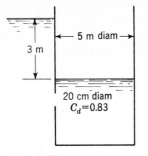

Fig. 8.36

8.44   Oxygen, $p_1 = 40$ psia, $t_1 = 120°F$, flows through a 1 by $\frac{1}{2}$ in venturi meter with a pressure drop of 6 psi. Find the mass per second flowing and the throat velocity.

8.45   Air flows through a 8-cm-diameter ISA flow nozzle in a 12-cm-diameter pipe. $p_1 = 1.5$ kg$_f$/cm² abs, $t_1 = 5°C$, and a differential manometer with liquid, sp gr 2.93, has a gage difference of 0.8 m when connected between the pressure taps. Calculate the mass rate of flow.

8.46   A 2.5-in-diameter ISA nozzle is used to measure flow of water at 40°F in a 6-in-diameter pipe. What gage difference on a water-mercury manometer is required for 300 gpm?

8.47   Determine the discharge in a 30-cm-diameter line with a 20-cm-diameter VDI orifice for water at 20°C when the gage difference is 30 cm on an acetylene tetrabromide (sp gr 2.94)–water differential manometer.

8.48   A $\frac{1}{2}$-in-diameter VDI orifice is installed in a 1-in-diameter pipe carrying nitrogen at $p_1 = 120$ psia, $t_1 = 120°F$. For a pressure drop of 20 psi across the orifice, calculate the mass flow rate.

8.49   Air at 1 atm, $t = 21°C$ flows through a 1-m-square duct that contains a 50-cm-diameter square-edged orifice. With a head loss of 8 cm H_2O across the orifice, compute the flow in cubic feet per minute.

8.50   A 6-in-diameter VDI orifice is installed in a 12-in-diameter oil line, $\mu = 6$ cP, $\gamma = 52$ lb/ft³. An oil-air differential manometer is used. For a gage difference of 22 in determine the flow rate in gallons per minute.

8.51   A rectangular sharp-crested weir 4 m long with end contractions suppressed is 1.3 m high. Determine the discharge when the head is 25 cm.

8.52   In Fig. 8.20, $L = 10$ ft, $P = 1.8$ ft, $H = 0.80$ ft. Estimate the discharge over the weir. $C = 3.33$.

8.53   A rectangular sharp-crested weir with end contractions is 1.5 m long. How high should it be placed in a channel to maintain an upstream depth of 2.25 m for 0.45 m³/s flow?

8.54   Determine the head on a 60° V-notch weir for discharge of 170 l/s.

8.55   Tests on a 90° V-notch weir gave the following results: $H = 0.60$ ft, $Q = 0.685$ cfs; $H = 1.35$ ft, $Q = 5.28$ cfs. Determine the formula for the weir.

8.56   A sharp-crested rectangular weir 3 ft long with end contractions suppressed and a 90° V-notch weir are placed in the same weir box, with the vertex of the 90° V-notch weir 6 in below the rectangular weir crest. Determine the head on the V-notch weir (*a*) when the discharges are equal and (*b*) when the rectangular weir discharges its greatest amount above the discharge of the V-notch weir.

8.57   A broad-crested weir 5 ft high and 10 ft long has a well-rounded upstream corner. What head is required for a flow of 100 cfs?

8.58   A circular disk 20 cm in diameter has a clearance of 0.3 mm from a flat plate. What torque is required to rotate the disk 800 rpm when the clearance contains oil, $\mu = 0.8$ P?

8.59   The concentric-cylinder viscometer (Fig. 8.27) has the following dimensions: $a = 0.012$ in; $b = 0.05$ in; $r_1 = 2.8$ in; $h = 6.0$ in. The torque is 24 lb·in when the speed is 160 rpm. What is the viscosity?

8.60   With the apparatus of Fig. 8.29, $D = 0.5$ mm, $L = 1$ m, $H = 0.75$ m, and 60 cm³ was discharged in 1 h 30 min. What is the viscosity in poises? $S = 0.83$.

8.61   The piezoelectric properties of quartz are used to measure
(*a*) temperature   (*b*) density   (*c*) velocity   (*d*) pressure
(*e*) none of these answers

8.62   A static tube is used to measure

(*a*) the pressure in a static fluid   (*b*) the velocity in a flowing stream
(*c*) the total pressure   (*d*) the dynamic pressure   (*e*) the undisturbed fluid pressure

8.63   A piezometer opening is used to measure

(*a*) the pressure in a static fluid   (*b*) the velocity in a flowing stream
(*c*) the total pressure   (*d*) the dynamic pressure   (*e*) the undisturbed fluid pressure

8.64   The simple pitot tube measures the

(*a*) static pressure
(*b*) dynamic pressure
(*c*) total pressure
(*d*) velocity at the stagnation point
(*e*) difference in total and dynamic pressure

8.65   A pitot-static tube ($C = 1$) is used to measure air speeds. With water in the differential manometer and a gage difference of 3 in, the air speed for $\gamma = 0.0624$ lb/ft³, in feet per second, is

(*a*) 4.01   (*b*) 15.8   (*c*) 24.06   (*d*) 127   (*e*) none of these answers

8.66 The pitot-static tube measures

(*a*) static pressure
(*b*) dynamic pressure
(*c*) total pressure
(*d*) difference in static and dynamic pressure
(*e*) difference in total and dynamic pressure

8.67 The temperature of a known flowing gas can be determined from measurement of

(*a*) static and stagnation pressure only   (*b*) velocity and stagnation pressure only
(*c*) velocity and dynamic pressure only
(*d*) velocity and stagnation temperature only
(*e*) none of these answers

8.68 The velocity of a known flowing gas may be determined from measurement of

(*a*) static and stagnation pressure only
(*b*) static pressure and temperature only
(*c*) static and stagnation temperature only
(*d*) stagnation temperature and stagnation pressure only
(*e*) none of these answers

8.69 The hot-wire anemometer is used to measure

(*a*) pressure in gases
(*b*) pressure in liquids
(*c*) wind velocities at airports
(*d*) gas velocities
(*e*) liquid discharges

8.70 A piston-type displacement meter has a volume displacement of 35 cm³ per revolution of its shaft. The discharge in liters per minute for 1000 rpm is

(*a*) 1.87   (*b*) 4.6   (*c*) 35   (*d*) 40.34   (*e*) none of these answers

8.71 Water for a pipeline was diverted into a weigh tank for exactly 10 min. The increased weight in the tank was 4765 lb. The average flow rate in gallons per minute was

(*a*) 66.1   (*b*) 57.1   (*c*) 7.95   (*d*) 0.13   (*e*) none of these answers

8.72 A rectangular tank with cross-sectional area of 8 m² was filled to a depth of 1.3 m by a steady flow of liquid for 12 min. The rate of flow in liters per second was

(*a*) 14.44   (*b*) 867   (*c*) 901   (*d*) 6471   (*e*) none of these answers

8.73 Which of the following measuring instruments is a rate meter?

(*a*) current meter   (*b*) disk meter   (*c*) hot-wire anemometer
(*d*) pitot tube   (*e*) venturi meter

8.74 The actual velocity at the *vena contracta* for flow through an orifice from a reservoir is expressed by

(*a*) $C_v \sqrt{2gH}$ (*b*) $C_c \sqrt{2gH}$ (*c*) $C_d \sqrt{2gH}$ (*d*) $\sqrt{2gH}$ (*e*) $C_v V_a$

8.75 A fluid jet discharging from a 2-in-diameter orifice has a diameter 1.75 in at its *vena contracta*. The coefficient of contraction is

(*a*) 1.31   (*b*) 1.14   (*c*) 0.875   (*d*) 0.766   (*e*) none of these answers

8.76 The ratio of actual discharge to theoretical discharge through an orifice is

(*a*) $C_c C_v$ (*b*) $C_c C_d$ (*c*) $C_v C_d$ (*d*) C_d / C_v (*e*) C_d / C_c

8.77 The losses in orifice flow are

(*a*) $\dfrac{1}{C_v^2} \left(\dfrac{V_{2a}^2}{2g} - 1 \right)$ (*b*) $\dfrac{V_{2t}^2}{2g} - \dfrac{V_{2a}^2}{2g}$ (*c*) $H(C_v^2 - 1)$

(*d*) $H - \dfrac{V_{2t}^2}{2g}$ (*e*) none of these answers

8.78 For a liquid surface to lower at a constant rate, the area of reservoir A_R must vary with head y on the orifice, as

(*a*) $\sqrt{y}$ (*b*) y (*c*) $1/\sqrt{y}$ (*d*) $1/y$ (*e*) none of these answers

8.79 A 5-cm-diameter Borda mouthpiece discharges 7.68 l/s under a head of 3 m. The velocity coefficient is

(*a*) 0.96   (*b*) 0.97   (*c*) 0.98   (*d*) 0.99   (*e*) none of these answers

8.80 The discharge coefficient for a 4 by 2 in venturi meter at a Reynolds number of 200,000 is

(*a*) 0.95   (*b*) 0.96   (*c*) 0.973   (*d*) 0.983   (*e*) 0.992

8.81 Select the correct statement:

(*a*) The discharge through a venturi meter depends upon Δp only and is independent of orientation of the meter.
(*b*) A venturi meter with a given gage difference R' discharges at a greater rate when the flow is vertically downward through it than when the flow is vertically upward.
(*c*) For a given pressure difference the equations show that the discharge of gas is greater through a venturi meter when compressibility is taken into account than when it is neglected.
(*d*) The coefficient of contraction of a venturi meter is unity.
(*e*) The overall loss is the same in a given pipeline whether a venturi meter or a nozzle with the same D_2 is used.

8.82 The expansion factor Y depends upon

(a) k, p_2/p_1, and A_2/A_1
(b) R, p_2/p_1, and A_2/A_1
(c) k, R, and p_2/p_1
(d) k, R, and A_2/A_1
(e) none of these answers

8.83 The discharge through a V-notch weir varies as

(a) $H^{-1/2}$ (b) $H^{1/2}$ (c) $H^{3/2}$ (d) $H^{5/2}$ (e) none of these answers

8.84 The discharge of a rectangular sharp-crested weir with end contractions is less than for the same weir with end contractions suppressed by

(a) 5%   (b) 10%   (c) 15%   (d) no fixed percentage
(e) none of these answers

8.85 A homemade viscometer of the Saybolt type is calibrated by two measurements with liquids of known kinematic viscosity. For $\nu = 0.461$ St, $t = 97$ s and for $\nu = 0.18$ St, $t = 46$ s. The coefficients C_1, C_2 in $\nu = C_1 t + C_2/t$ are

(a) $C_1 = 0.005$ (b) $C_1 = 0.0044$ (c) $C_1 = 0.0046$ (d) $C_1 = 0.00317$
      $C_2 = -2.3$ $C_2 = 3.6$ $C_2 = 1.55$ $C_2 = 14.95$
(e) none of these answers

REFERENCES

Dowden, R. Rosemary: "Fluid Flow Measurement: A Bibliography," BHRA Fluid Engineering, Cranfield, Bedford, England, 1972.

American Society of Mechanical Engineers: "Fluid Meters," 6th ed., New York, 1971.

ASME Symposium on Flow, Its Measurement and Control in Science and Industry, Pittsburgh, May 9–14, 1971.

TURBOMACHINERY

To turn a fluid stream or change the magnitude of its velocity requires that forces be applied. When a moving vane deflects a fluid jet and changes its momentum, forces are exerted between vane and jet and work is done by displacement of the vane. Turbomachines make use of this principle: the axial and centrifugal pumps, blowers, and compressors, by continuously doing work on the fluid, add to its energy; the impulse, Francis, and propeller turbines and steam and gas turbines continuously extract energy from the fluid and convert it into torque on a rotating shaft; the fluid coupling and the torque converter, each consisting of a pump and a turbine built together make use of the fluid to transmit power smoothly. Designing efficient turbomachines utilizes both theory and experimentation. A good design of given size and speed may be readily adapted to other speeds and other geometrically similar sizes by application of the theory of scaled models, as outlined in Sec. 4.5.

Similarity relationships are first discussed in this chapter by consideration of homologous units and specific speed. Elementary cascade theory is next taken up, before considering the theory of turbomachines. Water turbines and pumps are then considered, followed by blowers and centrifugal compressors. The chapter closes with a discussion of cavitation.

9.1 HOMOLOGOUS UNITS; SPECIFIC SPEED

In utilizing scaled models in designing turbomachines, geometric similitude is required as well as geometrically similar velocity vector diagrams at entrance to, or exit from, the impellers. Viscous effects must, unfortunately, be neglected, as it is generally impossible to satisfy the two above conditions and have equal Reynolds numbers in model and prototype. Two geometrically

498

similar units having similar velocity vector diagrams are *homologous*. They will also have geometrically similar streamlines.

The velocity vector diagram in Fig. 9.1 at exit from a pump impeller can be used to formulate the condition for similar streamline patterns. The blade angle is β, u is the peripheral speed of the impeller at the end of the vane or blade, v is the velocity of fluid *relative* to the vane, and V is the absolute velocity leaving the impeller, the vector sum of u and v; V_r is the radial component of V and is proportional to the discharge; α is the angle which the absolute velocity makes with u, the tangential direction. According to geometric similitude, β must be the same for two units, and for similar streamlines α must also be the same in each case.

It is convenient to express the fact that α is to be the same in any of a series of turbomachines, called homologous units, by relating the speed of rotation N, the impeller diameter (or other characteristic dimension) D, and the flow rate Q. For constant α, V_r is proportional to V ($V_r = V \sin \alpha$) and u is proportional to V_r. Hence the conditions for constant α in a homologous series of units may be expressed as

$$\frac{V_r}{u} = \text{const}$$

The discharge Q is proportional to $V_r D^2$, since any cross-sectional flow area is proportional to D^2. The speed of rotation N is proportional to u/D. When these values are inserted,

$$\frac{Q}{ND^3} = \text{const} \tag{9.1.1}$$

expresses the condition in which geometrically similar units are homologous.

The discharge Q through homologous units can be related to head H

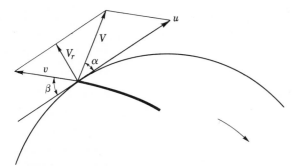

Fig. 9.1 Velocity vector diagram for exit from a pump impeller.

and a representative cross-sectional area A by the orifice formula

$$Q = C_d A \sqrt{2gH}$$

in which C_d, the discharge coefficient, varies slightly with the Reynolds number and so causes a small change in efficiency with size in a homologous series. The change in discharge with the Reynolds number is referred to as *scale effect*. The smaller machines, having smaller hydraulic radii of passages, have lower Reynolds numbers and correspondingly higher friction factors; hence they are less efficient. The change in efficiency from model to prototype may be from 1 to 4 percent. However, in the homologous theory, the scale effect must be neglected, and so an empirical correction for change in efficiency with size is used [see Eq. (9.5.1)]. As $A \sim D^2$, the discharge equation may be

$$\frac{Q}{D^2 \sqrt{H}} = \text{const} \qquad (9.1.2)$$

Eliminating Q between Eqs. (9.1.1) and (9.1.2) gives

$$\frac{H}{N^2 D^2} = \text{const} \qquad (9.1.3)$$

Equations (9.1.1) and (9.1.3) are most useful in determining performance characteristics for one unit from those of a homologous unit of different size and speed.[1]

---

[1] Application of dimensional analysis is illuminating. The variables appearing to be pertinent to the flow relations for similar units would be $F(H,Q,N,D,g) = 0$. There are two dimensions involved, L and T; N and D may be selected as the repeating variables, yielding

$$f\left(\frac{Q}{ND^3}, \frac{H}{D}, \frac{g}{N^2 D}\right) = 0$$

Solving for H gives

$$H = Df_1\left(\frac{Q}{ND^3}, \frac{g}{N^2 D}\right)$$

Experiment shows that the second dimensionless parameter actually occurs to the power -1; hence

$$H = \frac{N^2 D^2}{g} f_2\left(\frac{Q}{ND^3}\right) \qquad \text{or} \qquad \frac{gH}{N^2 D^2} = f_2\left(\frac{Q}{ND^3}\right)$$

The characteristic curve for a pump in dimensionless form is the plot of Q/ND^3 as abscissa against $H/(N^2 D^2/g)$ as ordinate. This curve, obtained from tests on one unit of the series, then applies to all homologous units, and may be converted to the usual characteristic curve by selecting desired values of N and D. As power is proportional to γQH, the dimensionless power term is

$$\frac{\gamma}{\gamma} \frac{Q}{ND^3} \frac{H}{N^2 D^2/g} = \frac{\text{power}}{\rho N^3 D^5}$$

EXAMPLE 9.1   A prototype test of a mixed-flow pump with a 72-in-diameter discharge opening, operating at 225 rpm, resulted in the following characteristics:

H, ft	Q, cfs	e, %	H, ft	Q, cfs	e, %	H, ft	Q, cfs	e, %
60	200	69	47.5	330	87.3	35	411	82
57.5	228	75	45	345	88	32.5	425	79
55	256	80	42.5	362	87.4	30	438	75
52.5	280	83.7	40	382	86.3	27.5	449	71
50	303	86	37.5	396	84.4	25	459	66.5

What size and synchronous speed (60 Hz) of homologous pump should be used to produce 200 cfs at 60 ft head at point of best efficiency? Find the characteristic curves for this case.

Subscript 1 refers to the 72-in pump. For best efficiency $H_1 = 45$, $Q_1 = 345$, $e = 88$ percent. With Eqs. (9.1.1) and (9.1.3),

$$\frac{H}{N^2 D^2} = \frac{H_1}{N_1^2 D_1^2} \qquad \frac{Q}{ND^3} = \frac{Q_1}{N_1 D_1^3}$$

or

$$\frac{60}{N^2 D^2} = \frac{45}{225^2 \times 72^2} \qquad \frac{200}{ND^3} = \frac{345}{225 \times 72^3}$$

After solving for N and D,

$$N = 366 \text{ rpm} \qquad D = 51.1 \text{ in}$$

The nearest synchronous speed (3600 divided by number of pairs of poles) is 360 rpm. To maintain the desired head of 60 ft, a new D is necessary. Its size can be computed:

$$D = \sqrt{\tfrac{60}{45}} \times \tfrac{225}{360} \times 72 = 52 \text{ in}$$

The discharge at best efficiency is then

$$Q = \frac{Q_1 ND^3}{N_1 D_1^3} = 345 \times \tfrac{360}{225} \left(\tfrac{52}{72} \right)^3 = 208 \text{ cfs}$$

which is slightly more capacity than required. With $N = 360$ and $D = 52$, equations for transforming the corresponding values of H and Q for any efficiency can be obtained:

$$H = H_1 \left(\frac{ND}{N_1 D_1}\right)^2 = H_1 (\tfrac{360}{225} \times \tfrac{52}{72})^2 = 1.335 H_1$$

and

$$Q = Q_1 \frac{ND^3}{N_1 D_1{}^3} = Q_1 (\tfrac{360}{225}) (\tfrac{52}{72})^3 = 0.603 Q_1$$

The characteristics of the new pump are

H, ft	Q, cfs	e, %	H, ft	Q, cfs	e, %	H, ft	Q, cfs	e, %
80	121	69	63.5	200	87.3	46.7	248	82
76.7	138	75	60	208	88	43.4	257	79
73.4	155	80	56.7	219	87.4	40	264	75
70	169	83.7	53.5	231	86.3	36.7	271	71
66.7	183	86	50	239	84.4	33.4	277	66.5

The efficiency of the 52-in pump might be a fraction of a percent less than that of the 72-in pump, as the hydraulic radii of flow passages are smaller, so that the Reynolds number would be less.

Specific speed

The specific speed of a homologous unit is a constant widely used in selecting the type of unit and in preliminary design. It is usually defined differently for a pump and a turbine.

The specific speed N_s of a homologous series of pumps is defined as the speed of some one unit of the series of such a size that it delivers unit discharge at unit head. It is obtained as follows. Eliminating D in Eqs. (9.1.1) and (9.1.3) and rearranging give

$$\frac{N\sqrt{Q}}{H^{3/4}} = \text{const} \tag{9.1.4}$$

By definition of specific speed, the constant is N_s, the speed of a unit for $Q = 1$,

$H = 1;$

$$N_s = \frac{N \sqrt{Q}}{H^{3/4}} \qquad (9.1.5)$$

The specific speed of a series is usually defined for the point of best efficiency, i.e., for the speed, discharge, and head that is most efficient.

The specific speed of a homologous series of turbines is defined as the speed of a unit of the series of such a size that it produces unit horsepower with unit head. Since power P is proportional to QH,

$$\frac{P}{QH} = \text{const} \qquad (9.1.6)$$

The terms D and Q may be eliminated from Eqs. (9.1.1), (9.1.3), and (9.1.6) to produce

$$\frac{N \sqrt{P}}{H^{5/4}} = \text{const} \qquad (9.1.7)$$

For unit power and unit head the constant of Eq. (9.1.7) becomes the speed, or the specific speed N_s of the series, so that

$$N_s = \frac{N \sqrt{P}}{H^{5/4}} \qquad (9.1.8)$$

The specific speed of a unit required for a given discharge and head can be estimated from Eqs. (9.1.5) and (9.1.8). For pumps handling large discharges at low heads a high specific speed is indicated; for a high-head turbine producing relatively low power (small discharge) the specific speed is low. Experience has shown that for best efficiency one particular type of pump or turbine is usually indicated for a given specific speed.

Because Eqs. (9.1.5) and (9.1.8) are not dimensionally correct (γ and g have been included in the constant term), the value of specific speed depends on the units involved. For example, in the United States Q is commonly expressed in gallons per minute, millions of gallons per day, or cubic feet per second when referring to specific speeds of pumps.

Centrifugal pumps have low specific speeds; mixed-flow pumps have medium specific speeds; and axial-flow pumps have high specific speeds. Impulse turbines have low specific speeds; Francis turbines have medium specific speeds; and propeller turbines have high specific speeds.

Fig. 9.2 Simple cascade system.

9.2 ELEMENTARY CASCADE THEORY

Turbomachines either do work on a fluid or extract work from it in a continuous manner by having it flow through a series of moving (and possibly fixed) vanes. By examination of flow through a series of similar blades or vanes, called a *cascade*, some of the requirements of an efficient system can be developed. Consider, first, flow through the simple fixed cascade system of Fig. 9.2. The velocity vector representing the fluid has been turned through an angle by the presence of the cascade system. A force has been exerted on the fluid, but (neglecting friction effects and turbulence) no work is done on the fluid. Section 3.11 deals with forces on a single vane.

Since turbomachines are rotational devices, the cascade system may be arranged symmetrically around the periphery of a circle, as in Fig. 9.3. If the fluid approaches the fixed cascade in a radial direction, it has its moment of momentum changed from zero to a value dependent upon the mass per

Fig. 9.3 Cascade arranged on the periphery of a circular cylinder.

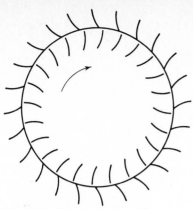

Fig. 9.4 Moving cascade within a fixed cascade.

unit time flowing, the tangential component of velocity V_t developed, and the radius, from Eq. (3.12.5),

$$T = \rho Q r V_t \tag{9.2.1}$$

Again, no work is done by the fixed-vane system.

Consider now another series of vanes (Fig. 9.4) rotating within the fixed-vane system at a speed ω. For efficient operation of the system it is important that the fluid flow onto the moving vanes with the least disturbance, i.e., in a tangential manner, as illustrated in Fig. 9.5. When the relative velocity is not tangent to the blade at its entrance, separation may occur, as shown in Fig. 9.6. The losses tend to increase rapidly (about as the square) with angle from the tangential and radically impair the efficiency of the machine. Separation also frequently occurs when the approaching relative velocity is tangential to the vane, owing to curvature of the vanes or to expansion of the flow passages, which causes the boundary layer to thicken and

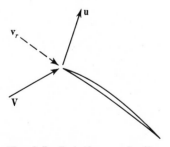

Fig. 9.5 Relative velocity tangent to blade.

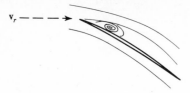

v_r

Fig. 9.6 Flow separation, or shock, from blade with relative velocity not tangent to leading edge.

come to rest. These losses are called *shock* or *turbulence* losses. When the fluid exits from the moving cascade, it will generally have its velocity altered in both magnitude and direction, thereby changing its moment of momentum and either doing work on the cascade or having work done on it by the moving cascade. In the case of a turbine it is desired to have the fluid leave with no moment of momentum. An old saying in turbine design is "have the fluid enter without shock and leave without velocity."

Turbomachinery design requires the proper arrangement and shaping of passages and vanes so that the purpose of the design can be met most efficiently. The particular design depends upon the purpose of the machine, the amount of work to be done per unit mass of fluid, and the fluid density.

9.3 THEORY OF TURBOMACHINES

Turbines extract useful work from fluid energy; and pumps, blowers, and turbocompressors add energy to fluids by means of a runner consisting of vanes rigidly attached to a shaft. Since the only displacement of the vanes is in the tangential direction, work is done by the displacement of the tangential components of force on the runner. The radial components of force on the runner have no displacement in a radial direction and, hence, can do no work.

In turbomachine theory, friction is neglected, and the fluid is assumed to have perfect guidance through the machine, i.e., an infinite number of thin vanes, and so the relative velocity of the fluid is always tangent to the vane. This yields circular symmetry and permits the moment-of-momentum equation, Sec. 3.12, to take the simple form of Eq. (3.12.5), for steady flow,

$$T = \rho Q [(rV_t)_{\text{out}} - (rV_t)_{\text{in}}] \tag{9.3.1}$$

in which T is the torque acting on the fluid within the control volume (Fig.

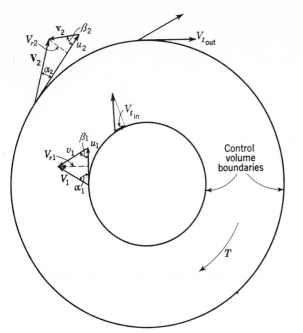

Fig. 9.7 Steady flow through control volume with circular symmetry.

9.7) and $\rho Q(rV_t)_{out}$ and $\rho Q(rV_t)_{in}$ represent the moment of momentum leaving and entering the control volume, respectively.

The polar vector diagram is generally used in studying vane relationships (Fig. 9.8), with subscript 1 for entering fluid and subscript 2 for exiting fluid. V is the absolute fluid velocity, u the peripheral velocity of the runner, and v the fluid velocity relative to the runner. The absolute velocities V, u are laid off from O, and the relative velocity connects them as shown. V_u is designated as the component of absolute velocity in the tangential direction. α is the angle the absolute velocity V makes with the peripheral velocity u, and β is the angle the relative velocity makes with $-u$, or it is the *blade angle*,

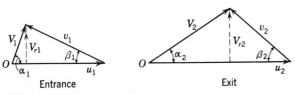

Fig. 9.8 Polar vector diagrams.

as perfect guidance is assumed. V_r is the absolute velocity component normal to the periphery. In this notation Eq. (9.3.1) becomes

$$T = \rho Q (r_2 V_2 \cos \alpha_2 - r_1 V_1 \cos \alpha_1)$$

$$= \rho Q (r_2 V_{u2} - r_1 V_{u1}) = \dot{m} (r_2 V_{u2} - r_1 V_{u1}) \qquad (9.3.2)$$

The mass per unit time flowing is $\dot{m} = \rho Q = (\rho Q)_{\text{out}} = (\rho Q)_{\text{in}}$. In the form above, when T is positive, the fluid moment of momentum increases through the control volume, as for a pump. For T negative, moment of momentum of the fluid is decreased, as for a turbine runner. When $T = 0$, as in passages where there are no vanes,

$$r V_u = \text{const}$$

This is *free-vortex* motion, with the tangential component of velocity varying inversely with radius. It is discussed in Sec. 7.9 and compared with the forced vortex in Sec. 2.9.

EXAMPLE 9.2 The wicket gates of Fig. 9.9 are turned so that the flow makes an angle of 45° with a radial line at section 1, where the speed is 2.5 m/s. Determine the magnitude of tangential velocity component V_u over section 2.

Since no torque is exerted on the flow between sections 1 and 2, the

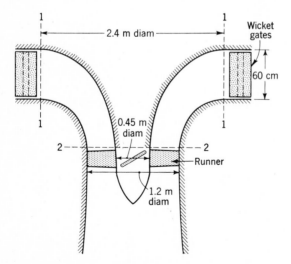

Fig. 9.9 Schematic view of propeller turbine.

moment of momentum is constant and the motion follows the free-vortex law

$$V_u r = \text{const}$$

At section 1

$$V_{u1} = 2.5 \cos 45° = 1.77 \text{ m/s}$$

Then

$$V_{u1} r_1 = (1.77 \text{ m/s})(1.2 \text{ m}) = 2.124 \text{ m}^2/\text{s}$$

Across section 2

$$V_{u2} = \frac{2.124 \text{ m}^2/\text{s}}{r \text{ m}}$$

At the hub $V_u = 2.124/0.225 = 9.44$ m/s, and at the outer edge $V_u = 2.124/0.6 = 3.54$ m/s.

Head and energy relations

By multiplying Eq. (9.3.2) by the rotational speed ω (rad/s) of runner,

$$T\omega = \rho Q(\omega r_2 V_{u2} - \omega r_1 V_{u1}) = \rho Q(u_2 V_{u2} - u_1 V_{u1}) \tag{9.3.3}$$

For no losses the power available from a turbine is $Q \Delta p = Q\gamma H$, in which H is the head on the runner, since $Q\gamma$ is the weight per unit time and H the potential energy per unit weight. Similarly a pump runner produces work $Q\gamma H$, in which H is the pump head. The power exchange is

$$T\omega = Q\gamma H \tag{9.3.4}$$

Solving for H, using Eq. (9.3.3) to eliminate T, gives

$$H = \frac{u_2 V_{u2} - u_1 V_{u1}}{g} \tag{9.3.5}$$

For turbines the sign is reversed in Eq. (9.3.5).
   For pumps the *actual* head H_p produced is

$$H_p = e_h H = H - H_L \tag{9.3.6}$$

and for turbines the actual head H_t is

$$H_t = \frac{H}{e_h} = H + H_L \qquad\qquad (9.3.7)$$

in which e_h is the hydraulic efficiency of the machine and H_L represents all the internal fluid losses in the machine. The overall efficiency of the machines is further reduced by bearing friction, by friction caused by fluid between runner and housing, and by leakage or flow that passes around the runner without going through it. These losses do not affect the head relations.

Pumps are generally designed so that the angular momentum of fluid entering the runner (impeller) is zero. Then

$$H = \frac{u_2 V_2 \cos \alpha_2}{g} \qquad\qquad (9.3.8)$$

Turbines are designed so that the angular momentum is zero at the exit section of the runner for conditions at best efficiency; hence,

$$H = \frac{u_1 V_1 \cos \alpha_1}{g} \qquad\qquad (9.3.9)$$

In writing the energy equation for a pump, with Eqs. (9.3.5) and (9.3.6),

$$H_p = \left(\frac{V_2^2}{2g} + \frac{p_2}{\gamma} + z_2\right) - \left(\frac{V_1^2}{2g} + \frac{p_1}{\gamma} + z_1\right)$$

$$= \frac{u_2 V_2 \cos \alpha_2 - u_1 V_1 \cos \alpha_1}{g} - H_L \qquad\qquad (9.3.10)$$

for which it is assumed that all streamlines through the pump have the same total energy. With the relations among the absolute velocity V, the velocity v relative to the runner, and the velocity u of runner, from the vector diagrams (Fig. 9.8) by the law of cosines,

$$u_1^2 + V_1^2 - 2u_1 V_1 \cos \alpha_1 = v_1^2 \qquad u_2^2 + V_2^2 - 2u_2 V_2 \cos \alpha_2 = v_2^2$$

Eliminating the absolute velocities V_1, V_2 in these relations and in Eq. (9.3.10) gives

$$H_L = \frac{u_2^2 - u_1^2}{2g} - \frac{v_2^2 - v_1^2}{2g} - \frac{p_2 - p_1}{\gamma} - (z_2 - z_1) \qquad\qquad (9.3.11)$$

or

$$H_L = \frac{u_2^2 - u_1^2}{2g} - \left[\left(\frac{v_2^2}{2g} + \frac{p_2}{\gamma} + z_2 \right) - \left(\frac{v_1^2}{2g} + \frac{p_1}{\gamma} + z_1 \right) \right] \tag{9.3.12}$$

The losses are the difference in centrifugal head, $(u_2^2 - u_1^2)/2g$, and in the head change in the relative flow. For no loss, the increase in pressure head, from Eq. (9.3.11), is

$$\frac{p_2 - p_1}{\gamma} + z_2 - z_1 = \frac{u_2^2 - u_1^2}{2g} - \frac{v_2^2 - v_1^2}{2g} \tag{9.3.13}$$

With no flow through the runner, v_1, v_2 are zero, and the head rise is as expressed in the relative equilibrium relationships [Eq. (2.9.7)]. When flow occurs, the head rise is equal to the centrifugal head minus the difference in relative velocity heads.

For the case of a turbine, exactly the same equations result.

EXAMPLE 9.3 A centrifugal pump with a 24-in-diameter impeller runs at 1800 rpm. The water enters without whirl, and $\alpha_2 = 60°$. The actual head produced by the pump is 50 ft. Find its hydraulic efficiency when $V_2 = 20$ ft/s.

From Eq. (9.3.8) the theoretical head is

$$H = \frac{u_2 V_2 \cos \alpha_2}{g} = \frac{1800 \times 2\pi \times 20 \times 0.50}{60 \times 32.2} = 58.6 \text{ ft}$$

The actual head is 50.0 ft; hence, the hydraulic efficiency is

$$e_h = \frac{50}{58.6} = 85.4\%$$

9.4 IMPULSE TURBINES

The impulse turbine is one in which all available energy of the flow is converted by a nozzle into kinetic energy at atmospheric pressure before the fluid contacts the moving blades. Losses occur in flow from the reservoir through the pressure pipe (penstock) to the base of the nozzle, which may be computed from pipe friction data. At the base of the nozzle the available energy, or total head, is

$$h_a = \frac{p_1}{\gamma} + \frac{V_1^2}{2g} \tag{9.4.1}$$

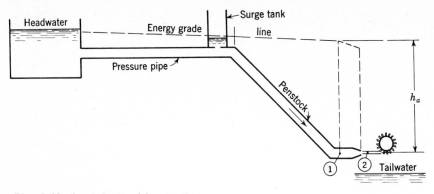

Fig. 9.10 Impulse turbine system.

from Fig. 9.10. With C_v the nozzle coefficient, the jet velocity V_2 is

$$V_2 = C_v \sqrt{2gh_a} = C_v \sqrt{2g\left(\frac{p_1}{\gamma} + \frac{V_1{}^2}{2g}\right)} \tag{9.4.2}$$

The head lost in the nozzle is

$$h_a - \frac{V_2{}^2}{2g} = h_a - C_v{}^2 h_a = h_a(1 - C_v{}^2) \tag{9.4.3}$$

and the efficiency of the nozzle is

$$\frac{V_2{}^2/2g}{h_a} = \frac{C_v{}^2 h_a}{h_a} = C_v{}^2 \tag{9.4.4}$$

The jet, with velocity V_2, strikes double-cupped buckets (Figs. 9.11 and 9.12) which split the flow and turn the relative velocity through the angle θ (Fig. 9.12).

The x component of momentum is changed by (Fig. 9.12)

$$F = \rho Q (v_r - v_r \cos \theta)$$

and the power exerted on the vanes is

$$Fu = \rho Q u v_r (1 - \cos \theta) \tag{9.4.5}$$

To maximize the power, theoretically, $\theta = 180°$, and uv_r must be a maximum; that is, $u(V_2 - u)$ must be a maximum. By differentiating with respect to u

Fig. 9.11 Southern California Edison, Big Creek 2A, 1948, $8\frac{1}{2}$-in-diameter jet impulse buckets and disk in process of being reamed; 56,000 hp, 2200 ft head, 300 rpm. (*Allis-Chalmers Mfg. Co.*)

and equating to zero,

$$(V_2 - u) + u(-1) = 0$$

$u = V_2/2$. After making these substitutions into Eq. (9.4.5),

$$Fu = \rho Q \frac{V_2}{2}\left(V_2 - \frac{V_2}{2}\right)(1 - -1) = \gamma Q \frac{V_2{}^2}{2g} \tag{9.4.6}$$

Fig. 9.12 Flow through bucket.

which accounts for the total kinetic energy of the jet. The velocity diagram for these values shows that the absolute velocity leaving the vanes is zero.

Practically, when vanes are arranged on the periphery of a wheel (Fig. 9.11), the fluid must retain enough velocity to move out of the way of the following bucket. Most of the practical impulse turbines are Pelton wheels. The jet is split in two and turned in a horizontal plane, and half is discharged from each side to avoid any unbalanced thrust on the shaft. There are losses due to the splitter and to friction between jet and bucket surface, which make the most economical speed somewhat less than $V_2/2$. It is expressed in terms of the *speed factor*

$$\phi = \frac{u}{\sqrt{2gh_a}} \tag{9.4.7}$$

For most efficient turbine operation ϕ has been found to depend upon specific speed as shown in Table 9.1. The angle θ of the bucket is usually 173 to 176°. If the diameter of the jet is d, and the diameter of the wheel is D at the center-line of the buckets, it has been found in practice that the diameter ratio D/d should be about $54/N_s$ (ft, hp, rpm), or $206/N_s$ (m, kW, rpm) for maximum efficiency.

In the majority of installations only one jet is used, which discharges horizontally against the lower periphery of the wheel as shown in Fig. 9.10. The wheel speed is carefully regulated for the generation of electric power. A governor operates a needle valve that controls the jet discharge by changing

Table 9.1 *Dependence of ϕ on specific speed**

Specific speed N_s		
(m, kW, rpm)	(ft, hp, rpm)	ϕ
7.62	2	0.47
11.42	3	0.46
15.24	4	0.45
19.05	5	0.44
22.86	6	0.433
26.65	7	0.425

* Modified from J. W. Daily, Hydraulic Machinery, in H. Rouse (ed.), "Engineering Hydraulics," p. 943, Wiley, New York, 1950.

its area. So V_0 remains practically constant for a wide range of positions of the needle valve.

The efficiency of the power conversion drops off rapidly with change in head (which changes V_0), as is evident when power is plotted against V_0 for constant u in Eq. (9.4.5). The wheel operates in atmospheric air although it is enclosed by a housing. It is therefore essential that the wheel be placed above the maximum floodwater level of the river into which it discharges. The head from nozzle to tail water is wasted. Because of their inefficiency at other than the design head and because of the wasted head, Pelton wheels usually are employed for high heads, e.g., from 200 m to more than 1 km. For high heads, the efficiency of the complete installation, from headwater to tail water, may be in the high 80s.

Impulse wheels with a single nozzle are most efficient in the specific speed range of 2 to 6, when P is in horsepower, H is in feet, and N is in revolutions per minute. Multiple nozzle units are designed in the specific speed range of 6 to 12.

EXAMPLE 9.4 A Pelton wheel is to be selected to drive a generator at 600 rpm. The water jet is 7.5 cm in diameter and has a velocity of 100 m/s. With the blade angle at 170°, the ratio of vane speed to initial jet speed at 0.47, and neglecting losses, determine (a) diameter of wheel to centerline of buckets (vanes), (b) power developed, and (c) kinetic energy per newton remaining in the fluid.

(a)   The peripheral speed of the wheel is

$$u = 0.47 \times 100 = 47 \text{ m/s}$$

Then

$$\frac{600}{60}\left(2\pi\frac{D}{2}\right) = 47 \text{ m/s} \qquad \text{or} \qquad D = 1.495 \text{ m}$$

(b)   From Eq. (9.4.5), the power, in kilowatts, is computed to be

$$(997.3 \text{ kg/m}^3)\frac{\pi}{4}(0.075 \text{ m})^2(100 \text{ m/s})(47 \text{ m/s})(100 - 47 \text{ m/s})$$

$$\times [1 - (-0.9848)]\frac{1 \text{ kW}}{1000 \text{ W}} = 2170 \text{ kW}$$

(c)   From Fig. 3.32, the absolute velocity components leaving the vane are

$$V_x = (100 - 47)(-0.9848) + 47 = -5.2 \text{ m/s}$$

$$V_y = (100 - 47)(0.1736) = 9.2 \text{ m/s}$$

The kinetic energy remaining in the jet is

$$\frac{5.2^2 + 9.2^2}{2 \times 9.806} = 5.69 \text{ m} \cdot \text{N/N}$$

EXAMPLE 9.5 A small impulse wheel is to be used to drive a generator for 60-Hz power. The head is 300 ft, and the discharge 1.40 cfs. Determine the diameter of the wheel at the centerline of the buckets and the speed of the wheel. $C_v = 0.98$. Assume efficiency of 80 percent.

The power is

$$P = \frac{\gamma QHe}{550} = \frac{62.4 \times 1.4 \times 300 \times 0.80}{550} = 38.1 \text{ hp}$$

Taking a trial value of N_s of 4 gives

$$N = \frac{N_s H^{5/4}}{\sqrt{P}} = \frac{4 \times 300^{5/4}}{\sqrt{38.1}} = 809 \text{ rpm}$$

For 60-Hz power the speed must be 3600 divided by the number of pairs of poles in the generator. For five pairs of poles the speed would be $3600/5 = 720$ rpm, and for four pairs of poles $3600/4 = 900$ rpm. The closer speed 720 is selected, although some engineers prefer an even number of pairs of poles in the generator. Then

$$N_s = \frac{N\sqrt{P}}{H^{5/4}} = \frac{720\sqrt{38.2}}{300^{5/4}} = 3.56$$

For $N_s = 3.56$, take $\phi = 0.455$,

$$u = \phi\sqrt{2gH} = 0.455\sqrt{2 \times 32.2 \times 300} = 63.2 \text{ ft/s}$$

and

$$\omega = \tfrac{720}{60} 2\pi = 75.4 \text{ rad/s}$$

The peripheral speed u and D and ω are related:

$$u = \frac{\omega D}{2} \qquad D = \frac{2u}{\omega} = \frac{2 \times 63.2}{75.4} = 1.676 \text{ ft} = 20.1 \text{ in}$$

The diameter d of the jet is obtained from the jet velocity V_2; thus

$$V_2 = C_v \sqrt{2gH} = 0.98 \sqrt{2 \times 32.2 \times 300} = 136 \text{ ft/s}$$

$$a = \frac{Q}{V_2} = \frac{1.40}{136} \times 144 = 1.482 \text{ in}^2 \qquad d = \sqrt{\frac{4a}{\pi}} = \sqrt{\frac{1.482}{0.7854}} = 1.374 \text{ in}$$

Hence the diameter ratio D/d is

$$\frac{D}{d} = \frac{20.1}{1.375} = 14.6$$

The desired diameter ratio for best efficiency is

$$\frac{D}{d} = \frac{54}{N_s} = \frac{54}{3.56} = 15.15$$

which is satisfactory. Hence the wheel diameter is 20.1 in and speed 720 rpm.

9.5 REACTION TURBINES

In the *reaction* turbine a portion of the energy of the fluid is converted into kinetic energy by the fluid's passing through adjustable gates (Fig. 9.13) before entering the runner, and the remainder of the conversion takes place through the runner. All passages are filled with liquid, including the passage (draft tube) from the runner to the downstream liquid surface. The static fluid pressure occurs on both sides of the vanes and hence does no work. The work done is entirely due to the conversion to kinetic energy.

The reaction turbine is quite different from the impulse turbine, discussed in Sec. 9.4. In an impulse turbine all the available energy of the fluid is converted into kinetic energy by a nozzle that forms a free jet. The energy is then taken from the jet by suitable flow through moving vanes. The vanes are partly filled, with the jet open to the atmosphere throughout its travel through the runner.

In contrast, in the reaction turbine the kinetic energy is appreciable as the fluid leaves the runner and enters the draft tube. The function of the

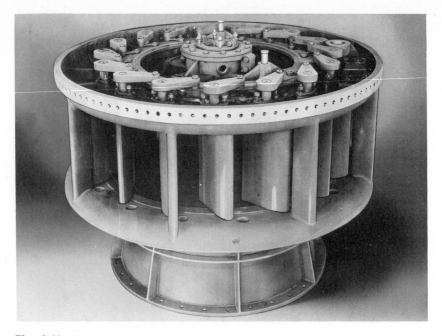

Fig. 9.13 Stay ring and wicket gates for reaction turbine. (*Allis-Chalmers Mfg. Co.*)

draft tube is to reconvert the kinetic energy to flow energy by a gradual expansion of the flow cross section. Application of the energy equation between the two ends of the draft tube shows that the action of the tube is to reduce the pressure at its upstream end to less than atmospheric pressure, thus increasing the effective head across the runner to the difference in elevation between headwater and tail water, less losses.

By referring to Fig. 9.14, the energy equation from 1 to 2 yields

$$z_s + \frac{V_1^2}{2g} + \frac{p_1}{\gamma} = 0 + 0 + 0 + \text{losses}$$

The losses include friction plus velocity head loss at the exit from the draft tube, both of which are quite small; hence

$$\frac{p_1}{\gamma} = -z_s - \frac{V_1^2}{2g} + \text{losses} \tag{9.5.1}$$

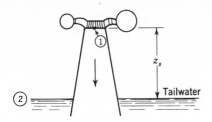

Fig. 9.14 Draft tube.

shows that considerable vacuum is produced at section 1, which effectively increases the head across the turbine runner. The turbine setting must not be too high, or cavitation occurs in the runner and draft tube (see Sec. 9.8).

EXAMPLE 9.6 A turbine has a velocity of 6 m/s at the entrance to the draft tube and a velocity of 1.2 m/s at its exit. For friction losses of 0.1 m and a tail water 5 m below the entrance to the draft tube, find the pressure head at the entrance.

From Eq. (9.5.1)

$$\frac{p_1}{\gamma} = -5 - \frac{6^2}{2 \times 9.806} + \frac{1.2^2}{2 \times 9.806} + 0.1 = -6.66 \text{ m}$$

as the kinetic energy at the exit from the draft tube is lost. Hence a suction head of 6.66 m is produced by the presence of the draft tube.

There are two forms of the reaction turbine in common use, the *Francis* turbine (Fig. 9.15) and the *propeller* (axial-flow) turbine (Fig. 9.16). In both, all passages flow full, and energy is converted to useful work entirely by changing the moment of momentum of the liquid. The flow passes first through the wicket gates, which impart a tangential and a radially inward velocity to the fluid. A space between the wicket gates and the runner permits the flow to close behind the gates and move as a free vortex, without external torque being applied.

In the Francis turbine (Fig. 9.15) the fluid enters the runner so that the relative velocity is tangent to the leading edge of the vanes. The radial component is gradually changed to an axial component, and the tangential component is reduced as the fluid traverses the vane, so that at the runner exit the flow is axial with very little whirl (tangential component) remaining. The pressure has been reduced to less than atmospheric, and most of the remaining kinetic energy is reconverted to flow energy by the time it dis-

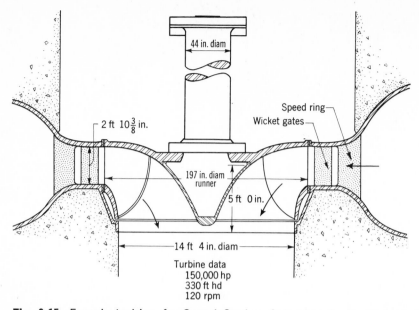

Fig. 9.15 Francis turbine for Grand Coulee, Columbia Basin Project. (*Newport News Shipbuilding and Dry Dock Co.*)

charges from the draft tube. The Francis turbine is best suited to medium head installations from 80 to 600 ft (25 to 180 m) and has an efficiency between 90 and 95 percent for the larger installations. Francis turbines are designed in the specific speed range of 10 to 110 (ft, hp, rpm) or 40 to 420 (m, kW, rpm) with best efficiency in the range 40 to 60 (ft, hp, rpm) or 150 to 230 (m, kW, rpm).

In the propeller turbine (Fig. 9.9), after passing through the wicket gates, the flow moves as a free vortex and has its radial component changed to axial component by guidance from the fixed housing. The moment of momentum is constant, and the tangential component of velocity is increased through the reduction in radius. The blades are few in number, relatively flat, with very little curvature, and placed so that the relative flow entering the runner is tangential to the leading edge of the blade. The relative velocity is high, as with the Pelton wheel, and changes slightly in traversing the blade. The velocity diagrams in Fig. 9.17 show how the tangential component of velocity is reduced. Propeller turbines are made with blades that pivot around the hub, thus permitting the blade angle to be adjusted for different gate openings and for changes in head. They are particularly suited for low-head installations, up to 30 m, and have top efficiencies around 94 percent. Axial-flow turbines

Fig. 9.16 Field view of installation of runner of 24,500-hp, 100-rpm, 41-ft-head, Kaplan adjustable-runner hydraulic turbine. Box Canyon Project, Public Utility District No. 1 of Pend Oreille County, Washington. Plant placed in operation in 1955. (*Allis-Chalmers Mfg. Co.*)

are designed in the specific speed range of 100 to 210 (ft, hp, rpm) or 380 to 800 (m, kW, rpm) with best efficiency from 120 to 160 (ft, hp, rpm) or 460 to 610 (m, kW, rpm).

The windmill is a form of axial-flow turbine. Since it has no fixed vanes to give an initial tangential component to the airstream, it must impart the tangential component to the air with the moving vanes. The airstream expands in passing through the vanes with a reduction in its axial velocity.

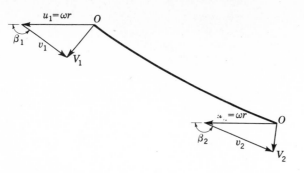

Fig. 9.17 Velocity diagrams for entrance and exit of a propeller turbine blade at fixed radial distance.

EXAMPLE 9.7 Assuming uniform axial velocity over section 2 of Fig. 9.9 and using the data of Example 9.2, determine the angle of the leading edge of the propeller at $r = 0.225$, 0.45, and 0.6 m, for a propeller speed of 240 rpm.

At $r = 0.225$ m,

$$u = \tfrac{240}{60}(2\pi)\,(0.225) = 5.66 \text{ m/s} \qquad V_u = 9.44 \text{ m/s}$$

At $r = 0.45$ m,

$$u = \tfrac{240}{60}(2\pi)\,(0.45) = 11.3 \text{ m/s} \qquad V_u = 4.72 \text{ m/s}$$

At $r = 0.6$ m,

$$u = \tfrac{240}{60}(2\pi)\,(0.6) = 15.06 \text{ m/s} \qquad V_u = 3.54 \text{ m/s}$$

The discharge through the turbine is, from section 1,

$$Q = (0.6 \text{ m})\,(2.4 \text{ m})\,(\pi)\,(2.5 \text{ m/s})\,(\cos 45°) = 8 \text{ m}^3\text{/s}$$

Hence, the axial velocity at section 2 is

$$V_a = \frac{8}{\pi(0.6^2 - 0.225^2)} = 8.24 \text{ m/s}$$

Figure 9.18 shows the initial vane angle for the three positions.

Moody[1] has developed a formula to estimate the efficiency of a unit of

[1] Lewis F. Moody, The Propeller Type Turbine, *Trans. ASCE*, vol. 89, p. 628, 1926.

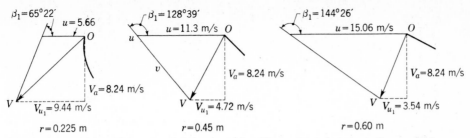

$\beta_1 = 65°22'$ $u = 5.66$ $V_a = 8.24$ m/s $V_{u_1} = 9.44$ m/s $r = 0.225$ m

$\beta_1 = 128°39'$ $u = 11.3$ m/s $V_a = 8.24$ m/s $V_{u_1} = 4.72$ m/s $r = 0.45$ m

$\beta_1 = 144°26'$ $u = 15.06$ m/s $V_a = 8.24$ m/s $V_{u_1} = 3.54$ m/s $r = 0.60$ m

Fig. 9.18 Velocity diagrams for angle of leading edge of a propeller turbine blade.

a homologous series of turbines when the efficiency of one of the series is known:

$$e = 1 - (1 - e_1)\left(\frac{D_1}{D}\right)^{1/4} \tag{9.5.2}$$

in which e_1 and D_1 are usually the efficiency and the diameter of a model.

9.6 PUMPS AND BLOWERS

Pumps add energy to liquids and blowers to gases. The procedure for designing them is the same for both, except when the density is appreciably increased. Turbopumps and -blowers are *radial-flow, axial-flow,* or a combination of the

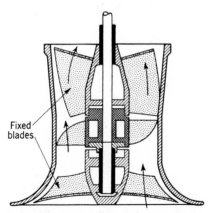

Fixed
blades

Fig. 9.19 Axial-flow pump. (*Ingersoll-Rand Co.*)

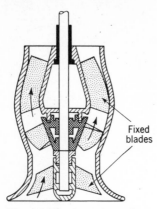

Fig. 9.20 A mixed-flow pump. (*Ingersoll-Rand Co.*)

two, called *mixed-flow*. For high heads the radial (centrifugal) pump, frequently with two or more stages (two or more impellers in series), is best adapted. For large flows under small heads the axial-flow pump or blower (Fig. 9.19) is best suited. The mixed-flow pump (Fig. 9.20) is used for medium head and medium discharge.

The equations developed in Sec. 9.2 apply just as well to pumps and blowers as to turbines. The usual centrifugal pump has a suction, or inlet,

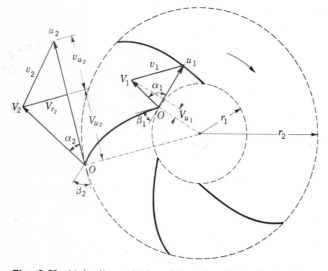

Fig. 9.21 Velocity relationships for flow through a centrifugal-pump impeller.

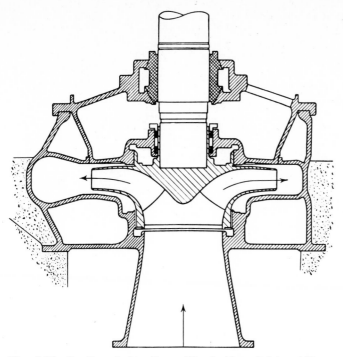

Fig. 9.22 Sectional elevation of Eagle Mountain and Hayfield pumps, Colorado River Aqueduct. (*Worthington Corp.*)

pipe leading to the center of the impeller, a radial outward-flow runner, as in Fig. 9.21, and a collection pipe or spiral casing that guides the fluid to the discharge pipe. Ordinarily, no fixed vanes are used, except for multistage units in which the flow is relatively small and the additional fluid friction is less than the additional gain in conversion of kinetic energy to pressure energy upon leaving the impeller.

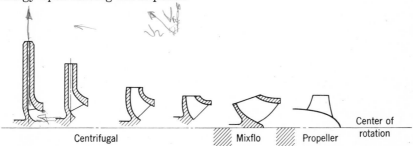

Fig. 9.23 Impeller types used in pumps and blowers. (*Worthington Corp.*)

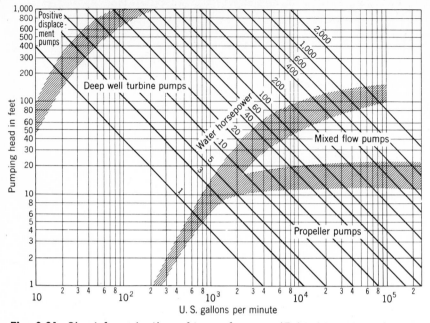

Fig. 9.24 Chart for selection of type of pump. (*Fairbanks, Morse & Co.*)

Figure 9.22 shows a sectional elevation of a large centrifugal pump. For lower heads and greater discharges (relatively) the impellers vary as shown in Fig. 9.23, from high head at left to low head at right with the axial-flow impeller. The specific speed increases from left to right. A chart for determining the types of pump for best efficiency is given in Fig. 9.24 for water.

Centrifugal and mixed-flow pumps are designed in the specific speed range 500 to 6500 and axial pumps from 5000 to 11,000; speed is expressed in revolutions per minute, discharge in gallons per minute, and head in feet.

Characteristic curves showing head, efficiency, and brake horsepower as a function of discharge for a typical centrifugal pump with backward-curved vanes are given in Fig. 9.25. Pumps are not so efficient as turbines, in general, owing to the inherently high losses that result from conversion of kinetic energy into flow energy.

Theoretical head-discharge curve

A theoretical head-discharge curve may be obtained by use of Eq. (9.3.8) and the vector diagrams of Fig. 9.8. From the exit diagram of Fig. 9.8

$$V_2 \cos \alpha_2 = V_{u2} = u_2 - V_{r2} \cot \beta_2$$

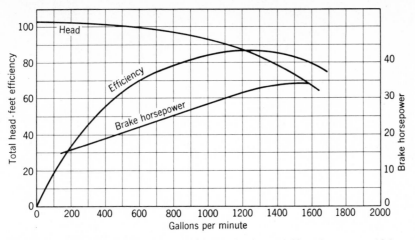

Fig. 9.25 Characteristic curves for typical centrifugal pump; 10-in impeller, 1750 rpm. (*Ingersoll-Rand Co.*)

From the discharge, if b_2 is the width of the impeller at r_2 and vane thickness is neglected,

$$Q = 2\pi r_2 b_2 V_{r2}$$

Eliminating V_{r2} and substituting these last two equations into Eq. (9.3.8) give

$$H = \frac{u_2^2}{g} - \frac{u_2 Q \cot \beta_2}{2\pi r_2 b_2 g} \qquad (9.6.1)$$

For a given pump and speed, H varies linearly with Q, as shown in Fig. 9.26. The usual design of centrifugal pump has $\beta_2 < 90°$, which gives decreasing

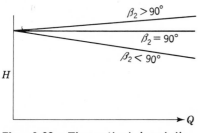

Fig. 9.26 Theoretical head-discharge curves.

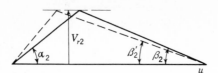

Fig. 9.27 Effect of circulatory flow.

head with increasing discharge. For blades radial at the exit, $\beta_2 = 90°$ and the theoretical head is independent of discharge. For blades curved forward, $\beta_2 > 90°$, and the head rises with discharge.

Actual head-discharge curve

By subtracting head losses from the theoretical head-discharge curve, the actual head-discharge curve is obtained. The most important subtraction is not an actual loss but a failure of the finite number of blades to impart the relative velocity with angle β_2 of the blades. Without perfect guidance (infinite number of blades) the fluid actually is discharged as if the blades had an angle β_2' which is less than β_2 (Fig. 9.27) for the same discharge. This inability of the blades to impart proper guidance reduces V_{u2} and hence decreases the actual head produced. This is called *circulatory flow* and is shown in Fig. 9.28. Fluid friction in flow through the fixed and moving passages causes losses that are proportional to the square of the discharge. They are shown in Fig. 9.28. The final head loss to consider is that of turbulence, the loss due to improper relative-velocity angle at the blade inlet. The pump can be designed for one discharge (at a given speed) at which the relative velocity is tangent to the blade at the inlet. This is the point of best efficiency, and shock or turbulence losses are negligible. For other discharges the loss varies

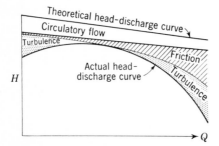

Fig. 9.28 Head-discharge relationships.

about as the square of the discrepancy in approach angle, as shown in Fig. 9.28. The final lower line then represents the actual head-discharge curve. Shutoff head is usually about $u_2^2/2g$, or half of the theoretical shutoff head.

In addition to the head losses and reductions, pumps and blowers have torque losses due to bearing- and packing-friction and disk-friction losses from the fluid between the moving impeller and housing. Internal leakage is also an important power loss, in that fluid which has passed through the impeller, with its energy increased, escapes through clearances and flows back to the suction side of the impeller.

EXAMPLE 9.8 A centrifugal water pump has an impeller (Fig. 9.21) with $r_2 = 12$ in, $r_1 = 4$ in, $\beta_1 = 20°$, $\beta_2 = 10°$. The impeller is 2 in wide at $r = r_1$ and $\frac{3}{4}$ in wide at $r = r_2$. For 1800 rpm, neglecting losses and vane thickness, determine (a) the discharge for shockless entrance when $\alpha_1 = 90°$; (b) α_2 and the theoretical head H; (c) the horsepower required; and (d) the pressure rise through the impeller.

(a) The peripheral speeds are

$$u_1 = \tfrac{1800}{60}(2\pi)\left(\tfrac{1}{3}\right) = 62.8 \text{ ft/s} \qquad u_2 = 3u_1 = 188.5 \text{ ft/s}$$

The vector diagrams are shown in Fig. 9.29. With u_1 and the angles α_1, β_1 known, the entrance diagram is determined, $V_1 = u_1 \tan 20° = 22.85$ ft/s; hence

$$Q = 22.85(\pi)\left(\tfrac{2}{3}\right)\left(\tfrac{2}{12}\right) = 7.97 \text{ cfs}$$

(b) At the exit the radial velocity V_{r2} is

$$V_{r2} = \frac{7.97 \times 12}{2\pi \times 0.75} = 20.3 \text{ ft/s}$$

By drawing u_2 (Fig. 9.29) and a parallel line distance V_{r2} from it, the vector

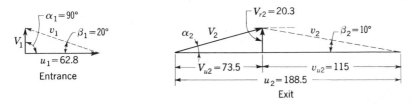

Fig. 9.29 Vector diagrams for entrance and exit of pump impeller.

triangle is determined when β_2 is laid off. Thus

$$v_{u2} = 20.3 \cot 10° = 115 \text{ ft/s} \qquad V_{u2} = 188.5 - 115 = 73.5 \text{ ft/s}$$

$$\alpha_2 = \tan^{-1} \frac{20.3}{73.5} = 15°26' \qquad V_2 = 20.3 \csc 15°26' = 76.2 \text{ ft/s}$$

From Eq. (9.3.8)

$$H = \frac{u_2 V_2 \cos \alpha_2}{g} = \frac{u_2 V_{u2}}{g} = \frac{188.5 \times 73.5}{32.2} = 430 \text{ ft}$$

$$(c) \quad \text{hp} = \frac{Q\gamma H}{550} = \frac{7.97 \times 62.4 \times 430}{550} = 388$$

(d) By applying the energy equation from entrance to exit of the impeller, including the energy H added (elevation change across impeller is neglected),

$$H + \frac{V_1^2}{2g} + \frac{p_1}{\gamma} = \frac{V_2^2}{2g} + \frac{p_2}{\gamma}$$

and

$$\frac{p_2 - p_1}{\gamma} = 430 + \frac{22.85^2}{64.4} - \frac{76.2^2}{64.4} = 348 \text{ ft}$$

or

$$p_2 - p_1 = 348 \times 0.433 = 151 \text{ psi}$$

9.7 CENTRIFUGAL COMPRESSORS

Centrifugal compressors operate according to the same principles as turbomachines for liquids. It is important for the fluid to enter the impeller without shock, i.e., with the relative velocity tangent to the blade. Work is done on the gas by rotation of the vanes, the moment-of-momentum equation relating torque to production of tangential velocity. At the impeller exit the high-velocity gas must have its kinetic energy converted in part to flow energy by suitable expanding flow passages. For adiabatic compression (no cooling of the gas) the actual work w_a of compression per unit mass is compared with the work w_{th} per unit mass to compress the gas to the same pressure isentropically. For cooled compressors the work w_{th} is based on the isothermal work of compression to the same pressure as the actual case. Hence

$$\eta = \frac{w_{th}}{w_a} \tag{9.7.1}$$

is the formula for efficiency of a compressor.

The efficiency formula for compression of a perfect gas is developed for the adiabatic compressor, assuming no internal leakage in the machine, i.e., no short-circuiting of high-pressure fluid back to the low-pressure end of the impeller. Centrifugal compressors are usually multistage, with pressure ratios up to 3 across a single stage. From the moment-of-momentum equation (9.3.2) with inlet absolute velocity radial, $\alpha_1 = 90°$, the theoretical torque T_{th} is

$$T_{th} = \dot{m} V_{u2} r_2 \tag{9.7.2}$$

in which $\dot{m}$ is the mass per unit time being compressed, V_{u2} is the tangential component of the absolute velocity leaving the impeller, and r_2 is the impeller radius at exit. The actual applied torque T_a is greater than the theoretical torque by the torque losses due to bearing and packing friction plus disk friction; hence

$$T_{th} = T_a \eta_m \tag{9.7.3}$$

if η_m is the mechanical efficiency of the compressor.

In addition to the torque losses, there are irreversibilities due to flow through the machine. The actual work of compression through the adiabatic machine is obtained from the steady-flow energy equation (3.7.1), neglecting elevation changes and replacing $u + p/\rho$ by h,

$$-w_a = \frac{V_{2a}^2 - V_1^2}{2} + h_2 - h_1 \tag{9.7.4}$$

The isentropic work of compression can be obtained from Eq. (3.7.1) in differential form, neglecting the z terms,

$$-dw_{th} = V\,dV + d\frac{p}{\rho} + du = V\,dV + \frac{dp}{\rho} + pd\frac{1}{\rho} + du$$

The last two terms are equal to $T\,ds$ from Eq. (3.7.6) which is zero for isentropic flow, so that

$$-dw_{th} = V\,dV + \frac{dp}{\rho} \tag{9.7.5}$$

By integrating for $p/\rho^k = $ const between sections 1 and 2,

$$-w_{th} = \frac{V_{2th}{}^2 - V_1{}^2}{2} + \frac{k}{k-1}\left(\frac{p_2}{\rho_{2th}} - \frac{p_1}{\rho_1}\right)$$

$$= \frac{V_{2th}{}^2 - V_1{}^2}{2} + c_p T_1\left[\left(\frac{p_2}{p_1}\right)^{(k-1)/k} - 1\right] \tag{9.7.6}$$

The efficiency may now be written

$$\eta = \frac{-w_{th}}{-w_a} = \frac{(V_{2th}{}^2 - V_1{}^2)/2 + c_p T_1[(p_2/p_1)^{(k-1)/k} - 1]}{(V_{2a}{}^2 - V_1{}^2)/2 + c_p(T_{2a} - T_1)} \tag{9.7.7}$$

since $h = c_p T$. In terms of Eqs. (9.7.2) and (9.7.3)

$$-w_a = \frac{T_a \omega}{\dot{m}} = \frac{T_{th}\omega}{\eta_m \dot{m}} = \frac{V_{u2}r_2\omega}{\eta_m} = \frac{V_{u2}u_2}{\eta_m} \tag{9.7.8}$$

then

$$\eta = \frac{\eta_m}{V_{u2}u_2}\left\{c_p T_1\left[\left(\frac{p_2}{p_1}\right)^{(k-1)/k} - 1\right] + \frac{V_{2th}{}^2 - V_1{}^2}{2}\right\} \tag{9.7.9}$$

Use of this equation is made in the following example.

EXAMPLE 9.9   An adiabatic turbocompressor has blades that are radial at the exit of its 15-cm-diameter impeller. It is compressing 0.5 kg/s air at 1 kg$_f$/cm^2 abs, $t = 15°C$, to 3 kg$_f$/cm^2 abs. The entrance area is 60 cm^2, and the exit area is 35 cm^2, $\eta = 0.75$; $\eta_m = 0.90$. Determine the rotational speed of the impeller and the actual temperature of air at the exit.

The density at the inlet is

$$\rho_1 = \frac{p_1}{RT_1} = \frac{9.806 \times 10^4 \text{ N/m}^2}{(287 \text{ J/kg·K})(273 + 15 \text{ K})} = 1.186 \text{ kg/m}^3$$

and the velocity at the entrance is

$$V_1 = \frac{\dot{m}}{\rho_1 A_1} = \frac{0.5 \text{ kg/s}}{(1.186 \text{ kg/m}^3)(0.006 \text{ m}^2)} = 70.26 \text{ m/s}$$

The theoretical density at the exit is

$$\rho_{2th} = \rho_1 \left(\frac{p_2}{p_1}\right)^{1/k} = 1.186 \times 3^{1/1.4} = 2.60 \text{ kg/m}^3$$

and the theoretical velocity at the exit is

$$V_{2th} = \frac{\dot{m}}{\rho_{2th} A_2} = \frac{0.5}{2.60 \times 0.0035} = 54.945 \text{ m/s}$$

For radial vanes at the exit, $V_{u2} = u_2 = \omega r_2$. From Eq. (9.7.9)

$$u_2{}^2 = \frac{\eta_m}{\eta} \left\{ c_p T_1 \left[\left(\frac{p_2}{p_1}\right)^{(k-1)/k} - 1 \right] + \frac{V_{2th}{}^2 - V_1{}^2}{2} \right\}$$

$$= \frac{0.90}{0.75} \left[(0.24 \times 4187)(273 + 15)(3^{0.4/1.4} - 1) + \frac{54.945^2 - 70.26^2}{2} \right]$$

and $u_2 = 359.56$ m/s. Then

$$\omega = \frac{u_2}{r_2} = \frac{359.56}{0.075} = 4794 \text{ rad/s}$$

and

$$N = \omega \left(\tfrac{60}{2\pi}\right) = 4794 \times \frac{60}{2\pi} = 45{,}781 \text{ rpm}$$

The theoretical work w_{th} is the term in the brackets in the expression for $u_2{}^2$. It is $-w_{th} = 0.1058 \times 10^6$ m·N/kg. Then from Eq. (9.7.1)

$$w_a = \frac{w_{th}}{\eta} = -\frac{1.058 \times 10^5}{0.75} = -1.411 \times 10^5 \text{ m·N/kg}$$

Since the kinetic energy term is small, Eq. (9.7.4) can be solved for $h_2 - h_1$

and a trial solution effected,

$$h_2 - h_1 = c_p(T_{2a} - T_1) = 1.411 \times 10^5 + \frac{70.26^2 - V_{2a}{}^2}{2}$$

As a first approximation, let $V_{2a} = V_{2th} = 54.945$; then

$$T_{2a} = 288 + \frac{1}{0.24 \times 4187}\left(1.411 \times 10^5 + \frac{70.26^2 - 54.945^2}{2}\right) = 429.4 \text{ K}$$

For this temperature the density at the exit is 2.387 kg/m³, and the velocity is 59.85 m/s. Insertion of this value in place of 54.945 reduces the temperature to $T_{2a} = 429.1$ K.

9.8 CAVITATION

When a liquid flows into a region where its pressure is reduced to vapor pressure, it boils and vapor pockets develop in the liquid. The vapor bubbles are carried along with the liquid until a region of higher pressure is reached, where they suddenly collapse. This process is called *cavitation*. If the vapor bubbles are near to (or in contact with) a solid boundary when they collapse, the forces exerted by the liquid rushing into the cavities create very high localized pressures that cause pitting of the solid surface. The phenomenon is accompanied by noise and vibrations that have been described as similar to gravel going through a centrifugal pump.

In a flowing liquid, the *cavitation parameter* σ is useful in characterizing the susceptibility of the system to cavitate. It is defined by

$$\sigma = \frac{p - p_v}{\rho V^2/2} \tag{9.8.1}$$

in which p is the absolute pressure at the point of interest, p_v is the vapor pressure of the liquid, ρ is the density of the liquid, and V is the undisturbed, or reference, velocity. The cavitation parameter is a form of pressure coefficient. In two geometrically similar systems, they would be equally likely to cavitate or would have the same degree of cavitation for the same value of σ. When $\sigma = 0$, the pressure is reduced to vapor pressure and boiling should occur.

Tests made on chemically pure liquids show that they will sustain high tensile stresses, of the order of thousands of pounds per square inch, which is in contradiction to the concept of cavities forming when pressure is reduced to vapor pressure. Since there is generally spontaneous boiling when vapor pressure is reached with commercial or technical liquids, it is generally accepted that nuclei must be present around which the vapor bubbles form and grow. The nature of the nuclei is not thoroughly understood, but they may be microscopic dust particles or other contaminants, which are widely dispersed through technical liquids.

Cavitation bubbles may form on nuclei, grow, then move into an area of higher pressure and collapse, all in a few thousandths of a second in flow within a turbomachine. In aerated water the bubbles have been photographed as they move through several oscillations, but this phenomenon does not seem to occur in nonaerated liquids. Surface tension of the vapor bubbles appears to be an important property accounting for the high-pressure pulses resulting from collapse of a vapor bubble. Recent experiments indicate pressures of the order of 200,000 psi based on the analysis of strain waves in a photoelastic specimen exposed to cavitation.[1] Pressures of this magnitude appear to be reasonable, in line with the observed mechanical damage caused by cavitation.

The formation and collapse of great numbers of bubbles on a surface subject that surface to intense local stressing, which appears to damage the surface by fatigue. Some ductile materials withstand battering for a period, called the *incubation period*, before damage is noticeable, while brittle materials may lose weight immediately. There may be certain electrochemical, corrosive, and thermal effects which hasten the deterioration of exposed surfaces. Rheingans[2] has collected a series of measurements made by magnetostriction-oscillator tests, showing weight losses of various metals used in hydraulic machines (see Table 9.2).

Protection against cavitation should start with the hydraulic design of the system in order to avoid the low pressures if practicable. Otherwise, use of special cavitation-resistant materials or coatings may be effective. Small amounts of air entrained into water systems have markedly reduced cavitation damage, and recent studies indicate that cathodic protection is helpful.

The formation of vapor cavities decreases the useful channel space for liquid and thus decreases the efficiency of a fluid machine. Cavitation causes three undesirable conditions: lowered efficiency, damage to flow passages, and noise and vibrations. Curved vanes are particularly susceptible to cavitation on their convex sides and may have localized areas where cavitation causes pitting or failure. Since all turbomachinery and ship propellers and many

[1] G. W. Sutton, A Photoelastic Study of Strain Waves Caused by Cavitation, *J. Appl. Mech.*, vol. 24, pt. 3, pp. 340–348, 1957.

[2] W. J. Rheingans, Selecting Materials to Avoid Cavitation Damage, *Mater. Des. Eng.*, pp. 102–106, 1958.

Table 9.2 *Weight loss in materials used in hydraulic machines*

Alloy	Weight loss after 2 h, mg
Rolled stellite†	0.6
Welded aluminum bronze‡	3.2
Cast aluminum bronze§	5.8
Welded stainless steel (two layers, 17% Cr, 7% Ni)	6.0
Hot-rolled stainless steel (26% Cr, 13% Ni)	8.0
Tempered, rolled stainless steel (12% Cr)	9.0
Cast stainless steel (18% Cr, 8% Ni)	13.0
Cast stainless steel (12% Cr)	20.0
Cast manganese bronze	80.0
Welded mild steel	97.0
Plate steel	98.0
Cast steel	105.0
Aluminum	124.0
Brass	156.0
Cast iron	224.0

† This material is not suitable for ordinary use, in spite of its high resistance, because of its high cost and difficulty in machining.
‡ Ampco-Trode 200: 83% Cu, 10.3% Al, 5.8% Fe.
§ Ampco 20: 83.1% Cu, 12.4% Al, 4.1% Fe.

hydraulic structures are subject to cavitation, special attention must be given to it in their design.

A *cavitation index* σ' is useful in the proper selection of turbomachinery and in its location with respect to suction or tail-water elevation. The minimum pressure in a pump or turbine generally occurs along the convex side of vanes near the suction side of the impeller. In Fig. 9.30, if e is the point of minimum pressure, Bernoulli's equation applied between e and the downstream liquid surface, neglecting losses between the two points, may be written

$$\frac{p_e}{\gamma} + \frac{V_e^2}{2g} + H_s = \frac{p_a}{\gamma} + 0 + 0$$

in which p_a is the atmospheric pressure and p_e the absolute pressure. For cavitation to occur at e, the pressure must be equal to or less than p_v, the

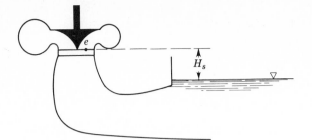

Fig. 9.30 Turbine or pump setting.

vapor pressure. If $p_e = p_v$,

$$\sigma' = \frac{V_e^2}{2gH} = \frac{p_a - p_v - \gamma H_s}{\gamma H} \qquad (9.8.2)$$

is the ratio of energy available at e to total energy H, since the only energy is kinetic energy. The ratio σ' is a cavitation index or number. The critical value σ_c may be determined by a test on a model of the homologous series. For cavitationless performance, the suction setting H_s for an impeller installation must be so fixed that the resulting value of σ' is greater than that of σ_c.

EXAMPLE 9.10 Tests on a pump model indicate a $\sigma_c = 0.10$. A homologous unit is to be installed at a location where $p_a = 13$ psi and $p_v = 0.50$ psi and is to pump water against a head of 80 ft. What is the maximum permissible suction head?

Solving Eq. (9.8.2) for H_s and substituting the values of σ_c, H, p_a, and p_v give

$$H_s = \frac{p_a - p_v}{\gamma} - \sigma'H = \frac{13 - 0.50}{0.433} - 0.10 \times 80 = 20.8 \text{ ft}$$

The less the value of H_s, the greater the value of the plant σ', and the greater the assurance against cavitation.

The *net positive suction head* (NPSH) is frequently used in the specification of minimum suction conditions for a turbomachine. It is

$$\text{NPSH} = \frac{V_e^2}{2g} = \frac{p_a - p_v - \gamma H_s}{\gamma} \qquad (9.8.3)$$

A test is run on the machine to determine the maximum value of H_s for operation of the machine with no impairment of efficiency and without objectionable noise or damage. Then, from this test NPSH is calculated from Eq. (9.8.3). Any setting of this machine where the suction lift is less than H_s, as found from Eq. (9.8.3), is then acceptable. Note that H_s is positive when the suction reservoir is below the turbomachine, as in Fig. 9.30.

A *suction specific speed* S for homologous units may be formulated. By elimination of D_e in the two equations

$$\text{NPSH} = \frac{V_e{}^2}{2g} \sim \frac{Q^2}{D^4} \qquad \frac{Q^2}{ND_e{}^3} = \text{const}$$

S is obtained,

$$S = \frac{N\sqrt{Q}}{(\text{NPSH})^{3/4}} \tag{9.8.4}$$

When different units of a series are operating under cavitating conditions, equal values of S indicate a similar degree of cavitation. When cavitation is not present, the equation is not valid.

PROBLEMS

9.1 By use of Eqs. (9.1.1) and (9.1.3) together with $P = \gamma QH$ for power, develop the homologous relationship for P in terms of speed and diameter.

9.2 A centrifugal pump is driven by an induction motor that reduces in speed as the pump load increases. A test determines several sets of values of N, Q, H for the pump. How is a characteristic curve of the pump for a constant speed determined from these data?

9.3 What is the specific speed of the pump of Example 9.1 at its point of best efficiency?

9.4 Plot the dimensionless characteristic curve of the pump of Example 9.1. On this same curve plot several points from the characteristics of the new (52-in) pump. Why are they not exactly on the same curve?

9.5 Determine the size and synchronous speed of a pump homologous to the 72-in pump of Example 9.1 that will produce 3 m³/s at 100 m head at its point of best efficiency.

9.6 Develop the characteristic curve for a homologous pump of the series of Example 9.1 for 18-in diameter discharge and 1800 rpm.

9.7 A pump with a 20-cm-diameter impeller discharges 100 l/s at 1140 rpm and 10 m head at its point of best efficiency. What is its specific speed?

9.8 A hydroelectric site has a head of 300 ft and an average discharge of 400 cfs. For a generator speed of 200 rpm, what specific speed turbine is needed? Assume an efficiency of 92 percent.

9.9 A model turbine, $N_s = 36$, with a 14-in-diameter impeller develops 27 hp at a head of 44 ft and an efficiency of 86 percent. What are the discharge and speed of the model?

9.10 What size and synchronous speed of homologous unit of Prob. 9.9 would be needed to discharge 600 cfs at 260 ft of head?

9.11 22 m³/s water flowing through the fixed vanes of a turbine has a tangential component of 2 m/s at a radius of 1.25 m. The impeller, turning at 180 rpm, discharges in an axial direction. What torque is exerted on the impeller?

9.12 In Prob. 9.11, neglecting losses, what is the head on the turbine?

9.13 A generator with speed $N = 240$ rpm is to be used with a turbine at a site where $H = 120$ m and $Q = 8$ m³/s. Neglecting losses, what tangential component must be given to the water at $r = 1$ m by the fixed vanes? What torque is exerted on the impeller? How much horsepower is produced?

9.14 A site for a Pelton wheel has a steady flow of 2 cfs with a nozzle velocity of 240 ft/s. With a blade angle of 174°, and $C_v = 0.98$, for 60-Hz power, determine (*a*) the diameter of wheel, (*b*) the speed, (*c*) the horsepower, (*d*) the energy remaining in the water. Neglect losses.

9.15 An impulse wheel is to be used to develop 50-Hz power at a site where $H = 120$ m, and $Q = 75$ l/s. Determine the diameter of the wheel and its speed. $C_v = 0.97$; $e = 82$ percent.

9.16 At what angle should the wicket gates of a turbine be set to extract 12,000 hp from a flow of 900 cfs? The diameter of the opening just inside the wicket gates is 12 ft, and the height is 3 ft. The turbine runs at 200 rpm, and flow leaves the runner in an axial direction.

9.17 For a given setting of wicket gates how does the moment of momentum vary with the discharge?

9.18 Assuming constant axial velocity just above the runner of the propeller turbine of Prob. 9.16, calculate the tangential-velocity components if the hub radius is 1 ft and the outer radius is 3 ft.

9.19 Determine the vane angles β_1, β_2 for entrance and exit from the propeller turbine of Prob. 9.18 so that no angular momentum remains in the flow. (Compute the angles for inner radius, outer radius, and midpoint.)

9.20 Neglecting losses, what is the head on the turbine of Prob. 9.16?

9.21 The hydraulic efficiency of a turbine is 95 percent, and its theoretical head is 80 m. What is the actual head required?

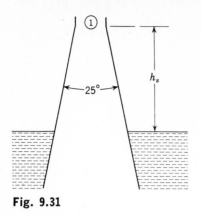

Fig. 9.31

9.22   A turbine model test with 25-cm-diameter impeller showed an efficiency of 90 percent. What efficiency could be expected from a 120-cm-diameter impeller?

9.23   A turbine draft tube (Fig. 9.31) expands from 6 to 18 ft diameter. At section 1 the velocity is 30 ft/s for vapor pressure of 1 ft and barometric pressure of 32 ft of water. Determine h_s for incipient cavitation (pressure equal to vapor pressure at section 1).

9.24   Construct a theoretical head-discharge curve for the following specifications of a centrifugal pump: $r_1 = 5$ cm, $r_2 = 10$ cm, $b_1 = 2.5$ cm, $b_2 = 2$ cm, $N = 1200$ rpm, and $\beta_2 = 30°$.

9.25   A centrifugal water pump (Fig. 9.21) has an impeller $r_1 = 2.5$ in, $b_1 = 1\frac{3}{8}$ in, $r_2 = 4.5$ in, $b_2 = \frac{3}{4}$ in, $\beta_1 = 30°$, $\beta_2 = 45°$ (b_1, b_2 are impeller width at r_1 and r_2, respectively). Neglect thickness of vanes. For 1800 rpm, calculate (*a*) the design discharge for no prerotation of entering fluid, (*b*) α_2 and the theoretical head at point of best efficiency, and (*c*) for hydraulic efficiency of 85 percent and overall efficiency of 78 percent, the actual head produced, losses in foot-pounds per pound, and brake horsepower.

9.26   A centrifugal pump has an impeller with dimensions $r_1 = 7.5$ cm, $r_2 = 15$ cm, $b_1 = 5$ cm, $b_2 = 3$ cm, $\beta_1 = \beta_2 = 30°$. For a discharge of 55 l/s and shockless entry to vanes compute (*a*) the speed, (*b*) the head, (*c*) the torque, (*d*) the power, and (*e*) the pressure rise across impeller. Neglect losses. $\alpha_1 = 90°$.

9.27   A centrifugal water pump with impeller dimensions $r_1 = 2$ in, $r_2 = 5$ in, $b_1 = 3$ in, $b_2 = 1$ in, $\beta_2 = 60°$ is to pump 5 cfs at 64 ft head. Determine (*a*) β_1, (*b*) the speed, (*c*) the horsepower, and (*d*) the pressure rise across the impeller. Neglect losses, and assume no shock at the entrance. $\alpha_1 = 90°$.

9.28   Select values of r_1, r_2, β_1, β_2, b_1, and b_2 of a centrifugal impeller to take 30 l/s water from a 10-cm-diameter suction line and increase its energy by 12 m·N/N. $N = 1200$ rpm; $\alpha_1 = 90°$. Neglect losses.

9.29 A pump has blade angles $\beta_1 = \beta_2$; $b_1 = 2b_2 = 1$ in; $r_1 = r_2/3 = 2$ in. For a theoretical head of 95.2 ft at a discharge at best efficiency of 1.052 cfs, determine the blade angles and speed of the pump. Neglect thickness of vanes and assume perfect guidance. (*Hint:* Write down every relation you know connecting β_1, β_2, b_1, b_2, r_1, r_2, u_1, u_2, H_{th}, Q, V_{r2}, V_{u2}, V_1, ω, and N from the two velocity vector diagrams, and by substitution reduce to one unknown.)

9.30 A mercury-water differential manometer, $R' = 65$ cm, is connected from the 10-cm-diameter suction pipe to the 8-cm-diameter discharge pipe of a pump. The centerline of the suction pipe is 30 cm below the discharge pipe. For $Q = 60$ l/s water, calculate the head developed by the pump.

9.31 The impeller for a blower (Fig. 9.32) is 18 in wide. It has straight blades and turns at 1200 rpm. For 10,000 ft³/min air, $\gamma = 0.08$ lb/ft³, calculate (*a*) entrance and exit blade angles ($\alpha_1 = 90°$), (*b*) the head produced in inches of water, and (*c*) the theoretical horsepower required.

9.32 An air blower is to be designed to produce pressure of 10 cm H₂O when operating at 3600 rpm. $\gamma = 11.5$ N/m³; $r_2 = 1.1r_1$; $\beta_2 = \beta_1$; width of impeller is 10 cm; $\alpha_1 = 90°$. Find r_1.

9.33 In Prob. 9.32 when $\beta_1 = 30°$, calculate the discharge in cubic meters per minute.

9.34 Develop the equation for efficiency of a cooled compressor,

$$\eta = \frac{\eta_m}{V_{u2}u_2}\left(\frac{V_{2th}^2 - V_1^2}{2} + \frac{p_1}{\rho_1}\ln\frac{p_2}{\rho_1}\right)$$

9.35 Find the rotational speed in Example 9.9 for a cooled compressor, using results of Prob. 9.34, with the actual air temperature at exit 15°C.

9.36 What is the cavitation parameter at a point in flowing water where $t = 68°F$, $p = 2$ psia and the velocity is 40 ft/s?

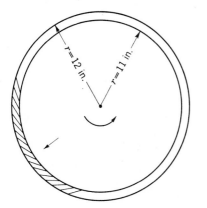

Fig. 9.32

9.37   A turbine with $\sigma_c = 0.08$ is to be installed at a site where $H = 60$ m and a water barometer stands at 8.3 m. What is the maximum permissible impeller setting above tail water?

9.38   Two units are homologous when they are geometrically similar and have

(*a*) similar streamlines   (*b*) the same Reynolds number
(*c*) the same efficiency   (*d*) the same Froude number
(*e*) none of these answers

9.39   The following two relationships are necessary for homologous units:

(*a*) $H/ND^3 = \text{const}; Q/N^2D^2 = \text{const}$ (*b*) $Q/D^2 \sqrt{H} = \text{const}; H/N^3D = \text{const}$
(*c*) $P/QH = \text{const}; H/N^2D^2 = \text{const}$
(*d*) $N \sqrt{Q}/H^{3/2} = \text{const}; N \sqrt{P}/H^{3/4} = \text{const}$ (*e*) none of these answers

9.40   The specific speed of a pump is defined as the speed of a unit

(*a*) of unit size with unit discharge at unit head
(*b*) of such a size that it requires unit power for unit head
(*c*) of such a size that it delivers unit discharge at unit head
(*d*) of such a size that it delivers unit discharge at unit power
(*e*) none of these answers

9.41   An impulse turbine

(*a*) always operates submerged
(*b*) makes use of a draft tube
(*c*) is most suited for low-head installations
(*d*) converts pressure head into velocity head throughout the vanes
(*e*) operates by initial complete conversion to kinetic energy

9.42   A Pelton 24-in-diameter wheel turns at 400 rpm. Select from the following the head, in feet, best suited for this wheel:

(*a*) 7   (*b*) 30   (*c*) 120   (*d*) 170   (*e*) 480

9.43   A shaft transmits 150 kW at 600 rpm. The torque in newton-meters is

(*a*) 26.2   (*b*) 250   (*c*) 2390   (*d*) 4780   (*e*) none of these answers

9.44   What torque is required to give 100 cfs water a moment of momentum so that it has a tangential velocity of 10 ft/s at a distance of 6 ft from the axis?

(*a*) 116 lb·ft   (*b*) 1935 lb·ft   (*c*) 6000 lb·ft   (*d*) 11,610 lb·ft
(*e*) none of these answers

9.45   The moment of momentum of water is reduced by 27,100 N·m in flowing through vanes on a shaft turning 400 rpm. The power developed on the shaft is, in kilowatts,

(*a*) 181.5   (*b*) 1134   (*c*) 10,800   (*d*) not determinable; insufficient data
(*e*) none of these answers

9.46 Liquid moving with constant angular momentum has a tangential velocity of 4.0 ft/s 10 ft from the axis of rotation. The tangential velocity 5 ft from the axis is, in feet per second,

(a) 2   (b) 4   (c) 8   (d) 16   (e) none of these answers

9.47 A reaction-type turbine discharges 34 m³/s under a head of 7.5 m and with an overall efficiency of 91 percent. The power developed is, in kilowatts,

(a) 2750   (b) 2500   (c) 2275   (d) 70.7   (e) none of these answers

9.48 The head developed by a pump with hydraulic efficiency of 80 percent, for $u_2 = 100$ ft/s, $V_2 = 60$ ft/s, $\alpha_2 = 45°$, $\alpha_1 = 90°$, is

(a) 52.6   (b) 105.3   (c) 132   (d) 165   (e) none of these answers

9.49 Select the correct relationship for pump vector diagrams.

(a) $\alpha_1 = 90°; v_1 = u_1 \cot \beta_1$ (b) $V_{u2} = u_2 - V_{r2} \cot \beta_2$
(c) $\omega_2 = r_2/u_2$ (d) $r_1 V_1 = r_2 V_{r2}$ (e) none of these answers

9.50 The cavitation parameter is defined by

(a) $\dfrac{p_v - p}{\rho V^2/2}$ (b) $\dfrac{p_{\text{atm}} - p_v}{\rho V^2/2}$ (c) $\dfrac{p - p_v}{\gamma V^2/2}$ (d) $\dfrac{p - p_v}{\rho V^2/2}$

(e) none of these answers

9.51 Cavitation is caused by

(a) high velocity   (b) low barometric pressure   (c) high pressure
(d) low pressure   (e) low velocity

REFERENCES

Church, A. H.: "Centrifugal Pumps and Blowers," Wiley, New York, 1944.
Daily, J. W.: Hydraulic Machinery, in H. Rouse (ed.), "Engineering Hydraulics," Wiley, New York, 1950.
Eisenberg, P., and M. P. Tulin: Cavitation, sec. 12 in V. L. Streeter (ed.), "Handbook of Fluid Dynamics," McGraw-Hill, New York, 1961.
Moody, L. F.: Hydraulic Machinery, in C. V. Davis (ed.), "Handbook of Applied Hydraulics," 2d ed., McGraw-Hill, New York, 1952.
Norrie, D. H.: "An Introduction to Incompressible Flow Machines," American Elsevier, New York, 1963.
Stepanoff, A. J.: "Centrifugal and Axial Flow Pumps," Wiley, New York, 1948.
Wislicenus, G. F.: "Fluid Mechanics of Turbomachinery," McGraw-Hill, New York, 1947.

STEADY CLOSED-CONDUIT FLOW

The basic procedures for solving problems in incompressible steady flow in closed conduits are presented in Sec. 5.10, where simple pipe-flow situations are discussed, including losses due to change in cross section or direction of flow. The great majority of practical problems deal with turbulent flow, and velocity distributions in turbulent pipe flow are discussed in Sec. 5.4. The Darcy-Weisbach equation is introduced in Chap. 5 to relate frictional losses to flow rate in pipes, with the friction factor determined from the Moody diagram. A number of exponential friction formulas commonly used in commercial and industrial applications are discussed in this chapter. The use of the hydraulic and energy grade lines in solving problems is reiterated before particular applications are developed. Complex flow problems are investigated, including hydraulic systems that incorporate various different elements such as pumps and piping networks. The use of the digital computer in analysis and design becomes particularly relevant when multielement systems are being investigated.

10.1 EXPONENTIAL PIPE-FRICTION FORMULAS

Industrial pipe-friction formulas are usually empirical, of the form

$$\frac{h_f}{L} = \frac{RQ^n}{D^m} \tag{10.1.1}$$

in which h_f/L is the head loss per unit length of pipe (slope of the energy grade line), Q the discharge, and D the inside pipe diameter. The resistance coefficient R is a function of pipe roughness only. An equation with specified exponents and coefficient R is valid only for the fluid viscosity for which it is developed, and it is normally limited to a range of Reynolds numbers and diameters. In its range of applicability such an equation is convenient, and nomographs are often used to aid problem solution.

The Hazen-Williams[1] formula for flow of water at ordinary temperatures through pipes is of this form, with R given by

$$R = \frac{4.727}{C^n} \qquad \text{English units} \tag{10.1.2}$$

$$R = \frac{10.675}{C^n} \qquad \text{SI units} \tag{10.1.3}$$

with $n = 1.852$, $m = 4.8704$, and C dependent upon roughness as follows:

$$
C = \begin{cases}
140 & \text{extremely smooth, straight pipes; asbestos-cement} \\
130 & \text{very smooth pipes; concrete; new cast iron} \\
120 & \text{wood stave; new welded steel} \\
110 & \text{vitrified clay; new riveted steel} \\
100 & \text{cast iron after years of use} \\
95 & \text{riveted steel after years of use} \\
60 \text{ to } 80 & \text{old pipes in bad condition}
\end{cases}
$$

One can develop a special-purpose formula for a particular application by using the Darcy-Weisbach equation and friction factors from the Moody diagram or, alternatively, by using experimental data if available. If P sets of values of D and Q are available and h_f/L values are obtained, either from an experimental program or by use of the Darcy-Weisbach equation, the resistance coefficient and exponents in Eq. (10.1.1) can be evaluated. The method of least squares is used to find R, n, and m in Eq. (10.1.1) as follows:

$$\ln \frac{h_f}{L} = \ln R + n \ln Q - m \ln D$$

Let $Z_i = \ln (h_{fi}/L)$, $B = \ln R$, $X_i = \ln Q_i$, and $Y_i = \ln D_i$; then

$$Z_i = B + nX_i - mY_i \tag{10.1.4}$$

[1] H. W. King and E. F. Brater, "Handbook of Hydraulics," p. **6**–11, McGraw-Hill, New York, 1954.

The left side of Eq. (10.1.4), Z_i, is computed from the Darcy-Weisbach equation at D_i and Q_i, or it is an experimental value. The right side of the equation is evaluated with the corresponding values of Q_i and D_i and represents the value of $\ln(h_f/L)$ as calculated from Eq. (10.1.1). The unknown values of R, n, and m that provide the best match between the P sets of values on the left and right side of Eq. (10.1.4) are determined by minimizing the sum of the squares of the differences:

$$F = \sum_{i=1}^{P} S^2 = \sum_{i=1}^{P} (Z_i - B - nX_i + mY_i)^2 \qquad (10.1.5)$$

The value of F, which is a function of the three unknowns B, n, and m for the given data set, will be minimized when

$$\frac{\partial F}{\partial B} = 0 \qquad \frac{\partial F}{\partial n} = 0 \qquad \frac{\partial F}{\partial m} = 0$$

These three conditions provide the three principal equations

$$\Sigma Z_i - PB - n\Sigma X_i + m\Sigma Y_i = 0$$

$$\Sigma Z_i X_i - B\Sigma X_i - n\Sigma X_i^2 + m\Sigma X_i Y_i = 0$$

$$\Sigma Z_i Y_i - B\Sigma Y_i - n\Sigma X_i Y_i + m\Sigma Y_i^2 = 0$$

in which the indices on the summation symbol have been dropped but are understood. A simultaneous solution of the three equations provides the values of B, n, and m, and since $R = e^B$, Eq. (10.1.1) is determined. The following example shows a calculation of the constants in Eq. (10.1.1) by use of the computer program in Fig. 10.1.

EXAMPLE 10.1 For the Reynolds number range, RMIN = 50,000 to RMAX = 10^6, diameter range DMIN = 0.667 ft to DMAX = 2.5 ft, and for pipe roughness, EPS = 0.0005 ft, and kinematic viscosity, VNU = 1.217×10^{-5} ft²/s, determine R, n, and m in Eq. (10.1.1).

Four (KDIV = 4) equally spaced values of **R** and of D are taken within the limiting values of **R** and D. Then since

$$Q = \frac{\pi}{4} D \mathbf{R} \nu$$

16 sets (PP = 16) of data, D and Q, are available to yield 16 values of f, by

```
      DIMENSION D(6),RE(6)
      NAMELIST/DIN/VNU,EPS,KDIV,DMAX,DMIN,RMIN,RMAX,G,EM,EN,R
      READ(5,DIN)
      DO 1 I=1,KDIV
      D(I)=DMIN+(I-1)*(DMAX-DMIN)/(KDIV-1)
1     RE(I)=RMIN+(I-1)*(RMAX-RMIN)/(KDIV-1)
      DATA J,SX,SY,SZ,SZX,SXX,SXY,SZY,SYY/0,.0,.0,.0,.0,.0,.0,.0,.0/
      CON=1./(.7854**2*2*G)
      DO 2 I=1,KDIV
      Y=ALOG(D(I))
      AW=.094*(EPS/D(I))**.225
      BW=88.*(EPS/D(I))**.44
      CW=1.62*(EPS/D(I))**.134
      DO 2 I1=1,KDIV
      J=J+1
      F=AW+BW*RE(I1)**(-CW)
      Q=.7854*D(I)*RE(I1)*VNU
      HL=F*CON*Q**2/D(I)**5
      Z=ALOG(HL)
      X=ALOG(Q)
      SX=SX+X
      SZ=SZ+Z
      SY=SY+Y
      SZX=SZX+Z*X
      SXX=SXX+X*X
      SXY=SXY+X*Y
      SZY=SZY+Z*Y
2     SYY=SYY+Y*Y
      PP=KDIV*KDIV
      C1=SXX/SX-SX/PP
      C4=SXY/SY-SXX/SX
      C5=SXY/SX-SYY/SY
      C6=SZX/SX-SZY/SY
      EM=(C6/C4-(SZ/PP-SZX/SX)/C1)/((SY/PP-SXY/SX)/C1-C5/C4)
      EN=-(C6+C5*EM)/C4
      R=EXP((SZ-SX*EN+SY*EM)/PP)
      WRITE(6,DIN)
      END
&DIN VNU=1.217E-5,EPS=.0005,KDIV=4,DMAX=2.5,DMIN=.667,RMAX=1E6,
RMIN=.5E5,G=32.174 &END
```

Fig. 10.1 Computer program for exponential equation, Example 10.1.

use of the Wood[1] approximation for friction factors, and 16 values of h_f/L from the Darcy-Weisbach equation. The computed results from the program are

$$R = 0.000542 \qquad n = 1.908 \qquad m = 5.041$$

If an equation is wanted in SI units, it is only necessary to use SI units in all input data to the program. These data include kinematic viscosity, diameters, absolute pipe roughness, and gravitational acceleration.

10.2 HYDRAULIC AND ENERGY GRADE LINES

The concepts of *hydraulic* and *energy* grade lines are useful in analyzing more complex flow problems. If, at each point along a pipe system, the term p/γ

---

[1] See p. 302.

is determined and plotted as a vertical distance above the center of the pipe, the locus of end points is the hydraulic grade line. More generally, the plot of the two terms

$$\frac{p}{\gamma} + z$$

for the flow, as ordinates, against length along the pipe as abscissas, produces the hydraulic grade line. The hydraulic grade line is the locus of heights to which liquid would rise in vertical glass tubes connected to piezometer openings in the line. When the pressure in the line is less than atmospheric, p/γ is negative and the hydraulic grade line is below the pipeline.

The energy grade line is a line joining a series of points marking the available energy in foot-pounds per pound for each point along the pipe as ordinate, plotted against distance along the pipe as the abscissa. It consists of the plot of

$$\frac{V^2}{2g} + \frac{p}{\gamma} + z$$

for each point along the line. By definition, the energy grade line is always vertically above the hydraulic grade line a distance of $V^2/2g$, neglecting the kinetic-energy correction factor.

The hydraulic and energy grade lines are shown in Fig. 10.2 for a simple pipeline containing a square-edged entrance, a valve, and a nozzle at the end of the line. To construct these lines when the reservoir surface is given, it is necessary first to apply the energy equation from the reservoir to the exit, including all minor losses as well as pipe friction, and to solve for the velocity head $V^2/2g$. Then, to find the elevation of hydraulic grade line at any point,

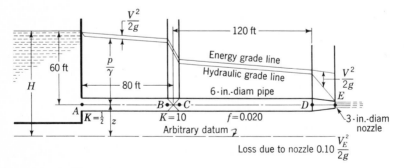

Fig. 10.2 Hydraulic and energy grade lines.

the energy equation is applied from the reservoir to that point, including all losses between the two points. The equation is solved for $p/\gamma + z$, which is plotted above the arbitrary datum. To find the energy grade line at the same point the equation is solved for $V^2/2g + p/\gamma + z$, which is plotted above the arbitrary datum.

The reservoir surface is the hydraulic grade line and is also the energy grade line. At the square-edged entrance the energy grade line drops by $0.5V^2/2g$ because of the loss there, and the hydraulic grade line drops $1.5V^2/2g$. This is made obvious by applying the energy equation between the reservoir surface and a point just downstream from the pipe entrance:

$$H + 0 + 0 = \frac{V^2}{2g} + z + \frac{p}{\gamma} + 0.5\frac{V^2}{2g}$$

Solving for $z + p/\gamma$,

$$z + \frac{p}{\gamma} = H - 1.5\frac{V^2}{2g}$$

shows the drop of $1.5V^2/2g$. The head loss due to the sudden entrance does not actually occur at the entrance itself, but over a distance of 10 or more diameters of pipe downstream. It is customary to show it at the fitting.

EXAMPLE 10.2 Determine the elevation of hydraulic and energy grade lines at points A, B, C, D, and E of Fig. 10.2.

If the arbitrary datum is selected as centerline of the pipe, both grade lines start at elevation 60 ft. First, solving for the velocity head is accomplished by applying the energy equation from the reservoir to E,

$$60 + 0 + 0 = \frac{V_E^2}{2g} + 0 + 0 + \frac{1}{2}\frac{V^2}{2g} + 0.020\frac{200}{0.50}\frac{V^2}{2g} + 10\frac{V^2}{2g} + 0.10\frac{V_E^2}{2g}$$

From the continuity equation, $V_E = 4V$. After simplifying,

$$60 = \frac{V^2}{2g}(16 + \tfrac{1}{2} + 8 + 10 + 16 \times 0.1) = 36.1\frac{V^2}{2g}$$

and $V^2/2g = 1.66$ ft. Applying the energy equation for the portion from the reservoir to A gives

$$60 + 0 + 0 = \frac{V^2}{2g} + \frac{p}{\gamma} + z + 0.5\frac{V^2}{2g}$$

Hence the hydraulic grade line at A is

$$\frac{p}{\gamma} + z \bigg|_A = 60 - 1.5 \frac{V^2}{2g} = 60 - 1.5 \times 1.66 = 57.51 \text{ ft}$$

The energy grade line for A is

$$\frac{V^2}{2g} + z + \frac{p}{\gamma} = 57.51 + 1.66 = 59.17 \text{ ft}$$

For B,

$$60 + 0 + 0 = \frac{V^2}{2g} + \frac{p}{\gamma} + z + 0.5\frac{V^2}{2g} + 0.02 \times \frac{80}{0.5}\frac{V^2}{2g}$$

and

$$\frac{p}{\gamma} + z \bigg|_B = 60 - (1.5 + 3.2)1.66 = 52.19 \text{ ft}$$

the energy grade line is at $52.19 + 1.66 = 53.85$ ft.

Across the valve the hydraulic grade line drops by $10V^2/2g$, or 16.6 ft. Hence, at C the energy and hydraulic grade lines are at 37.25 ft and 35.59 ft, respectively.

At point D,

$$60 = \frac{V^2}{2g} + \frac{p}{\gamma} + z + \left(10.5 + 0.02 \times \frac{200}{0.50}\right)\frac{V^2}{2g}$$

and

$$\frac{p}{\gamma} + z \bigg|_D = 60 - 19.5 \times 1.66 = 27.6 \text{ ft}$$

with the energy grade line at $27.6 + 1.66 = 29.96$ ft.

At point E the hydraulic grade line is at zero elevation, and the energy grade line is

$$\frac{V_E^2}{2g} = 16\frac{V^2}{2g} = 16 \times 1.66 = 26.6 \text{ ft}$$

The *hydraulic gradient* is the slope of the hydraulic grade line if the conduit is horizontal; otherwise, it is

$$\frac{d(z + p/\gamma)}{dL}$$

The *energy gradient* is the slope of the energy grade line if the conduit is horizontal; otherwise, it is

$$\frac{d(z + p/\gamma + V^2/2g)}{dL}$$

In many situations involving long pipelines the minor losses may be neglected (when less than 5 percent of the pipe friction losses) or they may be included as equivalent lengths of pipe which are added to actual length in solving the problem. For these situations the value of the velocity head $V^2/2g$ is small compared with $f(L/D)V^2/2g$ and is neglected.

In this special but very common case, when minor effects are neglected, the energy and hydraulic grade lines are superposed. The single grade line, shown in Fig. 10.3, is commonly referred to as the *hydraulic grade line*. No change in hydraulic grade line is shown for minor losses. For these situations with long pipelines the hydraulic gradient becomes h_f/L, with h_f given by the Darcy-Weisbach equation

$$h_f = f\frac{L}{D}\frac{V^2}{2g} \tag{10.2.1}$$

or by Eq. (10.1.1). Flow (except through a pump) is always in the direction of decreasing energy grade line.

Pumps add energy to the flow, a fact which may be expressed in the energy equation either by including a *negative loss* or by stating the energy per unit weight added as a positive term on the upstream side of the equation.

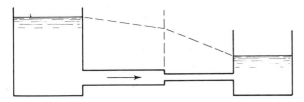

Fig. 10.3 Hydraulic grade line for long pipelines where minor losses are neglected or included as equivalent lengths of pipe.

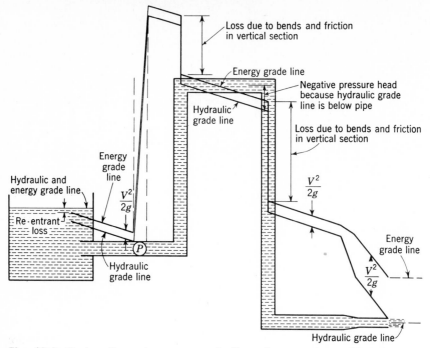

Fig. 10.4 Hydraulic and energy grade lines for a system with pump and siphon.

The hydraulic grade line rises sharply at a pump. Figure 10.4 shows the hydraulic and energy grade lines for a system with a pump and a siphon. The true slope of the grade lines can be shown only for horizontal lines.

EXAMPLE 10.3 A pump with a shaft input of 7.5 kW and an efficiency of 70 percent is connected in a waterline carrying 0.1 m³/s. The pump has a 15-cm-diameter suction line and a 12-cm-diameter discharge line. The suction line enters the pump 1 m below the discharge line. For a suction pressure of 70 kN/m², calculate the pressure at the discharge flange and the rise in the hydraulic grade line across the pump.

If the energy added in meter-newtons per newton is symbolized by E, the fluid power added is

$$Q\gamma E = 7500 \times 0.70 \qquad \text{or} \qquad E = \frac{7500 \times 0.7}{0.1 \times 9802} = 5.356 \text{ m}$$

Applying the energy equation from suction flange to discharge flange gives

$$\frac{V_s^2}{2g} + \frac{p_s}{\gamma} + 0 + 5.356 = \frac{V_d^2}{2g} + \frac{p_d}{\gamma} + 1$$

in which the subscripts s and d refer to the suction and discharge conditions, respectively. From the continuity equation

$$V_s = \frac{0.1 \times 4}{0.15^2\pi} = 5.66 \text{ m/s} \qquad V_d = \frac{0.1 \times 4}{0.12^2\pi} = 8.84 \text{ m/s}$$

Solving for p_d gives

$$\frac{p_d}{\gamma} = \frac{5.66^2}{2 \times 9.806} + \frac{70,000}{9802} + 5.356 - \frac{8.84^2}{2 \times 9.806} - 1 = 9.145 \text{ m}$$

and $p_d = 89.6 \text{ kN/m}^2$. The rise in hydraulic grade line is

$$\left(\frac{p_d}{\gamma} + 1\right) - \frac{p_s}{\gamma} = 9.145 + 1 - \frac{70,000}{9802} = 3.004 \text{ m}$$

In this example much of the energy was added in the form of kinetic energy, and the hydraulic grade line rises only 3.004 m for a rise of energy grade line of 5.356 m.

A turbine takes energy from the flow and causes a sharp drop in both the energy and the hydraulic grade lines. The energy removed per unit weight of fluid may be treated as a loss in computing grade lines.

10.3 THE SIPHON

A closed conduit, arranged as in Fig. 10.5, which lifts the liquid to an elevation higher than its free surface and then discharges it at a lower elevation is a *siphon*. It has certain limitations in its performance due to the low pressures that occur near the summit s.

Assuming that the siphon flows full, with a continuous liquid column throughout the siphon, the application of the energy equation for the portion from 1 to 2 produces the equation

$$H = \frac{V^2}{2g} + K\frac{V^2}{2g} + f\frac{L}{D}\frac{V^2}{2g}$$

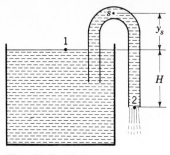

Fig. 10.5 Siphon.

in which K is the sum of all the minor-loss coefficients. Factoring out the velocity head gives

$$H = \frac{V^2}{2g}\left(1 + K + \frac{fL}{D}\right) \tag{10.3.1}$$

which is solved in the same fashion as the simple pipe problems of the first or second type. With the discharge known, the solution for H is straightforward, but the solution for velocity with H given is a trial solution started by assuming an f.

The pressure at the summit s is found by applying the energy equation for the portion between 1 and s after Eq. (10.3.1) is solved. It is

$$0 = \frac{V^2}{2g} + \frac{p_s}{\gamma} + y_s + K'\frac{V^2}{2g} + f\frac{L'}{D}\frac{V^2}{2g}$$

in which K' is the sum of the minor-loss coefficients between the two points and L' is the length of conduit upstream from s. Solving for the pressure gives

$$\frac{p_s}{\gamma} = -y_s - \frac{V^2}{2g}\left(1 + K' + \frac{fL'}{D}\right) \tag{10.3.2}$$

which shows that the pressure is negative and that it decreases with y_s and $V^2/2g$. If the solution of the equation should be a value of p_s/γ equal to or less than the vapor pressure[1] of the liquid, then Eq. (10.3.1) is not valid be-

---

[1] A liquid boils when its pressure is reduced to its vapor pressure. The vapor pressure is a function of temperature for a particular liquid. Water has a vapor pressure of 0.203 ft (0.0619 m) H$_2$O abs at 32°F (0°C), and 33.91 ft (10.33 m) H$_2$O abs at 212°F (100°C). See Appendix C.

cause the vaporization of portions of the fluid column invalidates the incompressibility assumption used in deriving the energy equation.

Although Eq. (10.3.1) is not valid for this case, theoretically there will be a discharge so long as y_s plus the vapor pressure is less than local atmospheric pressure expressed in length of the fluid column. When Eq. (10.3.2) yields a pressure less than vapor pressure at s, the pressure at s may be taken as vapor pressure. Then, with this pressure known, Eq. (10.3.2) is solved for $V^2/2g$, and the discharge is obtained therefrom. It is assumed that air does not enter the siphon at 2 and break at s the vacuum that produces the flow.

Practically a siphon does not work satisfactorily when the pressure intensity at the summit is close to vapor pressure. Air and other gases come out of solution at the low pressures and collect at the summit, thus reducing the length of the right-hand column of liquid that produces the low pressure at the summit. Large siphons that operate continuously have vacuum pumps to remove the gases at the summits.

The lowest pressure may not occur at the summit but somewhere downstream from that point, because friction and minor losses may reduce the pressure more than the decrease in elevation increases pressure.

EXAMPLE 10.4 Neglecting minor losses and considering the length of pipe equal to its horizontal distance, determine the point of minimum pressure in the siphon of Fig. 10.6.

When minor losses are neglected, the kinetic-energy term $V^2/2g$ is usually neglected also. Then the hydraulic grade line is a straight line connecting the two liquid surfaces. Coordinates of two points on the line are

$$x = -40 \text{ m} \qquad y = 4 \text{ m} \qquad \text{and} \qquad x = 56.57 \text{ m} \qquad y = 8 \text{ m}$$

The equation of the line is, by substitution into $y = mx + b$,

$$y = 0.0414x + 5.656 \text{ m}$$

The minimum pressure occurs where the distance between hydraulic grade

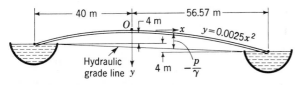

Fig. 10.6 Siphon connecting two reservoirs.

line and pipe is a maximum,

$$\frac{p}{\gamma} = 0.0025x^2 - 0.0414x - 5.656$$

To find minimum p/γ, set $d(p/\gamma)/dx = 0$, which yields $x = 8.28$, and $p/\gamma = -5.827$ m of fluid flowing. The minimum point occurs where the slopes of the pipe and of the hydraulic grade line are equal.

10.4 PIPES IN SERIES

When two pipes of different sizes or roughnesses are connected so that fluid flows through one pipe and then through the other, they are said to be connected in series. A typical series-pipe problem, in which the head H may be desired for a given discharge or the discharge wanted for a given H, is illustrated in Fig. 10.7. Applying the energy equation from A to B, including all losses, gives

$$H + 0 + 0 = 0 + 0 + 0 + K_e \frac{V_1^2}{2g} + f_1 \frac{L_1}{D_1} \frac{V_1^2}{2g} + \frac{(V_1 - V_2)^2}{2g}$$

$$+ f_2 \frac{L_2}{D_2} \frac{V_2^2}{2g} + \frac{V_2^2}{2g}$$

in which the subscripts refer to the two pipes. The last item is the head loss at exit from pipe 2. With the continuity equation

$$V_1 D_1^2 = V_2 D_2^2$$

V_2 is eliminated from the equations, so that

$$H = \frac{V_1^2}{2g} \left\{ K_e + \frac{f_1 L_1}{D_1} + \left[1 - \left(\frac{D_1}{D_2}\right)^2 \right]^2 + \frac{f_2 L_2}{D_2} \left(\frac{D_1}{D_2}\right)^4 + \left(\frac{D_1}{D_2}\right)^4 \right\}$$

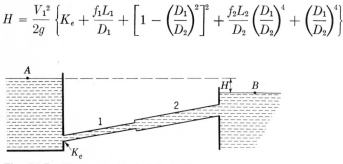

Fig. 10.7 Pipes connected in series.

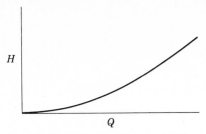

Fig. 10.8 Plot of calculated H for selected values of Q.

For known lengths and sizes of pipes this reduces to

$$H = \frac{V_1^2}{2g} (C_1 + C_2 f_1 + C_3 f_2) \tag{10.4.1}$$

in which C_1, C_2, C_3 are known. With the discharge given, the Reynolds number is readily computed, and the f's may be looked up in the Moody diagram. Then H is found by direct substitution. With H given, V_1, f_1, f_2 are unknowns in Eq. (10.4.1). By assuming values of f_1 and f_2 (they may be assumed equal), a trial V_1 is found from which trial Reynolds numbers are determined and values of f_1, f_2 looked up. With these new values, a better V_1 is computed from Eq. (10.4.1). Since f varies so slightly with the Reynolds number, the trial solution converges very rapidly. The same procedures apply for more than two pipes in series.

In place of the assumption of f_1 and f_2 when H is given, a graphical solution may be utilized in which several values of Q are assumed in turn, and the corresponding values of H are calculated and plotted against Q, as in Fig. 10.8. By connecting the points with a smooth curve, it is easy to read off the proper Q for the given value of H.

EXAMPLE 10.5 In Fig. 10.7, $K_e = 0.5$, $L_1 = 1000$ ft, $D_1 = 2$ ft, $\epsilon_1 = 0.005$ ft, $L_2 = 800$ ft, $D_2 = 3$ ft, $\epsilon_2 = 0.001$ ft, $\nu = 0.00001$ ft^2/s, and $H = 20$ ft. Determine the discharge through the system.

From the energy equation,

$$20 = \frac{V_1^2}{2g} \{0.5 + f_1(\tfrac{1000}{2}) + [1 - (\tfrac{2}{3})^2]^2 + f_2(\tfrac{800}{3})(\tfrac{2}{3})^4 + (\tfrac{2}{3})^4\}$$

After simplifying,

$$20 = \frac{V_1^2}{2g} (1.01 + 500f_1 + 52.6f_2)$$

From $\epsilon_1/D_1 = 0.0025$, $\epsilon_2/D_2 = 0.00033$, and Fig. 5.32 values of f's are assumed for the complete turbulence range,

$$f_1 = 0.025 \qquad f_2 = 0.015$$

By solving for V_1, with these values, $V_1 = 9.49$ ft/s, $V_2 = 4.21$ ft/s,

$$\mathbf{R}_1 = \frac{9.49 \times 2}{0.00001} = 1,898,000 \qquad \mathbf{R}_2 = \frac{4.21 \times 3}{0.00001} = 1,263,000$$

and from Fig. 5.32, $f_1 = 0.025$, $f_2 = 0.016$. By solving for V_1 again, $V_1 = 9.46$, and $Q = 9.46\pi = 29.8$ cfs.

Equivalent pipes

Series pipes can be solved by the method of equivalent lengths. Two pipe systems are said to be equivalent when the same head loss produces the same discharge in both systems. From Eq. (10.2.1)

$$h_{f_1} = f_1 \frac{L_1}{D_1} \frac{Q_1^2}{(D_1^2\pi/4)^2 2g} = \frac{f_1 L_1}{D_1^5} \frac{8Q_1^2}{\pi^2 g}$$

and for a second pipe

$$h_{f_2} = \frac{f_2 L_2}{D_2^5} \frac{8Q_2^2}{\pi^2 g}$$

For the two pipes to be equivalent,

$$h_{f_1} = h_{f_2} \qquad Q_1 = Q_2$$

After equating $h_{f_1} = h_{f_2}$ and simplifying,

$$\frac{f_1 L_1}{D_1^5} = \frac{f_2 L_2}{D_2^5}$$

Solving for L_2 gives

$$L_2 = L_1 \frac{f_1}{f_2} \left(\frac{D_2}{D_1}\right)^5 \tag{10.4.2}$$

which determines the length of a second pipe to be equivalent to that of the first pipe. For example, to replace 300 m of 25-cm pipe with an equivalent length of 15-cm pipe, the values of f_1 and f_2 must be approximated by selecting a discharge within the range intended for the pipes. Say $f_1 = 0.020$, $f_2 = 0.018$; then

$$L_2 = 300 \frac{0.020}{0.018} \left(\frac{15}{25}\right)^5 = 25.9 \text{ m}$$

For these assumed conditions 25.9 m of 15-cm pipe is equivalent to 300 m of 25-cm pipe.

Hypothetically two or more pipes composing a system may also be replaced by a pipe which has the same discharge for the same overall head loss.

EXAMPLE 10.6 Solve Example 10.5 by means of equivalent pipes.

First, by expressing the minor losses in terms of equivalent lengths, for pipe 1,

$$K_1 = 0.5 + [1 - (\tfrac{2}{3})^2]^2 = 0.809 \qquad L_{e1} = \frac{K_1 D_1}{f_1} = \frac{0.809 \times 2}{0.025} = 65 \text{ ft}$$

and for pipe 2,

$$K_2 = 1 \qquad L_{e2} = \frac{K_2 D_2}{f_2} = \frac{1 \times 3}{0.015} = 200 \text{ ft}$$

The values of f_1, f_2 are selected for the fully turbulent range as an approximation. The problem is now reduced to 1065 ft of 2-ft pipe and 1000 ft of 3-ft pipe. By expressing the 3-ft pipe in terms of an equivalent length of 2-ft pipe, by Eq. (10.4.2)

$$L_e = 1000 \times \frac{0.015}{0.025} \left(\frac{2}{3}\right)^5 = 79 \text{ ft}$$

By adding to the 2-ft pipe, the problem is reduced to the simple pipe problem of finding the discharge through $1065 + 79 = 1144$ ft of 2-ft diameter pipe,

$\epsilon = 0.005$ ft, for a head loss of 20 ft,

$$20 = f \frac{1144}{2} \frac{V^2}{2g}$$

With $f = 0.025$, $V = 9.5$ ft/s and $\mathbf{R} = 9.5 \times 2/0.00001 = 1,900,000$. For $\epsilon/D = 0.0025$, $f = 0.025$ and $Q = 9.5\pi = 29.9$ cfs.

10.5 PIPES IN PARALLEL

A combination of two or more pipes connected as in Fig. 10.9, so that the flow is divided among the pipes and then is joined again, is a *parallel-pipe* system. In series pipes the same fluid flows through all the pipes, and the head losses are cumulative; however, in parallel pipes the head losses are the same in any of the lines, and the discharges are cumulative.

In analyzing parallel-pipe systems, it is assumed that the minor losses are added into the lengths of each pipe as equivalent lengths. From Fig. 10.9 the conditions to be satisfied are

$$h_{f_1} = h_{f_2} = h_{f_3} = \frac{p_A}{\gamma} + z_A - \left(\frac{p_B}{\gamma} + z_B\right) \tag{10.5.1}$$

$$Q = Q_1 + Q_2 + Q_3$$

in which z_A, z_B are elevations of points A and B, and Q is the discharge through the approach pipe or the exit pipe.

Two types of problems occur: (1) with elevation of hydraulic grade line at A and B known, to find the discharge Q; (2) with Q known, to find the distribution of flow and the head loss. Sizes of pipe, fluid properties, and roughnesses are assumed to be known.

The first type is, in effect, the solution of simple pipe problems for dis-

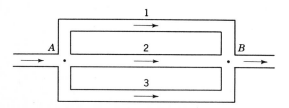

Fig. 10.9 Parallel-pipe system.

charge since the head loss is the drop in hydraulic grade line. These discharges are added to determine the total discharge.

The second type of problem is more complex, as neither the head loss nor the discharge for any one pipe is known. The recommended procedure is as follows:

1. Assume a discharge Q_1' through pipe 1.
2. Solve for h_{f1}', using the assumed discharge.
3. Using h_{f1}', find Q_2', Q_3'.
4. With the three discharges for a common head loss, now assume that the given Q is split up among the pipes in the same proportion as Q_1', Q_2', Q_3'; thus

$$Q_1 = \frac{Q_1'}{\Sigma Q'} Q \qquad Q_2 = \frac{Q_2'}{\Sigma Q'} Q \qquad Q_3 = \frac{Q_3'}{\Sigma Q'} Q \qquad (10.5.2)$$

5. Check the correctness of these discharges by computing h_{f1}, h_{f2}, h_{f3} for the computed Q_1, Q_2, Q_3.

This procedure works for any number of pipes. By judicious choice of Q_1', obtained by estimating the percent of the total flow through the system that should pass through pipe 1 (based on diameter, length, and roughness), Eq. (10.5.2) produces values that check within a few percent, which is well within the range of accuracy of the friction factors.

EXAMPLE 10.7   In Fig. 10.9, $L_1 = 3000$ ft, $D_1 = 1$ ft, $\epsilon_1 = 0.001$ ft; $L_2 = 2000$ ft, $D_2 = 8$ in, $\epsilon_2 = 0.0001$ ft; $L_3 = 4000$ ft, $D_3 = 16$ in, $\epsilon_3 = 0.0008$ ft; $\rho = 2.00$ slugs/ft^3, $\nu = 0.00003$ ft^2/s, $p_A = 80$ psi, $z_A = 100$ ft, $z_B = 80$ ft. For a total flow of 12 cfs, determine flow through each pipe and the pressure at B.

Assume $Q_1' = 3$ cfs; then $V_1' = 3.82$, $\mathbf{R}_1' = 3.82 \times 1/0.00003 = 127,000$, $\epsilon_1/D_1 = 0.001$, $f_1' = 0.022$, and

$$h_{f1}' = 0.022 \times \frac{3000}{1.0} \frac{3.82^2}{64.4} = 14.97 \text{ ft}$$

For pipe 2

$$14.97 = f_2' \frac{2000}{0.667} \frac{V_2'^2}{2g}$$

Then $\epsilon_2/D_2 = 0.00015$. Assume $f_2' = 0.020$; then $V_2' = 4.01$ ft/s, $\mathbf{R}_2' = 4.01 \times \frac{2}{3} \times 1/0.00003 = 89,000$, $f_2' = 0.019$, $V_2' = 4.11$ ft/s, $Q_2' = 1.44$ cfs.

For pipe 3

$$14.97 = f_3' \frac{4000}{1.333} \frac{V_3'^2}{2g}$$

Then $\epsilon_3/D_3 = 0.0006$. Assume $f_3' = 0.020$; then $V_3' = 4.01$ ft/s, $\mathbf{R}_3' = 4.01 \times 1.333/0.00003 = 178,000$, $f_3' = 0.020$, $Q_3' = 5.60$ cfs.
The total discharge for the assumed conditions is

$$\Sigma Q' = 3.00 + 1.44 + 5.60 = 10.04 \text{ cfs}$$

Hence

$$Q_1 = \frac{3.00}{10.04} \times 12 = 3.58 \text{ cfs} \qquad Q_2 = \frac{1.44}{10.04} \times 12 = 1.72 \text{ cfs}$$

$$Q_3 = \frac{5.60}{10.04} \times 12 = 6.70 \text{ cfs}$$

Check the values of h_1, h_2, h_3:

$$V_1 = \frac{3.58}{\pi/4} = 4.46 \qquad \mathbf{R}_1 = 152,000 \qquad f_1 = 0.021 \qquad h_{f_1} = 20.4 \text{ ft}$$

$$V_2 = \frac{1.72}{\pi/9} = 4.93 \qquad \mathbf{R}_2 = 109,200 \qquad f_2 = 0.019 \qquad h_{f_2} = 21.6 \text{ ft}$$

$$V_3 = \frac{6.70}{4\pi/9} = 4.80 \qquad \mathbf{R}_3 = 213,000 \qquad f_3 = 0.019 \qquad h_{f_3} = 20.4 \text{ ft}$$

f_2 is about midway between 0.018 and 0.019. If 0.018 had been selected, h_2 would be 20.4 ft.
To find p_B,

$$\frac{p_A}{\gamma} + z_A = \frac{p_B}{\gamma} + z_B + h_f$$

or

$$\frac{p_B}{\gamma} = \frac{80 \times 144}{62.4} + 100 - 80 - 20.8 = 183.5$$

in which the average head loss was taken. Then

$$p_B = \frac{183.5 \times 2 \times 32.2}{144} = 81.8 \text{ psi}$$

10.6 BRANCHING PIPES

A simple *branching-pipe* system is shown in Fig. 10.10. In this situation the flow through each pipe is wanted when the reservoir elevations are given. The sizes and types of pipes and fluid properties are assumed known. The Darcy-Weisbach equation must be satisfied for each pipe, and the continuity equation must be satisfied. It takes the form that the flow into the junction J must just equal the flow out of the junction. Flow must be out of the highest reservoir and into the lowest; hence, the continuity equation may be either

$$Q_1 = Q_2 + Q_3 \quad \text{or} \quad Q_1 + Q_2 = Q_3$$

If the elevation of hydraulic grade line at the junction is above the elevation of the intermediate reservoir, flow is *into* it; but if the elevation of hydraulic grade line at J is below the intermediate reservoir, the flow is *out* of it. Minor losses may be expressed as equivalent lengths and added to the actual lengths of pipe.

The solution is effected by assuming an elevation of hydraulic grade line at the junction, then computing Q_1, Q_2, Q_3 and substituting into the continuity equation. If the flow into the junction is too great, a higher grade-line elevation, which will reduce the inflow and increase the outflow, is assumed.

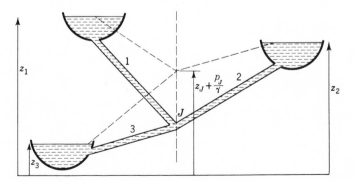

Fig. 10.10 Three interconnected reservoirs.

EXAMPLE 10.8 In Fig. 10.10, find the discharges for water at 20°C and with the following pipe data and reservoir elevations: $L_1 = 3000$ m, $D_1 = 1$ m, $\epsilon_1/D_1 = 0.0002$; $L_2 = 600$ m, $D_2 = 0.45$ m, $\epsilon_2/D_2 = 0.002$; $L_3 = 1000$ m, $D_3 = 0.6$ m, $\epsilon_3/D_3 = 0.001$; $z_1 = 30$ m, $z_2 = 18$ m, $z_3 = 9$ m.

Assume $z_J + p_J/\gamma = 23$ m. Then

$$7 = f_1 \frac{3000}{1} \frac{V_1^2}{2g} \qquad f_1 = 0.014 \qquad V_1 = 1.75 \text{ m/s} \qquad Q_1 = 1.380 \text{ m}^3/\text{s}$$

$$5 = f_2 \frac{600}{0.45} \frac{V_2^2}{2g} \qquad f_2 = 0.024 \qquad V_2 = 1.75 \text{ m/s} \qquad Q_2 = 0.278 \text{ m}^3/\text{s}$$

$$14 = f_3 \frac{1000}{0.60} \frac{V_3^2}{2g} \qquad f_3 = 0.020 \qquad V_3 = 2.87 \text{ m/s} \qquad Q_3 = 0.811 \text{ m}^3/\text{s}$$

so that the inflow is greater than the outflow by

$$1.380 - 0.278 - 0.811 = 0.291 \text{ m}^3/\text{s}$$

Assume $z_J + p_J/\gamma = 24.6$ m. Then

$$5.4 = f_1 \frac{3000}{1} \frac{V_1^2}{2g} \qquad f_1 = 0.015 \qquad V_1 = 1.534 \text{ m/s} \qquad Q_1 = 1.205 \text{ m}^3/\text{s}$$

$$6.6 = f_2 \frac{600}{0.45} \frac{V_2^2}{2g} \qquad f_2 = 0.024 \qquad V_2 = 2.011 \text{ m/s} \qquad Q_2 = 0.320 \text{ m}^3/\text{s}$$

$$15.6 = f_3 \frac{1000}{0.60} \frac{V_3^2}{2g} \qquad f_3 = 0.020 \qquad V_3 = 3.029 \text{ m/s} \qquad Q_3 = 0.856 \text{ m}^3/\text{s}$$

The inflow is still greater by 0.029 m³/s. By extrapolating linearly, $z_J + p_J/\gamma = 24.8$ m, $Q_1 = 1.183$, $Q_2 = 0.325$, $Q_3 = 0.862$ m³/s.

In pumping from one reservoir to two or more other reservoirs, as in Fig. 10.11, the characteristics of the pump must be known. Assuming that the pump runs at constant speed, its head depends upon the discharge. A suitable procedure is as follows:

1. Assume a discharge through the pump.
2. Compute the hydraulic-grade-line elevation at the suction side of the pump.

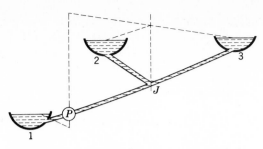

Fig. 10.11 Pumping from one reservoir to two other reservoirs.

3. From the pump characteristic curve find the head produced and add it to suction hydraulic grade line.
4. Compute drop in hydraulic grade line to the junction J and determine elevation of hydraulic grade line there.
5. For this elevation, compute flow into reservoirs 2 and 3.
6. If flow into J equals flow out of J, the problem is solved. If flow into J is too great, assume less flow through the pump and repeat the procedure.

This procedure is easily plotted on a graph, so that the intersection of two elevations vs. flow curves yields the answer.

   More complex branching-pipe problems may be solved with a similar approach by beginning with a trial solution. However, the network-analysis procedure in the next section is recommended for multibranch systems as well as for multi-parallel-loop systems. Such problems are most easily handled with a digital computer.

10.7 NETWORKS OF PIPES

Interconnected pipes through which the flow to a given outlet may come from several circuits are called a *network of pipes*, in many ways analogous to flow through electric networks. Problems on these in general are complicated and require trial solutions in which the elementary circuits are balanced in turn until all conditions for the flow are satisfied.

   The following conditions must be satisfied in a network of pipes:

1. The algebraic sum of the pressure drops around each circuit must be zero.
2. Flow into each junction must equal flow out of the junction.
3. The Darcy-Weisbach equation, or equivalent exponential friction formula,

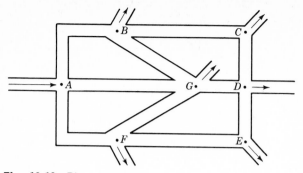

Fig. 10.12 Pipe network.

must be satisfied for each pipe; i.e., the proper relation between head loss and discharge must be maintained for each pipe.

The first condition states that the pressure drop between any two points in the circuit, for example, A and G (Fig. 10.12), must be the same whether through the pipe AG or through $AFEDG$. The second condition is the continuity equation.

Since it is impractical to solve network problems analytically, methods of successive approximations are utilized. The Hardy Cross method[1] is one in which flows are assumed for each pipe so that continuity is satisfied at every junction. A correction to the flow in each circuit is then computed in turn and applied to bring the circuits into closer balance.

Minor losses are included as equivalent lengths in each pipe. Exponential equations are commonly used, in the form $h_f = rQ^n$, where $r = RL/D^m$ in Eq. (10.1.1). The value of r is a constant in each pipeline (unless the Darcy-Weisbach equation is used) and is determined in advance of the loop-balancing procedure. The corrective term is obtained as follows.

For any pipe in which Q_0 is an assumed initial discharge

$$Q = Q_0 + \Delta Q \tag{10.7.1}$$

where Q is the correct discharge and ΔQ is the correction. Then for each pipe,

$$h_f = rQ^n = r(Q_0 + \Delta Q)^n = r(Q_0^n + nQ_0^{n-1}\Delta Q + \cdots)$$

If ΔQ is small compared with Q_0, all terms of the series after the second may

[1] Hardy Cross, Analysis of Flow in Networks of Conduits or Conductors, *Univ. Ill. Bull.* 286, November 1936.

be dropped. Now for a circuit,

$$\Sigma h_f = \Sigma r Q \,|\, Q \,|^{n-1} = \Sigma r Q_0 \,|\, Q_0 \,|^{n-1} + \Delta Q \,\Sigma r n \,|\, Q_0 \,|^{n-1} = 0$$

in which ΔQ has been taken out of the summation as it is the same for all pipes in the circuit and absolute-value signs have been added to account for the direction of summation around the circuit. The last equation is solved for ΔQ in each circuit in the network

$$\Delta Q = -\frac{\Sigma r Q_0 \,|\, Q_0 \,|^{n-1}}{\Sigma r n \,|\, Q_0 \,|^{n-1}} \tag{10.7.2}$$

When ΔQ is applied to each pipe in a circuit in accordance with Eq. (10.7.1), the directional sense is important; i.e., it adds to flows in the clockwise direction and subtracts from flows in the counterclockwise direction.

Steps in an arithmetic procedure may be itemized as follows:

1. Assume the best distribution of flows that satisfies continuity by careful examination of the network.
2. Compute the head loss $h_f = rQ_0{}^n$ in each pipe. Compute the net head loss around each elementary circuit: $\Sigma h_f = \Sigma r Q_0 \,|\, Q_0 \,|^{n-1}$ (should be zero for a balanced circuit).
3. Compute for each circuit $\Sigma n r \,|\, Q_0 \,|^{n-1}$.
4. Evaluate the corrective flow ΔQ in each circuit by use of Eq. (10.7.2).
5. Compute the revised flows in each pipe by use of Eq. (10.7.1).
6. Repeat the procedure, beginning with the revised flows, until the desired accuracy is obtained.

The values of r occur in both numerator and denominator; hence, values proportional to the actual r may be used to find the distribution. Similarly, the apportionment of flows may be expressed as a percent of the actual flows. To find a particular head loss, the actual values of r and Q must be used after the distribution has been determined.

EXAMPLE 10.9 The distribution of flow through the network of Fig. 10.13 is desired for the inflows and outflows as given. For simplicity n has been given the value 2.0.

The assumed distribution is shown in diagram *a*. At the upper left the term $\Sigma r Q_0 \,|\, Q_0 \,|^{n-1}$ is computed for the lower circuit number 1. Next to the diagram on the left is the computation of $\Sigma n r \,|\, Q_0 \,|^{n-1}$ for the same circuit. The same format is used for the second circuit in the upper right of the figure.

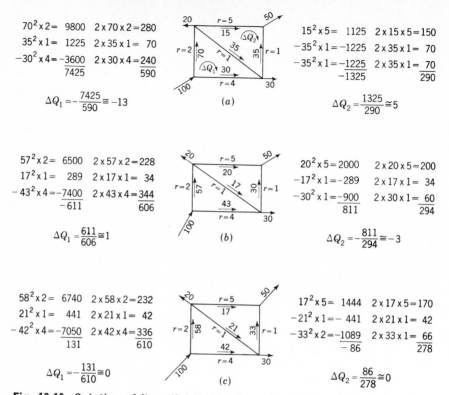

$$70^2 \times 2 = 9800 \quad 2 \times 70 \times 2 = 280$$
$$35^2 \times 1 = 1225 \quad 2 \times 35 \times 1 = 70$$
$$-30^2 \times 4 = -3600 \quad 2 \times 30 \times 4 = 240$$
$$\overline{7425} \qquad \overline{590}$$

$$\Delta Q_1 = -\frac{7425}{590} \cong -13$$

$$15^2 \times 5 = 1125 \quad 2 \times 15 \times 5 = 150$$
$$-35^2 \times 1 = -1225 \quad 2 \times 35 \times 1 = 70$$
$$-35^2 \times 1 = -1225 \quad 2 \times 35 \times 1 = 70$$
$$\overline{-1325} \qquad \overline{290}$$

$$\Delta Q_2 = \frac{1325}{290} \cong 5$$

(a)

$$57^2 \times 2 = 6500 \quad 2 \times 57 \times 2 = 228$$
$$17^2 \times 1 = 289 \quad 2 \times 17 \times 1 = 34$$
$$-43^2 \times 4 = -7400 \quad 2 \times 43 \times 4 = 344$$
$$\overline{-611} \qquad \overline{606}$$

$$\Delta Q_1 = \frac{611}{606} \cong 1$$

$$20^2 \times 5 = 2000 \quad 2 \times 20 \times 5 = 200$$
$$-17^2 \times 1 = -289 \quad 2 \times 17 \times 1 = 34$$
$$-30^2 \times 1 = -900 \quad 2 \times 30 \times 1 = 60$$
$$\overline{811} \qquad \overline{294}$$

$$\Delta Q_2 = -\frac{811}{294} \cong -3$$

(b)

$$58^2 \times 2 = 6740 \quad 2 \times 58 \times 2 = 232$$
$$21^2 \times 1 = 441 \quad 2 \times 21 \times 1 = 42$$
$$-42^2 \times 4 = -7050 \quad 2 \times 42 \times 4 = 336$$
$$\overline{131} \qquad \overline{610}$$

$$\Delta Q_1 = -\frac{131}{610} \cong 0$$

$$17^2 \times 5 = 1444 \quad 2 \times 17 \times 5 = 170$$
$$-21^2 \times 1 = -441 \quad 2 \times 21 \times 1 = 42$$
$$-33^2 \times 2 = -1089 \quad 2 \times 33 \times 1 = 66$$
$$\overline{-86} \qquad \overline{278}$$

$$\Delta Q_2 = \frac{86}{278} \cong 0$$

(c)

Fig. 10.13 Solution of flow distribution in a simple network.

The correction for the top horizontal pipe is determined as $15 + 5 = 20$ and for the diagonal as $35 + (-13) - 5 = 17$. Diagram b gives the distribution after both circuits have been corrected once. Diagram c shows the values correct to within about 1 percent of the distribution, which is more accurate than the exponential equations for head loss.

For networks larger than the previous example or for networks that contain multiple reservoirs, supply pumps, or booster pumps, the Hardy Cross loop-balancing method may be programmed for numerical solution on a digital computer. Such a program is provided in the next section.

A number of more general methods[1-3] are available, primarily based upon

[1] R. Epp and A. G. Fowler, Efficient Code for Steady-State Flows in Networks, *J. Hydraul. Div. ASCE*, vol. 96, no. HY1, pp. 43–56, January 1970.

[2] Uri Shamir and C. D. D. Howard, Water Distribution Systems Analysis, *J. Hydraul. Div., ASCE*, vol. 94, no. HY1, pp. 219–234, January 1968.

[3] Michael A. Stoner, A New Way to Design Natural Gas Systems, *Pipe Line Ind.*, vol. 32, no. 2, pp. 38–42, 1970.

the Hardy Cross loop-balancing or node-balancing schemes. In the more general methods the system is normally modeled with a set of simultaneous equations which are solved by the Newton-Raphson method. Some programmed solutions[2,3] are very useful as design tools since pipe sizes or roughnesses may be treated as unknowns in addition to junction pressures and flows.

10.8 COMPUTER PROGRAM FOR STEADY-STATE HYDRAULIC SYSTEMS

Hydraulic systems that contain components different from pipelines can be handled by replacing the component with an equivalent length of pipeline. When the additional component is a pump, special consideration is needed. Also, in systems that contain more than one fixed hydraulic-grade-line elevation, a special artifice must be introduced.

For systems with multiple fixed-pressure-head elevations, Fig. 10.14, *pseudo loops* are created to account for the unknown outflows and inflows at the reservoirs and to satisfy continuity conditions during balancing. A pseudo loop is created by using an imaginary pipeline that interconnects each pair of fixed pressure levels. These imaginary pipelines carry no flow but maintain a fixed drop in the hydraulic grade line equal to the difference in elevation of the reservoirs. If head drop is considered positive in an assumed positive direction in the imaginary pipe, then the correction in loop 3, Fig. 10.14, is

$$\Delta Q_3 = \frac{150 - 135 - r_4 Q_4 \, |\, Q_4 \, |^{n-1} - r_1 Q_1 \, |\, Q_1 \, |^{n-1}}{n r_4 \, |\, Q_4 \, |^{n-1} + n r_1 \, |\, Q_1 \, |^{n-1}}$$

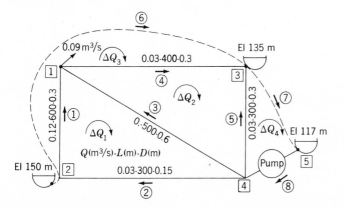

Fig. 10.14 Sample network.

This correction is applied to pipes 1 and 4 only. If additional real pipelines existed in a pseudo loop, each would be adjusted accordingly during each loop-balancing iteration.

A pump in a system may be considered as a flow element with a negative head loss equal to the head rise that corresponds to the flow through the unit. If the pump-head-discharge curve, element 8 in Fig. 10.14, were expressed by a cubic equation

$$H = A_0 + A_1 Q_8 + A_2 Q_8^2 + A_3 Q_8^3$$

where A_0 is the shutoff head of the pump, then the correction in loop 4 would be

$$\Delta Q_4 = \frac{135 - 117 - (A_0 + A_1 Q_8 + A_2 Q_8^2 + A_3 Q_8^3) + r_5 Q_5 \mid Q_5 \mid^{n-1}}{n r_5 \mid Q_5 \mid^{n-1} - (A_1 + 2 A_2 Q_8 + 3 A_3 Q_8^2)}$$

This correction is applied to pipe 5 and to pump 8 in the loop.

The FORTRAN IV program in Fig. 10.15 may be used to analyze a wide variety of liquid steady-state pipe-flow problems. The Hardy-Cross loop-balancing method is used. Pipeline flows described by the Hazen-Williams equation or laminar or turbulent flows analyzed with the Darcy-Weisbach equation can be handled; multiple reservoirs or fixed pressure levels, as in a sprinkler system, can be analyzed; and systems with booster pumps or supply pumps can be treated. Either English or SI units may be used by proper specification of input data.

A network is visualized as a combination of elements that are interconnected at junctions. These elements may include pipelines, pumps, and imaginary elements which are used to create pseudo loops in multiple-reservoir systems. All minor losses are handled by estimating equivalent lengths and adding them onto the actual pipe lengths. Each element in the system is numbered up to a maximum of 100, without duplication and not necessarily consecutively. A positive flow direction is assigned to each element, and, as in the arithmetic solution, an estimated flow is assigned to each element such that continuity is satisfied at each junction. The assigned positive flow direction in a pump must be in the intended direction of normal pump operation. Any solution with backward flow through a pump is invalid. The flow direction in the imaginary element that creates a pseudo loop indicates only the direction of fixed positive head drop, since the flow must be zero in this element. Each junction, which may represent the termination of a single element or the intersection of many elements, is numbered up to a maximum of 100, without duplication and not necessarily sequentially. An outflow or inflow at a junction is defined during the assignment of initial element flows.

```
C FARDY CROSS LOOP BALANCING INCLUDING MULTIPLE RESERVOIRS & PUMPS(PU)
C HAZEN-WILLIAMS(HW) CR CARCY-WEISBACH(DW) MAY BE USED FOR PIPES
C ENGLISH(EN) OR SI UNITS(SI) MAY BE USED, PCSITIVE DH IN ELEMENT IS FEAD DROP
      DIMENSICN ITY(4),IC(2),ITYPE(100),ELEM(500),IND(500),Q(10C),H(100)
     2,S(20),IX(24C)
      DATA ITY/'FW','DW','PS','PU'/,IE/'&&'/,ID/'EN','SI'/
 10   DO 12 J=1,50C
      IF (J.LE.1CC)ITYPE(J)=5
      IF (J.LE.100) H(J)=-1000.
      IF (J.LE.240)IX(J)=0
 12   IND(J)=C
C READ PARAMETERS FOR PROBLEM, AND ELEMENT CATA
      READ (5,15,END=99) NT,KK,TOL,VNU,DEF
 15   FORMAT (A2,I8,F10.4,F10.7,F10.5)
      IF (NT.EC.IC(2))GO TO 20
      WRITE (6,18) VNU
 18   FORMAT (' ENGLISH UNITS SPECIFIED, VISCOSITY IN FT**2/SEC=',F10.7)
      UNITS=4.727
      C=32.174
      GO TO 22
 20   WRITE (6,21) VNU
 21   FORMAT (' SI UNITS SPECIFIED, VISCOSITY IN M**2/SEC=',F10.7)
      UNITS=1C.674
      G=9.806
 22   WRITE (6,24) TCL,KK
 24   FORMAT (' CESIRED FLCW TCLERANCE=',F5.3,' NO. OF ITERATICNS=',I5//
     2' PIPE      Q(CFS CR M**3/S)  L(FT OR M)   C(FT OR M)  HW C CR EPS')
 26   READ (5,30) NT,I,CQ,X1,X2,X3,X4,X5
 30   FORMAT (A2,3X,I5,3F10.3,F10.5,2F10.3)
      IF (NT.EC.IE) GO TO 68
      Q(I)=QQ
      CO 32 NTY=1,4
      IF (NT.EQ.ITY(NTY))CC TO 33
 32   CONTINUE
 33   ITYPE(I)=NTY
      KP=5*(I-1)+1
      GO TO (41,42,53,64),NTY
 41   IF (X3.EC.0.)X3=DEF
      ELEM(KP)=UNITS*X1/(X3**1.852*X2**4.87C4)
      EX=1.852
      GO TO 43
 42   IF (X3.EQ.C.)X3=DEF
      EX=2.
      ELEM(KP)=X1/(2.*G*X2**5*.7854*.7854)
      ELEM(KP+1)=1./(.7854*X2*VNU)
      ED=X3/X2
      ELEM(KP+2)=.094*ED**.225+.53*ED
      ELEM(KP+3)=88.*ED**.44
      ELEM(KP+4)=1.62*ED**.134
 43   WRITE (6,45)I,Q(I),X1,X2,X3
 45   FORMAT (I5,F18.3,F12.1,F12.3,F14.5)
      EN=EX-1.
      GO TO 26
 53   ELEM(KP)=X1
      WRITE (6,55) I,X1
 55   FORMAT (I5,' RESERVCIR ELEV DIFFERENCE=',F10.2)
      GO TO 26
 64   ELEM(KP)=X2
      ELEM(KP+3)=(X5-3.*(X4-X3)-X2)/(6.*X1**3)
      ELEM(KP+2)=(X4-2.*X3+X2)/(2.*X1**2)-ELEM(KP+3)*3.*X1
      ELEM(KP+1)=(X3-X2)/X1-ELEM(KP+2)*X1-ELEM(KP+3)*X1*X1
      WRITE (6,66) I,X1,X2,X3,X4,X5,(ELEM(KF+J-1),J=1,4)
 66   FORMAT (I5,' PUMP CURVE, DQ=',F7.3,' H=',4F8.1/5X,
     2' COEF IN PUMP EQ=',4F11.3)
      GO TO 26
C READ LOOP INDEXING CATA, IND=NO. PIPES,PIFE,PIPE,ETC. CLOCKWISE+,CC-
 68   I1=1
 7C   I2=I1+14
      READ (5,75) NT,(INC(I),I=I1,I2)
 75   FORMAT (A2,2X,15I4)
      IF (NT.EC.IE) GO TC 78
      I1=I2+1
      GO TO 7C
 78   IF (I1.EC.1) GC TC 140
      WRITE (6,79) (IND(I),I=1,I1)
 79   FORMAT (' INC='/(15I4))
C BALANCE ALL LCCPS
      CO 130 K=1,KK
      CDQ=0.
      IP=1
```

Fig. 10.15 (Continued on pages 572 and 573)

571

```
 80   I1=IND(IP)
      IF(I1.EQ.0) GC TO 124
      DH=0.
      HDQ=0.
      DO 110 J=1,I1
      I=IND(IP+J)
      IF (I) 81,110,82
 81   S(J)=-1.
      I=-I
      GO TO 83
 82   S(J)=1.
 83   NTY=ITYPE(I)
      KP=5*(I-1)+1
      GO TO (91,92,103,104),NTY
 91   R=ELEM(KP)
      GO TO 95
 92   REY=ELEM(KP+1)*ABS(Q(I))
      IF(REY.LT.1.) REY=1.
      IF (REY-2000.) 93,94,94
 93   R=ELEM(KP)*64./REY
      GO TO 95
 94   R=ELEM(KP)*(ELEM(KP+2)+ELEM(KP+3)/REY**ELEM(KP+4))
 95   DH=DH+S(J)*R*Q(I)*AES(Q(I))**EN
     +DQ=HDQ+EN*R*AES(Q(I))**EN
      GO TO 110
 103  DH=DH+S(J)*ELEM(KP)
      GO TO 110
 104  DH=DH-S(J)*(ELEM(KP)+Q(I)*(ELEM(KP+1)+Q(I)*(ELEM(KP+2)+Q(I)*
     2ELEM(KP+3))))
     +DQ=HDQ-(ELEM(KP+1)+2.*ELEM(KP+2)*Q(I)+3.*ELEM(KP+3)*Q(I)**2)
 110  CONTINUE
      IF (ABS(HDQ).LT..0001) HDQ=1.
      DQ=-DH/HDQ
      DDQ=DDQ+ABS(DQ)
      DO 120 J=1,I1
      I=IABS(IND(IP+J))
      IF (ITYPE(I).EQ.3) GO TO 120
      Q(I)=Q(I)+S(J)*DQ
 120  CONTINUE
      IP=IP+I1+1
      GO TO 80
 124  WRITE (6,125) K,DDQ
 125  FORMAT (' ITERATION NO.',I4,' SUM OF FLOW CORRECTIONS=',F10.4)
      IF (DDQ.LT.TOL) GC TO 140
 130  CONTINUE
 140  WRITE (6,141)
 141  FORMAT (' ELEMENT    FLOW')
      CO 150 I=1,100
      NTY=ITYPE(I)
      GO TO (142,142,150,142,150),NTY
 142  WRITE (6,143) I,Q(I)
 143  FORMAT (I5,F10.3)
 150  CONTINUE
C     READ CATA FOR FGL CCMPUTATION, IX=JUNC,ELEMENT,JUNC,ELEM,JUNC,ETC.
 152  READ (5,155) NT,K,HH
 155  FORMAT (A2,I8,F10.3)
      IF (NT.EQ.IE) GO TO 160
      H(K)=HH
      GO TO 152
 160  I1=1
 162  I2=I1+14
      READ (5,75)NT,(IX(K),K=I1,I2)
      IF (NT.EQ.IE) GO TO 170
      I1=I2+1
      GO TO 162
 170  WRITE (6,171) (IX(I),I=1,I1)
 171  FORMAT ('  IX=',/(15I4))
      IP=1
 180  CO 200 J=1,238,2
      IF (J.EQ.1) I1=IX(IP)
      I=IX(IP+J)
      N=IX(IP+J+1)
      IF(I) 181,199,182
 181  SS=-1.
      I=-I
      GO TO 183
 182  SS=1.
 183  NTY=ITYPE(I)
      KP=5*(I-1)+1
      GO TO (184,185,189,190,199),NTY
```

Fig. 10.15

```
184 R=ELEM(KP)
    GO TO 188
185 REY=ELEM(KP+1)*ABS(C(I))
    IF (REY.LT.1.) REY=1.
    IF (REY-2000.) 186,187,187
186 R=ELEM(KP)*64./REY
    GO TO 188
187 R=ELEM(KP)*(ELEM(KP+2)+ELEM(KP+3)/REY**ELEM(KP+4))
188 H(N)=H(I1)-SS*R*Q(I)*ABS(Q(I))**EN
    GO TO 199
189 H(N)=H(I1)-SS*ELEM(KP)
    CO TO 199
190 H(N)=H(I1)+SS*(ELEM(KP)+Q(I)*(ELEM(KP+1)+C(I)*(ELEM(KP+2)+C(I)*
   2ELEM(KP+3))))
199 IF ((IX(J+IP+3).EC.0) GC TC 210
    IF ((IX(J+IP+2).EJ.0) GO TO 205
200 I1=N
205 IP=IP+J+3
    GO TO 180
210 WRITE (6,215)
215 FORMAT (' JUNCTICN  HEAC')
    DO 220 N=1,100
    IF (H(N).EC.-1000.) GO TO 220
    WRITE (6,143) N,H(N)
220 CONTINUE
    GO TO 10
99  STOP
    END
```

Fig. 10.15 FORTRAN program for hydraulic systems.

The operation of the program is best visualized in two major parts: the first performs the balancing of each loop in the system successively and then repeats in an iterative manner until the sum of all loop-flow corrections is less than a specified tolerance. At the end of this balancing process the element flows are computed and printed. The second part of an analysis involves the computation of the hydraulic-grade-line elevations at junctions in the system. Each of these parts requires a special indexing of the system configuration in the input data. The indexing of the system loops for balancing is placed in the vector IND. A series of integer values identifies each loop sequentially by the number of elements in the loop followed by the element number of each element in the loop. The directional sense of flow in each element is identified by a positive element number for the clockwise direction and a negative element number for counterclockwise. The second part of the program requires an identification of one or more junctions with known heads. Then a series of junction and element numbers indexes a continuous path through the system to all junctions where the hydraulic grade line is wanted. The path may be broken at any point by an integer zero followed by a new junction where the head is known. These data are stored in the vector IX by a junction number where the head is known followed by a contiguous element number and junction number. Again the positive element number is used in the assigned flow direction, and the negative element number is used when tracing a path against the assigned element-flow direction. Any continuous path may be broken by inserting a zero; then a new path is begun with a new initial junction, an element, and a node, etc. All junction hydraulic-grade-line elevations that are computed are printed.

As shown below, the type of each element is identified in the input data, and each element is identified in the program by the assignment of a unique numerical value in the vector ITYPE.

Element	Data	Program
Hazen-Williams pipeline	HW	1
Darcy-Weisbach pipeline	DW	2
Pseudo element	PS	3
Pump	PU	4

The physical data associated with each element are entered on separate cards. In the program the physical data that describe all elements in the system are stored in the vector ELEM, with five locations reserved for each element. As an example of the position of storage of element information, the data pertaining to element number 13 are located in positions 61 to 65 in ELEM.

Data preparation for the program is best visualized in four steps, as shown in Fig. 10.16 and described below. Formated input is used, as shown in Fig. 10.16 and the program.

Step 1: Parameter description card

The type of unit to be used in the analysis is defined by the characters EN for English units or SI for the SI units. An integer defines the maximum number of iterations to be allowed during the balancing scheme. An acceptable tolerance is set for the sum of the absolute values of the corrections in each loop during each iteration. The liquid kinematic viscosity must be specified if the Darcy-Weisbach equation is used for pipeline losses. If the Hazen-Williams equation is used, a default value for the coefficient C may be defined, or if the Darcy-Weisbach equation is used, a default value for absolute pipe roughness may be defined. If the default value is used on the parameter card, it need not be placed on the element cards; however, if it is, the element data override the default value.

Step 2: Element cards

Each element in the system requires a separate card. Pipeline elements require either HW or DW to indicate the equation for the problem solution, the element number, the estimated flow, the length, the inside diameter, and (if the default value is not used) either the Hazen-Williams coefficient or the pipe roughness for the Darcy-Weisbach equation. Pump elements require

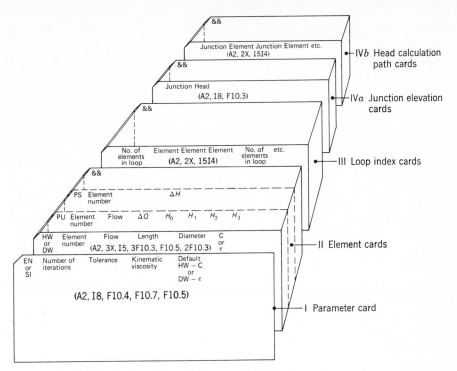

Fig. 10.16 Data cards for Hardy Cross program.

PU to indicate the element type, the element number, the estimated flow, a flow increment ΔQ at which values of pump head are specified, and four values of head from the pump-characteristic curve beginning at shutoff head and at equal flow intervals of ΔQ. The pseudo element for the pseudo loop requires PS to indicate the type, the element number, a zero or blank for the flow, and a difference in elevation between the interconnected fixed-pressure-head levels with head drop positive. The end of the element data is indicated by a card with "&&" in the first two columns.

Step 3: Loop index cards

These data are supplied with 15 integer numbers per card in the following order: the number of elements in a loop (maximum of 20) followed by the element number of each element in the loop with a negative sign to indicate counterclockwise flow direction. This information is repeated until all loops are defined. The end of step 3 data is indicated by a card with "&&" in columns 1 and 2.

Step 4: Head-calculation cards

Junctions with fixed elevations are identified on separate cards by giving the junction number and the hydraulic-grade-line elevation. There must be one or more of these cards followed by a card with "&&" in the first two columns to indicate the end of this type of data.

The path to be followed in computing the hydraulic-grade-line elevations is specified by supplying 15 integer values per card in the following order:

```
SI      30    .001    .000001    100.
HW       1    .12      600.       .3
HW       2    .03      300.       .15
HW       3    .0       500.       .6
HW       4    .03      400.       .3
HW       5    .03      300.       .3
PS       6             15.
PS       7             18.
PU       8    .06      .03        30.      29.      26.      20.
&&
         3    2    1   -3    3    4   -5    3    3    6   -4   -1    3    5    7
         8
&&
         5    117.
&&
         5    8    4    2    2    1    1    4    3
&&

SI UNITS SPECIFIED, VISCOSITY IN M**2/SEC=  0.0000010
DESIRED FLOW TOLERANCE=0.001  NO. OF ITERATIONS=     30
PIPE      Q(CFS OR M**3/S)  L(FT OR M)   D(FT OR M)    HW C OR EPS
  1              0.120         600.0         0.300       100.00000
  2              0.030         300.0         0.150       100.00000
  3              0.000         500.0         0.600       100.00000
  4              0.030         400.0         0.300       100.00000
  5              0.030         300.0         0.300       100.00000
  6  RESERVOIR ELEV DIFFERENCE=        15.00
  7  RESERVOIR ELEV DIFFERENCE=        18.00
  8  PUMP CURVE, CQ=  0.030 H=        30.0     29.0     26.0     20.0
     COEF IN PUMP EQ=         30.000     -11.111   -555.556   -6172.836
INC=
  3    2    1   -3    3    4   -5    3    3    6   -4   -1    3    5    7
  8    0    0    0    0    0    0    0    0    0    0    0    0    0    0
  C
ITERATION NO.    1 SUM OF FLOW CORRECTIONS=      0.1385
ITERATION NO.    2 SUM OF FLOW CORRECTIONS=      0.1040
ITERATION NO.    3 SUM OF FLOW CORRECTIONS=      0.0372
ITERATION NO.    4 SUM OF FLOW CORRECTIONS=      0.0034
ITERATION NO.    5 SUM OF FLOW CORRECTIONS=      0.0006
ELEMENT     FLOW
  1        0.143
  2       -0.034
  3        0.027
  4        0.080
  5        0.094
  6        0.087
IX=
  5    8    4    2    2    1    1    4    3    0    0    0    0    0    0
  C
JUNCTION    HEAD
  1       137.811
  2       150.044
  3       135.044
  4       137.797
  5       117.000
```

Fig. 10.17 Program input and output for Example 10.11.

junction number where the head is known, element number (with a negative sign to indicate a path opposite to the assumed flow direction), junction number, etc. If one wants a new path to begin at a junction different from the last listed junction, a single zero is added, followed by a junction where the head is known, element number, junction number, etc. The end of step 4 data is indicated by a card with "&&" in columns 1 and 2.

EXAMPLE 10.10 The program in Fig. 10.15 is used to solve the network problem displayed in Fig. 10.14. The pump data are as follows:

Q, m³/s	0	0.03	0.06	0.09
H, m	30	29	26	20

The Hazen-Williams pipeline coefficient for all pipes is 100. Figure 10.17 displays the input data and the computer output for this problem.

Figures 10.18 to 10.20 give input data for three systems which can be solved with this program.

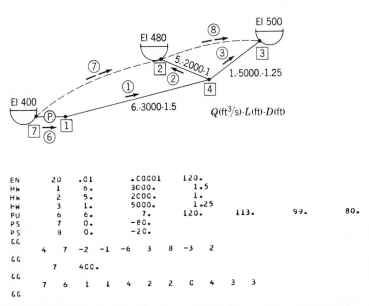

Fig. 10.18 Input data for branching pipe system in English units with Hazen-Williams formula.

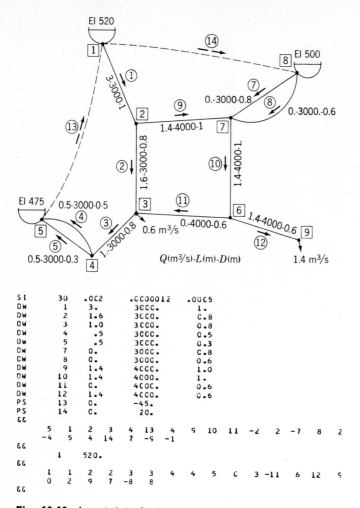

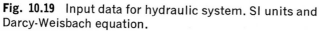

```
S I       30    .0C2     .CC00012   .00C5
DW        1     3.       3CCC.      1.
DW        2     1.6      3CCO.      C.8
DW        3     1.0      3CCO.      0.8
DW        4     .5       3CCC.      0.5
DW        5     .5       3CCC.      0.3
DW        7     0.       300C.      C.8
CW        8     0.       300C.      0.6
DW        9     1.4      4CCC.      1.0
DW        10    1.4      4000.      1.
DW        11    C.       4C0C.      0.6
DW        12    1.4      4CC0.      C.6
PS        13    0.       -45.
PS        14    C.       20.
&&
          5     1     2     3     4    13    4    9   10   11   -2    2   -7   8   2
         -4     5     4    14     7    -9   -1
&&
          1    520.
&&
          1     1     2     2     3     3    4    4    5    C    3  -11    6  12   9
          0     2     9     7    -8     8
&&
```

Fig. 10.19 Input data for hydraulic system. SI units and Darcy-Weisbach equation.

10.9 CONDUITS WITH NONCIRCULAR CROSS SECTIONS

In this chapter so far, only circular pipes have been considered. For cross sections that are noncircular, the Darcy-Weisbach equation may be applied if the term D can be interpreted in terms of the section. The concept of the *hydraulic radius R* permits circular and noncircular sections to be treated in

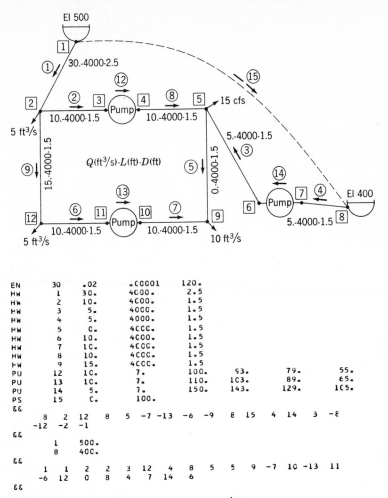

Fig. 10.20 Input for booster-pump system.

the same manner. The hydraulic radius is defined as the cross-sectional area divided by the *wetted perimeter*. Hence, for a circular section,

$$R = \frac{\text{area}}{\text{perimeter}} = \frac{\pi D^2/4}{\pi D} = \frac{D}{4} \tag{10.9.1}$$

and the diameter is equivalent to $4R$. Assuming that the diameter may be replaced by $4R$ in the Darcy-Weisbach equation, in the Reynolds number, and in the relative roughness,

$$h_f = f \frac{L}{4R} \frac{V^2}{2g} \qquad \mathbf{R} = \frac{V4R\rho}{\mu} \qquad \frac{\epsilon}{D} = \frac{\epsilon}{4R} \tag{10.9.2}$$

Noncircular sections may be handled in a similar manner. The Moody diagram applies as before. The assumptions in Eqs. (10.9.2) cannot be expected to hold for odd-shaped sections but should give reasonable values for square, oval, triangular, and similar types of sections.

EXAMPLE 10.11 Determine the head loss in inches of water required for flow of 10,000 ft³/min of air at 60°F and 14.7 psia through a rectangular galvanized-iron section 2 ft wide, 1 ft high, and 200 ft long.

$$R = \tfrac{2}{6} = 0.333 \text{ ft} \qquad \frac{\epsilon}{4R} = 0.00038$$

$$V = \frac{10,000}{60 \times 2} = 83.3 \text{ ft/s} \qquad V4R'' = 83.3 \times 4 \times 4 = 1330$$

$$f = 0.017$$

Then

$$h_f = f \frac{L}{4R} \frac{V^2}{2g} = \frac{0.017 \times 200}{4 \times 0.333} \frac{83.3^2}{64.4} = 275 \text{ ft}$$

The specific weight of air is $\gamma = (14.7 \times 144)/(53.3 \times 520) = 0.0762$ lb/ft³. In inches of water the head loss is

$$\frac{275 \times 0.0762 \times 12}{62.4} = 4.04 \text{ in}$$

10.10 AGING OF PIPES

The Moody diagram, with the values of absolute roughness shown there, is for new, clean pipe. With use, pipes become rougher, owing to corrosion, incrustations, and deposition of material on the pipe walls. The speed with which the friction factor changes with time depends greatly on the fluid being handled. Colebrook and White[1] found that the absolute roughness ϵ increases linearly with time,

$$\epsilon = \epsilon_0 + \alpha t \tag{10.10.1}$$

in which ϵ_0 is the absolute roughness of the new surface. Tests on a pipe are required to determine α.

The time variation of the Hazen-Williams coefficient has been summarized graphically[2] for water-distribution systems in seven major U.S. cities. Although it is not a linear variation, the range of values for the average rate of decline in C may typically be between 0.5 and 2 per year, with the larger values generally applicable in the first years following installation. The only sure way that accurate coefficients can be obtained for older water mains is through field tests.

PROBLEMS

10.1 Sketch the hydraulic and energy grade lines for Fig. 10.21. $H = 24$ ft.

10.2 Calculate the value of K for the valve of Fig. 10.21 so that the discharge of Prob. 10.1 is reduced by one-half. Sketch the hydraulic and energy grade lines.

10.3 Compute the discharge of the system in Fig. 10.22. Draw the hydraulic and energy grade lines.

10.4 What head is needed in Fig. 10.22 to produce a discharge of 0.3 m³/s?

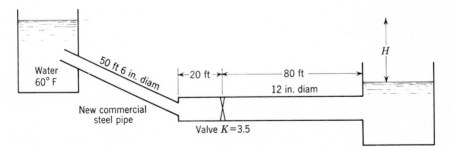

Fig. 10.21

[1] C. F. Colebrook and C. M. White, The Reduction of Carrying Capacity of Pipes with Age, *J. Inst. Civ. Eng. Lond.*, 1937.
[2] W. D. Hudson, Computerized Pipeline Design, *Transp. Eng. J. ASCE*, vol. 99, no. TE1, 1973.

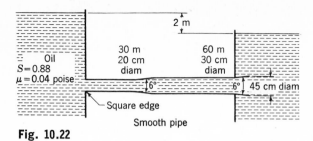

Fig. 10.22

10.5 Calculate the discharge through the siphon of Fig. 10.23 with the conical diffuser removed. $H = 4$ ft.

10.6 Calculate the discharge in the siphon of Fig. 10.23 for $H = 8$ ft. What is the minimum pressure in the system?

10.7 Find the discharge through the siphon of Fig. 10.24. What is the pressure at A? Estimate the minimum pressure in the system.

10.8 Neglecting minor losses other than the valve, sketch the hydraulic grade line for Fig. 10.25. The globe valve has a loss coefficient $K = 4.5$.

10.9 What is the maximum height of point A (Fig. 10.25) for no cavitation? Barometer reading is 29.5 in Hg.

10.10 Two reservoirs are connected by three clean cast-iron pipes in series; $L_1 = 300$ m, $D_1 = 20$ cm; $L_2 = 360$ m, $D_2 = 30$ cm; $L_3 = 1200$ m, $D_3 = 45$ cm. When $Q = 0.1$ m³/s water at 70°F, determine the difference in elevation of the reservoirs.

10.11 Solve Prob. 10.10 by the method of equivalent lengths.

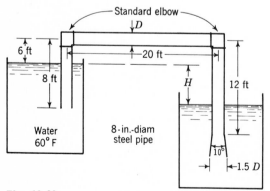

Fig. 10.23

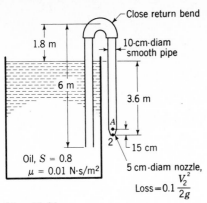

Fig. 10.24

10.12 For a difference in elevation of 10 m in Prob. 10.10, find the discharge by use of the Hazen-Williams equation.

10.13 By use of the program in Fig. 10.1, determine the exponential formula for flow of water at 21°C through clean cast-iron pipelines ranging in diameter from 10 to 30 cm for the Reynolds number range 10^5 to 5×10^5.

10.14 What diameter smooth pipe is required to convey 100 gpm kerosene at 90°F 500 ft with a head of 16 ft? There are a valve and other minor losses with total K of 7.6.

10.15 Air at atmospheric pressure and 60°F is carried through two horizontal pipes ($\epsilon = 0.06$) in series. The upstream pipe is 400 ft of 24 in diameter, and the downstream pipe is 100 ft of 36 in diameter. Estimate the equivalent length of 18-in ($\epsilon = 0.003$) pipe. Neglect minor losses.

10.16 What pressure drop, in inches of water, is required for flow of 6000 cfm in Prob. 10.15? Include losses due to sudden expansion.

10.17 Two pipes are connected in parallel between two reservoirs; $L_1 = 2500$ m, $D_1 = 1.2$-m-diameter old cast-iron pipe, $C = 100$; $L_2 = 2500$ m, $D_2 = 1$ m, $C = 90$. For a difference in elevation of 3.6 m determine the total flow of water at 20°C.

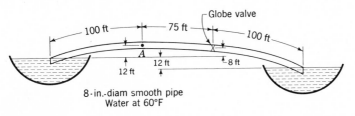

Fig. 10.25

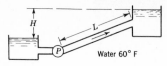

Water 60° F

Pump A				Pump B		
H_{ft}	Q_{cfs}	$e\%$		H_{ft}	Q_{cfs}	$e\%$
70	0	0		80	0	0
60	2.00	59		70	2.60	54
55	2.56	70		60	3.94	70
50	3.03	76		50	4.96	80
45	3.45	78		40	5.70	73
40	3.82	76.3		30	6.14	60
35	4.11	72		20	6.24	40
30	4.48	65				
25	4.59	56.5				
20	4.73	42				

Fig. 10.26

10.18 For 4.5 m³/s flow in the system of Prob. 10.17, determine the difference in elevation of reservoir surfaces.

10.19 Three smooth tubes are connected in parallel: $L_1 = 40$ ft, $D_1 = \frac{1}{2}$ in; $L_2 = 60$ ft, $D_2 = 1$ in; $L_3 = 50$ ft, $D_3 = \frac{3}{4}$ in. For total flow of 30 gpm oil, $\gamma = 55$ lb/ft³, $\mu = 0.65$ P, what is the drop in hydraulic grade line between junctions?

10.20 Determine the discharge of the system of Fig. 10.26 for $L = 2000$ ft, $D = 18$ in, $\epsilon = 0.0015$, and $H = 25$ ft, with the pump A characteristics given.

10.21 Determine the discharge through the system of Fig. 10.26 for $L = 4000$ ft, $D = 24$-in smooth pipe, $H = 40$ ft, with pump B characteristics.

10.22 Construct a head–discharge-efficiency table for pumps A and B (Fig. 10.26) connected in series.

10.23 Construct a head–discharge-efficiency table for pumps A and B (Fig. 10.26) connected in parallel.

10.24 Find the discharge through the system of Fig. 10.26 for pumps A and B in series; 5000 ft of 12-in clear cast-iron pipe, $H = 100$ ft.

10.25 Determine the horsepower needed to drive pumps A and B in Prob. 10.24.

10.26 Find the discharge through the system of Fig. 10.26 for pumps A and B in parallel; 5000 ft of 18-in steel pipe, $H = 30$ ft.

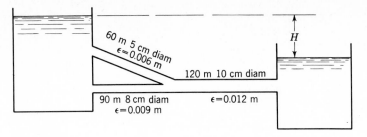

Fig. 10.27

10.27   Determine the horsepower needed to drive the pumps in Prob. 10.26.

10.28   For $H = 12$ m in Fig. 10.27, find the discharge through each pipe, $\mu = 8$ cP; sp gr $= 0.9$.

10.29   Find H in Fig. 10.27 for 0.03 m³/s flowing. $\mu = 5$ cP; sp gr $= 0.9$.

10.30   Find the equivalent length of 12-in-diameter clean cast-iron pipe to replace the system of Fig. 10.28. For $H = 30$ ft, what is the discharge?

10.31   With velocity of 4 ft/s in the 8-in-diameter pipe of Fig. 10.28 calculate the flow through the system and the head H required.

10.32   In Fig. 10.29 find the flow through the system when the pump is removed.

10.33   If the pump of Fig. 10.29 is delivering 3 cfs toward J, find the flow into A and B and the elevation of the hydraulic grade line at J.

10.34   The pump is adding 7500 W fluid power to the flow (toward J) in Fig. 10.29. Find Q_A and Q_B.

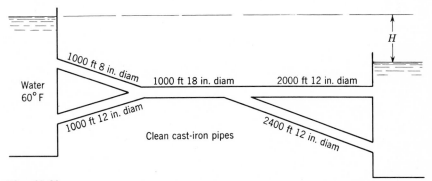

Fig. 10.28

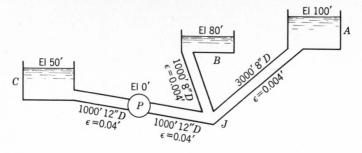

Fig. 10.29

10.35 With pump A of Fig. 10.26 in the system of Fig. 10.29, find Q_A, Q_B, and the elevation of the hydraulic grade line at J.

10.36 With pump B of Fig. 10.26 in the system of Fig. 10.29, find the flow into B and the elevation of the hydraulic grade line at J.

10.37 For flow of 1 cfs into B of Fig. 10.29, what head is produced by the pump? For pump efficiency of 70 percent, how much power is required?

10.38 Find the flow through the system of Fig. 10.30 for no pump in the system.

10.39 (a) With pumps A and B of Fig. 10.26 in parallel in the system of Fig. 10.30, find the flow into B, C, and D and the elevation of the hydraulic grade line at J_1 and J_2. (b) Assume the pipes in Fig. 10.30 are all cast iron. Prepare the data for solution to this problem by use of the program in Fig. 10.15.

10.40 Calculate the flow through each of the pipes of the network shown in Fig. 10.31. $n = 2$.

10.41 Determine the flow through each line of Fig. 10.32. $n = 2$.

10.42 By use of the program in Fig. 10.15, solve Prob. 10.35.

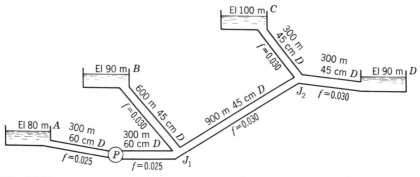

Fig. 10.30

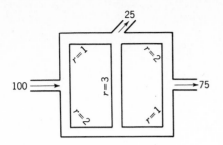

Fig. 10.31

10.43 By use of the program in Fig. 10.15 solve the problems given in (a) Fig. 10.18; (b) Fig. 10.19; (c) Fig. 10.20.

10.44 Determine the slope of the hydraulic grade line for flow of atmospheric air at 80°F through a rectangular 18 by 6 in galvanized-iron conduit. $V = 30$ ft/s.

10.45 What size square conduit is needed to convey 10 cfs water at 60°F with slope of hydraulic grade line of 0.001? $\epsilon = 0.003$.

10.46 Calculate the discharge of oil, $S = 0.85$, $\mu = 4$ cP, through 30 m of 5 by 10 cm sheet-metal conduit when the head loss is 60 cm. $\epsilon = 0.00015$ m.

10.47 A duct, with cross section an equilateral triangle 1 ft on a side, conveys 6 cfs water at 60°F. $\epsilon = 0.003$. Calculate the slope of the hydraulic grade line.

10.48 A clean 24-in-diameter cast-iron water pipe has its absolute roughness doubled in 5 years of service. Estimate the head loss per 1000 ft for a flow of 15 cfs when the pipe is 25 years old.

10.49 An 18-in-diameter pipe has an f of 0.020 when new for 5 ft/s water flow of 60°F. In 10 years $f = 0.029$ for $V = 3$ ft/s. Find f for 4 ft/s at end of 20 years.

10.50 The hydraulic grade line is

(a) always above the energy grade line   (b) always above the closed conduit

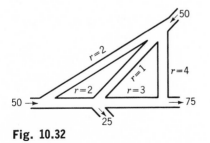

Fig. 10.32

(c) always sloping downward in the direction of flow
(d) the velocity head below the energy grade line
(e) upward in direction of flow when pipe is inclined downward

10.51   In solving a series-pipe problem for discharge, the energy equation is used along with the continuity equation to obtain an expression that contains a $V^2/2g$ and f_1, f_2, etc. The next step in the solution is to assume

(a) Q (b) V (c) **R**   (d) $f_1, f_2, \ldots$ (e) none of these quantities

10.52   One pipe system is said to be equivalent to another pipe system when the following two quantities are the same:

(a) h, Q (b) L, Q (c) L, D (d) f, D (e) V, D

10.53   In parallel-pipe problems

(a) the head losses through each pipe are added to obtain the total head loss
(b) the discharge is the same through all the pipes
(c) the head loss is the same through each pipe
(d) a direct solution gives the flow through each pipe when the total flow is known
(e) a trial solution is not needed

10.54   Branching-pipe problems are solved

(a) analytically by using as many equations as unknowns
(b) by assuming the head loss is the same through each pipe
(c) by equivalent lengths
(d) by assuming a distribution which satisfies continuity and computing a correction
(e) by assuming the elevation of hydraulic grade line at the junction point and trying to satisfy continuity

10.55   In networks of pipes

(a) the head loss around each elementary circuit must be zero
(b) the (horsepower) loss in all circuits is the same
(c) the elevation of hydraulic grade line is assumed for each junction
(d) elementary circuits are replaced by equivalent pipes
(e) friction factors are assumed for each pipe

10.56   The following quantities are computed by using $4R$ in place of diameter for noncircular sections:

(a) velocity, relative roughness
(b) velocity, head loss
(c) Reynolds number, relative roughness, head loss
(d) velocity, Reynolds number, friction factor
(e) none of these answers

10.57   Experiments show that in the aging of pipes

(a)   the friction factor increases linearly with time
(b)   a pipe becomes smoother with use
(c)   the absolute roughness increases linearly with time
(d)   no appreciable trends can be found
(e)   the absolute roughness decreases with time

STEADY FLOW IN OPEN CHANNELS

A broad coverage of topics in open-channel flow has been selected for this chapter. Steady uniform flow was discussed in Sec. 5.9, and application of the momentum equation to the hydraulic jump in Sec. 3.11. Weirs were introduced in Sec. 8.4. In this chapter open-channel flow is first classified and then the *shape* of optimum canal cross sections is discussed, followed by a section on flow through a floodway. The hydraulic jump and its application to stilling basins is then treated, followed by a discussion of specific energy and critical depth which leads into gradually varied flow. Water-surface profiles are classified and related to channel control sections. Transitions are next discussed, with one special application to the critical-depth meter.

The mechanics of flow in open channels is more complicated than closed-conduit flow owing to the presence of a free surface. The hydraulic grade line coincides with the free surface, and, in general, its position is unknown.

For laminar flow to occur, the cross section must be extremely small, the velocity very small, or the kinematic viscosity extremely high. One example of laminar flow is given by a thin film of liquid flowing down an inclined or vertical plane. This case is treated by the methods developed in Chap. 5 (see Prob. 5.10). Pipe flow has a lower critical Reynolds number of 2000, and this same value may be applied to an open channel when the diameter D is replaced by $4R$. R is the hydraulic radius, which is defined as the cross-sectional area of the channel divided by the wetted perimeter. In the range of Reynolds number, based on R in place of D, $\mathbf{R} = VR/\nu < 500$ flow is laminar, $500 < \mathbf{R} < 2000$ flow is *transitional* and may be either laminar or turbulent, and $\mathbf{R} > 2000$ flow is generally turbulent.

Most open-channel flows are turbulent, usually with water as the liquid. The methods for analyzing open-channel flow are not developed to the extent of those for closed conduits. The equations in use assume complete turbulence, with the head loss proportional to the square of the velocity. Although practically all data on open-channel flow have been obtained from experiments on the flow of water, the equations should yield reasonable values for other liquids of low viscosity. The material in this chapter applies to turbulent flow only.

11.1 CLASSIFICATION OF FLOW

Open-channel flow occurs in a large variety of forms, from flow of water over the surface of a plowed field during a hard rain to the flow at constant depth through a large prismatic channel. It may be classified as steady or unsteady, uniform or nonuniform. *Steady uniform flow* occurs in very long inclined channels of constant cross section, in those regions where *terminal velocity* has been reached, i.e., where the head loss due to turbulent flow is exactly supplied by the reduction in potential energy due to the uniform decrease in elevation of the bottom of the channel. The depth for steady uniform flow is called the *normal depth*. In steady uniform flow the discharge is constant, and the depth is everywhere constant along the length of the channel. Several equations are in common use for determining the relations between the average velocity, the shape of the cross section, its size and roughness, and the slope, or inclination, of the channel bottom (Sec. 5.9).

Steady nonuniform flow occurs in any irregular channel in which the discharge does not change with the time; it also occurs in regular channels when the flow depth and hence the average velocity change from one cross section to another. For gradual changes in depth or section, called *gradually varied flow*, methods are available, by numerical integration or step-by-step means, for computing flow depths for known discharge, channel dimensions and roughness, and given conditions at one cross section. For those reaches of a channel where pronounced changes in velocity and depth occur in a short distance, as in a transition from one cross section to another, model studies are frequently made. The *hydraulic jump* is one example of steady nonuniform flow; it is discussed in Secs. 3.11 and 11.4.

Unsteady uniform flow rarely occurs in open-channel flow. *Unsteady nonuniform flow* is common but is difficult to analyze. Wave motion is an example of this type of flow, and its analysis is complex when friction is taken into account. Positive and negative *surge waves* in a rectangular channel are analyzed, neglecting effects of friction, in Secs. 12.9 and 12.10.

The routing of unsteady flood flows in channels is discussed in Sec. 12.10.

Flow is also classified as *tranquil* or *rapid*. When flow occurs at low velocities so that a small disturbance can travel upstream and thus change upstream conditions, it is said to be tranquil flow[1] ($\mathbf{F} < 1$). Conditions up-stream are affected by downstream conditions, and the flow is controlled by the downstream conditions. When flow occurs at such high velocities that a small disturbance, such as an elementary wave is swept downstream, the flow is described as *shooting* or *rapid* ($\mathbf{F} > 1$). Small changes in downstream conditions do not effect any change in upstream conditions; hence, the flow is controlled by upstream conditions. When flow is such that its velocity is just equal to the velocity of an elementary wave, the flow is said to be critical ($\mathbf{F} = 1$).

The terms subcritical and supercritical are also used to classify flow velocities. *Subcritical* refers to tranquil flow at velocities less than critical, and *supercritical* corresponds to rapid flows when velocities are greater than critical.

Velocity distribution

The velocity at a solid boundary must be zero, and in open-channel flow it generally increases with distance from the boundaries. The maximum velocity does not occur at the free surface but is usually below the free surface a distance of 0.05 to 0.25 of the depth. The average velocity along a vertical line is sometimes determined by measuring the velocity at 0.6 of the depth, but a more reliable method is to take the average of the velocities at 0.2 and 0.8 of the depth, according to measurements of the U.S. Geological Survey.

11.2   BEST HYDRAULIC-CHANNEL CROSS SECTIONS

For the cross section of channel for conveying a given discharge for given slope and roughness factor, some shapes are more efficient than others. In general, when a channel is constructed, the excavation, and possibly the lining, must be paid for. From the Manning formula it is shown that when the area of cross section is a minimum, the wetted perimeter is also a minimum, and so both lining and excavation approach their minimum value for the same dimensions of channel. The *best hydraulic section* is one that has the least wetted perimeter or its equivalent, the least area for the type of section. The Manning formula is

$$Q = \frac{C_m}{n} AR^{2/3}S^{1/2} \tag{11.2.1}$$

---

[1] See Sec. 4.4 for definition and discussion of the Froude number $\mathbf{F}$.

in which Q is the discharge (L^3/T), A the cross-sectional flow area, R (area divided by wetted perimeter P) the hydraulic radius, S the slope of energy grade line, n the Manning roughness factor (Table 5.2), C_m an empirical constant $(L^{1/3}/T)$ equal to 1.49 in English units and to 1.0 in SI units. With Q, n, and S known, Eq. (11.2.1) may be written

$$A = cP^{2/5} \tag{11.2.2}$$

in which c is known. This equation shows that P is a minimum when A is a minimum. To find the best hydraulic section for a *rectangular* channel (Fig. 11.1) $P = b + 2y$, and $A = by$. Then

$$A = (P - 2y)y = cP^{2/5}$$

by elimination of b. The value of y is sought for which P is a minimum. Differentiating with respect to y gives

$$\left(\frac{dP}{dy} - 2\right)y + P - 2y = \tfrac{2}{5}cP^{-3/5}\frac{dP}{dy}$$

Setting $dP/dy = 0$ gives $P = 4y$, or since $P = b + 2y$,

$$b = 2y \tag{11.2.3}$$

Therefore, the depth is one-half the bottom width, independent of the size of rectangular section.

To find the best hydraulic *trapezoidal* section (Fig. 11.2) $A = by + my^2$, $P = b + 2y\sqrt{1 + m^2}$. After eliminating b and A in these equations and

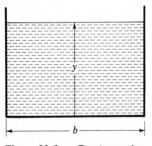

Fig. 11.1 Rectangular cross section.

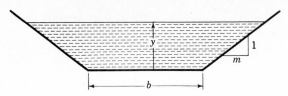

Fig. 11.2 Trapezoidal cross section.

Eq. (11.2.2),

$$A = by + my^2 = (P - 2y\sqrt{1 + m^2})y + my^2 = cP^{2/5} \tag{11.2.4}$$

By holding m constant and by differentiating with respect to y, $\partial P/\partial y$ is set equal to zero; thus

$$P = 4y\sqrt{1 + m^2} - 2my \tag{11.2.5}$$

Again, by holding y constant, Eq. (11.2.4) is differentiated with respect to m, and $\partial P/\partial m$ is set equal to zero, producing

$$\frac{2m}{\sqrt{1 + m^2}} = 1$$

After solving for m,

$$m = \frac{\sqrt{3}}{3}$$

and after substituting for m in Eq. (11.2.5),

$$P = 2\sqrt{3}\,y \qquad b = 2\frac{\sqrt{3}}{3}y \qquad A = \sqrt{3}\,y^2 \tag{11.2.6}$$

which shows that $b = P/3$ and hence the sloping sides have the same length as the bottom. As $\tan^{-1} m = 30°$, the best hydraulic section is one-half a hexagon. For trapezoidal sections with m specified (maximum slope at which wet earth will stand) Eq. (11.2.5) is used to find the best bottom-width-to-depth ratio.

The semicircle is the best hydraulic section of all possible open-channel cross sections.

EXAMPLE 11.1 Determine the dimensions of the most economical trapezoidal brick-lined channel to carry 8000 cfs with a slope of 0.0004.

With Eq. (11.2.6),

$$R = \frac{A}{P} = \frac{y}{2}$$

and by substituting into Eq. (11.2.1),

$$8000 = \frac{1.49}{0.016} \sqrt{3} \, y^2 \left(\frac{y}{2}\right)^{2/3} \sqrt{0.0004}$$

or

$$y^{8/3} = 3930 \qquad y = 22.3 \text{ ft}$$

and from Eq. (11.2.6), $b = 25.8$ ft.

11.3 STEADY UNIFORM FLOW IN A FLOODWAY

A practical open-channel problem of importance is the computation of discharge through a floodway (Fig. 11.3). In general the floodway is much rougher than the river channel, and its depth (and hydraulic radius) is much less. The slope of energy grade line must be the same for both portions. The discharge for each portion is determined separately, using the dashed line of Fig. 11.3 as the separation line for the two sections (but not as solid boundary), and then the discharges are added to determine the total capacity of the system.

Since both portions have the same slope, the discharge may be expressed

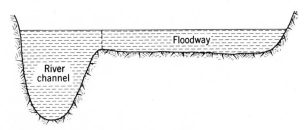

Fig. 11.3 Floodway cross section.

as

$$Q_1 = K_1 \sqrt{S} \qquad Q_2 = K_2 \sqrt{S}$$

or

$$Q = (K_1 + K_2)\sqrt{S} \qquad\qquad (11.3.1)$$

in which the value of K is

$$K = \frac{C_m}{n} A R^{2/3}$$

from Manning's formula and is a function of depth only for a given channel with fixed roughness. By computing K_1 and K_2 for different elevations of water surface, their sum may be taken and plotted against elevation. From this plot it is easy to determine the slope of energy grade line for a given depth and discharge from Eq. (11.3.1).

11.4 HYDRAULIC JUMP; STILLING BASINS

The relations among the variables V_1, y_1, V_2, y_2 for a hydraulic jump to occur in a horizontal rectangular channel are developed in Sec. 3.11. Another way of determining the conjugate depths for a given discharge is the $F + M$ method. The momentum equation applied to the free body of liquid between y_1 and y_2 (Fig. 11.4) is, for unit width ($V_1 y_1 = V_2 y_2 = q$),

$$\frac{\gamma y_1^2}{2} - \frac{\gamma y_2^2}{2} = \rho q (V_2 - V_1) = \rho V_2^2 y_2 - \rho V_1^2 y_1$$

Rearranging gives

$$\frac{\gamma y_1^2}{2} + \rho V_1^2 y_1 = \frac{\gamma y_2^2}{2} + \rho V_2^2 y_2 \qquad\qquad (11.4.1)$$

Fig. 11.4 Hydraulic jump in horizontal rectangular channel.

or

$$F_1 + M_1 = F_2 + M_2 \qquad (11.4.2)$$

in which F is the hydrostatic force at the section and M is the momentum per second passing the section. By writing $F + M$ for a given discharge q per unit width

$$F + M = \frac{\gamma y^2}{2} + \frac{\rho q^2}{y} \qquad (11.4.3)$$

a plot is made of $F + M$ as abscissa against y as ordinate, Fig. 11.5, for $q = 10$ cfs/ft. Any vertical line intersecting the curve cuts it at two points having the same value of $F + M$; hence, they are conjugate depths. The value of y for minimum $F + M$ [by differentiation of Eq. (11.4.3) with respect to y and setting $d(F + M)/dy$ equal to zero] is

$$y_c = \left(\frac{q^2}{g}\right)^{1/3} \qquad (11.4.4)$$

The jump must always occur from a depth less than this value to a depth greater than this value. This depth is the *critical* depth, which is shown in the following section to be the depth of minimum energy. Therefore, the

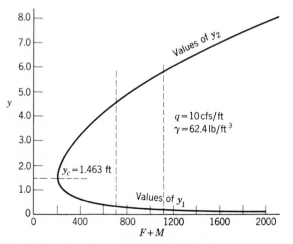

Fig. 11.5 $F + M$ curve for hydraulic jump.

jump always occurs from rapid flow to tranquil flow. The fact that available energy is lost in the jump prevents any possibility of its suddenly changing from the higher conjugate depth to the lower conjugate depth.

The conjugate depths are directly related to the Froude numbers before and after the jump,

$$\mathbf{F}_1 = \frac{V_1{}^2}{gy_1} \qquad \mathbf{F}_2 = \frac{V_2{}^2}{gy_2} \tag{11.4.5}$$

From the continuity equation

$$V_1{}^2 y_1{}^2 = g\mathbf{F}_1 y_1{}^3 = V_2{}^2 y_2{}^2 = g\mathbf{F}_2 y_2{}^3$$

or

$$\mathbf{F}_1 y_1{}^3 = \mathbf{F}_2 y_2{}^3 \tag{11.4.6}$$

From Eq. (11.4.1)

$$y_1{}^2 \left(1 + 2\frac{V_1{}^2}{gy_1} \right) = y_2{}^2 \left(1 + 2\frac{V_2{}^2}{gy_2} \right)$$

Substituting from Eqs. (11.4.5) and (11.4.6) gives

$$(1 + 2\mathbf{F}_1)\,\mathbf{F}_1{}^{-2/3} = (1 + 2\mathbf{F}_2)\,\mathbf{F}_2{}^{-2/3} \tag{11.4.7}$$

The value of $\mathbf{F}_2$ in terms of $\mathbf{F}_1$ is obtained from the hydraulic-jump equation (3.11.23),

$$y_2 = -\frac{y_1}{2} + \sqrt{\left(\frac{y_1}{2}\right)^2 + 2\frac{V_1{}^2 y_1}{g}} \qquad \text{or} \qquad 2\frac{y_2}{y_1} = -1 + \sqrt{1 + 8\frac{V_1{}^2}{gy_1}}$$

By Eqs. (11.4.5) and (11.4.6)

$$\mathbf{F}_2 = \frac{8\mathbf{F}_1}{(\sqrt{1 + 8\mathbf{F}_1} - 1)^3} \tag{11.4.8}$$

These equations apply only to a rectangular section.

The Froude number before the jump is always greater than unity, and after the jump it is always less than unity.

Stilling basins

A stilling basin is a structure for dissipating available energy of flow below a spillway, outlet works, chute, or canal structure. In the majority of existing installations a hydraulic jump is housed within the stilling basin and used as the energy dissipator. This discussion is limited to rectangular basins with horizontal floors although sloping floors are used in some cases to save excavation. An authoritative and comprehensive work[1] by personnel of the Bureau of Reclamation classified the hydraulic jump as an effective energy dissipator in terms of the Froude number $\mathbf{F}_1(V_1{}^2/gy_1)$ entering the basin as follows:

At $\mathbf{F}_1 = 1$ to 3. Standing wave. There is only a slight difference in conjugate depths. Near $\mathbf{F}_1 = 3$ a series of small rollers develops.

At $\mathbf{F}_1 = 3$ to 6. Pre-jump. The water surface is quite smooth, the velocity is fairly uniform, and the head loss is low. No baffles required if proper length of pool is provided.

At $\mathbf{F}_1 = 6$ to 20. Transition. Oscillating action of entering jet, from bottom of basin to surface. Each oscillation produces a large wave of irregular period that can travel downstream for miles and damage earth banks and riprap. If possible, it is advantageous to avoid this range of Froude numbers in stilling-basin design.

At $\mathbf{F}_1 = 20$ to 80. Range of good jumps. The jump is well balanced, and the action is at its best. Energy absorption (irreversibilities) ranges from 45 to 70 percent. Baffles and sills may be utilized to reduce length of basin.

At $\mathbf{F}_1 = 80$ upward. Effective but rough. Energy dissipation up to 85 percent. Other types of stilling basins may be more economical.

Baffle blocks are frequently used at the entrance to a basin to corrugate the flow. They are usually regularly spaced with gaps about equal to block widths. Sills, either triangular or dentated, are frequently employed at the downstream end of a basin to aid in holding the jump within the basin and to permit some shortening of the basin.

The basin should be paved with high-quality concrete to prevent erosion and cavitation damage. No irregularities in floor or training walls should be permitted. The length of the jump, about $6y_2$, should be within the paved basin, with good riprap downstream if the material is easily eroded.

[1] Research Study on Stilling Basins, Energy Dissipators, and Associated Appurtenances, Progress Report II, *U.S. Bur. Reclam. Hydraul. Lab. Rep.* Hyd-399, Denver, June 1, 1955. In this report the Froude number was defined as $V/\sqrt{gy}$.

EXAMPLE 11.2 A hydraulic jump occurs downstream from a 15-m-wide sluice gate. The depth is 1.5 m, and the velocity is 20 m/s. Determine (a) the Froude number and the Froude number corresponding to the conjugate depth; (b) the depth and velocity after the jump; and (c) the power dissipated by the jump.

$$(a) \quad \mathbf{F}_1 = \frac{V_1^2}{gy_1} = \frac{20^2}{9.806 \times 1.5} = 27.2$$

From Eq. (11.4.8)

$$\mathbf{F}_2 = \frac{8 \times 27.2}{(\sqrt{1 + 8 \times 27.2} - 1)^3} = 0.0831$$

$$(b) \quad \mathbf{F}_2 = \frac{V_2^2}{gy_2} = 0.0831 \qquad V_2 y_2 = V_1 y_1 = 1.5 \times 20 = 30$$

Then

$$V_2^3 = 30 \times 9.806 \times 0.0831$$

and $V_2 = 2.90$ m/s, $y_2 = 10.34$ m.

(c)   From Eq. (3.11.24), the head loss h_j in the jump is

$$h_j = \frac{(y_2 - y_1)^3}{4y_1 y_2} = \frac{(10.34 - 1.50)^3}{4 \times 1.5 \times 10.34} = 11.13 \text{ m} \cdot \text{N/N}$$

The power dissipated is

Power $= \gamma Q h_j = 9802 \times 15 \times 30 \times 11.13 = 49{,}093$ kW

11.5 SPECIFIC ENERGY; CRITICAL DEPTH

The energy per unit weight E with elevation datum taken as the bottom of the channel is called the *specific energy*. It is a convenient quantity to use in studying open-channel flow and was introduced by Bakhmeteff in 1911. It is plotted vertically above the channel floor,

$$E = y + \frac{V^2}{2g} \tag{11.5.1}$$

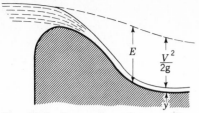

Fig. 11.6 Example of specific energy.

A plot of specific energy for a particular case is shown in Fig. 11.6. In a rectangular channel, in which q is the discharge per unit width, with $Vy = q$,

$$E = y + \frac{q^2}{2gy^2}$$
(11.5.2)

It is of interest to note how the specific energy varies with the depth for a constant discharge (Fig. 11.7). For small values of y the curve goes to infinity along the E axis, while for large values of y the velocity-head term is negligible and the curve approaches the 45° line $E = y$ asymptotically. The specific energy has a minimum value below which the given q cannot occur. The value of y for minimum E is obtained by setting dE/dy equal to

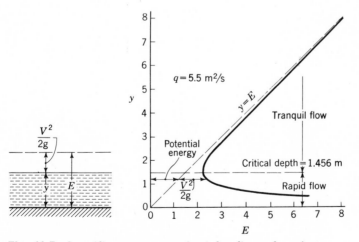

Fig. 11.7 Specific energy required for flow of a given discharge at various depths.

zero, from Eq. (11.5.2), holding q constant,

$$\frac{dE}{dy} = 0 = 1 - \frac{q^2}{gy^3}$$

or

$$y_c = \left(\frac{q^2}{g}\right)^{1/3} \qquad\qquad (11.5.3)$$

The depth for minimum energy y_c is called *critical depth*. Eliminating q^2 in Eqs. (11.5.2) and (11.5.3) gives

$$E_{\min} = \tfrac{3}{2} y_c \qquad\qquad (11.5.4)$$

showing that the critical depth is two-thirds of the specific energy. Eliminating E in Eqs. (11.5.1) and (11.5.4) gives

$$V_c = \sqrt{gy_c} \qquad\qquad (11.5.5)$$

The velocity of flow at critical condition V_c is $\sqrt{gy_c}$, which was used in Sec. 8.4 in connection with the broad-crested weir. Another method of arriving at the critical condition is to determine the maximum discharge q that could occur for a given specific energy. The resulting equations are the same as Eqs. (11.5.3) to (11.5.5).

For nonrectangular cross sections, as illustrated in Fig. 11.8, the specific-

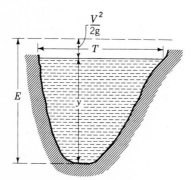

Fig. 11.8 Specific energy for a nonrectangular section.

energy equation takes the form

$$E = y + \frac{Q^2}{2gA^2} \tag{11.5.6}$$

in which A is the cross-sectional area. To find the critical depth,

$$\frac{dE}{dy} = 0 = 1 - \frac{Q^2}{gA^3}\frac{dA}{dy}$$

From Fig. 11.8, the relation between dA and dy is expressed by

$$dA = T\,dy$$

in which T is the width of the cross section at the liquid surface. With this relation,

$$\frac{Q^2}{gA_c^3}T_c = 1 \tag{11.5.7}$$

The critical depth must satisfy this equation. Eliminating Q in Eqs. (11.5.6) and (11.5.7) gives

$$E = y_c + \frac{A_c}{2T_c} \tag{11.5.8}$$

This equation shows that the minimum energy occurs when the velocity head is one-half the average depth A/T. Equation (11.5.7) may be solved by trial for irregular sections, by plotting

$$f(y) = \frac{Q^2T}{gA^3}$$

Critical depth occurs for that value of y which makes $f(y) = 1$.

EXAMPLE 11.3 Determine the critical depth for 300 cfs flowing in a trapezoidal channel with bottom width 8 ft and side slopes 1 horizontal to 2 vertical (1 on 2).

$$A = 8y + \frac{y^2}{2} \qquad T = 8 + y$$

Hence

$$f(y) = \frac{300^2(8+y)}{32.2(8y+y^2/2)^3} = \frac{2795(8+y)}{(8y+0.5y^2)^3}$$

By trial

$y = 2$	4	3	3.2	3.24	3.26	3.30	3.28
$f(y) = 4.8$	0.52	1.33	1.08	1.03	1.02	0.975	0.997

The critical depth is 3.28 ft.

In uniform flow in an open channel, the energy grade line slopes downward parallel to the bottom of the channel, thus showing a steady decrease in available energy. The specific energy, however, remains constant along the channel, since $y + V^2/2g$ does not change. In nonuniform flow, the energy grade line always slopes downward, or the available energy is decreased. The specific energy may either increase or decrease, depending upon the slope of the channel bottom, the discharge, the depth of flow, properties of the cross section, and channel roughness. In Fig. 11.6 the specific energy increases during flow down the steep portion of the channel and decreases along the horizontal channel floor.

The specific-energy and critical-depth relationships are essential in studying gradually varied flow and in determining control sections in open-channel flow.

The head loss in a hydraulic jump is easily displayed by drawing the $F + M$ curve (Fig. 11.5) and the specific-energy curve (Fig. 11.7) to the same vertical scale for the same discharge. Conjugate depths exist where any given vertical line intersects the $F + M$ curve. The specific energy at the upper depth may be observed to be always less than the specific energy at the corresponding lower conjugate depth.

11.6 GRADUALLY VARIED FLOW

Gradually varied flow is steady nonuniform flow of a special class. The depth area, roughness, bottom slope, and hydraulic radius change very slowly (if at all) along the channel. The basic assumption required is that the head-loss rate at a given section is given by the Manning formula for the same depth and discharge, regardless of trends in the depth. Solving Eq. (11.2.1) for the head loss per unit length of channel produces

$$S = -\frac{\Delta E}{\Delta L} = \left(\frac{nQ}{C_m A R^{2/3}}\right)^2 \tag{11.6.1}$$

in which S is now the slope of the energy grade line or, more specifically, the sine of the angle the energy grade line makes with the horizontal. In gradually varied flow the slopes of energy grade line, hydraulic grade line, and bottom are all different. Computations of gradually varied flow may be carried out either by the *standard-step method* or by *numerical integration*. Horizontal channels of great width are treated as a special case that may be integrated.

Standard-step method

Applying the energy equation between two sections a finite distance ΔL apart, Fig. 11.9, including the loss term, gives

$$\frac{V_1^2}{2g} + S_0\,\Delta L + y_1 = \frac{V_2^2}{2g} + y_2 + S\,\Delta L \tag{11.6.2}$$

Solving for the length of reach gives

$$\Delta L = \frac{(V_1^2 - V_2^2)/2g + y_1 - y_2}{S - S_0} \tag{11.6.3}$$

If conditions are known at one section, e.g., section 1, and the depth y_2 is wanted a distance ΔL away, a trial solution is required. The procedure is as follows:

1. Assume a depth y_2; then compute A_2, V_2.
2. For the assumed y_2 find an average y, P, and A for the reach [for prismatic

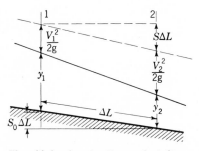

Fig. 11.9 Gradually varied flow.

channels $y = (y_1 + y_2)/2$ with A and R computed for this depth] and compute S.
3. Substitute in Eq. (11.6.3) to compute ΔL.
4. If ΔL is not correct, assume a new y_2 and repeat the procedure.

EXAMPLE 11.4 At section 1 of a canal the cross section is trapezoidal, $b_1 = 10$ m, $m_1 = 2$, $y_1 = 7$ m, and at section 2, downstream 200 m, the bottom is 0.08 m higher than at section 1, $b_2 = 15$ m, and $m_2 = 3$. $Q = 200$ m³/s, $n = 0.035$. Determine the depth of water at section 2.

$$A_1 = b_1 y_1 + m_1 y_1^2 = 10 \times 7 + 2 \times 7^2 = 168 \text{ m}^2 \qquad V_1 = \tfrac{200}{168} = 1.19 \text{ m/s}$$

$$P_1 = b_1 + 2y_1 \sqrt{m_1^2 + 1} = 10 + 2 \times 7 \sqrt{2^2 + 1} = 41.3 \text{ m}$$

$$S_0 = -\frac{0.08}{200} = -0.0004$$

Since the bottom has an adverse slope, i.e., it is rising in the downstream direction, and since section 2 is larger than section 1, y_2 is probably less than y_1. Assume $y_2 = 6.9$ m; then

$$A_2 = 15 \times 6.9 + 3 \times 6.9^2 = 246 \text{ m}^2 \qquad V_2 = \tfrac{200}{246} = 0.813 \text{ m/s}$$

and

$$P_2 = 15 + 2 \times 6.9 \sqrt{10} = 58.6 \text{ m}$$

The average $A = 207$ and average wetted perimeter $P = 50.0$ are used to find an average hydraulic radius for the reach, $R = 4.14$ m. Then

$$S = \left(\frac{nQ}{C_m A R^{2/3}}\right)^2 = \left(\frac{0.035 \times 200}{1.0 \times 207 \times 4.14^{2/3}}\right)^2 = 0.000172$$

Substituting into Eq. (11.6.3) gives

$$\Delta L = \frac{(1.19^2 - 0.813^2)/(2 \times 9.806) + 7 - 6.9}{0.000172 + 0.0004} = 242 \text{ m}$$

A larger y_2, for example, 6.92 m, would bring the computed value of length closer to the actual length.

Numerical-integration method

A more satisfactory procedure, particularly for flow through channels having a constant shape of cross section and constant bottom slope, is to obtain a differential equation in terms of y and L and then to perform the integration numerically. When ΔL is considered as an infinitesimal in Fig. 11.9, the rate of change of available energy is equal to the rate of head loss $-\Delta E/\Delta L$ given by Eq. (11.6.1), or

$$\frac{d}{dL}\left(\frac{V^2}{2g} + z_0 - S_0 L + y\right) = -\left(\frac{nQ}{C_m A R^{2/3}}\right)^2 \tag{11.6.4}$$

in which $z_0 - S_0 L$ is the elevation of bottom of channel at L, z_0 is the elevation of bottom at $L = 0$, and L is measured positive in the downstream direction. After performing the differentiation,

$$-\frac{V}{g}\frac{dV}{dL} + S_0 - \frac{dy}{dL} = \left(\frac{nQ}{C_m A R^{2/3}}\right)^2 \tag{11.6.5}$$

Using the continuity equation $VA = Q$ leads to

$$\frac{dV}{dL}A + V\frac{dA}{dL} = 0$$

and expressing $dA = T\,dy$, in which T is the liquid-surface width of the cross section, gives

$$\frac{dV}{dL} = -\frac{VT}{A}\frac{dy}{dL} = -\frac{QT}{A^2}\frac{dy}{dL}$$

Substituting for V in Eq. (11.6.5) yields

$$\frac{Q^2}{gA^3}T\frac{dy}{dL} + S_0 - \frac{dy}{dL} = \left(\frac{nQ}{C_m A R^{2/3}}\right)^2$$

and solving for dL gives

$$dL = \frac{1 - Q^2 T/gA^3}{S_0 - (nQ/C_m A R^{2/3})^2} \tag{11.6.6}$$

After integrating,

$$L = \int_{y_1}^{y_2} \frac{1 - Q^2 T/gA^3}{S_0 - (nQ/C_m AR^{2/3})^2} \, dy \tag{11.6.7}$$

in which L is the distance between the two sections having depths y_1 and y_2.

When the numerator of the integrand is zero, critical flow prevails; there is no change in L for a change in y (neglecting curvature of the flow and nonhydrostatic pressure distribution at this section). Since this is not a case of gradual change in depth, the equations are not accurate near critical depth. When the denominator of the integrand is zero, uniform flow prevails and there is no change in depth along the channel. The flow is at *normal depth*.

For a channel of prismatic cross section, constant n and S_0, the integrand becomes a function of y only,

$$F(y) = \frac{1 - Q^2 T/gA^3}{S_0 - (nQ/C_m AR^{2/3})^2}$$

and the equation can be integrated numerically by plotting $F(y)$ as ordinate against y as abscissa. The area under the curve (Fig. 11.10) between two values of y is the length L between the sections, since

$$L = \int_{y_1}^{y_2} F(y) \, dy$$

EXAMPLE 11.5 A trapezoidal channel, $b = 10$ ft, $m = 1$, $n = 0.014$, $S_0 = 0.001$, carries 1000 cfs. If the depth is 10 ft at section 1, determine the water-surface profile for the next 2000 ft downstream.

To determine whether the depth increases or decreases, the slope of

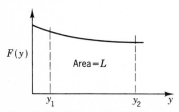

Fig. 11.10 Numerical integration of equation for gradually varied flow.

energy grade line at section 1 is computed [using Eq. (11.6.1)]:

$$A = by + my^2 = 10 \times 10 + 1 \times 10^2 = 200 \text{ ft}^2$$

$$P = b + 2y \sqrt{m^2 + 1} = 38.2 \text{ ft}$$

and

$$R = \frac{200}{38.2} = 5.24 \text{ ft}$$

Then

$$S = \left(\frac{0.014 \times 1000}{1.49 \times 200 \times 5.24^{2/3}} \right)^2 = 0.000243$$

The depth is greater than critical, and $S < S_0$; hence, the specific energy is increasing, and this can be accomplished only by increasing the depth downstream. Substituting into Eq. (11.6.7) gives

$$L = \int_{10}^{y} \frac{1 - 3.105 \times 10^4 T/A^3}{0.001 - 88.3/A^2 R^{4/3}} \, dy$$

The following table evaluates the terms in the integrand:

y	A	P	R	T	Numer-ator	$10^3 \times$ Denomi-nator	$F(y)$	L
10.0	200	38.2	5.24	30	0.8836	757	1167	0
10.5	215.2	39.8	5.41	31	0.9037	800	1129	574
11.0	232	41.1	5.64	32	0.9204	836	1101	1131
11.5	247.2	42.5	5.82	33	0.9323	862	1082	1677
12.0	264	43.9	6.01	34	0.9426	884	1067	2214

The integral $\int F(y) \, dy$ can be evaluated by plotting the curve and taking the area under it between $y = 10$ and the following values of y. As $F(y)$ does not vary greatly in this example, the average of $F(y)$ may be used for each reach and when it is multiplied by Δy, the length of reach is obtained. Between $y = 10$ and $y = 10.5$

$$\frac{1167 + 1129}{2} \times 0.50 = 574$$

Between $y = 10.5$ and $y = 11.0$

$$\frac{1129 + 1101}{2} \times 0.50 = 557$$

and so on. Five points on the water surface are known, so that it can be plotted.

Horizontal channels of great width

For channels of great width the hydraulic radius equals the depth; and for horizontal channel floors, $S_0 = 0$; hence, Eq. (11.6.7) can be simplified. The width may be considered as unity; that is, $T = 1$, $Q = q$ and $A = y$, $R = y$; thus

$$L = -\int_{y_1}^{y} \frac{1 - q^2/gy^3}{n^2q^2/C_m^2y^{10/3}} \, dy \tag{11.6.8}$$

or, after performing the integration,

$$L = -\frac{3}{13}\left(\frac{C_m}{nq}\right)^2 (y^{13/3} - y_1^{13/3}) + \frac{3}{4g}\left(\frac{C_m}{n}\right)^2 (y^{4/3} - y_1^{4/3}) \tag{11.6.9}$$

EXAMPLE 11.6 After contracting below a sluice gate water flows onto a wide horizontal floor with a velocity of 15 m/s and a depth of 0.7 m. Find the equation for the water-surface profile, $n = 0.015$.

From Eq. (11.6.9), with x replacing L as distance from section 1, where $y_1 = 0.7$, and with $q = 0.7 \times 15 = 10.5$ m²/s,

$$x = -\frac{3}{13}\left(\frac{1}{0.015 \times 10.5}\right)^2 (y^{13/3} - 0.7^{13/3})$$

$$+ \frac{3}{4 \times 9.806}\left(\frac{1}{0.015}\right)^2 (y^{4/3} - 0.7^{4/3})$$

$$= -209.3 - 9.30y^{13/3} + 340y^{4/3}$$

Critical depth occurs [Eq. (11.5.3)] at

$$y_c = \left(\frac{q^2}{g}\right)^{1/3} = \left(\frac{10.5^2}{9.806}\right)^{1/3} = 2.24 \text{ m}$$

The depth must increase downstream, since the specific energy decreases, and the depth must move toward the critical value for less specific energy. The equation does not hold near the critical depth because of vertical accelerations that have been neglected in the derivation of gradually varied flow. If the channel is long enough for critical depth to be attained before the end of the channel, the high-velocity flow downstream from the gate may be drowned or a jump may occur. The water-surface calculation for the subcritical flow must begin with critical depth at the downstream end of the channel.

The computation of water-surface profiles with the aid of a digital computer is discussed after the various types of gradually varied flow profiles are classified.

11.7 CLASSIFICATION OF SURFACE PROFILES

A study of Eq. (11.6.7) reveals many types of surface profiles, each of which has its definite characteristics. The bottom slope is classified as *adverse, horizontal, mild, critical,* and *steep*; and, in general, the flow can be above the normal depth or below the normal depth, and it can be above critical depth or below critical depth.

The various profiles are plotted in Fig. 11.11; the procedures used are discussed for the various classifications in the following paragraphs. A very wide channel is assumed in the reduced equations which follow, with $R = y$.

Adverse slope profiles

When the channel bottom rises in the direction of flow (S_0 is negative), the resulting surface profiles are said to be adverse. There is no normal depth, but the flow may be either below critical depth or above critical depth. Below critical depth the numerator is negative, and Eq. (11.6.6) has the form

$$dL = \frac{1 - C_1/y^3}{S_0 - C_2/y^{10/3}} \, dy$$

where C_1 and C_2 are positive constants. Here $F(y)$ is positive and the depth increases downstream. This curve is labeled **A₃** and shown in Fig. 11.11. For depths greater than critical depth, the numerator is positive, and $F(y)$ is negative; i.e., the depth decreases in the downstream direction. For y very large, $dL/dy = 1/S_0$, which is a horizontal asymptote for the curve. At $y = y_c$, dL/dy is 0, and the curve is perpendicular to the critical-depth line. This curve is labeled **A₂**.

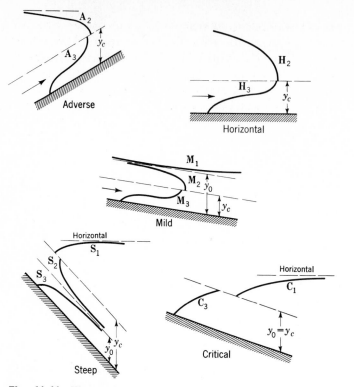

Fig. 11.11   The various typical liquid-surface profiles.

Horizontal slope profiles

For a horizontal channel $S_0 = 0$, the normal depth is infinite and flow may be either below critical depth or above critical depth.   The equation has the form

$$dL = -Cy^{1/3}(y^3 \doteq C_1)\ dy$$

For y less than critical, dL/dy is positive, and the depth increases downstream. It is labeled **H₃**.   For y greater than critical (**H₂** curve), dL/dy is negative, and the depth decreases downstream.   These equations are integrable analytically for very wide channels.

Mild slope profiles

A mild slope is one on which the normal flow is tranquil, i.e., where normal depth y_0 is greater than critical depth.   Three profiles may occur, **M₁, M₂, M₃,** for depth above normal, below normal and above critical, and below critical, respectively. For the $\mathbf{M}_1$ curve, dL/dy is positive and approaches $1/S_0$ for very large y; hence, the $\mathbf{M}_1$ curve has a horizontal asymptote downstream. As the denominator approaches zero as y approaches y_0, the normal depth is an asymptote at the upstream end of the curve. Thus, dL/dy is negative for the $\mathbf{M}_2$ curve, with the upstream asymptote the normal depth, and $dL/dy = 0$ at critical. The $\mathbf{M}_3$ curve has an increasing depth downstream, as shown.

Critical slope profiles

When the normal depth and the critical depth are equal, the resulting profiles are labeled $\mathbf{C}_1$ and $\mathbf{C}_3$ for depth above and below critical, respectively. The equation has the form

$$dL = \frac{1}{S_0} \frac{1 - b/y^3}{1 - b_1/y^{10/3}} \, dy$$

with both numerator and denominator positive for $\mathbf{C}_1$ and negative for $\mathbf{C}_3$. Therefore the depth increases downstream for both. For large y, dL/dy approaches $1/S_0$; hence, a horizontal line is an asymptote. The value of dL/dy at critical depth is $0.9/S_0$; hence, curve $\mathbf{C}_1$ is convex upward. Curve $\mathbf{C}_3$ is also convex upward, as shown.

Steep slope profiles

When the normal flow is rapid in a channel (normal depth less than critical depth), the resulting profiles $\mathbf{S}_1$, $\mathbf{S}_2$, $\mathbf{S}_3$ are referred to as steep profiles: $\mathbf{S}_1$ is above the normal and critical, $\mathbf{S}_2$ between critical and normal, and $\mathbf{S}_3$ below normal depth. For curve $\mathbf{S}_1$ both numerator and denominator are positive, and the depth increases downstream approaching a horizontal asymptote. For curve $\mathbf{S}_2$ the numerator is negative, and the denominator positive but approaching zero at $y = y_0$. The curve approaches the normal depth asymptotically. The $\mathbf{S}_3$ curve has a positive dL/dy as both numerator and denominator are negative. It plots as shown on Fig. 11.11.

It should be noted that a given channel may be classified as mild for one discharge, critical for another discharge, and steep for a third discharge, since normal depth and critical depth depend upon different functions of the discharge. The use of the various surface profiles is discussed in the next section.

11.8 CONTROL SECTIONS

A small change in downstream conditions cannot be relayed upstream when the depth is critical or less than critical; hence, downstream conditions do not control the flow. All rapid flows are controlled by upstream conditions, and computations of surface profiles must be started at the upstream end of a channel.

Tranquil flows are affected by small changes in downstream conditions and therefore are controlled by them. Tranquil-flow computations must start at the downstream end of a reach and be carried upstream.

Control sections occur at entrances and exits to channels and at changes in channel slopes, under certain conditions. A gate in a channel can be a control for both the upstream and downstream reaches. Three control sections are illustrated in Fig. 11.12. In *a* the flow passes through critical at the entrance to a channel, and depth can be computed there for a given discharge. The channel is steep; therefore, computations proceed downstream. In *b* a change in channel slope from mild to steep causes the flow to pass through critical at the break in grade. Computations proceed both upstream and downstream from the control section at the break in grade. In *c* a gate in a horizontal channel provides control both upstream and downstream from it. The various curves are labeled according to the classification in Fig. 11.11.

The hydraulic jump occurs whenever the conditions required by the momentum equation are satisfied. In Fig. 11.13, liquid issues from under a gate in rapid flow along a horizontal channel. If the channel were short enough, the flow could discharge over the end of the channel as an **H**₃ curve. With a longer channel, however, the jump occurs, and the resulting profile

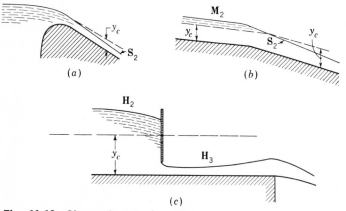

Fig. 11.12 Channel control sections.

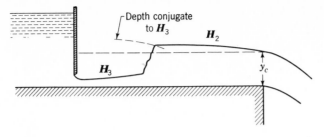

Fig. 11.13 Hydraulic jump between two control sections.

consists of pieces of **H₃** and **H₂** curves with the jump in between. In computing these profiles for a known discharge, the **H₃** curve is computed, starting at the gate (contraction coefficient must be known) and proceeding downstream until it is clear that the depth will reach critical before the end of the channel is reached. Then the **H₂** curve is computed, starting with critical depth at the end of the channel and proceeding upstream. The depths conjugate to those along **H₃** are computed and plotted as shown. The intersection of the conjugate-depth curve and the **H₂** curve locates the position of the jump. The channel may be so long that the **H₂** curve is everywhere greater than the depth conjugate to **H₃**. A *drowned jump* then occurs, with **H₂** extending to the gate.

All sketches are drawn to a greatly exaggerated vertical scale, since usual channels have small bottom slopes.

11.9 COMPUTER CALCULATION OF GRADUALLY VARIED FLOW

In Sec. 11.6 the standard-step and numerical-integration methods of computing water-surface profiles were introduced. The repetitious calculation in the latter method is easily handled by digital computer. The program, listed in Fig. 11.14, calculates the steady gradually varied water-surface profile in any prismatic rectangular, symmetric trapezoidal or triangular channel. The concepts of physical control sections in a channel must be understood in order to use the program successfully.

Input data include the specification of the system of units (SI or ENGLISH) in the first columns of the first data card, followed by the channel dimensions, discharge, and water-surface control depth on the second card. If the control depth is left blank or set to zero in data, it is automatically assumed to be the critical depth in the program. For subcritical flow the control is downstream, and distances are measured in the upstream direction.

```
C WATER SURFACE PROFILE IN RECT, TRAPEZOIDAL OR TRIANGULAR CHANNEL.
C XL=LENGTH, B=BOT WIDTH, Z=SIDE SLOPE,RN=MANNING N,SO=BOT SLOPE,Q=FLOW.
C YCCNT=CONTROL DEPTH. IF YCCNT=0. IN DATA, YCCNT IS SET EQUAL TC YC.
      DATA ISI/'SI'/
      AREA(YY)=YY*(B+Z*YY)
      PER(YY)=B+2.*YY*SQRT(1.+Z*Z)
      YCRIT(YY)=1.-Q*Q*(B+2.*Z*YY)/(G*AREA(YY)**3)
      YNORM(YY)=1.-Q*Q*CCN/(AREA(YY)**3.333/PER(YY)**1.333)
      DL(YY)=YCRIT(YY)/(YNORM(YY)*SO)
      FPM(YY)=GAM*(YY*YY*(B*.5+Z*YY/3.)+Q*Q/(G*AREA(YY)))
      ENERGY(YY)=YY+Q*Q/(2.*G*AREA(YY)**2)
    5 READ (5,7,END=99) IUNIT,XL,B,Z,RN,SO,Q,YCCNT
    7 FORMAT (A2/F10.1,3F10.4,F10.6,2F10.3)
      IF (IUNIT.EQ.ISI) GO TO 10
      GAM=62.4
      G=32.2
      CCN=(RN/1.486)**2/SO
      WRITE (6,9)
    9 FORMAT (' ENGLISH UNITS')
      GO TO 12
   10 G=9.806
      GAM=9802.
      CCN=RN**2/SO
      WRITE (6,11)
   11 FORMAT (' SI UNITS')
   12 WRITE (6,13) XL,Q,B,Z,RN,SO
   13 FORMAT (/' CHANNEL LENGTH=',F10.1,' DISCHARGE =',F12.3/
     2' B=',F8.2,' Z=',F6.3,' RN=',F6.4,' SO=',F9.6)
      NN=30
C DETERMINATION OF CRITICAL AND NORMAL DEPTHS
      UP=30.
      DN=0.
      YC=15.
      DO 20 I=1,15
      IF (YCRIT(YC)) 14,21,15
   14 DN=YC
      GO TO 20
   15 UP=YC
   20 YC=(UP+DN)*.5
   21 IF (YCCNT.EQ.0.) YCCNT=YC
      IF (SO.LE.0.) GO TO 33
      UP=40.
      DN=0.
      YN=20.
      DO 30 I=1,15
      IF (YNORM(YN)) 23,31,24
   23 DN=YN
      GO TO 30
   24 UP=YN
   30 YN=(UP+DN)*.5
   31 WRITE (6,32) YN,YC
   32 FORMAT (/' NORMAL DEPTH=',F7.3,' CRITICAL DEPTH=',F7.3)
      GO TO 35
   33 YN=3.*YC
   34 FORMAT (/' CRITICAL DEPTH=',F7.3)
      WRITE (6,34) YC
   35 IF (YN.LT.YC) GO TO 50
C MILD,ADVERSE, OR HORIZONTAL CHANNEL, YN. GT. YC
      IF (YCONT .LT.YC) GO TO 45
C SUBCRITICAL FLOW, YCCNT .GE. YC
      SIGN=-1.
      DY=(YCONT-YN)*.998/NN
      WRITE (6,44) YCCNT
   44 FORMAT (/' CONTROL IS DOWNSTREAM, DEPTH=',F7.3)
      GO TO 60
C SUPERCRITICAL FLOW
   45 SIGN=1.
      DY=(YC-YCCNT)/NN
      WRITE (6,46) YCCNT
   46 FORMAT (/' CONTROL IS UPSTREAM, DEPTH=',F7.3)
      GO TO 60
C STEEP CHANNEL, YN .LT. YC
   50 IF (YCONT.LE.YC) GO TO 55
C SUBCRITICAL FLOW, YCCNT .GT.YC
      SIGN=-1.
      DY=(YCONT-YC)/NN
      WRITE (6,44) YCCNT
      GO TO 60
```

Fig. 11.14

616

```
C SUPERCRITICAL FLCW, YCCNT .LE. YC
   55 SIGN=1.
      NN=NN*2
      DY=(YN-YCCNT)*.998/NN
      WRITE (6,46) YCCNT
   60 SL=0.
      Y=YCONT
      E=ENERGY(Y)
      FM=FPM(Y)
      WRITE (6,62)
   62 FORMAT (//'         DISTANCE      DEPTH      ENERGY          F+M')
      WRITE (6,64) SL,Y,E,FM
   64 FORMAT (5X,F10.1,2F12.3,F13.0)
C WATER SURFACE PRCFILE CALCULATION
      DO 80 I=1,NN,2
      Y2=YCONT+SIGN*CY*(I+1)
      CX=DY*(DL(Y)+DL(Y2)+4.*DL(YCCNT+SIGN*I*DY))/3.
      SL=SL+DX
      IF(SL.GT.XL) GC TC 82
      Y=Y2
      E=ENERGY(Y)
      FM=FPM(Y)
      IF (I.EC.NN-1.AND.SL.LT.C.) SL=XL
   80 WRITE (6,64)SL,Y,E,FM
      GO TO 5
   82 Y=Y2-SIGN*2.*CY*(SL-XL)/DX
      E=ENERGY(Y)
      FM=FPM(Y)
      WRITE (6,64)XL,Y,E,FM
      GO TO 5
   99 STOP
      END
ENCLISH UNITS, EXAMPLE 11.7
     600.      8.      0.8      0.012      C.025      873.      C.0
ENCLISH UNITS
    1800.      8.      C.8      0.C12      C.C002     873.      3.023
ENCLISH UNITS
    1800.      8.      C.E      0.C12      C.C002     873.      6.C
```

Fig. 11.14 FORTRAN program for water-surface profiles.

For supercritical flow the control depth is upstream, and distances are measured in the downstream direction.

The program begins with several line functions to compute the various variables and functions in the problem. After the necessary data input, critical depth is computed, followed by the normal-depth calculation if normal depth exists. The bisection method, Appendix E.3, is used in these calculations. The type of profile is then categorized, and finally the water-surface profile, specific energy, and $F + M$ are calculated and printed. Simpson's rule is used in the integration for the water-surface profile, Appendix E.1.

The program can be utilized for other channel sections, such as circular or parabolic, by simply changing the line functions at the beginning.

EXAMPLE 11.7 A trapezoidal channel, $B = 8$ ft, side slope $= 0.8$, has two bottom slopes. The upstream portion is 600 ft long, $S_0 = 0.025$, and the downstream portion, 1800 ft long, $S_0 = 0.0002$, $n = 0.012$. A discharge of 873 cfs enters at critical depth from a reservoir at the upstream end, and at the downstream end of the system the water depth is 6 ft. Determine the water-surface profiles throughout the system, including jump location.

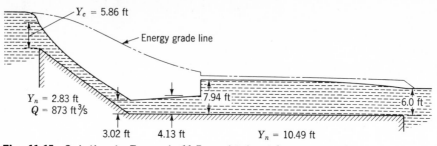

Fig. 11.15 Solution to Example 11.7 as obtained from computer.

Three separate sets of data, shown in Fig. 11.14, are needed to obtain the results used to plot the solution as shown in Fig. 11.15. The first set for the steep upstream channel has a control depth equal to zero since it will be at automatically assumed critical depth in the program. The second set is for the supercritical flow in the mild channel. It begins at a control depth equal to the end depth from the upstream channel and computes the water surface downstream to the critical depth. The third set of data used the 6-ft down-

```
ENGLISH UNITS
CHANNEL LENGTH=    1800.C CISCHARGE =      873.000
B=    8.00 Z= C.800 RN=0.0120 SO= 0.CC0200
NORMAL DEPTH= 10.495 CRITICAL DEPTH=  5.861
CCNTROL IS UPSTREAM, DEFTH=  3.C23
     DISTANCE        DEPTH       ENERGY            F+M
        0.C         3.023       14.954          49635.
       83.3         3.212       13.478          46627.
      165.0         3.401       12.301          44043.
      244.9         3.591       11.356          41821.
      322.4         3.780       10.596          39910.
      397.0         3.969        9.985          38271.
      468.4         4.158        9.493          36871.
      535.8         4.347        9.101          35683.
      598.6         4.536        8.789          34686.
      656.2         4.726        8.544          33860.
      707.7         4.915        8.356          33190.
      752.4         5.104        8.215          32662.
      789.1         5.293        8.115          32267.
      816.9         5.482        8.049          31994.
      834.6         5.672        8.012          31835.
      840.9         5.861        8.001          31784.
ENGLISH UNITS
CHANNEL LENGTH=    1800.0 CISCHARGE =      873.00C
B=    8.00 Z= C.8C0 RN=0.0120 SO= 0.CC0200
NORMAL DEPTH= 1C.495 CRITICAL CEPTH=  5.861
CCNTRCL IS DCWNSTREAM, DEPTH=  6.J00
     DISTANCE        DEPTH       ENERGY            F+M
        0.C         6.CC0        8.0C6          31811.
       34.7         6.299        8.053          32044.
      112.7         6.598        8.140          32504.
      243.7         6.897        8.259          33174.
      439.5         7.196        8.4C4          34046.
      715.8         7.495        8.571          35108.
     1092.8         7.794        8.756          36354.
     1597.9         8.C93        8.956          37779.
     1800.C         8.183        9.C19          38241.
```

Fig. 11.16 Computer output, Example 11.7.

stream depth as the control depth and computes in the upstream direction. Figure 11.16 shows the computer output from the last two data sets. The jump is located by finding the position of equal $F + M$ from the output of the last two data sets.

11.10 TRANSITIONS

At entrances to channels and at changes in cross section and bottom slope, the structure that conducts the liquid from the upstream section to the new section is a *transition*. Its purpose is to change the shape of flow and surface profile in such a manner that minimum losses result. A transition for tranquil flow from a rectangular channel to a trapezoidal channel is illustrated in Fig. 11.17. Applying the energy equation from section 1 to section 2 gives

$$\frac{V_1^2}{2g} + y_1 = \frac{V_2^2}{2g} + y_2 + z + E_1 \tag{11.10.1}$$

In general, the sections and depths are determined by other considerations, and z must be determined for the expected available energy loss E_1. By good design, i.e., with slowly tapering walls and flooring with no sudden changes in cross-sectional area, the losses can be held to about one-tenth the difference between velocity heads for accelerated flow and to about three-tenths the

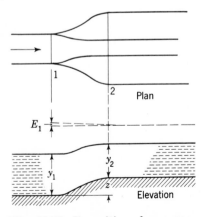

Fig. 11.17 Transition from rectangular channel to trapezoidal channel for tranquil flow.

difference between velocity heads for retarded flow. For rapid flow, wave mechanics is required in designing the transitions.[1]

EXAMPLE 11.8 In Fig. 11.17, 400 cfs flows through the transition; the rectangular section is 8 ft wide; and $y_1 = 8$ ft. The trapezoidal section is 6 ft wide at the bottom with side slopes 1:1, and $y_2 = 7.5$ ft. Determine the rise z in the bottom through the transition.

$$V_1 = \frac{400}{64} = 6.25 \qquad \frac{V_1^2}{2g} = 0.61 \qquad A_2 = 101.25 \text{ ft}^2$$

$$V_2 = \frac{400}{101.25} = 3.95 \qquad \frac{V_2^2}{2g} = 0.24 \qquad E_1 = 0.3\left(\frac{V_1^2}{2g} - \frac{V_2^2}{2g}\right) = 0.11$$

After substituting into Eq. (11.10.1),

$$z = 0.61 + 8 - 0.24 - 7.5 - 0.11 = 0.76 \text{ ft}$$

The *critical-depth meter*[2] is an excellent device for measuring discharge in an open channel. The relationships for determination of discharge are worked out for a rectangular channel of constant width, Fig. 11.18, with a raised floor over a reach of channel about $3y_c$ long. The raised floor is of such height that the restricted section becomes a control section with critical velocity occurring over it. By measuring only the upstream depth y_1, the discharge per foot of width is accurately determined. Applying the energy equation from section 1 to the critical section (exact location unimportant),

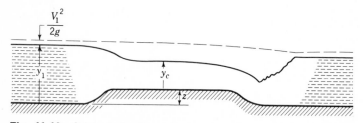

Fig. 11.18 Critical-depth meter.

[1] A. T. Ippen, Channel Transitions and Controls, in H. Rouse (ed.), "Engineering Hydraulics," Wiley, New York, 1950.

[2] H. W. King and E. F. Brater, "Handbook of Hydraulics," 5th ed., pp. **8**-14 to **8**-16, McGraw-Hill, New York, 1963.

including the transition-loss term, gives

$$\frac{V_1^2}{2g} + y_1 = z + y_c + \frac{V_c^2}{2g} + \frac{1}{10}\left(\frac{V_c^2}{2g} - \frac{V_1^2}{2g}\right)$$

Since

$$y_c + \frac{V_c^2}{2g} = E_c \qquad \frac{V_c^2}{2g} = \frac{E_c}{3}$$

in which E_c is the specific energy at critical depth,

$$y_1 + 1.1\frac{V_1^2}{2g} = z + 1.033E_c \qquad\qquad (11.10.2)$$

From Eq. (11.5.3)

$$y_c = \tfrac{2}{3}E_c = \left(\frac{q^2}{g}\right)^{1/3} \qquad\qquad (11.10.3)$$

In Eqs. (11.10.2) and (11.10.3) E_c is eliminated and the resulting equation solved for q,

$$q = 0.517g^{1/2}\left(y_1 - z + 1.1\frac{V_1^2}{2g}\right)^{3/2}$$

Since $q = V_1 y_1$, V_1 can be eliminated,

$$q = 0.517g^{1/2}\left(y_1 - z + \frac{0.55}{g}\frac{q^2}{y_1^2}\right)^{3/2} \qquad\qquad (11.10.4)$$

The equation is solved by trial. As y_1 and z are known and the right-hand term containing q is small, it may first be neglected for an approximate q. A value a little larger than the approximate q may be substituted on the right-hand side. When the two q's are the same the equation is solved. Once z and the width of channel are known, a chart or table may be prepared yielding Q for any y_1. Experiments indicate that accuracy within 2 to 3 percent may be expected.

With tranquil flow a jump occurs downstream from the meter and with rapid flow a jump occurs upstream from the meter.

EXAMPLE 11.9 In a critical-depth meter 2 m wide with $z = 0.3$ m the depth y_1 is measured to be 0.75 m. Find the discharge.

$$q = 0.517(9.806^{1/2})(0.45^{3/2}) = 0.489 \text{ m}^2/\text{s}$$

As a second approximation let q be 0.50,

$$q = 0.517(9.806^{1/2})\left(0.45 + \frac{0.55}{9.806} \times 0.5^2\right)^{3/2} = 0.512 \text{ m}^2/\text{s}$$

and as a third approximation, 0.513,

$$q = 0.517(9.806^{1/2})\left(0.45 + \frac{0.55}{9.806} \times 0.513^2\right)^{3/2} = 0.513 \text{ m}^2/\text{s}$$

Then

$$Q = 2 \times 0.513 = 1.026 \text{ m}^3/\text{s}$$

PROBLEMS

11.1 Show that for laminar flow to be ensured down an inclined surface, the discharge per unit width cannot be greater than 500ν. (See Prob. 5.10.)

11.2 Calculate the depth of laminar flow of water at 70°F down a plane surface making an angle of 30° with the horizontal for the lower critical Reynolds number. (See Prob. 5.10.)

11.3 Calculate the depth of turbulent flow at $\mathbf{R} = VR/\nu = 500$ for flow of water at 20°C down a plane surface making an angle θ of 30° with the horizontal. Use Manning's formula. $n = 0.01$; $S = \sin\theta$.

11.4 A rectangular channel is to carry 1.2 m³/s at a slope of 0.009. If the channel is lined with galvanized iron, $n = 0.011$, what is the minimum number of square meters of metal needed for each 100 m of channel? Neglect freeboard.

11.5 A trapezoidal channel, with side slopes 2 on 1 (2 horizontal to 1 vertical), is to carry 600 cfs with a bottom slope of 0.0009. Determine the bottom width, depth, and velocity for the best hydraulic section. $n = 0.025$.

11.6 A trapezoidal channel made out of brick, with bottom width 6 ft and with bottom slope 0.001, is to carry 600 cfs. What should the side slopes and depth of channel be for the least number of bricks?

11.7 What radius semicircular corrugated-metal channel is needed to convey 2.5 m³/s

Fig. 11.19

a distance of 1 km with a head loss of 2 m? Can you find another cross section that requires less perimeter?

11.8 Determine the best hydraulic trapezoidal section to convey 85 m³/s with a bottom slope of 0.001. The lining is finished concrete.

11.9 Calculate the discharge through the channel and floodway of Fig. 11.19 for steady uniform flow, with $S = 0.0009$ and $y = 8$ ft.

11.10 For 7000 cfs flow in the section of Fig. 11.19 when the depth over the floodway is 4 ft, calculate the energy gradient.

11.11 For 25,000 cfs flow through the section of Fig. 11.19, find the depth of flow in the floodway when the slope of the energy grade line is 0.0004.

11.12 Draw an $F + M$ curve for 2.5 m³/s per meter of width.

11.13 Draw the specific-energy curve for 2.5 m³/s per meter of width on the same chart as Prob. 11.12. What is the energy loss in a jump whose upstream depth is 0.5 m?

11.14 Prepare a plot of Eq. (11.4.7).

11.15 With $q = 100$ cfs/ft and $\mathbf{F}_1 = 12$, determine v_1, y_1, and the conjugate depth y_2.

11.16 Determine the two depths having a specific energy of 6 ft for 1 m³/s per meter of width.

11.17 What is the critical depth for flow of 18 cfs per foot of width?

11.18 What is the critical depth for flow of 0.3 m³/s through the cross section of Fig. 5.49?

11.19 Determine the critical depth for flow of 8.5 m³/s through a trapezoidal channel with a bottom width of 2.5 m and side slopes of 1 on 1.

11.20 An unfinished concrete rectangular channel 12 ft wide has a slope of 0.0009. It carries 480 cfs and has a depth of 7 ft at one section. By using the step method and taking one step only, compute the depth 1000 ft downstream.

11.21 Solve Prob. 11.20 by taking two equal steps. What is the classification of this water-surface profile?

11.22 A very wide gate (Fig. 11.20) admits water to a horizontal channel. Considering

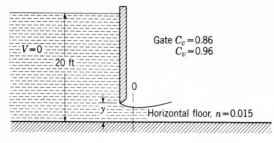

Gate $C_c = 0.86$
$C_v = 0.96$

Fig. 11.20

the pressure distribution hydrostatic at section 0, compute the depth at section 0 and the discharge per foot of width when $y = 3.0$ ft.

11.23   If the depth at section 0 of Fig. 11.20 is 2 ft and the discharge per foot of width is 65.2 cfs, compute the water-surface curve downstream from the gate.

11.24   Draw the curve of conjugate depths for the surface profile of Prob. 11.23.

11.25   If the very wide channel in Fig. 11.20 extends downstream 2000 ft and then has a sudden dropoff, compute the flow profile upstream from the end of the channel for $q = 65.2$ cfs/ft by integrating the equation for gradually varied flow.

11.26   Using the results of Probs. 11.24 and 11.25, determine the position of a hydraulic jump in the channel.

11.27   (*a*) In Fig. 11.21 the depth downstream from the gate is 0.6 m, and the velocity is 12 m/s. For a very wide channel, compute the depth at the downstream end of the adverse slope. (*b*) Solve part (*a*) by use of the computer program given in Fig. 11.14, or write a similar program to obtain the solution.

11.28   Sketch (without computation) and label all the liquid-surface profiles that can be obtained from Fig. 11.22 by varying z_1, z_2, and the lengths of the channels for $z_2 < z_1$, with a steep, inclined channel.

11.29   In Fig. 11.22 determine the possible combinations of control sections for various

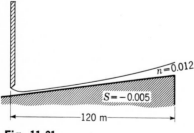

$n = 0.012$

$S = -0.005$

|← ——————— 120 m ——————— →|

Fig. 11.21

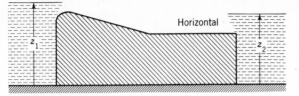

Fig. 11.22

values of z_1, z_2 and various channel lengths for $z_1 > z_2$, with the inclined channel always steep.

11.30 Sketch the various liquid-surface profiles and control sections for Fig. 11.22 obtained by varying channel length for $z_2 > z_1$.

11.31 Show an example of a channel that is mild for one discharge and steep for another discharge. What discharge is required for it to be critical?

11.32 Use the computer program in Fig. 11.14 or a similar program you have written to locate the hydraulic jump in a 90° triangular channel, 0.5 km long, that carries a flow of 1 m³/s, $n = 0.015$, $S_0 = 0.001$. The upstream depth is 0.2 m, and the downstream depth is 0.8 m.

11.33 Design a transition from a trapezoidal section, 8 ft bottom width and side slopes 1 on 1, depth 4 ft, to a rectangular section, 6 ft wide and 6 ft deep, for a flow of 250 cfs. The transition is to be 20 ft long, and the loss is one-tenth the difference between velocity heads. Show the bottom profile, and do not make any sudden changes in cross-sectional area.

11.34 A transition from a rectangular channel, 2.6 m wide and 2 m deep, to a trapezoidal channel, bottom width 4 m and side slopes 2 on 1, with depth 1.3 m has a loss four-tenths the difference between velocity heads. The discharge is 5.6 m³/s. Determine the difference between elevations of channel bottoms.

11.35 A critical-depth meter 20 ft wide has a rise in bottom of 2.0 ft. For an upstream depth of 3.52 ft determine the flow through the meter.

11.36 With flow approaching a critical-depth meter site at 6 m/s and a Froude number of 10, what is the minimum amount the floor must be raised?

11.37 In open-channel flow

(a) the hydraulic grade line is always parallel to the energy grade line
(b) the energy grade line coincides with the free surface
(c) the energy and hydraulic grade lines coincide
(d) the hydraulic grade line can never rise
(e) the hydraulic grade line and free surface coincide

11.38   Gradually varied flow is

(a) steady uniform flow   (b) steady nonuniform flow
(c) unsteady uniform flow   (d) unsteady nonuniform flow
(e) none of these answers

11.39   Tranquil flow must always occur

(a) above normal depth   (b) below normal depth   (c) above critical depth
(d) below critical depth   (e) on adverse slopes

11.40   Supercritical flow can never occur

(a) directly after a hydraulic jump   (b) in a mild channel
(c) in an adverse channel   (d) in a horizontal channel
(e) in a steep channel

11.41   Flow at critical depth occurs when

(a) changes in upstream resistance alter downstream conditions
(b) the specific energy is a maximum for a given discharge
(c) any change in depth requires more specific energy
(d) the normal depth and critical depth coincide for a channel
(e) the velocity is given by $\sqrt{2gy}$

11.42   The best hydraulic rectangular cross section occurs when (b = bottom width, y = depth)

(a) $y = 2b$ (b) $y = b$ (c) $y = b/2$ (d) $y = b^2$ (e) $y = b/5$

11.43   The best hydraulic canal cross section is defined as

(a) the least expensive canal cross section
(b) the section with minimum roughness coefficient
(c) the section that has a maximum area for a given flow
(d) the one that has a minimum perimeter
(e) none of these answers

11.44   The hydraulic jump always occurs from·

(a) an $\mathbf{M}_3$ curve to an $\mathbf{M}_1$ curve   (b) an $\mathbf{H}_3$ curve to an $\mathbf{H}_2$ curve
(c) an $\mathbf{S}_3$ curve to an $\mathbf{S}_1$ curve   (d) below normal depth to above normal depth
(e) below critical depth to above critical depth

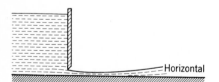

Fig. 11.23

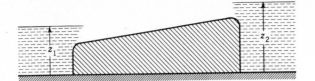

Fig. 11.24

11.45 Critical depth in a rectangular channel is expressed by

(a) $\sqrt{Vy}$ (b) $\sqrt{2gy}$ (c) $\sqrt{gy}$ (d) $\sqrt{q/g}$ (e) $(q^2/g)^{1/3}$

11.46 Critical depth in a nonrectangular channel is expressed by

(a) $Q^2T/gA^3 = 1$ (b) $QT^2/gA^2 = 1$ (c) $Q^2A^3/gT^2 = 1$ (d) $Q^2/gA^3 = 1$
(e) none of these answers

11.47 The specific energy for the flow expressed by $V = 4.43$ m/s $y = 1$ m, in meter-newtons per newton, is

(a) 2   (b) 3   (c) 5.43   (d) 9.86   (e) none of these answers

11.48 The minimum possible specific energy for a flow is 2.475 ft·lb/lb. The discharge per foot of width, in cubic feet per second, is

(a) 4.26   (b) 12.02   (c) 17   (d) 22.15   (e) none of these answers

11.49 The profile resulting from flow under the gate in Fig. 11.23 is classified as

(a) H_1 (b) H_2 (c) H_3 (d) A_2 (e) A_3

11.50 The number of different possible surface profiles that can occur for any variations of z_1, z_2, and length of channel in Fig. 11.24 is ($z_1 \neq z_2$)

(a) 2   (b) 3   (c) 4   (d) 5   (e) 6

11.51 The loss through a diverging transition is about

(a) $0.1 \dfrac{(V_1 - V_2)^2}{2g}$ (b) $0.1 \dfrac{(V_1^2 - V_2^2)}{2g}$ (c) $0.3 \dfrac{(V_1 - V_2)^2}{2g}$

(d) $0.3 \dfrac{V_1^2 - V_2^2}{2g}$ (e) none of these answers

11.52 A critical-depth meter

(a) measures the depth at the critical section
(b) is always preceded by a hydraulic jump
(c) must have a tranquil flow immediately upstream
(d) always has a hydraulic jump downstream
(e) always has a hydraulic jump associated with it

REFERENCES

Bakhmeteff, B. A.: "Hydraulics of Open Channels," McGraw-Hill, New York, 1932.

Chow, V. T.: "Open-channel Hydraulics," McGraw-Hill, New York, 1959.

Henderson, F. M.: "Open Channel Flow," Macmillan, New York, 1966.

12

UNSTEADY FLOW

Up to this point practically all flow cases examined have been for steady flow or have been reducible to a steady-flow situation. As technology advances and larger equipment is constructed or higher speeds employed, the problems of hydraulic transients become increasingly important. The hydraulic transients not only cause dangerously high pressures but produce excessive noise, fatigue, pitting due to cavitation, and disruption of normal control of circuits. Owing to the inherent period of certain systems of pipes, resonant vibrations may be incurred which can be destructive.

Hydraulic-transient analysis deals with the calculation of pressures and velocities during an unsteady-state mode of operation of a system. This may be caused by adjustment of a valve in a piping system, stopping a pump, or innumerable other possible changes in system operation.

The analysis of unsteady flow is much more complex than that of steady flow. Another independent variable enters, time, and equations may be partial differential equations rather than ordinary differential equations. The digital computer is ideally suited to the solution of such problems because of its large storage capacity and its ability to operate at very high computing rates. This chapter is in two parts: the first part dealing with closed-conduit transients, and the second part with open-channel unsteady flow.

First, oscillation of a U tube is studied, followed by its application to pipelines and reservoirs, the use of surge tanks, and the establishment of flow in a system. Equations are next developed for cases with more severe changes in velocity that require consideration of liquid compressibility and pipe-wall elasticity (usually called waterhammer). The open-channel cases are the positive and negative surge waves in a frictionless prismatic channel with instantaneous gate changes, the more general case of flood routing through a prismatic channel with friction, and flow over plane inclined surfaces due to rainfall.

FLOW IN CLOSED CONDUITS

The unsteady flow cases in closed conduits are treated as one-dimensional distributed-parameter problems. The equation of motion or the unsteady linear-momentum equation is used, and the unsteady continuity equation takes special forms. With nonlinear resistance terms for friction and other effects, the differential equations are frequently solved by numerical methods, usually by digital computer.

12.1 OSCILLATION OF LIQUID IN A U TUBE

Three cases of oscillations of liquid in a simple U tube are of interest: (1) frictionless liquid, (2) laminar resistance, and (3) turbulent resistance.

Frictionless liquid

For the frictionless case, Euler's equation of motion in unsteady form [Eq. (3.5.3)] may be applied. It is

$$\frac{1}{\rho}\frac{\partial p}{\partial s} + g\frac{\partial z}{\partial s} + v\frac{\partial v}{\partial s} + \frac{\partial v}{\partial t} = 0$$

When sections 1 and 2 are designated (Fig. 12.1) and the equation is integrated from 1 to 2, for incompressible flow

$$\frac{p_2 - p_1}{\rho} + g(z_2 - z_1) + \frac{v_2^2 - v_1^2}{2} + \int_1^2 \frac{\partial v}{\partial t}\,ds = 0 \qquad (12.1.1)$$

But $p_1 = p_2$ and $v_1 = v_2$; also $\partial v/\partial t$ is independent of s, hence

$$g(z_2 - z_1) = -L\frac{\partial v}{\partial t} \qquad (12.1.2)$$

in which L is the length of liquid column. By changing the elevation datum to the equilibrium position through the menisci, $g(z_2 - z_1) = 2gz$; since v is

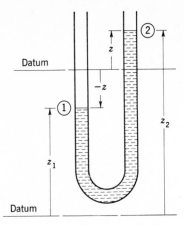

Fig. 12.1 Oscillation of liquid in a U tube.

a function of t only, $\partial v/\partial t$ may be written dv/dt, or d^2z/dt^2,

$$\frac{d^2z}{dt^2} = \frac{dv}{dt} = -\frac{2g}{L}z \qquad\qquad (12.1.3)$$

The general solution of this equation is

$$z = C_1 \cos \sqrt{\frac{2g}{L}}\, t + C_2 \sin \sqrt{\frac{2g}{L}}\, t$$

in which C_1 and C_2 are arbitrary constants of integration. The solution is readily checked by differentiating twice and substituting into the differential equation. To evaluate the constants, if $z = Z$ and $dz/dt = 0$ when $t = 0$, then $C_1 = Z$ and $C_2 = 0$, or

$$z = Z \cos \sqrt{\frac{2g}{L}}\, t \qquad\qquad (12.1.4)$$

This equation defines a simple harmonic motion of a meniscus, with a period for a complete oscillation of $2\pi \sqrt{L/2g}$. Velocity of the column may be obtained by differentiating z with respect to t.

EXAMPLE 12.1 A frictionless fluid column 2.18 m long has a speed of 2 m/s when $z = 0.5$ m. Find (a) the maximum value of z, (b) the maximum speed, and (c) the period.

(a)   Differentiating Eq. (12.1.4), after substituting for L, gives

$$\frac{dz}{dt} = -3Z \sin 3t$$

If t_1 is the time when $z = 0.5$ and $dz/dt = 2$,

$$0.5 = Z \cos 3t_1 \qquad -2 = -3Z \sin 3t_1$$

Dividing the second equation by the first equation gives

$$\tan 3t_1 = \tfrac{4}{3}$$

or $3t_1 = 0.927$ rad, $t_1 = 0.309$ s, $\sin 3t_1 = 0.8$, and $\cos 3t_1 = 0.6$. Then $Z = 0.5/(\cos 3t_1) = 0.5/0.6 = 0.833$ m, the maximum value of z.

(b)   The maximum speed occurs when

$$\sin 3t = 1, \qquad \text{or} \qquad 3Z = 3 \times 0.833 = 2.466 \text{ m/s}$$

(c)   The period is

$$2\pi \sqrt{\frac{L}{2g}} = 2.094 \text{ s}$$

Laminar resistance

When a shear stress τ_0 at the wall of the tube resists motion of the liquid column, it may be introduced into the Euler equation of motion along a streamline (Fig. 3.8). The resistance in length δs is $\tau_0 \pi D \, \delta s$. After dividing through by the mass of the particle $\rho A \, \delta s$, it is $4\tau_0/D$, and Eq. (12.1.1) becomes

$$\frac{1}{\rho} \frac{\partial p}{\partial s} + g \frac{\partial z}{\partial s} + v \frac{\partial v}{\partial s} + \frac{\partial v}{\partial t} + \frac{4\tau_0}{\rho D} = 0 \tag{12.1.5}$$

This equation is good for either laminar or turbulent resistance. The assumption is made that the frictional resistance in unsteady flow is the same as for the steady flow at the same velocity. From the Poiseuille equation the shear stress at the wall of a tube is

$$\tau_0 = \frac{8\mu v}{D} \tag{12.1.6}$$

After making the substitution for τ_0 in Eq. (12.1.5) and integrating with respect to s as before,

$$g(z_2 - z_1) + L \frac{\partial v}{\partial t} + \frac{32\nu v L}{D^2} = 0$$

Setting $2gz = g(z_2 - z_1)$, changing to total derivatives, and replacing v by dz/dt give

$$\frac{d^2 z}{dt^2} + \frac{32\nu}{D^2} \frac{dz}{dt} + \frac{2g}{L} z = 0 \qquad (12.1.7)$$

In effect, the column is assumed to have the average velocity dz/dt at any cross section.

By substitution

$$z = C_1 e^{at} + C_2 e^{bt}$$

can be shown to be the general solution of Eq. (12.1.7), provided that

$$a^2 + \frac{32\nu}{D^2} a + \frac{2g}{L} = 0 \qquad \text{and} \qquad b^2 + \frac{32\nu}{D^2} b + \frac{2g}{L} = 0$$

C_1 and C_2 are arbitrary constants of integration that are determined by given values of z and dz/dt at a given time. To keep a and b distinct, since the equations defining them are identical, they are taken with opposite signs before the radical term in solution of the quadratics; thus

$$a = -\frac{16\nu}{D^2} + \sqrt{\left(\frac{16\nu}{D^2}\right)^2 - \frac{2g}{L}} \qquad b = -\frac{16\nu}{D^2} - \sqrt{\left(\frac{16\nu}{D^2}\right)^2 - \frac{2g}{L}}$$

To simplify the formulas, if

$$m = \frac{16\nu}{D^2} \qquad n = \sqrt{\left(\frac{16\nu}{D^2}\right)^2 - \frac{2g}{L}}$$

then

$$z = C_1 \exp(-mt + nt) + C_2 \exp(-mt - nt)$$

When the initial condition is taken that $t = 0$, $z = 0$, $dz/dt = V_0$, then by substitution $C_1 = -C_2$, and

$$z = C_1 e^{-mt}(e^{nt} - e^{-nt}) \tag{12.1.8}$$

Since

$$\frac{e^{nt} - e^{-nt}}{2} = \sinh nt$$

Eq. (12.1.8) becomes

$$z = 2C_1 e^{-mt} \sinh nt$$

By differentiating with respect to t,

$$\frac{dz}{dt} = 2C_1(-me^{-mt} \sinh nt + ne^{-mt} \cosh nt)$$

and setting $dz/dt = V_0$ for $t = 0$ gives

$$V_0 = 2C_1 n$$

since $\sinh 0 = 0$ and $\cosh 0 = 1$. Then

$$z = \frac{V_0}{n} e^{-mt} \sinh nt \tag{12.1.9}$$

This equation gives the displacement z of one meniscus of the column as a function of time, starting with the meniscus at $z = 0$ when $t = 0$, and rising with velocity V_0.

Two principal cases[1] are to be considered. When

$$\frac{16\nu}{D^2} > \sqrt{\frac{2g}{L}}$$

n is a real number and the viscosity is so great that the motion is damped out in a partial cycle with z never becoming negative, Fig. 12.2 ($m/n = 2$). The time t_0 for maximum z to occur is found by differentiating z [Eq. (12.1.9)] with

---

[1] A third case, $16\nu/D^2 = \sqrt{2g/L}$, must be treated separately, yielding $z = V_0 t e^{-mt}$. The resulting oscillation is for a partial cycle only and is a limiting case of $16\nu/D^2 > \sqrt{2g/L}$.

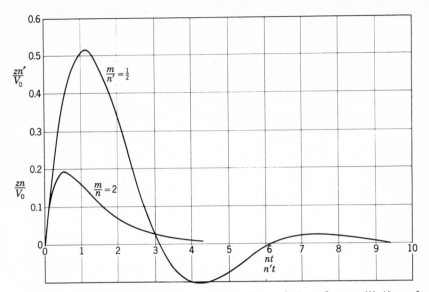

Fig. 12.2 Position of meniscus as a function of time for oscillation of liquid in a U tube with laminar resistance.

respect to t and equating to zero,

$$\frac{dz}{dt} = 0 = \frac{V_0}{n} \left(-me^{-mt} \sinh nt + ne^{-mt} \cosh nt \right)$$

or

$$\tanh nt_0 = \frac{n}{m} \tag{12.1.10}$$

Substitution of this value of t into Eq. (12.1.9) yields the maximum displacement Z

$$Z = \frac{V_0}{\sqrt{m^2 - n^2}} \left(\frac{m-n}{m+n} \right)^{m/2n} = V_0 \sqrt{\frac{L}{2g}} \left(\frac{m-n}{m+n} \right)^{m/2n} \tag{12.1.11}$$

The second case, when

$$\frac{16\nu}{D^2} < \sqrt{\frac{2g}{L}}$$

results in a negative term, within the radical,

$$n = \sqrt{-1\left[\frac{2g}{L} - \left(\frac{16\nu}{D^2}\right)^2\right]} = i\sqrt{\frac{2g}{L} - \left(\frac{16\nu}{D^2}\right)^2} = in'$$

in which $i = \sqrt{-1}$ and n' is a real number. Replacing n by in' in Eq. (12.1.9) produces the real function

$$z = \frac{V_0}{in'} e^{-mt} \sinh in't = \frac{V_0}{n'} e^{-mt} \sin n't \qquad (12.1.12)$$

since

$$\sin n't = \frac{1}{i} \sinh in't$$

The resulting motion of z is an oscillation about $z = 0$ with decreasing amplitude, as shown in Fig. 12.2 for the case $m/n' = \frac{1}{2}$. The time t_0 of maximum or minimum displacement is obtained from Eq. (12.1.12) by equating $dz/dt = 0$, producing

$$\tan n't_0 = \frac{n'}{m} \qquad (12.1.13)$$

There are an indefinite number of values of t_0 satisfying this expression, corresponding with all the maximum and minimum positions of a meniscus. By substitution of t_0 into Eq. (12.1.12)

$$Z = \frac{V_0}{\sqrt{n'^2 + m^2}} \exp\left(-\frac{m}{n'} \tan^{-1} \frac{n'}{m}\right)$$

$$= V_0 \sqrt{\frac{L}{2g}} \exp\left(-\frac{m}{n'} \tan^{-1} \frac{n'}{m}\right) \quad (12.1.14)$$

EXAMPLE 12.2 A 1.0-in-diameter U tube contains oil, $\nu = 1 \times 10^{-4}$ ft²/s, with a total column length of 120 in. Applying air pressure to one of the tubes makes the gage difference 16 in. By quickly releasing the air pressure the oil column is free to oscillate. Find the maximum velocity, the maximum Reynolds number, and the equation for position of one meniscus z, in terms of time.

The assumption is made that the flow is laminar, and the Reynolds number will be computed on this basis. The constants m and n are

$$m = \frac{16\nu}{D^2} = \frac{16 \times 10^{-4}}{(\frac{1}{12})^2} = 0.2302$$

$$n = \sqrt{\left(\frac{16\nu}{D^2}\right)^2 - \frac{2g}{L}} = \sqrt{0.2302^2 - \frac{2 \times 32.2}{10}} = \sqrt{-1}\, 2.527 = i2.527$$

or

$$n' = 2.527$$

Equations (12.1.12) to (12.1.14) apply to this case, as the liquid will oscillate above and below $z = 0$. The oscillation starts from the maximum position, that is, $Z = 0.667$ ft. By use of Eq. (12.1.14) the velocity (fictitious) when $z = 0$ at time t_0 before the maximum is determined to be

$$V_0 = Z\sqrt{\frac{2g}{L}} \exp\left(\frac{m}{n'} \tan^{-1}\frac{n'}{m}\right) = 0.667\sqrt{\frac{64.4}{10}} \exp\left(\frac{0.2302}{2.527} \tan^{-1}\frac{2.527}{0.2302}\right)$$

$$= 1.935 \text{ ft/s}$$

and

$$\tan n't_0 = \frac{n'}{m} \qquad t_0 = \frac{1}{2.527} \tan^{-1}\frac{2.527}{0.2302} = 0.586 \text{ s}$$

Hence by substitution into Eq. (12.1.12)

$$z = 0.766 \exp\left[-0.2302(t + 0.586)\right] \sin 2.527(t + 0.586)$$

in which $z = Z$ at $t = 0$. The maximum velocity (actual) occurs for $t > 0$. Differentiating with respect to t to obtain the expression for velocity,

$$V = \frac{dz}{dt} = -0.1763 \exp\left[-0.2302(t + 0.586)\right] \sin 2.57(t + 0.586)$$

$$+ 1.935 \exp\left[-0.2302(t + 0.586)\right] \cos 2.527(t + 0.586)$$

Differentiating again with respect to t and equating to zero to obtain maximum V produce

$$\tan 2.527(t + 0.586) = -0.1837$$

The solution in the second quadrant should produce the desired maximum, $t = 0.584$ s. Substituting this time into the expression for V produces $V = -1.48$ ft/s. The corresponding Reynolds number is

$$\mathbf{R} = \frac{VD}{\nu} = 1.48 \left(\tfrac{1}{12} \times 10^4 \right) = 1234$$

hence the assumption of laminar resistance is justified.

Turbulent resistance

In the majority of practical cases of oscillation, or surge, in pipe systems there is turbulent resistance. With large pipes and tunnels the Reynolds number is large except for those time periods when the velocity is very near to zero. The assumption of fluid resistance proportional to the square of the average velocity is made (constant f). It closely approximates true conditions, although it yields too small a resistance for slow motions, in which case resistance is almost negligible. The equations will be developed for $f = $ const for oscillation within a simple U tube. This case will then be extended to include oscillation of flow within a pipe or tunnel between two reservoirs, taking into account the minor losses. The assumption is again made that resistance in unsteady flow is given by steady-flow resistance at the same velocity.

Using Eq. (5.10.2) to substitute for τ_0 in Eq. (12.1.5) leads to

$$\frac{1}{\rho} \frac{\partial p}{\partial s} + g \frac{\partial z}{\partial s} + v \frac{\partial v}{\partial s} + \frac{\partial v}{\partial t} + \frac{fv^2}{2D} = 0 \tag{12.1.15}$$

When this equation is integrated from section 1 to section 2 (Fig. 12.1), the first term drops out as the limits are $p = 0$ in each case; the third term drops out as $\partial v / \partial s \equiv 0$; the fourth and fifth terms are independent of s, hence

$$g(z_2 - z_1) + \frac{\partial v}{\partial t} L + \frac{fv^2}{2D} L = 0$$

Since v is a function of t only the partial may be replaced with the total derivative

$$\frac{dv}{dt} + \frac{f}{2D} v \, | \, v \, | + \frac{2g}{L} z = 0 \tag{12.1.16}$$

The absolute-value sign on the velocity term is needed so that the resistance opposes the velocity, whether positive or negative. By expressing $v = dz/dt$,

$$\frac{d^2z}{dt^2} + \frac{f}{2D}\frac{dz}{dt}\left|\frac{dz}{dt}\right| + \frac{2g}{L}z = 0 \tag{12.1.17}$$

This is a nonlinear differential equation because of the v-squared term. It can be integrated once with respect to t, but no closed solution is known for the second integration. It is easily handled by the Runge-Kutta methods (Appendix E) with the digital computer when initial conditions are known: $t = t_0$, $z = z_0$, $dz/dt = 0$. Much can be learned from Eq. (12.1.17), however, by restricting the motion to the $-z$ direction; thus

$$\frac{d^2z}{dt^2} - \frac{f}{2D}\left(\frac{dz}{dt}\right)^2 + \frac{2g}{L}z = 0 \tag{12.1.18}$$

The equation may be integrated once,[1] producing

$$\left(\frac{dz}{dt}\right)^2 = \frac{4gD^2}{f^2L}\left(1 + \frac{fz}{D}\right) + Ce^{fz/D} \tag{12.1.19}$$

in which C is the constant of integration. To evaluate the constant, if $z = z_m$ for $dz/dt = 0$,

$$C = -\frac{4gD^2}{f^2L}\left(1 + \frac{fz_m}{D}\right)\exp\left(-\frac{fz_m}{D}\right)$$

and

$$\left(\frac{dz}{dt}\right)^2 = \frac{4gD^2}{f^2L}\left[1 + \frac{fz}{D} - \left(1 + \frac{fz_m}{D}\right)\exp\frac{f(z - z_m)}{D}\right] \tag{12.1.20}$$

---

[1] By substitution of

$$p = \frac{dz}{dt} \qquad \frac{d^2z}{dt^2} = \frac{dp}{dt} = \frac{dp}{dz}\frac{dz}{dt} = p\frac{dp}{dz}$$

then

$$p\frac{dp}{dz} - \frac{f}{2D}p^2 + \frac{2gz}{L} = 0$$

This equation can be made exact by multiplying by the integrating factor $e^{-fz/D}$. For the detailed method see Earl D. Rainville, "Elementary Differential Equations," 3d ed., Macmillan, New York, 1964.

Although this equation cannot be integrated again, numerical integration of particular situations yields z as a function of t. The equation, however, can be used to determine the magnitude of successive oscillations. At the instants of maximum or minimum z, say z_m and z_{m+1}, respectively, $dz/dt = 0$, and Eq. (12.1.20) simplifies to

$$\left(1 + \frac{fz_m}{D}\right) \exp\left(-\frac{fz_m}{D}\right) = \left(1 + \frac{fz_{m+1}}{D}\right) \exp\left(-\frac{fz_{m+1}}{D}\right) \tag{12.1.21}$$

Since Eq. (12.1.18), the original equation, holds only for decreasing z, z_m must be positive and z_{m+1} negative. To find z_{m+2} the other meniscus could be considered and z_{m+1} as a positive number substituted into the left-hand side of the equation to determine a minus z_{m+2} in place of z_{m+1} on the right-hand side of the equation.

EXAMPLE 12.3 A U tube consisting of 50-cm-diameter pipe with $f = 0.03$ has a maximum oscillation (Fig. 12.1) of $z_m = 6$ m. Find the minimum position of the surface and the following maximum.

From Eq. (12.1.21)

$$\left(1 + \frac{0.03 \times 6}{0.5}\right) \exp\left(-\frac{0.03 \times 6}{0.5}\right) = (1 + 0.06z_{m+1}) \exp\left(-0.06z_{m+1}\right)$$

or

$$(1 + 0.06z_{m+1}) \exp\left(-0.06z_{m+1}\right) = 0.9488$$

which is satisfied by $z_{m+1} = -4.84$ m. Using $z_m = 4.84$ m in Eq. (12.1.21),

$$(1 + 0.06z_{m+1}) \exp\left(-0.06z_{m+1}\right) = (1 + 0.06 \times 4.84) \exp\left(-0.06 \times 4.84\right)$$

$$= 0.9651$$

which is satisfied by $z_{m+1} = -4.05$ m. Hence, the minimum water surface is $z = -4.84$ m, and the next maximum is $z = 4.05$ m.

Equation (12.1.21) can be solved graphically. If $\phi = fz/D$, then

$$F(\phi) = (1 + \phi)e^{-\phi} \tag{12.1.22}$$

which is conveniently plotted with $F(\phi)$ as ordinate and both $-\phi$ and $+\phi$ on the same abscissa scale (Fig. 12.3). Successive values of ϕ are found as indicated by the dashed stepped line.

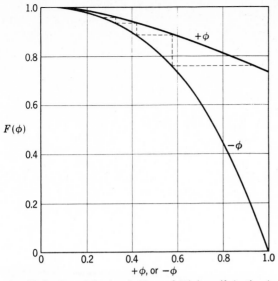

Fig. 12.3 Graphical solution of $F(\phi) = (1+\phi)e^{-\phi}$, yielding successive maximum and minimum displacements.

Although z cannot be found as a function of t from Eq. (12.1.20), V is given as a function of z, since $V = dz/dt$. The maximum value of V is found by equating $dV^2/dz = 0$ to find its position z'; thus

$$\frac{dV^2}{dz} = 0 = \frac{f}{D} - \left(1 + \frac{fz_m}{D}\right)\left[\exp\frac{f(z' - z_m)}{D}\right]\frac{f}{D}$$

After solving for z',

$$z' = z_m - \frac{D}{f}\ln\left(1 + \frac{fz_m}{D}\right)$$

and after substituting back into Eq. (12.1.20),

$$V_m{}^2 = \frac{4gD^2}{f^2L}\left[\frac{fz_m}{D} - \ln\left(1 + \frac{fz_m}{D}\right)\right] \tag{12.1.23}$$

Oscillation of two reservoirs

The equation for oscillation of two reservoirs connected by a pipeline is the same as that for oscillation of a U tube, except for value of constant terms.

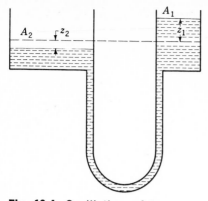

Fig. 12.4 Oscillation of two reservoirs.

If z_1 and z_2 represent displacements of the reservoir surfaces from their equilibrium positions (Fig. 12.4), and if z represents displacement of a water particle within the connecting pipe from its equilibrium position,

$$zA = z_1 A_1 = z_2 A_2$$

in which A_1 and A_2 are the reservoir areas, assumed to be constant in this derivation. Taking into account minor losses in the system by using the equivalent length L_e of pipe and fittings plus other minor losses, the Euler equation with resistance included is

$$-\gamma A\,(z_1 + z_2) + \frac{\gamma A f L_e}{2gD}\left(\frac{dz}{dt}\right)^2 = \frac{\gamma A L}{g}\frac{d^2z}{dt^2}$$

for z decreasing. After simplifying

$$\frac{d^2z}{dt^2} - \frac{f}{2D}\frac{L_e}{L}\left(\frac{dz}{dt}\right)^2 + \frac{gA}{L}\left(\frac{1}{A_1} + \frac{1}{A_2}\right)z = 0 \tag{12.1.24}$$

After comparing with Eq. (12.1.18), f is replaced by fL_e/L, and $2g/L$ by $gA\,(1/A_1 + 1/A_2)/L$. In Eq. (12.1.22)

$$\phi = f\frac{L_e}{L}\frac{z}{D}$$

EXAMPLE 12.4 In Fig. 12.4 a valve is opened suddenly in the pipeline when $z_1 = 40$ ft. $L = 2000$ ft, $A_1 = 200$ ft^2, $A_2 = 300$ ft^2, $D = 3.0$ ft, $f = 0.024$, and minor losses are $3.50V^2/2g$. Determine the subsequent maximum negative and positive surges in the reservoir A_1.

The equivalent length of minor losses is

$$\frac{KD}{f} = \frac{3.5 \times 3}{0.024} = 438 \text{ ft}$$

Then $L_e = 2000 + 438 = 2438$ and

$$z_m = \frac{z_1 A_1}{A} = \frac{40 \times 200}{2.25\pi} = 1132 \text{ ft}$$

The corresponding ϕ is

$$\phi = f\frac{L_e}{L}\frac{z_m}{D} = 0.024\left(\tfrac{2438}{2000}\right)\left(\tfrac{1132}{3}\right) = 11.04$$

and

$$F(\phi) = (1 + \phi)e^{-\phi} = (1 + 11.04)e^{-11.04} = 0.000193$$

which is satisfied by $\phi \approx -1.0$. Then

$$F(\phi) = (1 + 1)e^{-1} = 0.736 = (1 + \phi)e^{-\phi}$$

which is satisfied by $\phi = -0.593$. The values of z_m are, for $\phi = -1$,

$$z_m = \frac{\phi LD}{fL_e} = \frac{-1 \times 2000 \times 3}{0.024 \times 2438} = -102.6$$

and for $\phi = 0.593$,

$$z_m = \frac{0.593 \times 2000 \times 3}{0.024 \times 2438} = 60.9$$

The corresponding values of z_1 are

$$z_1 = z_m\frac{A}{A_1} = -102.6 \times \frac{2.25\pi}{200} = -3.63 \text{ ft}$$

and

$$z_1 = 60.9 \times \frac{2.25\pi}{200} = 2.15 \text{ ft}$$

12.2 ESTABLISHMENT OF FLOW

The problem of determination of time for flow to become established in a pipeline when a valve is suddenly opened is easily handled when friction and minor losses are taken into account. After a valve is opened (Fig. 12.5), the head H is available to accelerate the flow in the first instants, but as the velocity increases, the accelerating head is reduced by friction and minor losses. If L_e is the equivalent length of the pipe system, the final velocity V_0 is given by application of the energy equation

$$H = f \frac{L_e}{D} \frac{V_0^2}{2g} \tag{12.2.1}$$

The equation of motion is

$$\gamma A \left(H - f \frac{L_e}{D} \frac{V^2}{2g} \right) = \frac{\gamma A L}{g} \frac{dV}{dt}$$

Solving for dt and rearranging, with Eq. (12.2.1), give

$$\int_0^t dt = \frac{L V_0^2}{gH} \int_0^V \frac{dV}{V_0^2 - V^2}$$

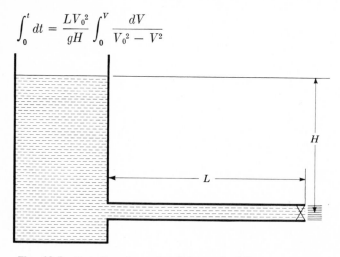

Fig. 12.5 Notation for establishment of flow.

After integration,

$$t = \frac{LV_0}{2gH} \ln \frac{V_0 + V}{V_0 - V} \tag{12.2.2}$$

The velocity V approaches V_0 asymptotically; i.e., mathematically it takes infinite time for V to attain the value V_0. Practically, for V to reach 0.99 V_0 takes

$$t = \frac{LV_0}{gH} \frac{1}{2} \ln \frac{1.99}{0.01} = 2.646 \frac{LV_0}{gH}$$

V_0 must be determined by taking minor losses into account, but Eq. (12.2.2) does not contain L_e.

EXAMPLE 12.5 In Fig. 12.5 the minor losses are $16V^2/2g$, $f = 0.030$, $L = 3000$ m, $D = 2.4$ m, and $H = 20$ m. Determine the time, after the sudden opening of a valve, for velocity to attain nine-tenths the final velocity.

$$L_e = 3000 + \frac{16 \times 2.4}{0.03} = 3128 \text{ m}$$

From Eq. (12.2.1)

$$V_0 = \sqrt{\frac{2gHD}{fL_e}} = \sqrt{\frac{19.612 \times 20 \times 2.4}{0.030 \times 3128}} = 3.17 \text{ m/s}$$

Substituting $V = 0.9V_0$ into Eq. (12.2.2) gives

$$t = \frac{3000 \times 3.17}{19.612 \times 20} \ln \frac{1.90}{0.10} = 71.3 \text{ s}$$

12.3 SURGE CONTROL

The oscillation of flow in pipelines, when compressibility effects are not important, is referred to as *surge*. For sudden deceleration of flow due to closure of the flow passage, compressibility of the liquid and elasticity of the pipe walls must be considered; this phenomenon, known as *waterhammer*, is discussed in Secs. 12.4 to 12.7. Oscillations in a U tube are special cases of surge. As one means of eliminating waterhammer, provision is made to permit the

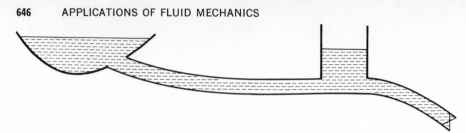

Fig. 12.6 Surge tank on a long pipeline.

liquid to surge into a tank (Fig. 12.6). The valve at the end of a pipeline may be controlled by a turbine governor, and may rapidly stop the flow if the generator loses its load. To destroy all momentum in the long pipe system quickly would require high pressure, which in turn would require a very costly pipeline. With a surge tank as near the valve as feasible, although surge will occur between the reservoir and surge tank, development of high pressure in this reach is prevented. It is still necessary to design the pipeline between surge tank and valve to withstand waterhammer.

Surge tanks may be classified as *simple, orifice,* and *differential.* The simple surge tank has an unrestricted opening into it and must be large enough not to overflow (unless a spillway is provided) or not to be emptied, allowing air to enter the pipeline. It must also be of a size that will not fluctuate in resonance with the governor action on the valve. The period of oscillation of a simple surge tank is relatively long.

The orifice surge tank has a restricted opening, or orifice, between pipeline and tank and hence allows more rapid pressure changes in the pipeline than the simple surge tank. The more rapid pressure change causes a more rapid adjustment of flow to the new valve setting, and losses through the orifice aid in dissipating excess available energy resulting from valve closure.

The differential surge tank (Fig. 12.7) is in effect a combination of an orifice surge tank and a simple surge tank of small cross-sectional area. In case of rapid valve opening a limited amount of liquid is directly available

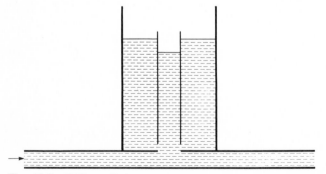

Fig. 12.7 Differential surge tank.

from the central riser, and flow from the large tank supplements this flow. For sudden valve closures the central riser may be designed so that it overflows into the outside tank.

Surge tanks operating under air pressure are utilized in certain circumstances, e.g., after a reciprocating pump. They are generally uneconomical for large pipelines.

Detailed analysis of surge tanks entails a numerical integration of the equation of motion for the liquid in the pipeline, taking into account the particular rate of valve closure, together with the continuity equation. The particular type of surge tank to be selected for a given situation depends upon a detailed study of the economics of the pipeline system. High-speed digital computers are most helpful in their design.

Another means of controlling surge and waterhammer is to supply a quick-opening bypass valve that opens when the control valve closes. The quick-opening valve has a controlled slow closure at such a rate that excessive pressure is not developed in the line. The bypass valve wastes liquid, however, and does not provide relief from surge due to opening of the control valve or starting of a pump.

The following sections on closed-conduit flow take into account compressibility of the liquid and elasticity of the pipe walls. Waterhammer calculations may be accomplished in several ways; the characteristics method, recommended for general use in computer solutions, is presented here.

12.4 DESCRIPTION OF THE WATERHAMMER PHENOMENON

Waterhammer may occur in a closed conduit flowing full when there is either a retardation or acceleration of the flow, such as with the change in opening of a valve in the line. If the changes are gradual, the calculations may be carried out by surge methods, considering the liquid incompressible and the conduit rigid. When a valve is rapidly closed in a pipeline during flow, the flow through the valve is reduced. This increases the head on the upstream side of the valve and causes a pulse of high pressure to be propagated upstream. The action of this pressure pulse is to decrease the velocity of flow. On the downstream side of the valve the pressure is reduced, and a wave of lowered pressure travels downstream, which also reduces the velocity. If the closure is rapid enough and the steady pressure low enough, a vapor pocket may be formed downstream from the valve. When this occurs, the cavity will eventually collapse and produce a high-pressure wave downstream.

Before undertaking the derivation of equations for solution of waterhammer, a description of the sequence of events following sudden closure of a valve at the downstream end of a pipe leading from a reservoir (Fig. 12.8) is given. Friction is neglected in this case. At the instant of valve closure

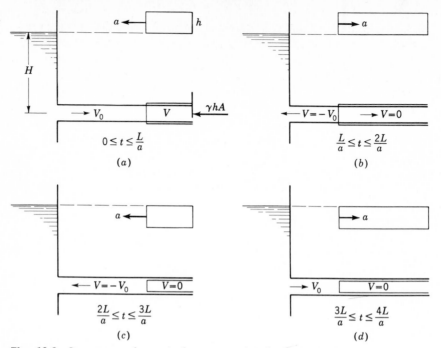

Fig. 12.8 Sequence of events for one cycle of sudden closure of a valve.

($t = 0$) the fluid nearest the valve is compressed and brought to rest, and the pipe wall is stretched (Fig. 12.8a). As soon as the first layer is compressed the process is repeated for the next layer. The fluid upstream from the valve continues to move downstream with undiminished speed until successive layers have been compressed back to the source. The high pressure moves upstream as a wave, bringing the fluid to rest as it passes, compressing it, and expanding the pipe. When the wave reaches the upstream end of the pipe ($t = L/a$ s), all the fluid is under the extra head h, all the momentum has been lost, and all the kinetic energy has been converted into elastic energy.

There is an unbalanced condition at the upstream (reservoir) end at the instant of arrival of the pressure wave, as the reservoir pressure is unchanged. The fluid starts to flow backward, beginning at the upstream end. This flow returns the pressure to the value which was normal before closure, the pipe wall returns to normal, and the fluid has a velocity V_0 in the backward sense. This process of conversion travels downstream toward the valve at the speed of sound a in the pipe. At the instant $2L/a$ the wave arrives at the valve, pressures are back to normal along the pipe, and the velocity is everywhere V_0 in the backward direction.

Since the valve is closed, no fluid is available to maintain the flow at the valve and a low pressure develops ($-h$) such that the fluid is brought to rest.

This low-pressure wave travels upstream at speed a and everywhere brings the fluid to rest, causes it to expand because of the lower pressure, and allows the pipe walls to contract. (If the static pressure in the pipe is not sufficiently high to sustain head $-h$ above vapor pressure, the liquid vaporizes in part and continues to move backward over a longer period of time.)

At the instant the negative pressure wave arrives at the upstream end of the pipe, $3L/a$ s after closure, the fluid is at rest but uniformly at head $-h$ less than before closure. This leaves an unbalanced condition at the reservoir, and fluid flows into the pipe, acquiring a velocity V_0 forward and returning the pipe and fluid to normal conditions as the wave progresses downstream at speed a. At the instant this wave reaches the valve, conditions are exactly the same as at the instant of closure, $4L/a$ s earlier.

This process is then repeated every $4L/a$ s. The action of fluid friction and imperfect elasticity of fluid and pipe wall, neglected heretofore, is to damp out the vibration and eventually cause the fluid to come permanently to rest. Closure of a valve in less than $2L/a$ is called *rapid closure*; *slow closure* refers to times of closure greater than $2L/a$.

The sequence of events taking place in a pipe may be compared with the sudden stopping of a freight train when the engine hits an immovable object. The car behind the engine compresses the spring in its forward coupling and stops as it exerts a force against the engine, and each car in turn keeps moving at its original speed until the preceding one suddenly comes to rest. When the caboose is at rest all the energy is stored in compressing the coupling springs (neglecting losses). The caboose has an unbalanced force exerted on it, and starts to move backward, which in turn causes an unbalanced force on the next car, setting it in backward motion. This action proceeds as a wave toward the engine, causing each car to move at its original speed in a backward direction. If the engine is immovable, the car next to it is stopped by a tensile force in the coupling between it and the engine, analogous to the low-pressure wave in waterhammer. The process repeats itself car by car until the train is again at rest, with all couplings in tension. The caboose is then acted upon by the unbalanced tensile force in its coupling and is set into forward motion, followed in turn by the rest of the cars. When this wave reaches the engine all cars are in motion as before the original impact. Then the whole cycle is repeated again. Friction acts to reduce the energy to zero in a very few cycles.

12.5 DIFFERENTIAL EQUATIONS FOR CALCULATION OF WATERHAMMER

Two basic mechanics equations are applied to a short segment of fluid in a pipe to obtain the differential equations for transient flow: Newton's second law of motion and the continuity equation. The dependent variables are the elevation of hydraulic grade line H above a fixed datum and the average velocity V at a cross section. The independent variables are distance x along the pipe measured from the upstream end and time t; hence, $H = H(x,t)$, $V = V(x,t)$. Poisson's ratio effect is not taken into account in this derivation. For pipelines with expansion joints it does not enter into the derivation. Friction is considered to be proportional to the square of the velocity.

Equation of motion

The fluid element between two parallel planes δx apart, normal to the pipe axis, is taken as a free body for application of Newton's second law of motion in the axial direction (Fig. 12.9). In equation form

$$pA - \left[pA + \frac{\partial}{\partial x}(pA)\,\delta x\right] + p\,\frac{\partial A}{\partial x}\,\delta x + \gamma A\,\delta x\,\sin\theta - \tau_0\pi D\,\delta x = \rho A\,\delta x\,\frac{dV}{dt}$$

Dividing through by the mass of the element $\rho A\,\delta x$ and simplifying give

$$-\frac{1}{\rho}\frac{\partial p}{\partial x} + g\sin\theta - \frac{4\tau_0}{\rho D} = \frac{dV}{dt} \tag{12.5.1}$$

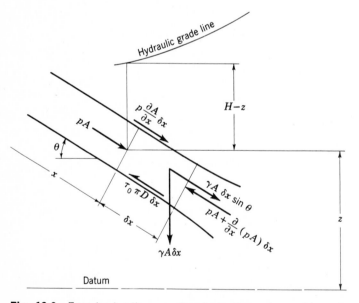

Fig. 12.9 Free-body diagram for derivation of equation of motion.

Introducing the hydraulic-grade-line elevation, from $p = \rho g(H - z)$, leads to

$$\frac{\partial p}{\partial x} \approx \rho g \left(\frac{\partial H}{\partial x} - \frac{\partial z}{\partial x} \right) \tag{12.5.2}$$

But

$$\frac{\partial z}{\partial x} = -\sin \theta$$

and

$$g \frac{\partial H}{\partial x} + \frac{4\tau_0}{\rho D} + \frac{dV}{dt} = 0 \tag{12.5.3}$$

For steady turbulent flow, $\tau_0 = \rho f V^2 / 8$ [Eq. (5.10.2)]. The assumption is made that the friction factor in unsteady flow is the same as in steady flow. Hence, the equation of motion becomes

$$g \frac{\partial H}{\partial x} + \frac{dV}{dt} + \frac{f V^2}{2D} = 0 \tag{12.5.4}$$

Since friction must oppose the motion, V^2 is written as $V \mid V \mid$ to provide the proper sign. After expanding the acceleration term,

$$L_2 = g \frac{\partial H}{\partial x} + V \frac{\partial V}{\partial x} + \frac{\partial V}{\partial t} + \frac{f V \mid V \mid}{2D} = 0 \tag{12.5.5}$$

The equation is indicated by L_2 to distinguish it from the equation of continuity L_1, which is next derived.

Equation of continuity

The unsteady continuity equation (3.2.1) is applied to the control volume of Fig. 12.10,

$$-\frac{\partial}{\partial x} (\rho A V) \, \delta x = \frac{\partial}{\partial t} (\rho A \, \delta x) \tag{12.5.6}$$

in which δx is not a function of t. Expanding the equation and dividing

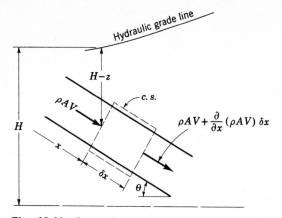

Fig. 12.10 Control volume for derivation of continuity equation.

through by the mass $\rho A \, \delta x$ give

$$\frac{V}{A}\frac{\partial A}{\partial x} + \frac{1}{A}\frac{\partial A}{\partial t} + \frac{V}{\rho}\frac{\partial \rho}{\partial x} + \frac{1}{\rho}\frac{\partial \rho}{\partial t} + \frac{\partial V}{\partial x} = 0 \tag{12.5.7}$$

The first two terms are the total derivative[1] $(1/A) \, dA/dt$, and the next two terms the total derivative $(1/\rho) \, d\rho/dt$, yielding

$$\frac{1}{A}\frac{dA}{dt} + \frac{1}{\rho}\frac{d\rho}{dt} + \frac{\partial V}{\partial x} = 0 \tag{12.5.8}$$

The first term deals with the elasticity of the pipe wall and its rate of deformation with pressure; the second term takes into account the compressibility of the liquid. For the wall elasticity the rate of change of tensile force per unit length (Fig. 12.11) is $(D/2) \, dp/dt$; when divided by the wall thickness t', it is the rate of change of unit stress $(D/2t') \, dp/dt$; when this is divided by Young's modulus of elasticity for the wall material, the rate of increase of unit strain is obtained, $(D/2t'E) \, dp/dt$. After multiplying this by the radius $D/2$, the rate of radial extension is obtained; finally, by multiplying by the perimeter πD the rate of area increase is obtained:

$$\frac{dA}{dt} = \frac{D}{2t'E}\frac{dp}{dt}\frac{D}{2}\,\pi D$$

[1] See Appendix B.

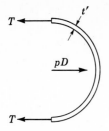

Fig. 12.11 Tensile force in pipe wall.

and hence

$$\frac{1}{A}\frac{dA}{dt} = \frac{D}{t'E}\frac{dp}{dt} \tag{12.5.9}$$

From the definition of bulk modulus of elasticity of fluid (Chap. 1),

$$K = -\frac{dp}{d\mathbb{v}/\mathbb{v}} = \frac{dp}{d\rho/\rho}$$

and the rate of change of density divided by density yields

$$\frac{1}{\rho}\frac{d\rho}{dt} = \frac{1}{K}\frac{dp}{dt} \tag{12.5.10}$$

By Eqs. (12.5.9) and (12.5.10), Eq. (12.5.8) becomes

$$\frac{1}{K}\frac{dp}{dt}\left(1 + \frac{K}{E}\frac{D}{t'}\right) + \frac{\partial V}{\partial x} = 0 \tag{12.5.11}$$

It is convenient to express the constants in this equation in the form

$$a^2 = \frac{K/\rho}{1 + (K/E)(D/t')c_1} \tag{12.5.12}$$

in which c_1 is unity for the pipeline with expansion joints. Equation (12.5.11) now becomes

$$\frac{1}{\rho}\frac{dp}{dt} + a^2\frac{\partial V}{\partial x} = 0 \tag{12.5.13}$$

Now, since $p = \rho g (H - z)$ (Fig. 12.10),

$$\frac{dp}{dt} = V \frac{\partial p}{\partial x} + \frac{\partial p}{\partial t} = V \rho g \left(\frac{\partial H}{\partial x} - \frac{\partial z}{\partial x} \right) + \rho g \left(\frac{\partial H}{\partial t} - \frac{\partial z}{\partial t} \right)$$

The change of ρ with respect to x or t is very much less than the change of H with respect to x or t, and so ρ was considered constant in the preceding equation. If the pipe is at rest, $\partial z / \partial t = 0$, and $\partial z / \partial x = -\sin \theta$; hence

$$\frac{1}{\rho} \frac{dp}{dt} = V g \left(\frac{\partial H}{\partial x} + \sin \theta \right) + g \frac{\partial H}{\partial t}$$

and Eq. (12.5.13) becomes

$$L_1 = \frac{a^2}{g} \frac{\partial V}{\partial x} + V \frac{\partial H}{\partial x} + \frac{\partial H}{\partial t} + V \sin \theta = 0 \qquad (12.5.14)$$

which is the continuity equation for a compressible liquid in an elastic pipe. L_1 and L_2 provide two nonlinear partial differential equations in V and H in terms of the independent variables x and t. No general solution to these equations is known, but they can be solved by the method of characteristics for a convenient finite-difference solution with the digital computer.

12.6 THE METHOD-OF-CHARACTERISTICS SOLUTION

Equations L_1 and L_2 in the preceding section contain two unknowns. These equations may be combined with an unknown multiplier as $L = L_1 + \lambda L_2$. Any two real, distinct values of λ yield two equations in V and H that contain all the physics of the original two equations L_1 and L_2 and may replace them in any solution. It may happen that great simplification will result if two particular values of λ are found. L_1 and L_2 are substituted into the equation for L, with some rearrangement.

$$L = \left[\frac{\partial H}{\partial x} (V + \lambda g) + \frac{\partial H}{\partial t} \right] + \lambda \left[\frac{\partial V}{\partial x} \left(V + \frac{a^2}{g\lambda} \right) + \frac{\partial V}{\partial t} \right]$$

$$+ V \sin \theta + \lambda \frac{fV \, | \, V \, |}{2D} = 0 \quad (12.6.1)$$

This expression is arranged so that the first term in brackets would be the total derivative dH/dt if

$$\frac{dx}{dt} = V + \lambda g \tag{12.6.2}$$

and the second term in brackets would be dV/dt if

$$\frac{dx}{dt} = V + \frac{a^2}{g\lambda} \tag{12.6.3}$$

since

$$\frac{dH}{dt} = \frac{\partial H}{\partial x}\frac{dx}{dt} + \frac{\partial H}{\partial t} \qquad \frac{dV}{dt} = \frac{\partial V}{\partial x}\frac{dx}{dt} + \frac{\partial V}{\partial t}$$

from calculus. Equations (12.6.2) and (12.6.3) must be equivalent,

$$V + \lambda g = V + \frac{a^2}{g\lambda} \tag{12.6.4}$$

Solving for λ gives

$$\lambda = \pm \frac{a}{g} \tag{12.6.5}$$

Therefore, two real, distinct values of λ have been found that convert the two partial differential equations into a pair of total differential equations restricted by Eqs. (12.6.2) and (12.6.3).

These equations, with λ substituted, become

$$\left.\begin{array}{l} \dfrac{dH}{dt} + \dfrac{a}{g}\dfrac{dV}{dt} + V \sin \theta + \dfrac{afV \mid V \mid}{2gD} = 0 \qquad (12.6.6) \\[4mm] \dfrac{dx}{dt} = V + a \qquad (12.6.7) \end{array}\right\} C^+$$

$$\left.\begin{array}{l} \dfrac{dH}{dt} - \dfrac{a}{g}\dfrac{dV}{dt} + V \sin \theta - \dfrac{afV \mid V \mid}{2gD} = 0 \qquad (12.6.8) \\[4mm] \dfrac{dx}{dt} = V - a \qquad (12.6.9) \end{array}\right\} C^-$$

To understand the significance of these four equations, it is convenient

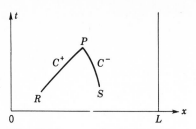

Fig. 12.12 xt plot of characteristics along which solution is obtained.

to consider the solution being carried out on an xt plot (Fig. 12.12). Consider that V and H are known at the two known locations R and S in the figure. The curve labeled C^+ is a plot of Eq. (12.6.7); Eq. (12.6.6) is valid only along a C^+ characteristic. The curve labeled C^- is a plot of Eq. (12.6.9); Eq. (12.6.8) is valid only along a C^- characteristic. Each equation, (12.6.6) and (12.6.8), contains two unknowns for a known point on its characteristic, but at the intersection of C^+ and C^- at P both equations may be solved to yield V_P and H_P. At this point Eqs. (12.6.7) and (12.6.9) may also be solved for x and t. Hence the solution is carried out along the characteristics, starting from known conditions and by finding new intersections so that heads and velocities are found for later times.

In waterhammer calculations in metal pipes, the subject of this treatment, V is very small compared with a and may be dropped from Eqs. (12.6.7) and (12.6.9). The characteristic lines are now straight, with slopes $\pm a$. a is the speed with which the pressure-pulse wave is transmitted along the pipe. The pipe is considered to be made up of N equal reaches (Fig. 12.13), and H and V are initially known at each of the dividing sections. The solution to the waterhammer problem can then be carried out at the intersections of the characteristic lines, as shown by the solid dots. It is to be noted that the solution can be carried over only a limited region, unless information is given at $x = 0$ and $x = L$, of some external condition as a function of time. (See under "Boundary conditions," page 658.)

From the grid of Fig. 12.13, it is seen that the time step of the calculation is $\Delta t = \Delta x/a$. By using the grid, x and t are known at each intersection, and Eqs. (12.6.7) and (12.6.9) need be considered no further. By multiplying Eq. (12.6.6) by dt and integrating along the C^+ characteristic, Fig. 12.13,

$$H_{P_i} - H_{i-1} + \frac{a}{gA}(Q_{P_i} - Q_{i-1}) + \frac{Q_{i-1}\,\Delta t}{A}\sin\theta + \frac{f\,\Delta x}{2gDA^2}Q_{i-1}\left|\,Q_{i-1}\,\right| = 0$$

$$(12.6.10)$$

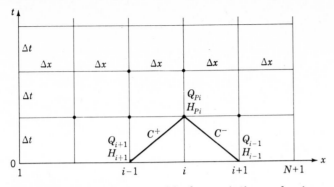

Fig. 12.13 Rectangular grid for solution of characteristics equations.

where a first-order approximation has been used in the integration of the last two terms and, for convenience, the equation has been written in terms of discharge Q in place of velocity V. Similarly for Eq. (12.6.8)

$$H_{P_i} - H_{i+1} - \frac{a}{gA}(Q_{P_i} - Q_{i+1}) + \frac{Q_{i+1}\,\Delta t}{A}\sin\theta - \frac{f\,\Delta x}{2gDA^2}Q_{i+1}\left| Q_{i+1} \right| = 0$$

$$(12.6.11)$$

By adding the last two equations, Q_{P_i} is eliminated:

$$H_{P_i} = 0.5\left[H_{i-1} + H_{i+1} + \frac{a}{gA}(Q_{i-1} - Q_{i+1}) - \frac{\Delta t \sin\theta}{A}(Q_{i-1} + Q_{i+1}) \right.$$

$$\left. - \frac{f\,\Delta x}{2gDA^2}(Q_{i-1}\,| Q_{i-1} | - Q_{i+1}\,| Q_{i+1} |) \right] \quad (12.6.12)$$

With H_{P_i} known, either Eq. (12.6.10) or (12.6.11) can be used to solve for Q_{P_i}. With N reaches in a pipe there are $N + 1$ sections along the pipe, including the two ends. The unknowns H_P and Q_P are determined at each section 2 to N by using Eq. (12.6.12) and either (12.6.10) or (12.6.11). The two variables at each of the two ends of the pipeline are determined by using the appropriate one of Eqs. (12.6.10) or (12.6.11) together with the equation for the boundary condition. When the variables H_P and Q_P have been evaluated at each section in the pipeline, time is incremented by Δt and the procedure is repeated. At the end of each time step the values of H and Q at each section are replaced by the newly computed values of H_P and Q_P.

Boundary conditions

At the upstream end of a pipe, Eq. (12.6.11) for the C^- characteristic provides one equation in the two unknowns Q_{P_1} and H_{P_1} (Fig. 12.14a). One condition is needed exterior to the pipe to relate the pipeline response to the boundary-condition behavior. This condition may be a constant value of one of the variables, such as a constant-head reservoir, a specified variation of one of the variables as a function of time, an algebraic relationship between the two variables, or a relationship in the form of a differential equation. Some boundary conditions may involve additional variables, e.g., pump speed in the case of a centrifugal pump connected at the upstream end of the line. In this case two independent equations must be available to combine with Eq. (12.6.11) to solve for the three unknowns at each time step. The simplest boundary condition is one in which one of the variables is given as a function of time. A direct solution of Eq. (12.6.11) for the other variable at each time step provides a complete solution of the interaction of the fluid in the pipeline and the particular boundary. This includes the appropriate reflection and transmission of transient pressure and flow waves that arrive at the pipe end.

At the downstream end of the pipe (Fig. 12.14b), Eq. (12.6.10) for the C^+ characteristic provides one equation in the two variables $H_{P_{N+1}}$ and $Q_{P_{N+1}}$. One external condition is needed that either specifies one of the variables to be constant or a known function of time or provides a relationship between the variables in algebraic or differential equation form. The simplest end condition is one in which a variable is held constant, as at a dead end where $Q_{P_{N+1}}$ is zero. Then Eq. (12.6.10) provides a direct solution for $H_{P_{N+1}}$ at each time step.

A computer program is provided for solution of the next example, which involves simple boundary conditions on a single pipeline. Complex systems can be visualized as a combination of single pipelines that are handled as described above, with boundary conditions at the pipe ends to transfer the transient response from one pipeline to another and to provide interaction with the system terminal conditions. Thus it may be noted that a compli-

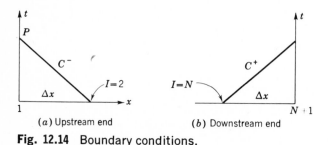

(a) Upstream end   (b) Downstream end

Fig. 12.14 Boundary conditions.

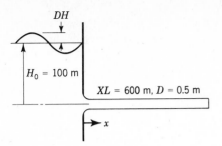

Fig. 12.15 Example 12.6.

cated system can be treated by a combination of a common solution procedure for the interior of each pipeline, together with a systematic coverage of each terminal and interconnection point in the system. The primary focus in the treatment of a variety of transient liquid-flow problems is on the handling of boundary conditions, which are discussed in the next section.

```
C BASIC WATERHAMMER PROGRAM. WAVE ON RESERVOIR UPSTREAM, DEAD END D.S.
      DIMENSICN HP(21),QP(21),H(21),Q(21)
   10 READ (5,15,END=99) A,XL,D,F,G,QO, HO,DH,CMEGA,TMAX,N,IPR
   15 FORMAT (3F10.2,3F10.4/4F10.2/2I5)
      NS=N+1
      R=F*XL/(2.*G*D**5*.7854**2*N)
      CH=A/(G*.7854*D*D)
      DT=XL/(A*N)
      T=0.
      K=0
      DO 20 I=1,NS
      H(I)=HO-(I-1)*R*QO*QO
   20 Q(I)=QO
      WRITE (6,25) A,XL,D,F,G,QC,HO,DH,CMEGA,TMAX,DT,N,IPR
   25 FORMAT(' A, XL, D & F=',2FE.1,2F8.4/' G, QO, HO & DH=',2F8.3,F8.1,
     2F8.3/' CMEGA, TMAX, CT=',F8.3,F8.1,F8.3/' N & IPR=',2I4//'     HEADS
     3 AND CISCHARGES ALONG THE PIPE'//'    TIME X/L=     0.     .2
     4   .4      .6      .8     1.')
   30 WRITE (6,35) T,(H(I),I=1,NS,2),(Q(I),I=1,NS,2)
   35 FORMAT (1H0,F7.3,5H   H=,6F8.2/10X,3H Q=,6F8.3)
   40 T=T+DT
      IF (T.GT.TMAX) GO TC 10
      K=K+1
C COMPUTATICN OF INTERICR PCINTS
      DO 50 I=2,N
      CP=H(I-1)+Q(I-1)*(CH-R*ABS(Q(I-1)))
      HP(I)=.5*(CP+H(I+1)+Q(I+1)*(R*AES(Q(I+1))-CH))
   50 QP(I)=(CP-HP(I))/CH
C BCUNDARY CCNDITIONS
      HP(1)=HC+DH*SIN(OMEGA*T)
      QP(1)=(HP(1)-H(2)-Q(2)*(R*ABS(Q(2))-CH))/CH
      QP(NS)=0.
      HP(NS)=H(N)+Q(N)*(CH-R*ABS(Q(N)))
      DO 60 I=1,NS
      H(I)=HP(I)
   60 Q(I)=QP(I)
      IF (K/IPR*IPR-K) 40,30,40
   99 STOP
      END
 1200.      600.       .5        .018      9.806     0.
  1CC.       3.       3.1416     6.
   1C     2
```

Fig. 12.16 Basic waterhammer program.

EXAMPLE 12.6 Determine the transient response in a single horizontal dead-end pipeline, Fig. 12.15, that is connected to a reservoir in which the water surface is oscillating, $H = H_0 + DH \sin \omega t$, where H_0 is the steady hydraulic-grade-line elevation, DH the amplitude of the wave, and ω the frequency.

```
        A, XL, D & F=  1200.0    6C0.0   C.5000  0.0180
        G, C0, H0 & CH=   9.806    0.00C   100.0    3.000
     CMEGA, TMAX, CT=    3.142    6.C    0.050
     N & IPR=  10    2
     HEADS AND DISCHARGES ALCNG THE PIPE
        TIME X/L=   0.       .2       .4       .6       .8       1.
       0.000    H=  100.00   100.00   1C0.00   100.00   10C.CC   100.00
                Q=    0.000     0.000    C.000    0.000    0.000    0.000
       0.1C0    H=  100.93   100.CC   1C0.0C   1C0.00   10C.C0   100.00
                C=    0.001     0.000    C.000    0.000    0.0CC    0.000
       0.200    H=  101.76   100.93   1C0.0C   100.00   100.00   100.00
                Q=    0.003     0.001    C.00C    0.000    0.000    0.000
       0.300    H=  102.43   101.76   100.93   100.00   10C.00   100.00
                Q=    0.004     0.003    0.001    0.000    0.000    0.000
       0.4C0    H=  102.85   102.43   1C1.76   100.93   10C.CC   1C0.C0
                C=    0.005     0.004    0.0C3    0.001    0.0C0    C.000
       C.500    H=  103.00   102.85   1C2.43   101.76   10C.93   100.00
                C=    0.C05     0.005    C.0C4    0.003    0.CC1    0.000
       0.600    H=  102.85   103.0C   1C2.85   102.43   101.76   101.85
                C=    0.0C5     0.005    C.005    C.004    0.0C3    0.000
       C.7C0    H=  102.43   102.85   1C3.00   102.85   103.35   103.53
                Q=    0.0C4     0.005    C.0C5    0.005    0.002    0.000
       0.800    H=  101.76   102.43   102.85   1C3.93   104.62   104.85
                Q=    0.0C3     0.004    C.005    0.005    0.0C2    0.000
       C.9C0    H=  100.93   101.76   103.35   1C4.62   105.43   105.71
                Q=    C.001     0.003    0.002    0.002    0.0C1    0.000
       1.000    H=  100.00   101.85   103.53   1C4.85   105.71   106.00
                Q=   -0.00C   -0.000   -C.000    0.000    C.0C0    0.000
       1.100    H=   99.07   101.76   103.35   1C4.62   105.43   105.71
                Q=   -0.C04   -0.003   -0.002   -0.002   -0.0C1    0.000
       1.2C0    H=   98.24   100.57   102.85   1C3.93   104.62   104.85
                C=   -0.008   -0.007   -C.005   -0.003   -0.0C2    0.000
       1.300    H=   97.57    95.33   101.15   102.85   103.35   103.53
                C=   -0.012   -0.01C   -0.J08   -0.005   -0.0C2    0.000
       1.400    H=   97.15    98.15    99.33   100.57   101.76   101.85
                Q=   -0.014   -0.C13   -0.010   -0.007   -0.0C3    0.000
       1.500    H=   97.00    97.15    97.57    98.24    99.C7   1C0.00
                Q=   -0.014   -0.014   -0.012   -0.008   -0.004    0.000
       1.600    H=   97.15    96.43    96.06    96.07    96.47    96.29
                Q=   -0.C14   -0.014   -C.012   -0.009   -0.0C6    0.000
       1.700    H=   97.57    96.06    94.93    94.30    93.29    92.95
                Q=   -0.012   -0.012   -C.011   -0.009   -0.005    0.000
       1.800    H=   98.24    96.07    94.30    92.15    90.77    90.30
                Q=   -0.C08    -0.009   -0.009   -0.007   -0.0C3    0.000
       1.900    H=   99.07    96.47    93.29    9C.77    89.15    88.59
                Q=   -0.004   -0.006   -0.J05   -0.003   -0.0C2    0.00C
       2.000    H=  100.00    96.29    92.95    90.30    88.59    88.01
                Q=    C.000    C.000    C.000    0.000    0.0CC    0.0C0
       2.1CC    H=  100.93    96.48    93.29    90.77    89.15    88.59
                Q=    0.C07     0.006    0.0C5    0.003    0.0C2    0.000
       2.200    H=  101.76    97.93    94.30    92.15    90.77    90.30
                Q=    0.014     0.012    C.009    0.007    0.003    0.000
       2.300    H=  102.43    99.58    96.78    94.30    93.29    92.95
                Q=    0.019     0.018    C.014    0.009    0.0C5    0.0C0
       2.4C0    H=  1C2.85   101.28    99.58    97.93    96.47    96.29
                C=    0.023     0.C21    C.018    0.012    C.CC6    0.000
       2.500    H=  103.00   102.85   1C2.42   101.76   100.93   100.00
                U=    0.C24     0.023    0.C19    0.014    C.0C7    0.000
       2.600    H=  102.85   104.14   1C5.C3   105.42   105.29   105.56
                C=    0.023     0.022    0.019    0.015    0.0C8    0.000
       2.700    H=  102.43   105.03   1C7.14   108.55   11C.C5   110.57
                Q=    0.C19     0.019    C.017    C.014    0.0C7    0.000
       2.8C0    H=  101.76   105.42   1C8.55   111.77   113.84   114.55
                Q=    0.C14    U.C15    C.014    0.010    C.0C5    0.00C
       2.9CC    H=  100.93   105.29   110.05   113.84   116.27   117.10
                Q=    C.C07     0.008    C.007    0.005    C.0C3    0.000
```

Fig. 12.17 Computer solution to Example 12.6.

Other conditions include $a = 1200$ m/s, $f = 0.018$, $Q_0 = 0.0$ m³/s, $H_0 = 100$ m, $DH = 3$ m, $\omega = 3.1416$ rad/s, $T_{\max} = 6$ s, $N = 10$ reaches in pipeline, and $IPR = 2$ time increments between printout. The value of g is also needed in the input data, and in this case it is 9.806 m/s² since SI units are used.

The computer program in Fig. 12.16 is used to solve the problem. The important steps in the program include (1) data input, (2) setup of initial conditions, (3) printout, (4) compute H_P and Q_P at interior points, (5) compute H_P and Q_P at the boundaries, (6) substitute H_P and Q_P into H and Q at each section, and (7) return to step 3. The formated input data are shown in Fig. 12.16, and the computed output appears in Fig. 12.17. It is of interest to follow the wave of the pressure pulse through the pipe. Since the excitation happens to be at the pipeline natural period, a pressure-head amplification may also be noted.

12.7   BOUNDARY CONDITIONS

The term boundary condition refers to the end condition on each pipeline. It may be a system terminal at a reservoir, valve, etc., or it may be a pipeline connection to another pipeline or a different type of element, e.g., a pump or a storage volume. In each of the many options at the downstream end of the pipe the equation along the C^+ characteristic is used to interface with the particular end condition. It is convenient to write Eq. (12.6.10) in an abbreviated form for section NS at the downstream end,

$$H_{P_{NS}} = C_P - C_H Q_{P_{NS}} \tag{12.7.1}$$

where $C_H = a/gA$ is a constant for the pipeline and C_P is a combination of known quantities at each time step

$$C_P = H_N + Q_N \left(C_H - \frac{\Delta t}{A} \sin\theta - \frac{f\,\Delta x}{2gDA^2} \,|\, Q_N \,| \right) \tag{12.7.2}$$

Similarly at the upstream end of the pipe the equation along the C^- characteristic is used to relate the pipeline behavior to the end condition. Equation (12.6.11) may be expressed

$$H_{P_1} = C_M + C_H Q_{P_1} \tag{12.7.3}$$

with

$$C_M = H_2 - Q_2 \left(C_H + \frac{\Delta t}{A} \sin\theta - \frac{f\,\Delta x}{2gDA^2} \,|\, Q_2 \,| \right) \tag{12.7.4}$$

A few common boundary conditions follow, and in each case either Eq. (12.7.1) or (12.7.3) is used to represent the pipeline response.

Valve at downstream end

For steady-state flow through the valve, considered as an orifice,

$$Q_0 = (C_d A_v)_0 \sqrt{2gH_0} \tag{12.7.5}$$

with Q_0 the steady-state flow, H_0 the head across the valve, and $(C_d A_v)_0$ the area of the opening times the discharge coefficient. For another opening, in general,

$$Q_P = C_d A_v \sqrt{2gH_P} \tag{12.7.6}$$

Dividing the second equation by the first and rearranging give

$$Q_P = \frac{Q_0}{\sqrt{H_0}} \tau \sqrt{H_P} \tag{12.7.7}$$

in which τ is the dimensionless valve opening. $\tau = 1$ for steady flow Q_0 and head drop H_0, and $\tau = 0$ for the closed position of the valve. When the subscript NS is added to the variables Q_P and H_P in Eq. (12.7.7) and this equation is solved simultaneously with Eq. (12.7.1), the value of $Q_{P_{NS}}$ is determined as a function of the valve position including the transient response from the pipeline

$$Q_{P_{NS}} = -\frac{Q_0^2 \tau^2 C_H}{2H_0} + \sqrt{\left(\frac{Q_0^2 \tau^2 C_H}{2H_0}\right)^2 + \frac{Q_0^2 \tau^2 C_P}{H_0}} \tag{12.7.8}$$

The corresponding value of $H_{P_{NS}}$ can be determined from either Eq. (12.7.1) or (12.7.7).

Minor loss

In some problems it may be important to use the energy equation at the boundary and to include a minor loss. The pipe entrance from a reservoir is discussed as an example. Figure 12.18, left shows the energy and hydraulic grade lines for flow into the pipe. The energy equation between the reservoir

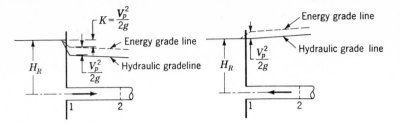

Fig. 12.18   Minor loss.

surface and section 1 in the pipeline is

$$H_R = H_{P_1} + K\frac{Q_{P_1}^2}{2gA^2} + \frac{Q_{P_1}^2}{2gA^2} \tag{12.7.9}$$

Simultaneous solution of this equation with Eq. (12.7.3) yields

$$Q_{P_1} = -\frac{gC_HA^2}{K+1} + \sqrt{\left(\frac{gC_HA^2}{K+1}\right)^2 + (H_R - C_M)\frac{2gA^2}{K+1}} \tag{12.7.10}$$

which is valid as long as $H_R - C_M$ is positive. For flow into the reservoir, Fig. 12.18, right, $H_{P_1} = H_R$, and a direct solution for Q_{P_1} is possible by use of Eq. (12.7.3).

EXAMPLE 12.7   Consider a single horizontal pipeline with reservoir upstream and an orifice a short distance (10 diameters) downstream from the entrance. At the downstream end a valve is to be closed linearly in 1.5 s. The steady flow Q_0 discharges to the atmosphere at the valve. Write the equations for both boundary conditions so they are ready to be programmed. The kinetic-energy term should be included at the upstream orifice. Assume the loss coefficient to be the same for flow in either direction.

The steady-state head loss at the valve must first be determined by writing the energy equation for the system. If H_R is the reservoir head, the head H_0 just upstream from the valve is

$$H_0 = H_R - \frac{Q_0^2}{2gA^2}\left(1 + K_e + K + \frac{fL}{D}\right)$$

K_e is the entrance-loss coefficient and K is the orifice-loss coefficient.

For the downstream boundary,

$$C_H = \frac{a}{gA}$$

$$C_P = H_N + Q_N \left(C_H - \frac{f\,\Delta x}{2gDA^2} |Q_N| \right)$$

If $t < 1.5$,

$$\tau = 1 - \frac{t}{1.5} \qquad C_4 = \frac{\tau^2 Q_0^2}{H_0} \qquad Q_{PNS} = -\frac{C_4 C_H}{2} + \sqrt{\left(\frac{C_4 C_H}{2}\right)^2 + C_4 C_P}$$

If $t \geq 1.5$,

$$\tau = 0.0 \qquad Q_{PNS} = 0.0 \qquad H_{PNS} = C_P$$

For the upstream boundary, when flow is negative, the energy equation gives

$$H_{P_1} + \frac{Q_{P_1}^2}{2gA^2} = H_R + \frac{Q_{P_1}^2}{2gA^2} + K \frac{Q_{P_1}^2}{2gA^2}$$

The second term on the right-hand side is the exit loss from the pipe. By solving simultaneously with Eq. (12.7.3)

$$Q_{P_1} = \frac{C_5}{K} - \sqrt{\left(\frac{C_5}{K}\right)^2 - \frac{(H_R - C_M)2gA^2}{K}}$$

in which

$$C_M = H_2 - Q_2 \left(C_H - \frac{f\,\Delta x}{2gDA^2} |Q_2| \right) \qquad C_5 = gC_H A^2$$

The equation for Q_{P_1} is valid only for negative flow; hence $H_R - C_M$ must be negative. For positive flow, the energy equation yields

$$H_{P_1} + \frac{Q_{P_1}^2}{2gA^2} = H_R - \frac{Q_{P_1}^2}{2gA^2} (K_e + K)$$

Combined with Eq. (12.7.3), it gives

$$Q_{P_1} = -\frac{C_5}{1 + K_e + K} + \sqrt{\left(\frac{C_5}{1 + K_e + K}\right)^2 + \frac{2gA^2(H_R - C_M)}{1 + K_e + K}}$$

which is valid only if $H_R - C_M \geq 0$.

Junction of two or more pipes

At a connection of pipelines of different properties, the continuity equation must be satisfied at each instant of time, and a common hydraulic-grade-line elevation may be assumed at the end of each pipe. These statements implicitly assume that there is no storage at the junction, and they also neglect all minor effects. In multipipe systems it is necessary either to use double-subscript notation, the first subscript referring to the pipe number and the second to the pipe section number, or to use continuous sectioning in the entire system. If the former scheme is used to handle the three-pipe junction in Fig. 12.19 and Eqs. (12.7.1) and (12.7.3) are written in the following form, a summation provides a simple solution for the common head:

$$Q_{P_{1,NS}} = -\frac{H_{P_{1,NS}}}{C_{H1}} + \frac{C_{P1}}{C_{H1}}$$

$$-Q_{P_{2,1}} = -\frac{H_{P_{1,NS}}}{C_{H2}} + \frac{C_{M2}}{C_{H2}}$$

$$-Q_{P_{3,1}} = -\frac{H_{P_{1,NS}}}{C_{H3}} + \frac{C_{M3}}{C_{H3}}$$

$$\Sigma Q_P = 0 = -H_{P_{1,NS}}\Sigma\frac{1}{C_H} + \frac{C_{P1}}{C_{H1}} + \frac{C_{M2}}{C_{H2}} + \frac{C_{M3}}{C_{H3}}$$

or

$$H_{P_{1,NS}} = \frac{C_{P1}/C_{H1} + C_{M2}/C_{H2} + C_{M3}/C_{H3}}{\Sigma(1/C_H)} \tag{12.7.11}$$

With the common head computed, the equations above can be used to determine the flow in each pipe.

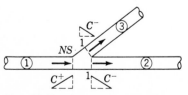

Fig. 12.19 Pipeline junction.

Valve in line

A valve or orifice between two different pipelines or within a given line must be treated simultaneously with the contiguous end sections of the pipelines. It is assumed that the orifice equation (12.7.7) is valid at any instant for the control volume shown in Fig. 12.20. This assumption neglects any inertia effects in accelerating or decelerating flow through the valve opening and also implies that there is a constant volume of fluid within the indicated control volume. At any instant the flow rate in the end sections are equal, $Q_{P1,NS} = Q_{P2,1}$, and the orifice equation for positive flow, written with the subscript notation to identify the pipeline as well as the section, becomes

$$Q_{P2,1} = Q_{P1,NS} = \frac{Q_0 \tau}{\sqrt{\Delta H_0}} \sqrt{H_{P1,NS} - H_{P2,1}} \tag{12.7.12}$$

where ΔH_0 is the steady-state drop in hydraulic grade line across the valve at the flow Q_0 when $\tau = 1$. When Eqs. (12.7.1) and (12.7.3) are written with the same notation

$$H_{P1,NS} = C_{P1} - C_{H1} Q_{P1,NS} \tag{12.7.13}$$

$$H_{P2,1} = C_{M2} + C_{H2} Q_{P2,1} \tag{12.7.14}$$

and are combined with Eq. (12.7.12), a quadratic equation results:

$$Q_{P1,NS}{}^2 + C_4(C_{H1} + C_{H2})Q_{P1,NS} - C_4(C_{P1} - C_{M2}) = 0$$

In this equation $C_4 = \tau^2 Q_0^2 / \Delta H_0$. The solution for flow in the positive direction is

$$Q_{P1,NS} = \frac{C_4}{2}\left[-(C_{H1} + C_{H2}) + \sqrt{(C_{H1} + C_{H2})^2 + \frac{4(C_{P1} - C_{M2})}{C_4}} \right]$$

$$\tag{12.7.15}$$

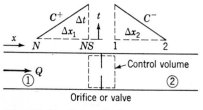

Fig. 12.20 Valve in line.

Equations (12.7.13) and (12.7.14) may be used to find the hydraulic grade line when the flow is known. For flow in the negative direction the orifice equation must be written

$$Q_{P_2,1} = Q_{P_1,NS} = -\frac{Q_0 \tau}{\sqrt{\Delta H_0}} \sqrt{H_{P_2,1} - H_{P_1,NS}} \tag{12.7.16}$$

When the equations that are valid along the characteristic lines are combined with this equation, the solution is

$$Q_{P_1,NS} = \frac{C_4}{2}\left[(C_{H_1} + C_{H_2}) - \sqrt{(C_{H_1} + C_{H_2})^2 - 4\frac{(C_{P_1} - C_{M_2})}{C_4}} \right] \tag{12.7.17}$$

This equation yields a valid answer only for negative flow. Examining the equation shows that a negative result is possible only if $C_{P_1} - C_{M_2} < 0$. Thus, Eq. (12.7.15) is used if $C_{P_1} - C_{M_2} \geq 0$, and Eq. (12.7.17) is used if $C_{P_1} - C_{M_2} < 0$.

Multiparallel pipes

If a number of identical parallel pipes exist in a system, as in a cooling-water condenser, they may be correctly treated as one element with a flow area equal to the sum of the areas of the individual pipes but with a resistance factor that is appropriate to the individual pipe diameters. Thus the multiplier for the resistance term in Eqs. (12.7.2) and (12.7.4) becomes $f_s \Delta x/2gD_s A_T^2$, where f_s and D_s refer to the smaller or individual pipes while A_T refers to the total flow area. As long as the individual pipelines are the same in every respect, the transient conditions in each line will be identical and this special element, which accommodates the total flow, will accurately portray the physical behavior of this portion of the system. The pressure-pulse wave speed must be appropriate to the smaller-diameter lines.

Centrifugal pump with speed known

If a pump is operating at a constant speed, or if the unit is started and the pump and motor come up to speed in a known manner, the interaction of the pump and the fluid in the connecting pipelines can be handled by a fairly simple boundary condition. The homologous conditions, Eqs. (9.1.1) and (9.1.3), when the transient behavior of a given pump is investigated, may be written

$$\frac{H}{N^2} = \text{const} \qquad \frac{Q}{N} = \text{const} \qquad\qquad (12.7.18)$$

where H is the head rise across the pump and N is the speed. If the pump characteristic curve is expressed by a parabola, then in the homologous form it may be written

$$\frac{H}{N^2} = B_1 + B_2 \frac{Q}{N} + B_3 \left(\frac{Q}{N}\right)^2 \qquad\qquad (12.7.19)$$

The compressibility of the fluid in the pump is assumed to be negligible compared with the rest of the system. When Eq. (12.7.19) is applied to the pump in Fig. 12.21, it takes the form

$$H_{P_{2,1}} - H_{P_{1,NS}} = N^2 B_1 + B_2 N Q_{P_{1,NS}} + B_3 Q_{P_{1,NS}}{}^2 \qquad\qquad (12.7.20)$$

When this equation is combined with Eqs. (12.7.13) and (12.7.14), the discharge may be determined as

$$Q_{P_{1,NS}} = \frac{C_{H_1} + C_{H_2} - B_2 N}{2B_3} \left\{ 1 - \left[1 + \frac{4B_3(N^2 B_1 + C_{P_1} - C_{M_2})}{(C_{H_1} + C_{H_2} - B_2 N)^2} \right]^{1/2} \right\}$$

$$(12.7.21)$$

During a pump start-up a linear speed rise is often assumed for the speed variation. If the speed of the pump is constant, N may be combined with the pump-curve constants. Also if the pump is operating directly from a suction reservoir, the equation may be simplified by the elimination of the equation along the C^+ characteristic in the suction pipeline.

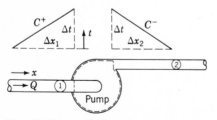

Fig. 12.21 Centrifugal pump.

EXAMPLE 12.8 Develop the necessary boundary-condition equations for the pump in Fig. 12.22. The pump is to be started with a linear speed rise to N_R in t_0 s. A check valve exists in the discharge pipe. The initial no-flow steady-state head on the downstream side of the check valve is H_C. For a steady flow of Q_0 there is a loss of ΔH_0 across the open check valve. Assume the check valve opens instantaneously when the pump has developed enough head to exceed H_C.

The equation for the hydraulic grade line across the pump and check valve (after the check valve is open) is

$$H_P = N^2 B_1 + B_2 N Q_P + B_3 Q_P^2 - \frac{Q_P^2}{C_4}$$

where $C_4 = \tau^2 Q_0 / \Delta H_0$ and N is the speed. When Eq. (12.7.3) is introduced on the left side of this equation, the quadratic may be solved to determine the flow:

$$Q_P = \frac{C_H - B_2 N}{2(B_3 - 1/C_4)} \left\{ 1 - \left[1 + \frac{4(B_3 - 1/C_4)(N^2 B_1 - C_M)}{(C_H - B_2 N)^2} \right]^{1/2} \right\}$$

The equations for the boundary condition are

$$C_H = \frac{a}{gA}$$

$$C_M = H_2 - Q_2 \left(C_H - \frac{f \, \Delta x}{2gDA^2} \, |Q_2| \right)$$

$$N = \begin{cases} N_R \left(\dfrac{t}{t_0} \right) & t \leq t_0 \\[2mm] N_R & t > t_0 \end{cases}$$

If $N^2 B_1 < H_C$,

$$Q_P = 0 \quad \text{and} \quad H_P = H_C$$

If $N^2 B_1 > H_C$,

Q_P is defined by the above solution to the quadratic equation, and

$$H_P = C_M + C_H Q_P$$

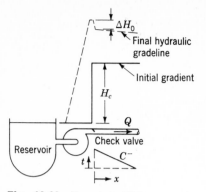

Fig. 12.22 Example 12.8.

Accumulator

Many different types of air or gas accumulators are used to help reduce pressure transients in liquid systems. In an analysis of the behavior of the accumulator shown in Fig. 12.23 the pressure is visualized as being the same throughout the indicated control volume at any instant. It is assumed to be frictionless and inertialess. The gas is assumed to follow the reversible polytropic relation

$$H_A \mathcal{V}^n = C \tag{12.7.22}$$

where H_A is the absolute head equal to the gage plus barometric pressure heads, $\mathcal{V}$ is the gas volume, n is the polytropic exponent, and C is a constant.

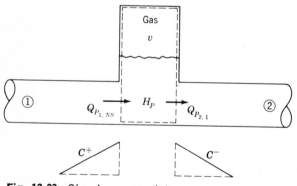

Fig. 12.23 Simple accumulator.

The derivative of this equation with respect to time leads to

$$\frac{dH_A}{dt} = -\frac{nC}{\mho^{n+1}}\frac{d\mho}{dt}$$

The continuity equation applied to the control volume in Fig. 12.23 yields

$$\frac{d\mho}{dt} = Q_{P_2} - Q_{P_1}$$

When these two equations are combined and placed in finite-difference form for the time increment Δt,

$$H_P = H + C_6(Q_{1,NS} - Q_{2,1}) + C_6(Q_{P_1,NS} - Q_{P_2,1}) \tag{12.7.23}$$

where $C_6 = 0.5nC\,\Delta t/\mho^{n+1}$. In this equation it is assumed that the volume $\mho$ at the beginning of the time step adequately represents the gas volume during the time step. The new volume is computed before proceeding to the next time step by using a finite-difference representation of the above continuity equation. Equation (12.7.23) can be combined with Eqs. (12.7.13) and (12.7.14) to evaluate the three unknowns $Q_{P_1,NS}$, $Q_{P_2,1}$, and H_P. For those cases where the change in $\mho$ is significant during Δt, a second-order method should be used.

An accumulator with inertia and friction is shown in Fig. 12.24. The equation of motion written for connecting line 3 yields

$$\gamma A_3 \left(\frac{H_P + H}{2} - \frac{H_{P_4 + H_4}}{2} - \frac{fL_3}{D_3 A_3{}^2}\frac{Q_3\,|\,Q_3\,|}{2g} \right) = \frac{\gamma A_3 L_3}{g}\frac{dV_3}{dt}$$

or

$$H_P - H_{P_4} = C_7 + C_8 Q_{P_3} \tag{12.7.24}$$

where

$$C_8 = \frac{2L_3}{gA_3\,\Delta t} \quad \text{and} \quad C_7 = H_4 - H + \frac{fL_3}{D_3 g A_3{}^2}Q_3\,|\,Q_3\,| - C_8 Q_3$$

Equation (12.7.23) written for the air volume with the notation in Fig. 12.24 yields

$$H_{P_4} = H_4 + C_6 Q_3 + C_6 Q_{P_3}$$

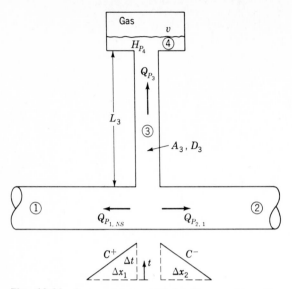

Fig. 12.24 Accumulator with friction and inertia.

The continuity equation at the bottom of pipeline 3 yields

$$Q_{P_1, NS} = Q_{P_{2,1}} + Q_{P_3}$$

The five unknowns may be determined at each time step by a simultaneous solution of Eqs. (12.7.13), (12.7.14), (12.7.24), and the last two equations.

Vapor-column separation

If the pressure level at a section in a pipeline drops below vapor pressure for the liquid, vaporization occurs and a vapor pocket forms. When this happens, the pressure level is fixed at vapor pressure at the section and the flows must be computed using this pressure. When the flows have been computed, a local continuity balance may be used to identify the size of cavity. The vapor column may grow and subsequently collapse, giving rise to a substantial over-pressure. The calculation of the pressure rise at the collapse of a vapor cavity is one of the most important calculations in many transient water-flow problems. An accurate description of the phenomenon is beyond the scope of this treatment; however, the following example provides a reasonable description of the behavior and yields a conservative design, in that pressure levels are likely to be less than predicted with the model.

```
C COLUMN SEPARATION STATEMENTS FOR INTERNAL SECTIONS
C ASSUMPTIONS: 1) COLUMN OPENS ONLY AT SECTIONS, 2) CAV. VOLUME
C SMALL COMPARED WITH LIQUID VOLUME IN REACH, 3) NO NEGATIVE
C PRESSURES, 4) WAVE SPEED REMAINS CONSTANT.  HV=VAPOR PRESSURE
C IN UNITS OF LENGTH OF FLUID, GAGE.

         DO 11 I=2,N
         VCAV(I)=.0
   11    ICAV(I)=0

         DO 34 I=2,N
         CP=H(I-1)+Q(I-1)*(CH-R*ABS(Q(I-1)))
         IF (ICAV(I+1).EQ.1)CM=H(I+1)-QPX(I+1)*(CH-R*ABS(QPX(I+1)))
         IF (ICAV(I+1).EQ.0)CM=H(I+1)-Q(I+1)*(CH-R*ABS(Q(I+1)))
         IF (ICAV(I).EQ.1) GO TO 33
   31    HP(I)=.5*(CP+CM)
         IF ((HP(I)-EL(I)).LT.HV) GO TO 32
         QP(I)=(CP-HP(I))/CH
         GO TO 34
   32    HP(I)=EL(I)+HV
         ICAV(I)=1
         QPX(I)=Q(I)
   33    QP(I)=(HP(I)-CM)/CH
         QPP(I)=(CP-HP(I))/CH
         VCAV(I)=VCAV(I)+.5*DT*(Q(I)+QP(I)-QPP(I)-QPX(I))
         QPX(I)=QPP(I)
         IF(VCAV(I).GT.0.) GO TO 34
         ICAV(I)=0
         VCAV(I)=.0
         HP(I)=.5*(CP+CM)
         QP(I)=(CP-HP(I))/CH
   34    CONTINUE
```

Fig. 12.25 Column-separation statements.

EXAMPLE 12.9 Prepare a set of FORTRAN statements that will be able to recognize and to calculate vapor cavities at all interior points in a single pipeline.

The set of statements is shown in Fig. 12.25. Two variables are defined to identify the existence of a vapor pocket at a section: VCAV(I) identifies the volume of the vapor cavity, and ICAV(I) is an integer that has the value 0 if no cavity exists at the section and 1 if a cavity exists. The comments in Fig. 12.25 set forth the assumptions; the first three statements are inserted above the time-increment iteration loop; and the rest of the statements replace all interior-point calculations in the pipeline. The size of the cavity at each section is accumulated, and when it collapses, the pressure rise is computed by use of the normal equations along the characteristic lines. When a cavity exists, the flows at the upstream and downstream sides of the cavity are identified as QPP(I) and QP(I) at the current time and QPX(I) and Q(I) at the previous time, respectively.

OPEN-CHANNEL FLOW

In general, open-channel transients are more complex to handle than closed-conduit transients. Surface-wave motion is an example of open-channel and unsteady flow. The subject is too vast to attempt to cover as part of a chapter. A few special topics are discussed that use about the same approach as the waterhammer equations: frictionless positive and negative surge waves, flood routing, and a case of rainfall and runoff from a plane area.

12.8 FRICTIONLESS POSITIVE SURGE WAVE IN A RECTANGULAR CHANNEL

In this section the surge wave resulting from a sudden change in flow (due to a gate or other mechanism) that increases the depth is studied. A rectangular channel is assumed, and friction is neglected. Such a situation is shown in Fig. 12.26 shortly after a sudden, partial closure of a gate. The problem is analyzed by reducing it to a steady-state problem, as in Fig. 12.27. The continuity equation yields, per unit width,

$$(V_1 + c)y_1 = (V_2 + c)y_2 \tag{12.8.1}$$

and the momentum equation for the control volume $1 - 2$, neglecting shear stress on the floor, per unit width, is

$$\frac{\gamma}{2}(y_1{}^2 - y_2{}^2) = \frac{\gamma}{g} y_1(V_1 + c)(V_2 + c - V_1 - c) \tag{12.8.2}$$

By elimination of V_2 in the last two equations,

$$V_1 + c = \sqrt{gy_1} \left[\frac{y_2}{2y_1} \left(1 + \frac{y_2}{y_1} \right) \right]^{1/2} \tag{12.8.3}$$

In this form the speed of an elementary wave is obtained by letting y_2 approach y_1, yielding

$$V_1 + c = \sqrt{gy} \tag{12.8.4}$$

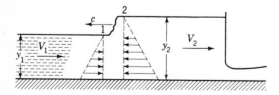

Fig. 12.26 Positive surge wave in a rectangular channel.

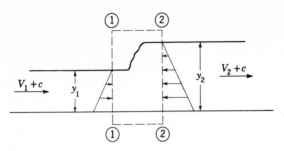

Fig. 12.27 Surge problem reduced to a steady-state problem by superposition of surge velocity.

For propagation through still liquid $V_1 \to 0$, and the wave speed is $c = \sqrt{gy}$ when the problem is converted back to the unsteady form by superposition of $V = -c$.

In general, Eqs. (12.8.1) and (12.8.2) have to be solved by trial. The hydraulic-jump formula results from setting $c = 0$ in the two equations (3.11.23).

EXAMPLE 12.10 A rectangular channel 3 m wide and 2 m deep, discharging 18 m³/s, suddenly has the discharge reduced to 12 m³/s at the downstream end. Compute the height and speed of the surge wave.

$V_1 = 3$, $y_1 = 2$, $V_2 y_2 = 4$. With Eqs. (12.8.1) and (12.8.2),

$$6 = 4 + c(y_2 - 2) \qquad \text{and} \qquad y_2^2 - 4 = \frac{2 \times 2}{9.806}(c + 3)(3 - V_2)$$

Eliminating c and V_2 gives

$$y_2^2 - 4 = \frac{4}{9.806}\left(\frac{2}{y_2 - 2} + 3\right)\left(3 - \frac{4}{y_2}\right)$$

or

$$\left(\frac{y_2 - 2}{3y_2 - 4}\right)^2 (y_2 + 2)y_2 = \frac{4}{9.806} = 0.407$$

After solving for y_2 by trial, $y_2 = 2.75$ m. Hence $V_2 = 4/2.75 = 1.455$ m/s.

The height of surge wave is 0.75 m, and the speed of the wave is

$$c = \frac{2}{y_2 - 2} = \frac{2}{0.75} = 2.667 \text{ m/s}$$

12.9 FRICTIONLESS NEGATIVE SURGE WAVE IN A RECTANGULAR CHANNEL

The negative surge wave appears as a gradual flattening and lowering of a liquid surface. It occurs, for example, in a channel downstream from a gate that is being closed or upstream from a gate that is being opened. Its propagation is accomplished by a series of elementary negative waves superposed on the existing velocity, each wave traveling at less speed than the one at next greater depth. Application of the momentum equation and the continuity equation to a small depth change produces simple differential expressions relating wave speed c, velocity V, and depth y. Integration of the equations yields liquid-surface profile as a function of time, and velocity as a function of depth or as a function of position along the channel and time (x and t). The fluid is assumed to be frictionless, and vertical accelerations are neglected.

In Fig. 12.28a an elementary disturbance is indicated in which the flow upstream has been slightly reduced. For application of the momentum and continuity equations it is convenient to reduce the motion to a steady one, as in Fig. 12.28b, by imposing a uniform velocity c to the left. The continuity

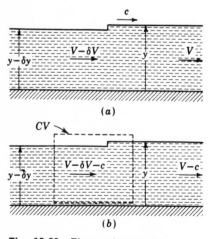

Fig. 12.28 Elementary wave.

equation is

$$(V - \delta V - c)(y - \delta y) = (V - c)y$$

or, by neglecting the product of small quantities,

$$(c - V)\,\delta y = y\delta V \qquad\qquad (12.9.1)$$

The momentum equation produces

$$\frac{\gamma}{2}\,(y - \delta y)^2 - \frac{\gamma}{2}\,y^2 = \frac{\gamma}{g}\,(V - c)y[V - c - (V - \delta V - c)]$$

After simplifying,

$$\delta y = \frac{c - V}{g}\,\delta V \qquad\qquad (12.9.2)$$

Equating $\delta V/\delta y$ in Eqs. (12.9.1) and (12.9.2) gives

$$c - V = \pm\sqrt{gy} \qquad\qquad (12.9.3)$$

or

$$c = V \pm\sqrt{gy}$$

The speed of an elementary wave in still liquid at depth y is $\sqrt{gy}$ and with flow the wave travels at the speed $\sqrt{gy}$ *relative* to the flowing liquid.
Eliminating c from Eqs. (12.9.1) and (12.9.2) gives

$$\frac{dV}{dy} = \pm\sqrt{\frac{g}{y}}$$

and integrating leads to

$$V = \pm 2\sqrt{gy} + \text{const}$$

For a negative wave forming downstream from a gate, Fig. 12.29, by using the plus sign, after an instantaneous partial closure, $V = V_0$ when $y = y_0$, and

$$V_0 = 2\sqrt{gy_0} + \text{const}$$

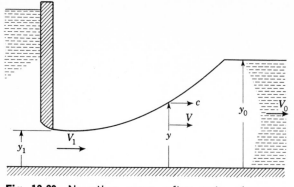

Fig. 12.29 Negative wave after gate closure.

After eliminating the constant,

$$V = V_0 - 2\sqrt{g}\left(\sqrt{y_0} - \sqrt{y}\right) \tag{12.9.4}$$

The wave travels in the $+x$ direction, so that

$$c = V + \sqrt{gy} = V_0 - 2\sqrt{gy_0} + 3\sqrt{gy} \tag{12.9.5}$$

If the gate motion occurs at $t = 0$, the liquid-surface position is expressed by $x = ct$, or

$$x = (V_0 - 2\sqrt{gy_0} + 3\sqrt{gy})t \tag{12.9.6}$$

Eliminating y from Eqs. (12.9.5) and (12.9.6) gives

$$V = \frac{V_0}{3} + \frac{2}{3}\frac{x}{t} - \tfrac{2}{3}\sqrt{gy_0} \tag{12.9.7}$$

which is the velocity in terms of x and t.

EXAMPLE 12.11 In Fig. 12.29 find the Froude number of the undisturbed flow such that the depth y_1 at the gate is just zero when the gate is suddenly closed. For $V_0 = 20$ ft/s, find the liquid-surface equation.

It is required that $V_1 = 0$ when $y_1 = 0$ at $x = 0$ for any time after $t = 0$. In Eq. (12.9.4), with $V = 0$, $y = 0$,

$$V_0 = 2\sqrt{gy_0} \quad \text{or} \quad \mathbf{F}_0 = \frac{V_0^2}{gy_0} = 4$$

For $V_0 = 20$,

$$y_0 = \frac{V_0^2}{4g} = \frac{20^2}{4g} = 3.11 \text{ ft}$$

By Eq. (12.9.6)

$$x = (20 - 2\sqrt{32.2 \times 3.11} + 3\sqrt{32.2y})t = 17.04\sqrt{y}\,t$$

The liquid surface is a parabola with vertex at the origin and surface concave upward.

EXAMPLE 12.12 In Fig. 12.29 the gate is partially closed at the instant $t = 0$ so that the discharge is reduced by 50 percent. $V_0 = 6$ m/s, $y_0 = 3$ m. Find V_1, y_1, and the surface profile.

The new discharge is

$$q = \frac{6 \times 3}{2} = 9 = V_1 y_1$$

By Eq. (12.9.4)

$$V_1 = 6 - 2\sqrt{9.806}\,(\sqrt{3} - \sqrt{y_1})$$

Then V_1 and y_1 are found by trial from the last two equations, $V_1 = 4.24$ m/s, $y_1 = 2.12$ m. The liquid-surface equation, from Eq. (12.9.6), is

$$x = (6 - 2\sqrt{3g} + 3\sqrt{gy})t \qquad \text{or} \qquad x = (9.39\sqrt{y} - 4.84)t$$

which holds for the range of values of y between 2.12 and 3 m.

Dam break

An idealized dam-break water-surface profile, Fig. 12.30, can be obtained from Eqs. (12.9.4) to (12.9.7). From a frictionless, horizontal channel with depth of water y_0 on one side of a gate and no water on the other side of the gate, the gate is suddenly removed. Vertical accelerations are neglected. $V_0 = 0$ in the equations, and y varies from y_0 to 0. The velocity at any section, from Eq. (12.9.4), is

$$V = -2\sqrt{g}\,(\sqrt{y_0} - \sqrt{y}) \tag{12.9.8}$$

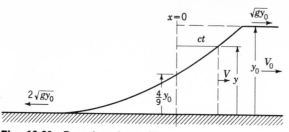

Fig. 12.30 Dam-break profile.

always in the downstream direction. The water-surface profile is, from Eq. (12.9.6),

$$x = (3\sqrt{gy} - 2\sqrt{gy_0})t \tag{12.9.9}$$

At $x = 0$, $y = 4y_0/9$, the depth remains constant and the velocity past the section $x = 0$ is, from Eq. (12.9.8),

$$V = -\tfrac{2}{3}\sqrt{gy_0}$$

also independent of time. The leading edge of the wave feathers out to zero height and moves downstream at $V = c = -2\sqrt{gy_0}$. The water surface is a parabola with vertex at the leading edge, concave upward.

With an actual dam break, ground roughness causes a positive surge, or wall of water, to move downstream; i.e., the feathered edge is retarded by friction.

12.10 FLOOD ROUTING IN PRISMATIC CHANNELS

In the two preceding sections instantaneous changes in frictionless rectangular channels were considered. In this section friction is taken into account, and conditions may change gradually at inlet or outlet sections. With prismatic channels the area is a function of depth of flow. The assumption is made that the channel slope α is small enough for cos α to approximate 1 and hydrostatic conditions to prevail along any vertical line in the fluid. Any known flow as a function of time may be added to the flow or taken from it at the upstream or downstream section of the channel.

In Fig. 12.31 an element of the flow is taken as control volume, the x direction is taken parallel to the bottom of the channel, and the depth y is measured normal to the bottom. The unsteady-momentum equation (3.11.2)

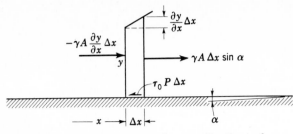

Fig. 12.31 Control volume for application of unsteady momentum equation.

is applied,

$$-\frac{\partial y}{\partial x} \gamma \, \Delta x \, A - \tau_0 P \, \Delta x + \gamma A \, \Delta x \sin \alpha = \frac{\partial}{\partial x}(\rho V^2 A) \, \Delta x + \frac{\partial}{\partial t}(\rho A V \, \Delta x)$$

Expanding and dividing through by the mass of the element $\rho A \, \Delta x$ gives

$$g\frac{\partial y}{\partial x} + \frac{\tau_0}{\rho R} - g\sin\alpha + 2V\frac{\partial V}{\partial x} + \frac{V^2}{A}\frac{\partial A}{\partial x} + \frac{V}{A}\frac{\partial A}{\partial t} + \frac{\partial V}{\partial t} = 0 \qquad (12.10.1)$$

P is the wetted perimeter of the cross section, and R is the hydraulic radius.

The continuity equation (3.2.1) applied to the control volume of Fig. 12.31 yields

$$-\frac{\partial}{\partial x}(\rho A V) \, \Delta x = \frac{\partial}{\partial t}(\rho A \, \Delta x)$$

After expanding and dividing by the mass of the element,

$$\frac{V}{A}\frac{\partial A}{\partial x} + \frac{1}{A}\frac{\partial A}{\partial t} + \frac{\partial V}{\partial x} = 0 \qquad (12.10.2)$$

Equation (12.10.1) may be simplified by subtracting Eq. (12.10.2), when multiplied by V, from it:

$$L_1 = g\frac{\partial y}{\partial x} + \frac{\tau_0}{\rho R} - g\sin\alpha + V\frac{\partial V}{\partial x} + \frac{\partial V}{\partial t} = 0 \qquad (12.10.3)$$

Equation (12.10.2) may be written

$$L_2 = \frac{A}{T}\frac{\partial V}{\partial x} + V\frac{\partial y}{\partial x} + \frac{\partial y}{\partial t} = 0 \qquad (12.10.4)$$

since

$$\frac{\partial A}{\partial x} = \frac{\partial A}{\partial y}\frac{\partial y}{\partial x} = T\frac{\partial y}{\partial x}$$

with T the top width. L_1 and L_2 are now in suitable form for solution by the method of characteristics. Combining, as in Sec. 12.6,

$$L_1 + \lambda L_2 = \left[\frac{\partial V}{\partial x}\left(V + \frac{\lambda A}{T}\right) + \frac{\partial V}{\partial t}\right] + \lambda\left[\frac{\partial y}{\partial x}\left(V + \frac{g}{\lambda}\right) + \frac{\partial y}{\partial t}\right]$$

$$+ \frac{\tau_0}{\rho R} - g\sin\alpha = 0 \quad (12.10.5)$$

For the term within the first pair of brackets to be a total derivative

$$\frac{dx}{dt} = V + \frac{\lambda A}{T}$$

and for the term within the second pair of brackets to be a total derivative

$$\frac{dx}{dt} = V + \frac{g}{\lambda}$$

After equating the two latter expressions and solving for λ,

$$\lambda = \pm\sqrt{\frac{gT}{A}} \qquad (12.10.6)$$

and

$$\frac{dx}{dt} = V \pm \sqrt{\frac{gA}{T}} \qquad (12.10.7)$$

For a rectangular cross section it should be noted that this expression for speed of a wave (surface wave) is given by

$$\frac{dx}{dt} = V \pm \sqrt{gy} \qquad (12.10.8)$$

Equation (12.10.5) reduces to

$$\frac{dV}{dt} + \lambda\frac{dy}{dt} + \frac{\tau_0}{\rho R} - g\sin\alpha = 0 \qquad (12.10.9)$$

subject to Eqs. (12.10.6) and (12.10.7).

For small slopes $\sin \alpha \approx S_0$, where S_0 is the bottom slope, and $\tau_0/\rho R$ may be expressed as gS, where S is the slope of the energy grade line as defined by the Manning or Chézy equation. This assumes that viscous losses in unsteady flow are described in the same manner as losses in steady flow at the same depth and discharge. After making these substitutions and those of the values of λ from Eq. (12.10.6), the four differential equations become, in finite-difference form,

$$V_P - V_R + \frac{g}{c_R}(y_P - y_R) + g(S_R - S_0)\,\Delta t = 0 \tag{12.10.10}$$

$$x_P - x_R = (V_R + c_R)\,\Delta t \tag{12.10.11}$$

C^+

$$V_P - V_S - \frac{g}{c_S}(y_P - y_S) + g(S_S - S_0)\,\Delta t = 0 \tag{12.10.12}$$

$$x_P - x_S = (V_S - c_S)\,\Delta t \tag{12.10.13}$$

C^-

where

$$c = \sqrt{\frac{gA}{T}} \tag{12.10.14}$$

and the subscripts R and S indicate evaluation of the quantity at points R and S in Fig. 12.32.

In Fig. 12.32 it is assumed that initially y and V are known at equidistant sections along the channel, Δx apart. To find y_P and V_P at one of the sections at the new time $t + \Delta t$, the variables y, V, and c must be evaluated at R and S. This is accomplished by linear interpolation from known values at A, C, and B. Then Eqs. (12.10.10) and (12.10.12) may be solved for V_P

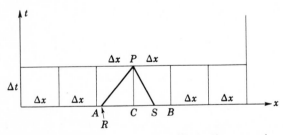

Fig. 12.32 Solution for specified time and distance interval.

and y_P. The linear interpolation for V_R is expressed by the proportion

$$\frac{V_C - V_R}{V_C - V_A} = \frac{x_C - x_R}{\Delta x} = \frac{x_P - x_R}{\Delta x} = \theta(V_R + c_R) \tag{12.10.15}$$

where $\theta = \Delta t/\Delta x$ and use has been made of Eq. (12.10.11). Similarly

$$\frac{c_C - c_R}{c_C - c_A} = \theta(V_R + c_R) \tag{12.10.16}$$

Simultaneous solution of these two equations yields

$$V_R = \frac{V_C + \theta(-V_C c_A + c_C V_A)}{1 + \theta(V_C - V_A + c_C - c_A)} \tag{12.10.17}$$

The value of c_R can be determined from Eq. (12.10.16) and for the rectangular section y_R from (12.10.14). For the general cross section it is easier to find y_R from

$$\frac{y_C - y_R}{y_C - y_A} = \theta(V_R + c_R) \tag{12.10.18}$$

by using the computed values of V_R and c_R; then find a new compatible value of c_R from Eq. (12.10.14).

For subcritical flow, which is most common for flood-routing analysis, point S lies between C and B in Fig. 12.32. A similar linear-interpolation procedure using Eq. (12.10.13) in place of Eq. (12.10.11) leads to

$$V_S = \frac{V_C - \theta(V_C c_B - c_C V_B)}{1 - \theta(V_C - V_B - c_C + c_B)} \tag{12.10.19}$$

$$c_S = \frac{c_C + V_S \theta(c_C - c_B)}{1 + \theta(c_C - c_B)} \tag{12.10.20}$$

$$y_S = y_C + \theta(V_S - c_S)(y_C - y_B) \tag{12.10.21}$$

Again in the rectangular section Eqs. (12.10.19), (12.10.20), and (12.10.14) are used, and for the general cross section the above three equations are used, with a final value of c_S determined from Eq. (12.10.14) using y_S from Eq. (12.10.21). In supercritical flow, which is not covered here, the control of the unsteady flow shifts to the upstream end only, and point S lies to the left of point C.

With the variables determined at points R and S, Eqs. (12.10.10) and

(12.10.12) are solved simultaneously for y_P and V_P:

$$y_P = \frac{1}{c_R + c_S} \left\{ y_S c_R + y_R c_S + c_R c_S \left[\frac{V_R - V_S}{g} - \Delta t (S_R - S_S) \right] \right\} \qquad (12.10.22)$$

$$V_P = V_R - g \frac{y_P - y_R}{c_R} - g \Delta t (S_R - S_0) \qquad (12.10.23)$$

For the end sections of the channel Eqs. (12.10.10) and (12.10.12) each yield one linear equation in two unknowns V_P, y_P, which, together with a known condition at the end, permits the solution to be carried forward. This analysis is limited to subcritical flow where one boundary condition is needed at each end of the channel.

In solving the equations it is absolutely essential that R and S always stay within AB (Fig. 12.32); otherwise the solution method is unstable.

EXAMPLE 12.13 A trapezoidal channel 3000 m long, 7 m wide, side slopes 0.8 horizontal to 1 vertical, 0.016 Manning roughness, and 0.001 bottom slope is discharging under steady uniform flow conditions at $y_n = 1.8$ m. At time $t = 0$ the flow begins to increase linearly to a flow of 60 m³/s in 20 min. The flow then decreases linearly to 15 m³/s in another 10 min. The downstream boundary condition is given as a gage-height discharge curve of the form $Q = C_W(y - y_{00})^{1.5}$.

A computer program is presented in Fig. 12.33 for finding the velocity and depth in the channel. The discharge Q_0 at steady uniform flow is first computed using the given normal depth in Manning's equation. The upstream boundary is

$$Q = \begin{cases} Q_0 + \dfrac{(Q_M - Q_0)t}{t_1} & 0 \le t < t_1 \\[2ex] Q_M + \dfrac{Q_F - Q_M}{t_F - t_1}(t - t_1) & t_1 \le t < t_F \\[2ex] Q_F & t \ge t_F \end{cases}$$

where $Q_M = 60$, $Q_F = 15$, $t_1 = 1200$ s, $t_F = 1800$ s. By continuity the discharge in a trapezoidal section is equal to $(By + Z_1 y^2) V$, where B is the bottom width and Z_1 the side slope. A nonlinear equation results which must be solved simultaneously with the C^- compatibility equation, Eq. (12.10.12). Newton's method is used in the solution (see Appendix E). Similarly the downstream gage-height discharge equation, when combined with the continuity equa-

```
C FLCCD ROUTING IN TRAPEZCICAL CHANNEL. D.S. BCUNCARY Q=CW*(Y-YOO)**1.5
C LINEAR RISE & FALL CF INFLOW HYDRUGRAPH, FLOW AT YN AT INITIAL T,QM AT
C T1, ANC QF AT TF. WIDTH B, SIDE SLCPE Z1, RUUGHNESS RN, SLOPE SO.
      DIMENSICN Y(21),V(21),YP(21),VP(21),C(21),IC(2)
      CATA ID/'EN','SI'/
      AREA(YY)=YY*(B+YY*Z1)
      PER(YY)=B+2.*YY*SCRT(Z1*Z1+1.)
      CEL(YY)=SQRT(G*YY*(B+YY*Z1)/(B+2.*Z1*YY))
      SLCPE(YY,VV)=(VV*RN/CMA)**2*(PER(YY)/(YY*(B+YY*Z1)))**1.3333
10    READ (5,15,END=99)XL,B,Z1,RN,SU,CW ,YN,CM,CF,T1,TF,TMAX, NT,N,IPR
15    FORMAT(3F10.3,2F10.5,F1C.2/6F10.2/A2,3X,2I5)
      IF (NT.EC.ID(2)) GO TO 16
      CMA=1.49
      G=32.174
      GO TO 17
16    CMA=1.
      G=S.8C6
17    WRITE (6,18) NT,XL,B,Z1,RN,SU,CW,YN,CM,QF,T1,TF,TMAX,N,IPR
18    FCRMAT(1HO,A2,' UNITS SPECIFIED'/' XL=',F9.1,' B=',F8.2,' Z1=',
     2F8.2,' RN=',F6.4,' SO=',F7.5,' CW=',F7.2/' YN=',F8.2,' CM=',F8.2
     3,' QF=',F8.2,' T1,TF,& TMAX=',3F8.1,' SEC'/' N=',I5,' IPR=',I5)
      QO=CMA*AREA(YN)**1.6666*SQRT(SO)/(RN*PER(YN)**C.6666)
      C6=(CM-CO)/T1
      C7=(QF-CM)/(TF-T1)
      YOO=YN-(CO/CW)**(2./3.)
      NS=N+1
      VO=QO/AREA(YN)
      CC 20 I=1,NS
      V(I)=VO
      C(I)=CEL(YN)
      YP(I)=YN
20    Y(I)=YN
      DX=XL/N
      Q=QO
      T=C.
      K=0
      CT=.9*CX/(VO+C(1))
      WRITE (6,25) QC,YOO
25    FORMAT (' QC=',F10.2,' YCC=',F8.3/' TIME IN MINUTES'//'      TIME
     2 U.S.C X/L=   .0     .2     .4     .6     .8    1.   Q')
30    TM=T/6C.
      CW=AREA(Y(NS))*V(NS)
      WRITE (6,35) TM,Q,(V(I),I=1,NS,2),(Y(I),I=1,NS,2),QW
35    FCRMAT(1H ,F9.3,F9.2,3H V=,6F8.3/19X,3H Y=,6F8.3,F8.2)
40    T=T+DT
      K=K+1
      IF (T.GT.TMAX) GO TC 10
      IX=0
      DXX=0.
      DO 45 I=1,NS
      DXI=(V(I)+C(I))*DT
      IF (CXI.GT.DXX) DXX=DXI
      IF (DXI.GT.DX.AND.IX.EQ.0) GO TO 43
      GO TO 45
43    T=T-.1*CT
      CT=.9*DT
      IX=1
45    CONTINUE
      TH=DT/DX
C INTERIOR PCINTS
      DO 50 I=2,N
      CA=C(I)-C(I-1)
      VR=(V(I)+TH*(C(I)*V(I-1)-V(I)*C(I-1)))/(1.+TH*(V(I)-V(I-1)+CA))
      CR=(C(I)-VR*TH*CA)/(1.+TH*CA)
      YR=Y(I)-TH*(VR+CR)*(Y(I)-Y(I-1))
      CR=CEL(YR)
      SR=SLOPE(YR,VR)
      CB=C(I)-C(I+1)
      VS=(V(I)-TH*(V(I)*C(I+1)-C(I)*V(I+1)))/(1.-TH*(V(I)-V(I+1)-CB))
      CS=(C(I)+VS*TH*CB)/(1.+TH*CB)
      YS=Y(I)+TH*(VS-CS)*(Y(I)-Y(I+1))
      CS=CEL(YS)
      YP(I)=(YS*CR+YR*CS+CR*CS*((VR-VS)/G-DT*(SR-SLCPE(YS,VS))))/(CR+CS)
50    VP(I)=VR-G*((YP(I)-YR)/CR+CT*(SR-SO))
C UPSTREAM BCUNCARY
      CB=C(1)-C(2)
      VS=(V(1)-TH*(V(1)*C(2)-C(1)*V(2)))/(1.-TH*(V(1)-V(2)-CB))
      CS=(C(1)+VS*TH*CB)/(1.+TH*CB)
      YS=Y(1)+TH*(VS-CS)*(Y(1)-Y(2))
      IF (T.LT.T1) Q=QO+C6*T
```

Fig. 12.33

```
      IF (T.GE.T1.AND.T.LT.TF) Q=CM+C7*(T-T1)
      IF (T.GT.TF) Q=QF
      C2=G/CEL(YS)
      CM=VS-C2*YS-G*DT*(SLOPE(YS,VS)-S0)
      E=CM*Z1+C2*B
      DO 60 J=1,3
      F=Q-YP(1)*(YP(1)*(YP(1)*C2*Z1+E)+CM*B)
   60 YP(1)=YP(1)+F/(YP(1)*(3.*C2*Z1*YP(1)+2.*E)+CM*B)
      VP(1)=CM+C2*YP(1)
C DOWNSTREAM BOUNDARY
      CA=C(NS)-C(N)
      VR=(V(NS)+TH*(C(NS)*V(N)-V(NS)*C(N)))/(1.+TH*(V(NS)-V(N)+CA))
      CR=(C(NS)-VR*TH*CA)/(1.+TF*CA)
      YR=Y(NS)-TH*(VR+CR)*(Y(NS)-Y(N))
      C4=G/CEL(YR)
      CP=VR*C4*YR-G*DT*(SLOPE(YR,VR)-S0)
      E=-CP*Z1+B*C4
      DO 70 J=1,3
      W=YP(NS)-Y00
      F=YP(NS)*(YP(NS)*(YP(NS)*C4*Z1+E)-B*CP)+CW*W**1.5
   70 YP(NS)=YP(NS)+F/(YP(NS)*(-YP(NS)*C4*Z1*3.-E*2.)+B*CP-1.5*CW*W**.5)
      VP(NS)=CP-C4*YP(NS)
      DO 80 I=1,NS
      V(I)=VP(I)
      C(I)=CEL(YP(I))
   80 Y(I)=YP(I)
      IF (DXX.LT.0.8*DX) DT=1.15*DT
      IF (K/IPR*IPR.EQ.K) GO TO 30
      GO TO 40
   99 STOP
      END
   3000.       7.      .8      .016     .0010      20.
      1.8      60.     15.    1200.    1800.     2400.
SI       10      2
```

Fig. 12.33 FORTRAN IV program for flood routing.

tion and Eq. (12.10.10), yields a nonlinear equation which is also solved iteratively using Newton's method.

Computed results are given in Fig. 12.34. It may be noted that formatted input is used in the program and that either SI or English units may be used by specifying SI or EN in the first two columns of the last card of input data. The two integers on this card specify the number of reaches the channel is divided into and the number of time-increment iterations between printout of computed results. The program is able to handle prismatic triangular, rectangular, or trapezoidal sections. It also contains an adjustment of the size of the time increment during the computations in order to minimize the approximation due to interpolations. Some caution should be used in applying the program to continually increasing flows or sudden flows, as no provision is made for handling hydraulic bores.

12.11 MECHANICS OF RAINFALL-RUNOFF RELATIONS FOR SLOPING PLANE AREAS

An interesting, simplified characteristics problem[1] is that of the relation between rainfall and runoff on a plane sloping surface. Percolation rate may be

[1] F. M. Henderson, and R. A. Wooding, Overland Flow and Groundwater Flow from a Steady Rainfall of Finite Duration, *J. Geophys. Res.*, vol. 69, no. 8, 1964.

```
'SI UNITS SPECIFIED
XL=    3000.0  B=     7.00  Z1=     C.80  RN=0.0160  S0=0.00100  CW=   20.00
YN=     1.80  CM=    60.00  QF=    15.00  T1,TF,& TMAX=  1200.0  1800.0  2400.0 SEC
N=     10  IPR=     2
CC=    35.92  Y00=    0.322
TIME IN MINUTES
```

TIME	L.S.Q	X/L=	.0	.2	.4	.6	.8	1.	Q
C.000	35.92	V=	2.364	2.364	2.364	2.364	2.364	2.364	
		Y=	1.800	1.800	1.800	1.800	1.800	1.800	35.92
1.441	37.65	V=	2.432	2.364	2.364	2.364	2.364	2.364	
		Y=	1.830	1.800	1.800	1.800	1.800	1.800	35.92
2.881	39.39	V=	2.488	2.411	2.364	2.364	2.364	2.364	
		Y=	1.865	1.820	1.800	1.800	1.800	1.800	35.92
4.322	41.12	V=	2.536	2.460	2.396	2.364	2.364	2.364	
		Y=	1.903	1.849	1.814	1.800	1.800	1.800	35.92
5.762	42.86	V=	2.577	2.505	2.437	2.386	2.364	2.364	
		Y=	1.944	1.882	1.836	1.809	1.800	1.800	35.92
7.203	44.59	V=	2.615	2.546	2.478	2.419	2.379	2.364	
		Y=	1.985	1.919	1.865	1.827	1.806	1.800	35.92
8.643	46.33	V=	2.649	2.585	2.518	2.454	2.404	2.369	
		Y=	2.028	1.958	1.897	1.850	1.819	1.806	36.13
10.084	48.06	V=	2.681	2.621	2.556	2.492	2.435	2.381	
		Y=	2.071	1.998	1.933	1.878	1.839	1.819	36.63
11.525	49.80	V=	2.711	2.654	2.592	2.529	2.468	2.398	
		Y=	2.114	2.040	1.971	1.911	1.863	1.839	37.35
12.965	51.53	V=	2.739	2.686	2.627	2.565	2.502	2.418	
		Y=	2.156	2.082	2.011	1.946	1.892	1.864	38.27
14.334	53.18	V=	2.764	2.714	2.658	2.598	2.534	2.441	
		Y=	2.157	2.122	2.049	1.981	1.923	1.892	39.32
15.630	54.74	V=	2.788	2.740	2.686	2.629	2.564	2.465	
		Y=	2.235	2.160	2.087	2.017	1.955	1.921	40.44
16.927	56.30	V=	2.810	2.764	2.713	2.658	2.593	2.490	
		Y=	2.272	2.198	2.124	2.053	1.989	1.953	41.66
18.223	57.86	V=	2.832	2.788	2.739	2.686	2.622	2.517	
		Y=	2.309	2.236	2.162	2.090	2.024	1.987	42.95
19.520	59.42	V=	2.853	2.810	2.764	2.712	2.650	2.543	
		Y=	2.346	2.274	2.200	2.128	2.061	2.022	44.32
20.816	56.33	V=	2.739	2.832	2.788	2.738	2.676	2.570	
		Y=	2.321	2.311	2.238	2.165	2.098	2.058	45.74
22.113	50.49	V=	2.561	2.759	2.811	2.763	2.702	2.597	
		Y=	2.242	2.305	2.276	2.203	2.136	2.095	47.21
23.409	44.66	V=	2.390	2.618	2.766	2.787	2.728	2.624	
		Y=	2.144	2.250	2.283	2.241	2.174	2.133	48.72
24.706	38.82	V=	2.219	2.480	2.657	2.762	2.752	2.650	
		Y=	2.029	2.176	2.248	2.257	2.212	2.171	50.25
26.002	32.99	V=	2.041	2.340	2.544	2.681	2.743	2.676	
		Y=	1.898	2.086	2.194	2.238	2.235	2.209	51.81
27.299	27.15	V=	1.847	2.195	2.428	2.589	2.683	2.691	
		Y=	1.750	1.982	2.124	2.201	2.229	2.231	52.73
28.595	21.32	V=	1.628	2.044	2.309	2.493	2.610	2.683	
		Y=	1.584	1.865	2.042	2.149	2.206	2.220	52.28
29.892	15.49	V=	1.368	1.881	2.184	2.392	2.532	2.663	
		Y=	1.395	1.735	1.948	2.084	2.167	2.190	51.04
31.188	15.00	V=	1.415	1.701	2.054	2.287	2.452	2.633	
		Y=	1.316	1.592	1.843	2.008	2.115	2.146	49.25
32.485	15.00	V=	1.486	1.633	1.915	2.179	2.370	2.594	
		Y=	1.261	1.486	1.728	1.923	2.051	2.091	47.04
33.781	15.00	V=	1.546	1.636	1.822	2.067	2.286	2.547	
		Y=	1.217	1.413	1.620	1.828	1.978	2.027	44.49
35.078	15.00	V=	1.597	1.650	1.785	1.975	2.199	2.492	
		Y=	1.182	1.353	1.545	1.736	1.896	1.955	41.71
36.375	15.00	V=	1.637	1.665	1.768	1.920	2.121	2.429	
		Y=	1.156	1.303	1.478	1.656	1.811	1.877	38.77
37.671	15.00	V=	1.670	1.680	1.759	1.887	2.065	2.363	
		Y=	1.136	1.261	1.420	1.586	1.733	1.798	35.86

Fig. 12.34 Computer solution of Example 12.13.

subtracted from rainfall rate to yield more meaningful results. One-dimensional flow is assumed (Fig. 12.35) on a mild slope so that $\cos\theta \approx 1$ and $\sin\theta \approx S_0$.

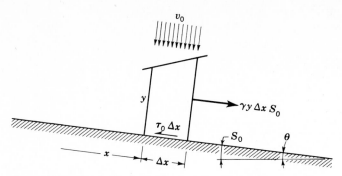

Fig. 12.35 Control volume for continuity and momentum equations for rainfall and runoff.

The momentum equation, for an element of unit width, yields

$$\gamma y \, \Delta x \, S_0 - \gamma y \frac{\partial y}{\partial x} \Delta x - \tau_0 \, \Delta x = \rho \frac{\partial}{\partial x} (V^2 y) \, \Delta x + \rho \frac{\partial}{\partial t} (Vy) \, \Delta x$$

After replacing τ_0 by γSy, where S is the slope of the energy grade line, and simplifying

$$gy \frac{\partial y}{\partial x} + gyS - gyS_0 + V^2 \frac{\partial y}{\partial x} + 2Vy \frac{\partial V}{\partial x} + y \frac{\partial V}{\partial t} + V \frac{\partial y}{\partial t} = 0 \qquad (12.11.1)$$

The continuity equation yields

$$\frac{\partial}{\partial x} (Vy) \, \Delta x = v_0 \, \Delta x - \frac{\partial y}{\partial t} \Delta x$$

or

$$y \frac{\partial V}{\partial x} + V \frac{\partial y}{\partial x} + \frac{\partial y}{\partial t} - v_0 = 0 \qquad (12.11.2)$$

Multiplying Eq. (12.11.2) by V and subtracting from Eq. (12.11.1) lead to

$$gy \frac{\partial y}{\partial x} + gyS - gyS_0 + Vy \frac{\partial V}{\partial x} + y \frac{\partial V}{\partial t} + v_0 V = 0 \qquad (12.11.3)$$

or

$$S = S_0 - \left(\frac{\partial y}{\partial x} + \frac{V}{g} \frac{\partial V}{\partial x} + \frac{1}{g} \frac{\partial V}{\partial t} + \frac{v_0 V}{gy} \right) \qquad (12.11.4)$$

For one class of problems the quantity in parenthesis is small compared with S_0 and may be dropped:

$$S \approx S_0 \qquad (12.11.5)$$

If Manning's equation is used to describe frictional losses

$$q = Vy = \frac{C_m}{n} \sqrt{S_0}\, y^{5/3} \qquad (12.11.6)$$

or, by generalizing, for resistance formulas other than Manning,

$$q = \alpha y^m \qquad (12.11.7)$$

The continuity equation in terms of q is

$$\frac{\partial q}{\partial x} + \frac{\partial y}{\partial t} = v_0 \qquad (12.11.8)$$

These last two equations, one algebraic, one differential, may be solved along a characteristic. From Eq. (12.11.7)

$$\frac{\partial q}{\partial x} = \frac{dq}{dy} \frac{\partial y}{\partial x} = m\alpha y^{m-1} \frac{\partial y}{\partial x} \qquad (12.11.9)$$

which, combined with Eq. (12.11.8), yields

$$m\alpha y^{m-1} \frac{\partial y}{\partial x} + \frac{\partial y}{\partial t} = v_0 \qquad (12.11.10)$$

Now, if

$$\frac{dx}{dt} = m\alpha y^{m-1} \qquad (12.11.11)$$

Eq. (12.11.10) becomes

$$\frac{dy}{dt} = v_0 \tag{12.11.12}$$

which is valid along the xt path given by Eq. (12.11.11).

For constant rate of rainfall starting from $t = 0$, Eq. (12.11.12) integrates to

$$y = v_0 t \tag{12.11.13}$$

which states that along any xt characteristic the depth of liquid increases linearly. By substituting Eq. (12.11.13) into Eq. (12.11.11) and integrating,

$$\frac{dx}{dt} = m\alpha(v_0 t)^{m-1}$$

or

$$x = x_0 + \alpha v_0^{m-1} t^m \tag{12.11.14}$$

with x_0 the starting point on the characteristic for $t = 0$. For $x_0 = 0$, the line plots on the xt plane of Fig. 12.36a as shown. Since all the characteristics issuing from the $t = 0$ line are parallel at a given t, and since $y = v_0 t$ along each characteristic, the build-up on the plane occurs with the surface parallel to the plane, as indicated in Fig. 12.36b. The steady-state profile is determined from Eq. (12.11.14) with $x_0 = 0$,

$$x = \alpha v_0^{m-1} t^m = \frac{\alpha}{v_0}(v_0 t)^m = \frac{\alpha y^m}{v_0}$$

or

$$x_{s_1} = \frac{\alpha y_s^m}{v_0} \tag{12.11.15}$$

The time required for steady state is given by Eq. (12.11.14) for $x = L$, $x_0 = 0$

$$t_s = \left(\frac{L}{\alpha v_0^{m-1}}\right)^{1/m} \tag{12.11.16}$$

which is also the time at which maximum depth and flow begin to occur. The

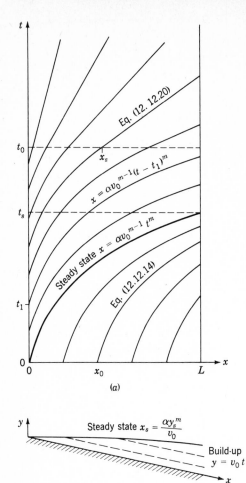

(a)

(b)

Fig. 12.36 (a) Characteristics on xt plane and (b) water surface on physical plane.

discharge at the downstream end of the plane is

$$q_{x=L} = \alpha y^m = \alpha(v_0 t)^m \qquad 0 < t < t_s \tag{12.11.17}$$

with steady-state discharge

$$q_{x=L} = Lv_0 = \alpha(v_0 t_s)^m \qquad t \geq t_s \tag{12.11.18}$$

For subsidence of the runoff following the cessation of the rainfall, let $v_0 = 0$, for $t > t_0 > t_s$. Then $dy/dt = 0$, which shows y to be a constant along the characteristic

$$\frac{dx}{dt} = \alpha m y_s{}^{m-1} \qquad (12.11.19)$$

Figure 12.36a shows the characteristics drawn from conditions for steady state at time t_0. The profile conditions as a function of time may be found by integration of Eq. (12.11.19)

$$x = x_s + \alpha m y_s{}^{m-1}(t - t_0) \qquad (12.11.20)$$

where x_s is the distance to the steady-state depth y_s. By substituting Eq. (12.11.15)

$$x = \frac{\alpha y_s{}^m}{v_0} + \alpha m y_s{}^{m-1}(t - t_0) \qquad (12.11.21)$$

Since $q = \alpha y_s{}^m$, the discharge during subsidence may be determined at any position x,

$$x = \frac{q}{v_0} + \alpha m \left(\frac{q}{\alpha}\right)^{(m-1)/m} (t - t_0) \qquad t > t_0 \qquad (12.11.22)$$

For $x = L$,

$$L = \frac{q}{v_0} + \alpha m \left(\frac{q}{\alpha}\right)^{(m-1)/m} (t - t_0) \qquad t > t_0 \qquad (12.11.23)$$

from which the subsidence hydrograph at the downstream end can be calculated. The discharge hydrograph for the complete rainfall period is schematically shown in Fig. 12.37 for $t_0 > t_s$.

EXAMPLE 12.14 A paved parking lot 200 yd square on a slope of 0.0016 is subjected to rain of 2 in/h for 30 min. Resistance is given by Manning's $n = 0.025$. Determine the hydrograph for this storm from this area.

$$v_0 = \frac{2.0}{12 \times 3600} = \frac{1}{21,600} \text{ ft/s}$$

$$\alpha = \frac{C_m}{n} \sqrt{S_0} = \frac{1.49}{0.025} \times \sqrt{0.0016} = 2.4$$

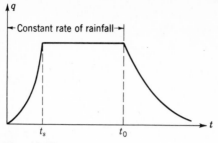

Fig. 12.37 Hydrograph for runoff from an inclined plane.

From Eq. (12.11.16), for $m = \frac{5}{3}$,

$$t_s = \left(\frac{L}{\alpha v_0{}^{m-1}}\right)^{1/m} = \left(\frac{600}{2.4 \times 0.0000463^{2/3}}\right)^{3/5} = 1488 \text{ s}$$

From Eq. (12.11.17)

$$Q = 600\alpha(v_0 t)^m = 600 \times 2.4(0.0000463t)^{5/3}$$

$$= 8.596 \times 10^{-5}t^{5/3} \qquad 0 < t < 1488$$

$$Q_{max} = 16.67 \text{ cfs} \qquad 1488 < t < 1800$$

From Eq. (12.11.23), for $q = Q/600$,

$$t - t_0 = \frac{600 - 36Q}{0.218Q^{2/5}} \qquad t > 1800$$

The hydrograph is plotted in Fig. 12.38.

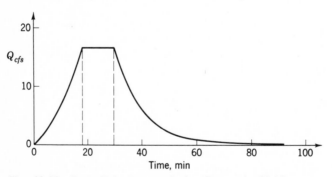

Fig. 12.38 Runoff hydrograph for Example 12.14.

PROBLEMS

12.1 Determine the period of oscillation of a U tube containing $\frac{1}{2}$ l of water. The cross-sectional area is 2.4 cm^2. Neglect friction.

12.2 A U tube containing alcohol is oscillating with maximum displacement from equilibrium position of 5.0 in. The total column length is 40 in. Determine the maximum fluid velocity and the period of oscillation. Neglect friction.

12.3 A liquid, $\nu = 0.002$ ft^2/s, is in a 0.50-in-diameter U tube. The total liquid column is 70 in long. If one meniscus is 15 in above the other meniscus when the column is at rest, determine the time for one meniscus to move to within 1.0 in of its equilibrium position.

12.4 Develop the equations for motion of a liquid in a U tube for laminar resistance when $16\nu/D^2 = \sqrt{2g/L}$. *Suggestion:* Try $z = e^{-mt}(c_1 + c_2 t)$.

12.5 A U tube contains liquid oscillating with a velocity 2 m/s at the instant the menisci are at the same elevation. Find the time to the instant the menisci are next at the same elevation and determine the velocity then. $\nu = 10$ μm^2/s, $D = 0.6$ cm, $L = 75$ cm.

12.6 A 10-ft-diameter horizontal tunnel has 10-ft-diameter vertical shafts spaced 1 mi apart. When valves are closed isolating this reach of tunnel, the water surges to a depth of 50 ft in one shaft when it is 20 ft in the other shaft. For $f = 0.022$ find the height of the next two surges.

12.7 Two standpipes 6 m in diameter are connected by 900 m of 2.5-m-diameter pipe; $f = 0.020$, and minor losses are 4.5 velocity heads. One reservoir level is 9 m above the other one when a valve is rapidly opened in the pipeline. Find the maximum fluctuation in water level in the standpipe.

12.8 A valve is quickly opened in a pipe 1200 m long, $D = 0.6$ m, with a 0.3-m-diameter nozzle on the downstream end. Minor losses are $4V^2/2g$, with V the velocity in the pipe, $f = 0.024$, $H = 9$ m. Find the time to attain 95 percent of the steady-state discharge.

12.9 A globe valve $(K = 10)$ at the end of a pipe 2000 ft long is rapidly opened. $D = 3.0$ ft, $f = 0.018$, minor losses are $2V^2/2g$, and $H = 75$ ft. How long does it take for the discharge to attain 80 percent of its steady-state value?

12.10 A steel pipeline with expansion joints is 90 cm in diameter and has a 1-cm wall thickness. When it is carrying water, determine the speed of a pressure wave.

12.11 Benzine $(K = 150{,}000$ psi, $S = 0.88)$ flows through $\frac{3}{4}$-in-ID steel tubing with $\frac{1}{8}$-in wall thickness. Determine the speed of a pressure wave.

12.12 Determine the maximum time for rapid valve closure on a pipeline: $L = 3000$ ft, $D = 4$ ft, $t' = \frac{1}{2}$ in steel pipe, $V_0 = 10$ ft/s, water flowing.

12.13 A valve is closed in 5 s at the downstream end of a 3000-m pipeline carrying water at 2 m/s. $a = 1000$ m/s. What is the peak pressure developed by the closure?

12.14   Determine the length of pipe in Prob. 12.13 subjected to the peak pressure.

12.15   A valve is closed at the downstream end of a pipeline in such a manner that only one-third of the line is subjected to maximum pressure. During what proportion of the time $2L/a$ is it closed?

12.16   A pipe, $L = 2000$ m, $a = 1000$ m/s, has a valve on its downstream end, $v_0 = 2.5$ m/s and $h_0 = 20$ m. It closes in three increments, spaced 1 s apart, each area reduction being one-third of the original opening. Find the pressure at the gate and at the midpoint of the pipeline at 1-s intervals for 5 s after initial closure.

12.17   A pipeline, $L = 2000$ ft, $a = 4000$ ft/s, has a valve at its downstream end, $V_0 = 6$ ft/s and $h_0 = 100$ ft. Determine the pressure at the valve for closure:

A_v/A_{v0}	1	0.75	0.60	0.45	0.30	0.15	0
t, s	0	0.5	1.0	1.5	2.0	2.5	3.0

12.18   In Prob. 12.17 determine the peak pressure at the valve for uniform area reduction in 3.0 s.

12.19   Find the maximum area reduction for $\frac{1}{2}$-s intervals for the pipeline of Prob. 12.17 when the maximum head at the valve is not to exceed 160 ft.

12.20   Derive the characteristics-method solution for waterhammer with the pressure p and the discharge Q as dependent variables.

12.21   Alter the reservoir boundary condition in Example 12.6 to include the minor loss and velocity head at the pipe inlet. Assume a square-edged pipe entrance.

12.22   Develop a single-pipeline waterhammer program to handle a valve closure at the downstream end of the pipe with a reservoir at the upstream end. The valve closure is given by $\tau = (1 - t/t_c)^m$, where t_c is the time of closure and is 6.2 s, and $m = 3.2$; $L = 5743.5$ ft, $a = 3927$ ft/s, $D = 4$ ft, $f = 0.019$, $V_0 = 3.6$ ft/s, and $H_0 = 300$ ft.

12.23   In Prob. 12.22 place a wave on the reservoir at a period of 1.95 s and obtain a solution with the aid of a computer.

12.24   Develop the computer program to solve Example 12.8 using your own particular input data.

12.25   The valve closure data for the series system shown in Fig. 12.39 is

τ	1	0.73	0.5	0.31	0.16	0.05	0.0
t, s	0	1	2	3	4	5	6

Develop a characteristics-method computer program to determine the pressure head and flow at the valve and at the series pipeline connection.

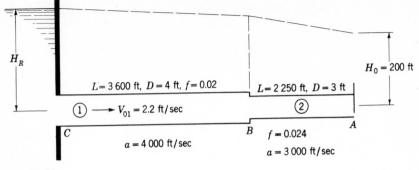

Fig. 12.39

12.26 In Prob. 12.25 place the valve at the connection between the two pipelines and an orifice at the downstream end with a steady-state head drop of 100 ft. Develop a program to analyze the valve motion given in Prob. 12.25.

12.27 In Example 12.6 reduce H_0 to 20 m and execute the program until vapor pressure is reached at the dead end. Modify the program to account for the vapor-column separation and subsequent collapse at the dead end.

12.28 A rectangular channel is discharging 50 cfs per foot of width at a depth of 10 ft when the discharge upstream is suddenly increased to 70 cfs/ft. Determine the speed and height of the surge wave.

12.29 In a rectangular channel with velocity 2 m/s flowing at a depth of 2 m, a surge wave 0.3 m high travels upstream. What is the speed of the wave, and how much is the discharge reduced per meter of width?

12.30 A rectangular channel 10 ft wide and 6 ft deep discharges 28 m³/s when the flow is completely stopped downstream by closure of a gate. Compute the height and speed of the resulting positive surge wave.

12.31 Determine the depth downstream from the gate of Prob. 12.30 after it closes.

12.32 Find the downstream water surface of Prob. 12.30 3 s after closure.

12.33 Determine the water surface 2 s after an ideal dam breaks. Original depth is 100 ft.

12.34 Work Example 12.13, by computer, for linear increase in flow to 15,000 cfs in 20 min followed by a decrease to 6000 cfs in an additional 40 min. $B = 100$ ft, $Z_1 = 0$, $n = 0.014$, $S_0 = 0.001$, $y_n = 8.143$ ft, $C_W = 132$.

12.35 Calculate the runoff hydrograph for a 1.5 in/h rainfall for 40 min on an impervious flat surface 400 ft long and 200 ft wide. The surface slopes downward 1 ft along the 400-ft length. $n = 0.03$.

12.36 Neglecting friction, the maximum difference in elevation of the two menisci of an oscillating U tube is 1.0 ft, $L = 3.0$ ft. The period of oscillation is, in seconds,

(*a*) 0.52   (*b*) 1.92   (*c*) 3.27   (*d*) 20.6   (*e*) none of these answers

12.37 The maximum speed of the liquid column in Prob. 12.36, in feet per second, is

(*a*) 0.15   (*b*) 0.31   (*c*) 1.64   (*d*) 3.28   (*e*) none of these answers

12.38 In frictionless oscillation of a U tube, $L = 2.179$ m, $z = 0$, $V = 6$ m/s. The maximum value of z is, in meters,

(*a*) 1.57   (*b*) 2.179   (*c*) 2.00   (*d*) 13.074   (*e*) none of these answers

12.39 In analyzing the oscillation of a U tube with laminar resistance, the assumption is made that the

(*a*) motion is steady   (*b*) resistance is constant
(*c*) Darcy-Weisbach equation applies
(*d*) resistance is a linear function of the displacement
(*e*) resistance is the same at any instant as if the motion were steady

12.40 When $16\nu/D^2 = 5$ and $2g/L = 12$ in oscillation of a U tube with laminar resistance,

(*a*) the resistance is so small that it may be neglected
(*b*) the menisci oscillate about the $z = 0$ axis
(*c*) the velocity is a maximum when $z = 0$
(*d*) the velocity is zero when $z = 0$
(*e*) the speed of column is a linear function of z

12.41 In laminar resistance to oscillation in a U tube, $m = 1$, $n = \frac{1}{2}$, $V_0 = 3$ ft/s when $t = 0$ and $z = 0$. The time of maximum displacement of meniscus is, in seconds,

(*a*) 0.46   (*b*) 0.55   (*c*) 0.93   (*d*) 1.1   (*e*) none of these answers

12.42 In Prob. 12.41 the maximum displacement, in feet, is

(*a*) 0.53   (*b*) 1.06   (*c*) 1.16   (*d*) 6.80   (*e*) none of these answers

12.43 In analyzing the oscillation of a U tube with turbulent resistance, the assumption is made that

(*a*) the Darcy-Weisbach equation applies
(*b*) the Hagen-Poiseuille equation applies   (*c*) the motion is steady
(*d*) the resistance is a linear function of velocity
(*e*) the resistance varies as the square of the displacement

12.44 The maximum displacement is $z_m = 20$ ft for $f = 0.020$, $D = 1.0$ ft in oscillation of a U tube with turbulent flow. The minimum displacement $(-z_{m+1})$ of the same fluid column is

(*a*) -13.3 (*b*) -15.7 (*c*) -16.5 (*d*) -20 (*e*) none of these answers

12.45 When a valve is suddenly opened at the downstream end of a long pipe connected at its upstream end with a water reservoir,

(a) the velocity attains its final value instantaneously if friction is neglected
(b) the time to attain nine-tenths of its final velocity is less with friction than without friction
(c) the value of f does not affect the time to acquire a given velocity
(d) the velocity increases exponentially with time
(e) the final velocity is attained in less than $2L/a$ s

12.46 Surge may be differentiated from waterhammer by

(a) the time for a pressure wave to traverse the pipe
(b) the presence of a reservoir at one end of the pipe
(c) the rate of deceleration of flow
(d) the relative compressibility of liquid to expansion of pipe walls
(e) the length-diameter ratio of pipe

12.47 Waterhammer occurs only when

(a) $2L/a > 1$ (b) $V_0 > a$ (c) $2L/a = 1$ (d) $K/E < 1$
(e) compressibility effects are important

12.48 Valve closure is *rapid* only when

(a) $2L/a \geq t_c$ (b) $L/a \geq t_c$ (c) $L/2a \geq t_c$ (d) $t_c = 0$
(e) none of these answers

12.49 The head rise at a valve due to sudden closure is

(a) $a^2/2g$ (b) $V_0 a/g$ (c) $V_0 a/2g$ (d) $V_0^2/2g$ (e) none of these answers

12.50 The speed of a pressure wave through a pipe depends upon

(a) the length of pipe   (b) the original head at the valve
(c) the viscosity of fluid   (d) the initial velocity   (e) none of these answers

12.51 When the velocity in a pipe is suddenly reduced from 3 to 2 m/s by downstream valve closure, for $a = 980$ m/s the head rise in meters is

(a) 100   (b) 200   (c) 300   (d) 980   (e) none of these answers

12.52 When $t_c = L/2a$, the proportion of pipe length subjected to maximum heads is, in percent,

(a) 25   (b) 50   (c) 75   (d) 100   (e) none of these answers

12.53 When the steady-state value of head at a valve is 120 ft, the valve is given a sudden partial closure such that $\Delta h = 80$ ft. The head at the valve at the instant this reflected wave returns is

(a) -80 (b) 40   (c) 80   (d) 200   (e) none of these answers

12.54 An elementary wave can travel upstream in a channel, $y = 4$ ft, $V = 8$ ft/s, with a velocity of

(a) 3.35 ft/s   (b) 11.35 ft/s   (c) 16.04 ft/s   (d) 19.35 ft/s
(e) none of these answers

12.55 The speed of an elementary wave in a still liquid is given by

(a) $(gy^2)^{1/3}$ (b) $2y/3$ (c) $\sqrt{2gy}$ (d) $\sqrt{gy}$ (e) none of these answers

12.56 A negative surge wave

(a) is a positive surge wave moving backward
(b) is an inverted positive surge wave
(c) can never travel upstream
(d) can never travel downstream
(e) is none of the above

REFERENCES

Bergeron, L.: "Water Hammer in Hydraulics and Wave Surges in Electricity," translated under the sponsorship of the ASME, Wiley, New York, 1961.

Halliwell, A. R.: Velocity of a Water-Hammer Wave in an Elastic Pipe, *J. Hydraul. Div. ASCE*, vol. 89, no. HY4, pp. 1–21, July 1963.

Parmakian, J.: "Waterhammer Analysis," Prentice-Hall, Englewood Cliffs, N.J., 1955 (also Dover, New York, 1963).

Streeter, V. L.: Unsteady Flow Calculations by Numerical Methods, *Trans. ASME J. Basic Eng.*, June 1972.

Streeter, V. L., and E. B. Wylie: "Hydraulic Transients," McGraw-Hill, New York, 1967.

FORCE SYSTEMS, MOMENTS, AND CENTROIDS

The material in this appendix has been assembled to aid in working with force systems. Simple force systems are briefly reviewed, and first and second moments, including the product of inertia, are discussed. Centroids and centroidal axes are defined.

SIMPLE FORCE SYSTEMS

A free-body diagram for an object or portion of an object shows the action of all other bodies on it. The action of the earth on the object is called a *body force* and is proportional to the mass of the object. In addition, forces and couples may act on the object by contact with its surface. When the free body is at rest or is moving in a straight line with uniform speed, it is said to be in *equilibrium*. By Newton's second law of motion, since there is no acceleration of the free body, the summation of all force components in any direction must be zero and the summation of all moments about any axis must be zero.

Two force systems are equivalent if they have the same value for summation of forces in every direction and the same value for summation of moments about every axis. The simplest equivalent force system is called the *resultant* of the force system. Equivalent force systems always cause the same motion (or lack of motion) of a free body.

In coplanar force systems the resultant is either a force or a couple. In noncoplanar parallel force systems the resultant is either a force or a couple. In general noncoplanar systems the resultant may be a force, a couple, or a force and a couple.

The action of a fluid on any surface may be replaced by the resultant force system that causes the same external motion or reaction as the distributed fluid-force system. In this situation the fluid may be considered to be completely removed, the resultant acting in its place.

FIRST AND SECOND MOMENTS; CENTROIDS

The moment of an area, volume, weight, or mass may be determined in a manner analogous to that of determining the moments of a force about an axis.

First moments

The moment of an area A about the y axis (Fig. A.1) is expressed by

$$\int_A x \, dA$$

in which the integration is carried out over the area. To determine the moment about a parallel axis, for example, $x = k$, the moment becomes

$$\int_A (x - k) \, dA = \int_A x \, dA - kA \tag{A.1}$$

which shows that there will always be a parallel axis $x = k = \bar{x}$, about which the moment is zero. This axis is called a *centroidal* axis and is obtained from

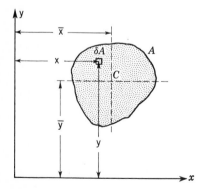

Fig. A.1 Notation for first and second moments.

Eq. (A.1) by setting it equal to zero and solving for $\bar{x}$,

$$\bar{x} = \frac{1}{A} \int_A x \, dA \tag{A.2}$$

Another centroidal axis may be determined parallel to the x axis,

$$\bar{y} = \frac{1}{A} \int_A y \, dA \tag{A.3}$$

The point of intersection of centroidal axes is called the *centroid* of the area. It may easily be shown, by rotation of axes, that the first moment of the area is zero about any axis through the centroid. When an area has an axis of symmetry, it is a centroidal axis because the moments of corresponding area elements on each side of the axis are equal in magnitude and opposite in sign. When location of the centroid is known, the first moment for any axis may be obtained without integration by taking the product of area and distance from centroid to the axis,

$$\int_A z \, dA = \bar{z}A \tag{A.4}$$

The centroidal axis of a triangle, parallel to one side, is one-third the altitude from that side; the centroid of a semicircle of radius a is $4a/3\pi$ from the diameter.

By taking the first moment of a volume $\mathcal{V}$ about a plane, say the yz plane, the distance to its centroid is similarly determined,

$$\bar{x} = \frac{1}{\mathcal{V}} \int_{\mathcal{V}} x \, d\mathcal{V} \tag{A.5}$$

The mass center of a body is determined by the same procedure,

$$x_m = \frac{1}{M} \int_M x \, dm \tag{A.6}$$

in which dm is an element of mass and M is the total mass of the body. For practical engineering purposes the *center of gravity* of a body is at its mass center.

Second moments

The second moment of an area A (Fig. A.1) about the y axis is

$$I_y = \int_A x^2 \, dA \tag{A.7}$$

It is called the *moment of inertia* of the area and is always positive since dA is always considered positive. After transferring the axis to a parallel axis through the centroid C of the area,

$$I_c = \int_A (x - \bar{x})^2 \, dA = \int_A x^2 \, dA - 2\bar{x} \int_A x \, dA + \bar{x}^2 \int_A dA$$

Since

$$\int_A x \, dA = \bar{x}A \qquad \int_A x^2 \, dA = I_y \qquad \int_A dA = A$$

therefore

$$I_c = I_y - \bar{x}^2 A \qquad \text{or} \qquad I_y = I_c + \bar{x}^2 A \tag{A.8}$$

In words, the moment of inertia of an area about any axis is the sum of the moment of inertia about a parallel axis through the centroid and the product of the area and square of distance between axes. Figure A.2 shows moments of inertia for four simple areas.

The *product of inertia* I_{xy} of an area is expressed by

$$I_{xy} = \int_A xy \, dA \tag{A.9}$$

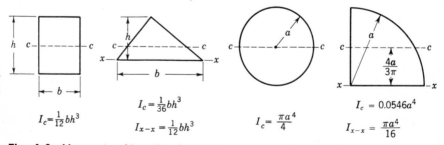

Fig. A.2 Moments of inertia of simple areas about centroidal axes.

with the notation of Fig. A.1. It may be positive or negative. Writing the expression for product of inertia $\bar{I}_{xy}$ about centroidal axes parallel to the xy axes produces

$$\bar{I}_{xy} = \int_A (x - \bar{x})(y - \bar{y})\, dA = \int_A xy\, dA - \bar{x} \int_A y\, dA - \bar{y} \int_A x\, dA + \bar{x}\bar{y}A$$

After simplifying, and solving for I_{xy},

$$I_{xy} = \bar{I}_{xy} + \bar{x}\bar{y}A \qquad\qquad\qquad\qquad\qquad\qquad (A.10)$$

Whenever either axis is an axis of symmetry of the area, the product of inertia is zero. The product of inertia I_{xy} of a triangle having sides b and h along the positive coordinate axis is $b^2h^2/24$.

PARTIAL DERIVATIVES AND TOTAL DIFFERENTIALS

PARTIAL DERIVATIVES

A partial derivative is an expression of the rate of change of one variable with respect to another variable when all other variables are held constant. When one sees a partial derivative, he should determine which variables are considered constant. For example, the temperature T at any point throughout a plane might be expressed as an equation containing space coordinates and time, x, y, and t. To determine how the temperature changes at some point, for example, (x_0, y_0), with the time, the actual numbers for coordinates are substituted, and the equation becomes a relation between T and t only. The rate of change of temperature with respect to time is dT/dt, which is written as a total differential because T and t are the only two variables in the equation. When one wants an expression for rate of change of temperature with time at any point (x, y), these are considered to be constants and the derivative of the equation with respect to t is taken. This is written $\partial T/\partial t$, to indicate that the other variables x, y have been held constant. Substitution of particular values of x, y into the expression yields $\partial T/\partial t$, in terms of t. As a specific case, if

$$T = x^2 + xyt + \sin t$$

then

$$\frac{\partial T}{\partial t} = xy + \cos t$$

For the point (1,2)

$$\frac{\partial T}{\partial t} = 2 + \cos t$$

which could have been obtained by first substituting (1,2) into the equation for T,

$$T = 1 + 2t + \sin t$$

and then, by taking the total derivative,

$$\frac{dT}{dt} = 2 + \cos t$$

If one wants to know the variation of temperature along any line parallel to the x axis at a given instant of time, then $\partial T/\partial x$ is taken and the specific y coordinates of the line and the time are substituted later; thus

$$\frac{\partial T}{\partial x} = 2x + yt$$

in which y, t have been considered constant. For the line through $y = 2$, at time $t = 4$,

$$\frac{\partial T}{\partial x} = 2x + 8$$

and the rate of change of T with respect to x at this instant can be found at any point x along the particular line.

In the function

$$u = f(x,y)$$

x and y are independent variables, and u is the dependent variable. If y is held constant, u becomes a function of x alone and its derivative may be determined as if u were a function of one variable. It is denoted by

$$\frac{\partial f}{\partial x} \quad \text{or} \quad \frac{\partial u}{\partial x}$$

and is called the partial derivative of f with respect to x or the partial derivative of u with respect to x. Similarly, if x is held constant, u becomes a function of y alone and $\partial u/\partial y$ is the partial of u with respect to y. These partials are defined by

$$\frac{\partial u}{\partial x} = \frac{\partial f(x,y)}{\partial x} = \lim_{\Delta x \to 0} \frac{f(x + \Delta x, y) - f(x,y)}{\Delta x}$$

$$\frac{\partial u}{\partial y} = \frac{\partial f(x,y)}{\partial y} = \lim_{\Delta y \to 0} \frac{f(x, y + \Delta y) - f(x,y)}{\Delta y}$$

EXAMPLES

(a) $u = x^3 + x^2 y^3 + 3y^2$

$$\frac{\partial u}{\partial x} = 3x^2 + 2xy^3 \qquad \frac{\partial u}{\partial y} = 3x^2 y^2 + 6y$$

(b) $u = \sin(ax + by^2)$

$$\frac{\partial u}{\partial x} = a \cos(ax + by^2) \qquad \frac{\partial u}{\partial y} = 2by \cos(ax + by^2)$$

(c) $u = x \ln y$

$$\frac{\partial u}{\partial x} = \ln y \qquad \frac{\partial u}{\partial y} = \frac{x}{y}$$

TOTAL DIFFERENTIALS

When u is a function of one variable only, $u = f(x)$,

$$\frac{du}{dx} = \lim_{\Delta x \to 0} \frac{\Delta u}{\Delta x} = \lim_{\Delta x \to 0} \frac{f(x + \Delta x) - f(x)}{\Delta x} = f'(x)$$

and

$$\Delta u = f'(x)\, \Delta x + \epsilon\, \Delta x$$

in which

$$\lim_{\Delta x \to 0} \epsilon = 0$$

After applying the limiting process to Δu,

$$du = f'(x)\,\Delta x \equiv f'(x)\,dx$$

is the differential du.

When $u = f(x,y)$, the differential du is defined in a similar manner. If x and y take on increments, Δx, Δy, then

$$\Delta u = f(x + \Delta x, y + \Delta y) - f(x,y)$$

in which Δx, Δy may approach zero in any manner. If Δu approaches zero regardless of the way in which Δx and Δy approach zero, $u = f(x,y)$ is called a *continuous* function of x and y. In the following it is assumed that $f(x,y)$ is continuous and that $\partial f/\partial x$ and $\partial f/\partial y$ are also continuous.

Adding and subtracting $f(x,y + \Delta y)$ to the expression for Δu give

$$\Delta u = f(x + \Delta x, y + \Delta y) - f(x,y + \Delta y) + f(x,y + \Delta y) - f(x,y)$$

Then

$$f(x + \Delta x, y + \Delta y) - f(x,y + \Delta y) = \frac{\partial f(x,y + \Delta y)}{\partial x}\,\Delta x + \epsilon_1\,\Delta x$$

in which $\lim_{\Delta y \to 0} \epsilon_1 = 0$, because

$$\lim_{\Delta x \to 0} \frac{f(x + \Delta x, y + \Delta y) - f(x,y + \Delta y)}{\Delta x} = \frac{\partial f(x,y + \Delta y)}{\partial x}$$

Furthermore

$$\lim_{\Delta y \to 0} \frac{\partial f(x,y + \Delta y)}{\partial x} = \frac{\partial f(x,y)}{\partial x}$$

as the derivative is continuous, and

$$\frac{\partial f(x,y + \Delta y)}{\partial x} = \frac{\partial f(x,y)}{\partial x} + \epsilon_2$$

in which $\lim_{\Delta y \to 0} \epsilon_2 = 0$. Similarly

$$f(x,y + \Delta y) - f(x,y) = \frac{\partial f(x,y)}{\partial y}\,\Delta y + \epsilon_3\,\Delta y$$

in which $\lim\limits_{\Delta y \to 0} \epsilon_2 = 0$. By substituting into the expression for Δu,

$$\Delta u = \frac{\partial f(x,y)}{\partial x} \Delta x + \frac{\partial f(x,y)}{\partial y} \Delta y + (\epsilon_1 + \epsilon_2) \Delta x + \epsilon_3 \Delta y$$

If the limit is taken as Δx and Δy approach zero, the last two terms drop out since they are the product of two infinitesimals and, hence, are of a higher order of smallness. The total differential of u is obtained,

$$du = \frac{\partial f}{\partial x} dx + \frac{\partial f}{\partial y} dy = \frac{\partial u}{\partial x} dx + \frac{\partial u}{\partial y} dy$$

If x and y in $u = f(x,y)$ are functions of one independent variable, for example, t, then u becomes a function of t alone and has a derivative with respect to t if the functions $x = f_1(t)$, $y = f_2(t)$ are assumed differentiable. An increment in t results in increments Δx, Δy, Δu which approach zero with Δt. By dividing the expression for Δu by Δt,

$$\frac{\Delta u}{\Delta t} = \frac{\partial u}{\partial x} \frac{\Delta x}{\Delta t} + \frac{\partial u}{\partial y} \frac{\Delta y}{\Delta t} + (\epsilon_1 + \epsilon_2) \frac{\Delta x}{\Delta t} + \epsilon_3 \frac{\Delta y}{\Delta t}$$

and by taking the limit as Δt approaches zero,

$$\frac{du}{dt} = \frac{\partial u}{\partial x} \frac{dx}{dt} + \frac{\partial u}{\partial y} \frac{dy}{dt}$$

The same general form results for additional variables, namely,

$$u = f(x,y,t)$$

in which x, y are functions of t; then

$$\frac{du}{dt} = \frac{\partial u}{\partial x} \frac{dx}{dt} + \frac{\partial u}{\partial y} \frac{dy}{dt} + \frac{\partial u}{dt}$$

PHYSICAL PROPERTIES OF FLUIDS

Table C.1 *Physical properties of water in English units†*

Temp, °F	Specific weight γ, lb/ft³	Density ρ, slugs/ft³	Viscosity μ, lb·s/ft² $10^5\,\mu =$	Kine-matic viscosity ν, ft²/s $10^5\,\nu =$	Surface tension σ, lb/ft $100\,\sigma =$	Vapor-pressure head p_v/γ, ft	Bulk modulus of elasticity K, lb/in² $10^{-3}\,K =$
32	62.42	1.940	3.746	1.931	0.518	0.20	293
40	62.43	1.940	3.229	1.664	0.514	0.28	294
50	62.41	1.940	2.735	1.410	0.509	0.41	305
60	62.37	1.938	2.359	1.217	0.504	0.59	311
70	62.30	1.936	2.050	1.059	0.500	0.84	320
80	62.22	1.934	1.799	0.930	0.492	1.17	322
90	62.11	1.931	1.595	0.826	0.486	1.61	323
100	62.00	1.927	1.424	0.739	0.480	2.19	327
110	61.86	1.923	1.284	0.667	0.473	2.95	331
120	61.71	1.918	1.168	0.609	0.465	3.91	333
130	61.55	1.913	1.069	0.558	0.460	5.13	334
140	61.38	1.908	0.981	0.514	0.454	6.67	330
150	61.20	1.902	0.905	0.476	0.447	8.58	328
160	61.00	1.896	0.838	0.442	0.441	10.95	326
170	60.80	1.890	0.780	0.413	0.433	13.83	322
180	60.58	1.883	0.726	0.385	0.426	17.33	313
190	60.36	1.876	0.678	0.362	0.419	21.55	313
200	60.12	1.868	0.637	0.341	0.412	26.59	308
212	59.83	1.860	0.593	0.319	0.404	33.90	300

† This table was compiled primarily from Hydraulic Models, *ASCE Man. Eng. Pract.* 25, 1942.

Table C.2 *Physical properties of water in SI units*

Temp, °C	Specific weight γ, N/m^3	Density ρ, kg/m^3	Viscosity μ, kg/m·s $10^3\,\mu=$	Kinematic viscosity ν, m^2/s $10^6\,\nu=$	Surface tension σ, N/m $100\,\sigma=$	Vapor-pressure head p_v/γ, m	Bulk modulus of elasticity K, N/m^2 $10^{-7}\,K=$
0	9805	999.9	1.792	1.792	7.62	0.06	204
5	9806	1000.0	1.519	1.519	7.54	0.09	206
10	9803	999.7	1.308	1.308	7.48	0.12	211
15	9798	999.1	1.140	1.141	7.41	0.17	214
20	9789	998.2	1.005	1.007	7.36	0.25	220
25	9779	997.1	0.894	0.897	7.26	0.33	222
30	9767	995.7	0.801	0.804	7.18	0.44	223
35	9752	994.1	0.723	0.727	7.10	0.58	224
40	9737	992.2	0.656	0.661	7.01	0.76	227
45	9720	990.2	0.599	0.605	6.92	0.98	229
50	9697	988.1	0.549	0.556	6.82	1.26	230
55	9679	985.7	0.506	0.513	6.74	1.61	231
60	9658	983.2	0.469	0.477	6.68	2.03	228
65	9635	980.6	0.436	0.444	6.58	2.56	226
70	9600	977.8	0.406	0.415	6.50	3.20	225
75	9589	974.9	0.380	0.390	6.40	3.96	223
80	9557	971.8	0.357	0.367	6.30	4.86	221
85	9529	968.6	0.336	0.347	6.20	5.93	217
90	9499	965.3	0.317	0.328	6.12	7.18	216
95	9469	961.9	0.299	0.311	6.02	8.62	211
100	9438	958.4	0.284	0.296	5.94	10.33	207

Table C.3 *Properties of gases at low pressures and 80°F*

Gas	Chemical formula	Molecular weight	Gas constant R m·N/Kg·K	Gas constant R ft·lb/lb$_m$·°R	Specific heat, Btu/lb$_m$·°R or kcal/kg·K c_p	Specific heat, Btu/lb$_m$·°R or kcal/kg·K c_v	Specific-heat ratio k
Air	—	29.0	287	53.3	0.240	0.171	1.40
Carbon monoxide	CO	28.0	297	55.2	0.249	0.178	1.40
Helium	He	4.00	2077	386	1.25	0.753	1.66
Hydrogen	H$_2$	2.02	4121	766	3.43	2.44	1.40
Nitrogen	N$_2$	28.0	297	55.2	0.248	0.177	1.40
Oxygen	O$_2$	32.0	260	48.3	0.219	0.157	1.40
Water vapor	H$_2$O	18.0	462	85.8	0.445	0.335	1.33

Table C.4 *One-dimensional isentropic relations (for a perfect gas with constant specific heat; $k = 1.4$)†*

M	A/A^*	p/p_0	ρ/ρ_0	T/T_0	M	A/A^*	p/p_0	ρ/ρ_0	T/T_0
0.00	.....	1.000	1.000	1.000	0.78	1.05	0.669	0.750	0.891
0.01	57.87	0.9999	0.9999	0.9999	0.80	1.04	0.656	0.740	0.886
0.02	28.94	0.9997	0.9999	0.9999	0.82	1.03	0.643	0.729	0.881
0.04	14.48	0.999	0.999	0.9996	0.84	1.02	0.630	0.719	0.876
0.06	9.67	0.997	0.998	0.999	0.86	1.02	0.617	0.708	0.871
0.08	7.26	0.996	0.997	0.999	0.88	1.01	0.604	0.698	0.865
0.10	5.82	0.993	0.995	0.998	0.90	1.01	0.591	0.687	0.860
0.12	4.86	0.990	9.993	0.997	0.92	1.01	0.578	0.676	0.855
0.14	4.18	0.986	0.990	0.996	0.94	1.00	0.566	0.666	0.850
0.16	3.67	0.982	0.987	0.995	0.96	1.00	0.553	0.655	0.844
0.18	3.28	0.978	0.984	0.994	0.98	1.00	0.541	0.645	0.839
0.20	2.96	0.973	0.980	0.992	1.00	1.00	0.528	0.632	0.833
0.22	2.71	0.967	0.976	0.990	1.02	1.00	0.516	0.623	0.828
0.24	2.50	0.961	0.972	0.989	1.04	1.00	0.504	0.613	0.822
0.26	2.32	0.954	0.967	0.987	1.06	1.00	0.492	0.602	0.817
0.28	2.17	0.947	0.962	0.985	1.08	1.01	0.480	0.592	0.810
0.30	2.04	0.939	0.956	0.982	1.10	1.01	0.468	0.582	0.805
0.32	1.92	0.932	0.951	0.980	1.12	1.01	0.457	0.571	0.799
0.34	1.82	0.923	0.944	0.977	1.14	1.02	0.445	0.561	0.794
0.36	1.74	0.914	0.938	0.975	1.16	1.02	0.434	0.551	0.788
0.38	1.66	0.905	0.931	0.972	1.18	1.02	0.423	0.541	0.782
0.40	1.59	0.896	0.924	0.969	1.20	1.03	0.412	0.531	0.776
0.42	1.53	0.886	0.917	0.966	1.22	1.04	0.402	0.521	0.771
0.44	1.47	0.876	0.909	0.963	1.24	1.04	0.391	0.512	0.765
0.46	1.42	0.865	0.902	0.959	1.26	1.05	0.381	0.502	0.759
0.48	1.38	0.854	0.893	0.956	1.28	1.06	0.371	0.492	0.753
0.50	1.34	0.843	0.885	0.952	1.30	1.07	0.361	0.483	0.747
0.52	1.30	0.832	0.877	0.949	1.32	1.08	0.351	0.474	0.742
0.54	1.27	0.820	0.868	0.945	1.34	1.08	0.342	0.464	0.736
0.56	1.24	0.808	0.859	0.941	1.36	1.09	0.332	0.455	0.730
0.58	1.21	0.796	0.850	0.937	1.38	1.10	0.323	0.446	0.724
0.60	1.19	0.784	0.840	0.933	1.40	1.11	0.314	0.437	0.718
0.62	1.17	0.772	0.831	0.929	1.42	1.13	0.305	0.429	0.713
0.64	1.16	0.759	0.821	0.924	1.44	1.14	0.297	0.420	0.707
0.66	1.13	0.747	0.812	0.920	1.46	1.15	0.289	0.412	0.701
0.68	1.12	0.734	0.802	0.915	1.48	1.16	0.280	0.403	0.695
0.70	1.09	0.721	0.792	0.911	1.50	1.18	0.272	0.395	0.690
0.72	1.08	0.708	0.781	0.906	1.52	1.19	0.265	0.387	0.684
0.74	1.07	0.695	0.771	0.901	1.54	1.20	0.257	0.379	0.678
0.76	1.06	0.682	0.761	0.896	1.56	1.22	0.250	0.371	0.672

† From Irving Shames, "Mechanics of Fluids," copyright © 1962 by McGraw-Hill, Inc. Used with permission of McGraw-Hill Book Company.

Table C.4 *One-dimensional isentropic relations (Continued)*

M	A/A^*	p/p_0	ρ/ρ_0	T/T_0	M	A/A^*	p/p_0	ρ/ρ_0	T/T_0
1.58	1.23	0.242	0.363	0.667	2.30	2.19	0.080	0.165	0.486
1.60	1.25	0.235	0.356	0.661	2.32	2.23	0.078	0.161	0.482
1.62	1.27	0.228	0.348	0.656	2.34	2.27	0.075	0.157	0.477
1.64	1.28	0.222	0.341	0.650	2.36	2.32	0.073	0.154	0.473
1.66	1.30	0.215	0.334	0.645	2.38	2.36	0.071	0.150	0.469
1.68	1.32	0.209	0.327	0.639	2.40	2.40	0.068	0.147	0.465
1.70	1.34	0.203	0.320	0.634	2.42	2.45	0.066	0.144	0.461
1.72	1.36	0.197	0.313	0.628	2.44	2.49	0.064	0.141	0.456
1.74	1.38	0.191	0.306	0.623	2.46	2.54	0.062	0.138	0.452
1.76	1.40	0.185	0.300	0.617	2.48	2.59	0.060	0.135	0.448
1.78	1.42	0.179	0.293	0.612	2.50	2.64	0.059	0.132	0.444
1.80	1.44	0.174	0.287	0.607	2.52	2.69	0.057	0.129	0.441
1.82	1.46	0.169	0.281	0.602	2.54	2.74	0.055	0.126	0.437
1.84	1.48	0.164	0.275	0.596	2.56	2.79	0.053	0.123	0.433
1.86	1.51	0.159	0.269	0.591	2.58	2.84	0.052	0.121	0.429
1.88	1.53	0.154	0.263	0.586	2.60	2.90	0.050	0.118	0.425
1.90	1.56	0.149	0.257	0.581	2.62	2.95	0.049	0.115	0.421
1.92	1.58	0.145	0.251	0.576	2.64	3.01	0.047	0.113	0.418
1.94	1.61	0.140	0.246	0.571	2.66	3.06	0.046	0.110	0.414
1.96	1.63	0.136	0.240	0.566	2.68	3.12	0.044	0.108	0.410
1.98	1.66	0.132	0.235	0.561	2.70	3.18	0.043	0.106	0.407
2.00	1.69	0.128	0.230	0.556	2.72	3.24	0.042	0.103	0.403
2.02	1.72	0.124	0.225	0.551	2.74	3.31	0.040	0.101	0.400
2.04	1.75	0.120	0.220	0.546	2.76	3.37	0.039	0.099	0.396
2.06	1.78	0.116	0.215	0.541	2.78	3.43	0.038	0.097	0.393
2.08	1.81	0.113	0.210	0.536	2.80	3.50	0.037	0.095	0.389
2.10	1.84	0.109	0.206	0.531	2.82	3.57	0.036	0.093	0.386
2.12	1.87	0.106	0.201	0.526	2.84	3.64	0.035	0.091	0.383
2.14	1.90	0.103	0.197	0.522	2.86	3.71	0.034	0.089	0.379
2.16	1.94	0.100	0.192	0.517	2.88	3.78	0.033	0.087	0.376
2.18	1.97	0.097	0.188	0.513	2.90	3.85	0.032	0.085	0.373
2.20	2.01	0.094	0.184	0.508	2.92	3.92	0.031	0.083	0.370
2.22	2.04	0.091	0.180	0.504	2.94	4.00	0.030	0.081	0.366
2.24	2.08	0.088	0.176	0.499	2.96	4.08	0.029	0.080	0.363
2.26	2.12	0.085	0.172	0.495	2.98	4.15	0.028	0.078	0.360
2.28	2.15	0.083	0.168	0.490	3.00	4.23	0.027	0.076	0.357

Table C.5 *One-dimensional normal-shock relations (for a perfect gas with k = 1.4)*†

M_1	M_2	$\dfrac{p_2}{p_1}$	$\dfrac{T_2}{T_1}$	$\dfrac{(p_0)_2}{(p_0)_1}$	M_1	M_2	$\dfrac{p_2}{p_1}$	$\dfrac{T_2}{P_1}$	$\dfrac{(p_0)_2}{(p_0)_1}$
1.00	1.000	1.000	1.000	1.000	1.72	0.635	3.285	1.473	0.847
1.02	0.980	1.047	1.013	1.000	1.74	0.631	3.366	1.487	0.839
1.04	0.962	1.095	1.026	1.000	1.76	0.626	3.447	1.502	0.830
1.06	0.944	1.144	1.039	1.000	1.78	0.621	3.530	1.517	0.821
1.08	0.928	1.194	1.052	0.999	1.80	0.617	3.613	1.532	0.813
1.10	0.912	1.245	1.065	0.999	1.82	0.612	3.698	1.547	0.804
1.12	0.896	1.297	1.078	0.998	1.84	0.608	3.783	1.562	0.795
1.14	0.882	1.350	1.090	0.997	1.86	0.604	3.869	1.577	0.786
1.16	0.868	1.403	1.103	0.996	1.88	0.600	3.957	1.592	0.777
1.18	0.855	1.458	1.115	0.995	1.90	0.596	4.045	1.608	0.767
1.20	0.842	1.513	1.128	0.993	1.92	0.592	4.134	1.624	0.758
1.22	0.830	1.570	1.140	0.991	1.94	0.588	4.224	1.639	0.749
1.24	0.818	1.627	1.153	0.988	1.96	0.584	4.315	1.655	0.740
1.26	0.807	1.686	1.166	0.986	1.98	0.581	4.407	1.671	0.730
1.28	0.796	1.745	1.178	0.983	2.00	0.577	4.500	1.688	0.721
1.30	0.786	1.805	1.191	0.979	2.02	0.574	4.594	1.704	0.711
1.32	0.776	1.866	1.204	0.976	2.04	0.571	4.689	1.720	0.702
1.34	0.766	1.928	1.216	0.972	2.06	0.567	4.784	1.737	0.693
1.36	0.757	1.991	1.229	0.968	2.08	0.564	4.881	1.754	0.683
1.38	0.748	2.055	1.242	0.963	2.10	0.561	4.978	1.770	0.674
1.40	0.740	2.120	1.255	0.958	2.12	0.558	5.077	1.787	0.665
1.42	0.731	2.186	1.268	0.953	2.14	0.555	5.176	1.805	0.656
1.44	0.723	2.253	1.281	0.948	2.16	0.553	5.277	1.822	0.646
1.46	0.716	2.320	1.294	0.942	2.18	0.550	5.378	1.839	0.637
1.48	0.708	2.389	1.307	0.936	2.20	0.547	5.480	1.857	0.628
1.50	0.701	2.458	1.320	0.930	2.22	0.544	5.583	1.875	0.619
1.52	0.694	2.529	1.334	0.923	2.24	0.542	5.687	1.892	0.610
1.54	0.687	2.600	1.347	0.917	2.26	0.539	5.792	1.910	0.601
1.56	0.681	2.673	1.361	0.910	2.28	0.537	5.898	1.929	0.592
1.58	0.675	2.746	1.374	0.903	2.30	0.534	6.005	1.947	0.583
1.60	0.668	2.820	1.388	0.895	2.32	0.532	6.113	1.965	0.575
1.62	0.663	2.895	1.402	0.888	2.34	0.530	6.222	1.984	0.566
1.64	0.657	2.971	1.416	0.880	2.36	0.527	6.331	2.003	0.557
1.66	0.651	3.048	1.430	0.872	2.38	0.525	6.442	2.021	0.549
1.68	0.646	3.126	1.444	0.864	2.40	0.523	6.553	2.040	0.540
1.70	0.641	3.205	1.458	0.856	2.42	0.521	6.666	2.060	0.532

† From Irving Shames, "Mechanics of Fluids," copyright © 1962 by McGraw-Hill, Inc. Used with permission of McGraw-Hill Book Company.

Table C.5 *One-dimensional normal-shock relations (continued)*

M_1	M_2	$\dfrac{p_2}{p_1}$	$\dfrac{T_2}{T_1}$	$\dfrac{(p_0)_2}{(p_0)_1}$	M_1	M_2	$\dfrac{p_2}{p_1}$	$\dfrac{T_2}{T_1}$	$\dfrac{(p_0)_2}{(p_0)_1}$
2.44	0.519	6.779	2.079	0.523	2.76	0.491	8.721	2.407	0.403
2.46	0.517	6.894	2.098	0.515	2.78	0.490	8.850	2.429	0.396
2.48	0.515	7.009	2.118	0.507	2.80	0.488	8.980	2.451	0.389
2.50	0.513	7.125	2.138	0.499	2.82	0.487	9.111	2.473	0.383
2.52	0.511	7.242	2.157	0.491	2.84	0.485	9.243	2.496	0.376
2.54	0.509	7.360	2.177	0.483	2.86	0.484	9.376	2.518	0.370
2.56	0.507	7.479	2.198	0.475	2.88	0.483	9.510	2.541	0.364
2.58	0.506	7.599	2.218	0.468	2.90	0.481	9.645	2.563	0.358
2.60	0.504	7.720	2.238	0.460	2.92	0.480	9.781	2.586	0.352
2.62	0.502	7.842	2.260	0.453	2.94	0.479	9.918	2.609	0.346
2.64	0.500	7.965	2.280	0.445	2.96	0.478	10.055	2.632	0.340
2.66	0.499	8.088	2.301	0.438	2.98	0.476	10.194	2.656	0.334
2.68	0.497	8.213	2.322	0.431	3.00	0.475	10.333	2.679	0.328
2.70	0.496	8.338	2.343	0.424					
2.72	0.494	8.465	2.364	0.417					
2.74	0.493	8.592	2.396	0.410					

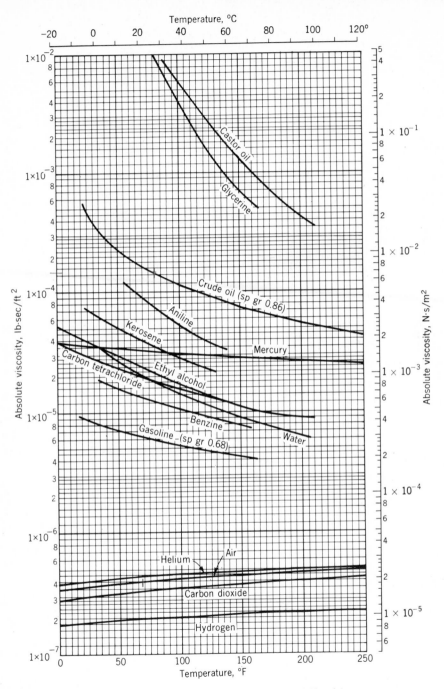

Fig. C.1 Absolute viscosities of certain gases and liquids.

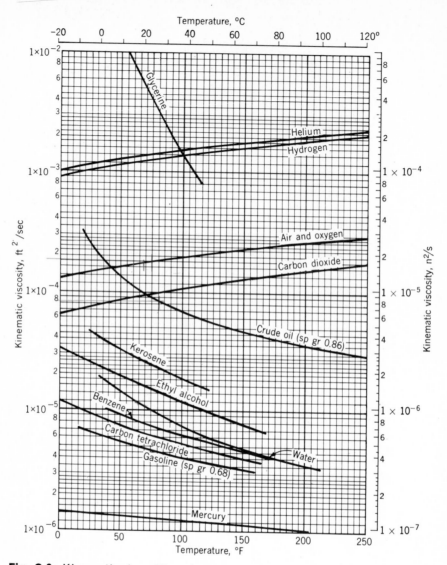

Fig. C.2 Kinematic viscosities of certain gases and liquids. The gases are at standard pressure.

NOTATION

Symbol	Quantity	Unit SI	Unit English	Dimensions (MLT)
a	Constant, pulse wave speed	m/s	ft/s	LT^{-1}
a	Acceleration	m/s²	ft/s²	LT^{-2}
a	Acceleration vector	m/s²	ft/s²	LT^{-2}
a^*	Velocity	m/s	ft/s	LT^{-1}
A	Area	m²	ft²	L^2
A	Adverse slope	none	none	
b	Distance	m	ft	L
b	Constant			
c	Speed of surge wave	m/s	ft/s	LT^{-1}
c	Speed of sound	m/s	ft/s	LT^{-1}
c_p	Specific heat, constant pressure	J/kg·K	ft·lb/slug·°R	
c_v	Specific heat, constant volume	J/kg·K	ft·lb/slug·°R	
C	Concentration	No./m³	No./ft³	L^{-3}
C	Coefficient		none	
C	Stress	Pa	lb/ft²	$ML^{-1}T^{-2}$
C_m	Empirical constant	$m^{1/3}/s$	ft$^{1/3}$/s	$L^{1/3}T^{-1}$
c	Critical slope		none	
D'	Volumetric displacement	m³	ft³	L^3
D	Diameter	m	ft	L
e	Efficiency		none	
e	Internal energy per unit mass	J/kg	ft·lb/slug	L^2T^{-2}
E	Internal energy	J	ft·lb	ML^2T^{-2}
E	Specific energy	m·N/N	ft·lb/lb	L
E	Losses per unit weight	m·N/N	ft·lb/lb	L
E	Modulus of elasticity	Pa	lb/ft²	$ML^{-1}T^{-2}$
f	Friction factor		none	

| Symbol | Quantity | Unit | | Dimensions (MLT) |
		SI	English	
F	Force	N	lb	MLT^{-2}
F	Force vector	N	lb	MLT^{-2}
F	Froude number		none	
F_B	Buoyant force	N	lb	MLT^{-2}
g	Acceleration of gravity	m/s²	ft/s²	LT^{-2}
g_0	Gravitation constant	kg·m/ N·s²	lb$_m$·ft/lb·s²	
G	Mass flow rate per unit area	kg/s·m²	slug/s·ft²	$ML^{-2}T^{-1}$
h	Head, vertical distance	m	ft	L
h	Enthalpy per unit mass	J/kg	ft·lb/slug	L^2T^{-2}
H	Head, elevation of hydraulic grade line	m	ft	L
H	Horizontal slope	none	none	
I	Moment of inertia	m⁴	ft⁴	L^4
J	Junction point	none	none	
k	Specific-heat ratio	none	none	
K	Bulk modulus of elasticity	Pa	lb/ft²	$ML^{-1}T^{-2}$
K	Minor loss coefficient	none	none	
L	Length	m	ft	L
L	Lift	N	lb	MLT^{-2}
l	Length, mixing length	m	ft	L
ln	Natural logarithm	none	none	
m	Mass	kg	slug	M
m	Form factor, constant	none	none	
m	Strength of source	m³/s	ft³/s	L^3T^{-1}
$\dot{m}$	Mass per unit time	kg/s	slug/s	MT^{-1}
M	Molecular weight			
M	Momentum per unit time	N	lb	MLT^{-2}
M	Mild slope	none	none	
M	Mach number	none	none	
$\overline{MG}$	Metacentric height	m	ft	L
n	Exponent, constant	none	none	
n	Normal direction	m	ft	L
n	Manning roughness factor			
n	Number of moles			
n$_1$	Normal unit vector			
N	Rotation speed	1/s	1/s	T^{-1}
$NPSH$	Net positive suction head	m	ft	L
p	Pressure	Pa	lb/ft²	$ML^{-1}T^{-2}$
p	Force	N	lb	MLT^{-2}
P	Height of weir	m	ft	L
P	Wetted perimeter	m	ft	L
q	Discharge per unit width	m²/s	ft²/s	L^2T^{-1}
q	Velocity	m/s	ft/s	LT^{-1}
q	Velocity vector	m/s	ft/s	LT^{-1}
q_H	Heat transfer per unit mass	J/kg	ft·lb/slug	L^2T^{-2}
Q	Discharge	m³/s	ft³/s	L^3T^{-1}

Symbol	Quantity	Unit SI	Unit English	Dimensions (MLT)
Q_H	Heat transfer per unit time	J/s	ft·lb/s	ML^2T^{-3}
r	Coefficient			
r	Radial distance	m	ft	L
$\mathbf{r}$	Position vector	m	ft	L
R	Hydraulic radius	m	ft	L
R	Gas constant	J/kg·K	ft·lb/slug·°R	
R, R'	Gage difference	m	ft	L
$\mathbf{R}$	Reynolds number	none	none	
s	Distance	m	ft	L
s	Entropy per unit mass	J/Kg·K	ft·lb/slug·°R	
s	Slip	none	none	
S	Entropy	J/K	ft·lb/°R	
S	Specific gravity, slope	none	none	
$\mathbf{S}$	Steep slope	none	none	
t	Time	s	s	T
t, t'	Distance, thickness	m	ft	L
T	Temperature	K	°R	
T	Torque	N·m	lb·ft	ML^2T^{-2}
T	Tensile force/ft	N/m	lb/ft	MT^{-2}
T	Top width	m	ft	L
u	Velocity, velocity component	m/s	ft/s	LT^{-1}
u	Peripheral speed	m/s	ft/s	LT^{-1}
u	Intrinsic energy	J/kg	ft·lb/slug	L^2T^{-2}
u_*	Shear stress velocity	m/s	ft/s	LT^{-1}
U	Velocity	m/s	ft/s	LT^{-1}
v	Velocity, velocity component	m/s	ft/s	LT^{-1}
v_s	Specific volume	m³/kg	ft³/slug	$M^{-1}L^3$
$\mathcal{U}$	Volume	m³	ft³	L^3
$\mathbf{V}$	Velocity vector	m/s	ft/s	LT^{-1}
V	Velocity	m/s	ft/s	LT^{-1}
w	Velocity component	m/s	ft/s	LT^{-1}
w	Work per unit mass	J/kg	ft·lb/slug	L^2T^{-2}
W	Work per unit time	J/s	ft·lb/s	ML^2T^{-3}
W	Work of expansion	m·N	ft·lb	ML^2T^{-2}
W_s	Shaft work	m·N	ft·lb	ML^2T^{-2}
W	Weight	N	lb	MLT^{-2}
$\mathbf{W}$	Weber number	none	none	
x	Distance	m	ft	L
x_p	Distance to pressure center	m	ft	L
X	Body-force component per unit mass	N/kg	lb/slug	LT^{-2}
y	Distance, depth	m	ft	L
y_p	Distance to pressure center	m	ft	L
Y	Expansion factor	none	none	
Y	Body-force component per unit mass	N/kg	lb/slug	LT^{-2}
z	Vertical distance	m	ft	L
Z	Vertical distance	m	ft	L

| Symbol | Quantity | Unit | | Dimensions (MLT) |
		SI	English	
Z	Body-force component per unit mass	N/kg	lb/slug	LT^{-2}
α	Kinetic-energy correction factor	none	none	
α	Angle, coefficient	none	none	
β	Momentum correction factor	none	none	
β	Blade angle	none	none	
Γ	Circulation	m²/s	ft²/s	L^2T^{-1}
∇	Vector operator	1/m	1/ft	L^{-1}
γ	Specific weight	N/m³	lb/ft³	$ML^{-2}T^{-2}$
δ	Boundary-layer thickness	m	ft	L
ϵ	Kinematic eddy viscosity	m²/s	ft²/s	L^2T^{-1}
ϵ	Roughness height	m	ft	L
η	Eddy viscosity	N·s/m²	lb·s/ft²	$ML^{-1}T^{-1}$
η	Head ratio	none	none	
η	Efficiency			
θ	Angle	none	none	
κ	Universal constant	none	none	
λ	Scale ratio, undetermined multiplier	none	none	
μ	Viscosity	N·s/m²	lb·s/ft²	$ML^{-1}T^{-1}$
μ	Constant			
ν	Kinematic viscosity	m²/s	ft²/s	L^2T^{-1}
ϕ	Velocity potential	m²/s	ft²/s	L^2T^{-1}
ϕ	Function			
π	Constant	none	none	
Π	Dimensionless parameter	none	none	
ρ	Density	kg/m³	slug/ft³	ML^{-3}
σ	Surface tension	N/m	lb/ft	MT^{-2}
σ	Cavitation index	none	none	
τ	Shear stress	Pa	lb/ft²	$ML^{-1}T^{-2}$
ψ	Stream function, two dimensions	m/s	ft/s	L^2T^{-1}
ψ	Stokes' stream function	m³/s	ft³/s	L^3T^{-1}
ω	Angular velocity	rad/s	rad/s	T^{-1}

APPENDIX E

COMPUTER PROGRAMMING AIDS

FORTRAN IV COMPILER

The programs listed in the text are in FORTRAN IV compiler language (G level). They have been executed on an IBM 360-67 system, using the NAMELIST as a convenient means of data input. The NAMELIST is described in the Organick[1] primer. It is assumed that the reader is acquainted with FORTRAN IV. Several useful programming techniques for engineering calculations are not given in first computing courses. These include quadratures, numerical integration using Simpson's rule, parabolic interpolation, solution of algebraic and transcendental equations by the bisection and Newton-Raphson methods, and the Runge-Kutta methods of solving systems of ordinary simultaneous differential equations. These techniques are discussed in this appendix.

E.1 QUADRATURES; NUMERICAL INTEGRATION BY SIMPSON'S RULE

The integral $V = \int_{y_0}^{y_1} F(y)\, dy$ is to be evaluated between the known limits y_0

and y_1 for the known, finite, continuous function $F(y)$. By dividing the interval between y_0 and y_1 into N equal reaches (N even) (Fig. E.1), Simpson's rule may be applied to find the area under the curve. A statement sequence

---

[1] E. I. Organick, "A FORTRAN IV Primer," Addison-Wesley, Reading, Mass., 1966.

723

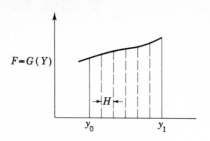

Fig. E.1 Determining area under
a curve by Simpson's rule.

to find V when we let $G(Y) = F(Y)$ is

```
      H = (Y1 − Y0)/N
      V = G(Y0) + G(Y1)
      DO 2  I = 1,N,2
2     V = V + 4.*G(Y0 + I*H)
      N1 = N − 1
      DO 3  I = 2,N1,2
3     V = V + 2.*G(Y0 + I*H)
      V = V*H/3.
      WRITE(6,1) V
1     FORMAT ('0 V = ', F 20.4)
      IF intermediate values are desired:
      V = .0
      N2 = N − 2
      DO 4  I = 0, N2, 2
      V = V + H*(G(Y0 + I*H) + G(Y0 + (I + 2)*H)
     1+ 4.*G(Y0 + (I + 1)*H))/3.
      Y = Y0 + (I + 2)*H
4     WRITE(6,5) Y,V
5     FORMAT ('0 Y = ', F 20.4, 5X, 'V = ', F20.4)
```

E.2 PARABOLIC INTERPOLATION

It is frequently desirable to use experimental data in computer programs.
For example, consider that A, the area of a reservoir, is known for 10-ft
intervals of elevation z, starting with z_0. Then for any elevation within the
range of data, the area of reservoir is desired. More generally (Fig. E.2),
values of y are known for equal increments in x. If the value of y is desired

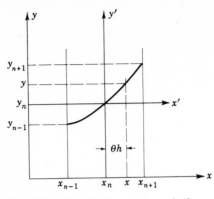

Fig. E.2 Parabolic   interpolation.

for the x shown, a parabola through the three points, with axis vertical, is first found by transferring the origin to x', y'.

$$y' = ax'^2 + bx'$$

Let $x_{n+1} - x_n = x_n - x_{n-1} = h$; then

$$y_{n+1} - y_n = ah^2 + bh \qquad y_{n-1} - y_n = ah^2 - bh$$

from which

$$a = \frac{1}{2h^2}\,(y_{n+1} + y_{n-1} - 2y_n) \qquad b = \frac{1}{2h}\,(y_{n+1} - y_{n-1})$$

and

$$y = y_n + \frac{\theta^2}{2}\,(y_{n+1} + y_{n-1} - 2y_n) + \frac{\theta}{2}\,(y_{n+1} - y_{n-1})$$

in which θh has been substituted for x'.

EXAMPLE E.1 The area of reservoir given for each 10 ft of elevation above $z_0 = 6320$ is

A	348	692	1217
z	6320	6330	6340

and so forth. The program is, for $dz = 10$:

```
   I = (Z − Z0)/DZ + 2
   TH = (Z − Z0 − (I − 1)*DZ)/DZ
   AREA = A(I) + .5*TH*(A(I + 1) − A(I − 1) + TH*(A(I + 1)
   2+ A(I − 1) − 2.*A(I)))
   WRITE (6,1)Z, AREA
1  FORMAT('0 Z = ', F10.2, 5X, 'AREA = ',F10.2)
```

In this program TH takes the place of θ, and the interpolation is taken so that TH is always negative.

E.3 SOLUTION OF ALBEBRAIC OR TRANSCENDENTAL EQUATIONS BY THE BISECTION METHOD

In the algebraic expression $F(x) = 0$, when a range of values of x is known that contains only one root, the *bisection* method is a practical way to obtain it. It is best shown by an example. The critical depth in a trapezoidal channel is wanted for given flow Q and channel dimensions (Fig. E.3). The formula

$$GG = 1 - \frac{Q^2T}{gA^3} = 0$$

must be satisfied by some positive depth YCR greater than 0 and less than, say, 100 ft. T is the top width $(B + 2.*M*YCR)$. The interval is bisected and this value of YCR tried. If the value of GG is positive, as with the solid line in Fig. E.3b, then the root is less than the midpoint and the upper limit is moved to the midpoint and the remaining half bisected, etc.

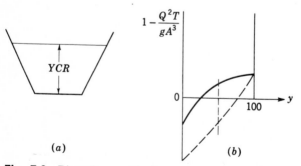

(a)

(b)

Fig. E.3 Bisection method.

In program form

```
REAL M
F(YY) = 1.   −Q**2*(B + 2.*M*YY)/((YY*(B + M*YY))**3*G)
YMAX = 100.
YMIN = .0
YCR = .5*(YMAX + YMIN)
DO  2   J = 0,13
X = F(YCR)
IF(X.GT..0) YMAX = YCR
IF(X.LE..0) YMIN = YCR
2   YCR = .5*(YMAX + YMIN)
WRITE (6,1) YCR
```

The 14 iterations reduce the interval within which the root must be to about 0.01 ft.

E.4 SOLUTION OF TRANSCENDENTAL OR ALGEBRAIC EQUATIONS BY THE NEWTON-RAPHSON METHOD

The Newton-Raphson method[1] is particularly convenient for solving easily differentiable equations when the value of the desired root is known approximately. Let $y = F(x)$ be the equation, Fig. E.4, with $x = x_0$ an approximate value of the root. The root is at B; that is, for $x = B$, $F(x) = 0$. Starting at $x = x_0$ and drawing the tangent to the curve at A give

$$F'(x_0) = \frac{F(x_0)}{x_0 - x_1} \qquad \text{or} \qquad x_1 = x_0 - \frac{F(x_0)}{F'(x_0)}$$

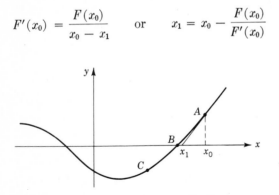

Fig. E.4 Newton-Raphson root-finding method.

[1] L. A. Pipes, "Applied Mathematics for Engineers and Physicists," 2d ed., pp. 115–118, McGraw-Hill, New York, 1958.

It is evident that x_1 is a better approximation to the root than x_0 if there is no point of inflection between A and B and if the slope of the curve does not become zero. If one were to apply these procedures starting at point C (with no inflection points or zero slopes), the first application would yield an x_1 to the right of B. The procedure may be repeated three or four times to obtain an accurate value of the root.

EXAMPLE E.2   It is desired to find the root near $x = x_0$ in the equation $y = a_0 + a_1 x + a_2 x^2 + a_3 \sin \omega x$.

The programming sequence for this solution follows ($\omega = $ OMEGA):

```
    NAMELIST/DIN/A0,A1,A2,A3,OMEGA,X0,X
    READ (5,DIN)
    X = X0
    DO 1   I = 1, 4
  1 X = X − (A0 + A1*X + A2*X*X + A3*SIN(OMEGA*X))/
  1 (A1 + 2.*A2*X + A3*OMEGA*COS(OMEGA*X))
    WRITE (6,DIN)
    END
    &DIN A0 = −1.55,A1 = 1.,A2 = −5,A3 = 1.,OMEGA = 1.5708,
    X0 = .95   &END
```

E.5   RUNGE-KUTTA SOLUTION OF DIFFERENTIAL EQUATIONS

The family of Runge-Kutta solutions is for various orders of accuracy, but they are alike in that the differential equation has its solution extended forward from known conditions by an increment of the independent variable without using information outside of this increment.

First order

In the equation

$$\frac{dy}{dt} = F(y,t)$$

$y = y_n$ when $t = t_n$ and y_{n+1} is desired when $t = t_n + h$. In Fig. E.5,

$$u_1 = hF(y_n,t_n) \qquad y_{n+1} = y_n + u_1 \qquad t_{n+1} = t_n + h$$

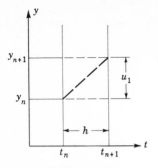

Fig. E.5 First-order Runge-Kutta method.

the equation is evaluated at the initial known conditions and the extension is taken as the tangent to the curve at this point.

Second order

The equation is evaluated at the end points of the interval h, as shown in Fig. E.6:

$$u_1 = hF(y_n, t_n) \qquad u_2 = hF(y_n + u_1, t_n + h)$$

$$y_{n+1} = y_n + \tfrac{1}{2}(u_1 + u_2) \qquad t_{n+1} = t_n + h$$

u_2 is evaluated for the point found by the first-order method.

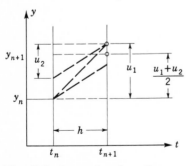

Fig. E.6 Second-order Runge-Kutta method.

Third order

The slope of the curve is evaluated at the initial point, the $\frac{1}{3}$ point, and the $\frac{2}{3}$ point, as follows (Fig. E.7):

$$u_1 = hF(y_n, t_n)$$

$$u_2 = hF\left(y_n + \frac{u_1}{3}, t_n + \frac{h}{3}\right)$$

$$u_3 = hF\left(y_n + \frac{2u_2}{3}, t_n + \frac{2h}{3}\right)$$

$$y_{n+1} = y_n + \frac{u_1}{4} + \frac{3u_3}{4}$$

$$t_{n+1} = t_n + h$$

Differential equations of higher order and degree may be simplified by expressing them as simultaneous first-order differential equations.

EXAMPLE E.3   Put Eq. (12.1.17) into suitable form for solution by the third-order Runge-Kutta method.

In Eq. (12.1.17)

$$\frac{d^2z}{dt^2} + \frac{f}{2D}\frac{dz}{dt}\left|\frac{dz}{dt}\right| + \frac{2g}{L}z = 0$$

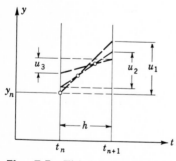

Fig. E.7   Third-order Runge-Kutta method.

Let $y = dz/dt$, then $dy/dt = d^2z/dt^2$, and

$$\frac{dy}{dt} = -\frac{f}{2D} y \,|\, y \,| - \frac{2g}{L} z = F_1(y,z,t)$$

$$\frac{dz}{dt} = y = F_2(y,z,t)$$

The two equations are solved simultaneously, from known initial conditions y_n, z_n, t_n:

$$u_{11} = hF_1(y_n, z_n, t_n) = h\left(-\frac{f}{2D} y_n \,|\, y_n \,| - \frac{2g}{L} z_n\right)$$

$$u_{12} = hF_2(y_n, z_n, t_n) = hy_n$$

$$u_{21} = hF_1\left(y_n + \frac{u_{11}}{3}, z_n + \frac{u_{12}}{3}, t_n + \frac{h}{3}\right)$$

$$= h\left[-\frac{f}{2D}\left(y_n + \frac{u_{11}}{3}\right)\left|y_n + \frac{u_{11}}{3}\right| - \frac{2g}{L}\left(z_n + \frac{u_{12}}{3}\right)\right]$$

$$u_{22} = hF_2\left(y_n + \frac{u_{11}}{3}, z_n + \frac{u_{12}}{3}, t_n + \frac{h}{3}\right) = h\left(y_n + \frac{u_{11}}{3}\right)$$

$$u_{31} = hF_1(y_n + \tfrac{2}{3}u_{21}, z_n + \tfrac{2}{3}u_{22}, t_n + \tfrac{2}{3}h)$$

$$= h\left[-\frac{f}{2D}\left(y_n + \tfrac{2}{3}u_{21}\right)\,|\,y_n + \tfrac{2}{3}u_{21}\,| - \frac{2g}{L}\left(z_n + \tfrac{2}{3}u_{22}\right)\right]$$

$$u_{32} = hF_2(y_n + \tfrac{2}{3}u_{21}, z_n + \tfrac{2}{3}u_{22}, t_n + \tfrac{2}{3}h) = h(y_n + \tfrac{2}{3}u_{21})$$

$$y_{n+1} = y_n + \frac{u_{11}}{4} + \tfrac{3}{4}u_{31}$$

$$z_{n+1} = z_n + \frac{u_{12}}{4} + \tfrac{3}{4}u_{32}$$

$$t_{n+1} = t_n + h$$

The equations for simultaneous solution have been written for a general case as well as for the specific case of solution of Eq. (12.1.17).

ANSWERS TO EVEN-NUMBERED PROBLEMS

Chapter 1

1.6 $W = 95.1$ lb **1.8** $g_0 = 1000$ slugs$\cdot$ft/kip$\cdot$s^2 **1.10** $g = 10.72$ ft/s^2

1.12 $\mu = 184.6$ slugs/ft$\cdot$s **1.14** $\mu = 0.004$ N$\cdot$s/m^2 **1.16** $\mu = 1.217$ N$\cdot$s/m^2

1.18 88.68% **1.20** $\nu = 4.992$ cSt **1.22** $t = 3.34 \times 10^{-3}$ in

1.24 $\nu = 7.618$ mSt **1.26** $v_s = 0.02136$ ft^3/lb$_m$ or 0.687 ft^3/slug

1.28 (a) $S = 2.94$; (b) $v_s = 0.34$ cm^3/g; (c) $\gamma = 2.94$ g$_f$/cm^3

1.30 $\rho = 3.187$ kg/m^3, $m = 127.5$ g **1.32** $\rho = 3.634$ kg/m^3

1.34 $p = 9.59$ MN/m^2 **1.36** Exponentially **1.38** 1500 psi

1.40 4 kg$_f$/cm^2 abs **1.42** 15.53 psia **1.44** 0.14 in **1.46** 0.045 in

Chapter 2

2.2 $p_A = 249.6$ psf $= 11,950$ Pa; $p_B = p_C = -62.4$ psf $= -2.99$ kPa; $p_D = -374.4$ psf $= -17.93$ kPa

2.4 $p_A = -0.866$ psi; $p_B = p_C = 0.866$ psi; $p_D = 3.207$ psi

2.8 0.683 kg$_f$/cm^2 abs; $\rho = 855$ g/m^3 **2.10** (a) 16.30 in; (b) 18.48 ft;

(c) 6.284 ft; (d) 55.2 kPa **2.12** 20.84 m **2.14** 2.449 m; 2.95 m; 83.3 cm

2.16 2.458 m **2.18** 4.645 mm; 10.23 m abs **2.20** -41.5 cm

2.22 (a) 11.50 psi; (b) 3.217 m   **2.24** 38.81 cm   **2.26** 50.01 kPa

2.28 16.76 cm   **2.30** 3600 kg$_f$ **2.32** -1129 N

2.34 (a) 45.82 kN; (b) 54.29 kN   **2.36** $\gamma bh^2/3$ **2.38** 1125 b^2h^2 N·m

2.40 0.774 ft below AB   **2.42** 46.19 kN   **2.44** 5990 lb 3 ft below A

2.46 410 kN   **2.48** 0.4506 ft   **2.50** $\frac{3}{4}h$; $\frac{3}{8}b$ **2.52** 7.429 ft

2.54 0.433b **2.56** 0.7143 m   **2.60** 0.58 ft   **2.62** 0.3334 m

2.66 (a) $x = 11.59$ m; (b) $\sigma_{max} = 69.44\gamma$, $\sigma_{min} = 5.12\gamma$ **2.68** 3994 lb·ft

2.70 (a) $W = 1083$ lb in air; (b) stable   **2.72** 1.44 in   **2.74** 1.839 cm

2.76 253.7 kg$_f$

2.78 (a) 156.8 kN, 4.083 m down; (b) 179.2 kN, 0.948 m left of hinge;

(c) 0; (d) 0   **2.80** -150 lb   **2.82** 12,000 N; $x_p = 0.30$ m

2.84 $100 - 20\rho/3$ lb   **2.86** 27,600 N·m   **2.88** 30.79 kN

2.92 16.53 mm   **2.94** $\bar{y} = 7.16$ ft   **2.96** 1.334 ft   **2.98** 10 cm; 980.2 N

2.100 126.9 lb   **2.102** 4.77 m   **2.104** No   **2.106** Unstable

2.108 86.31 ft/s^2 **2.110** $\theta = 14.05°$; $p_B = 0$; $p_C = -0.866$ psi; $p_D = 0.953$ psi;

$p_E = 0.693$ psi   **2.112** $\theta = 9.47°$; $p_B = 0$; $p_C = -0.26$ psi; $p_D = 1.30$ psi;

$p_E = 1.04$ psi   **2.114** $p_A = 0$; $p_B = 17.89$ kPa; $p_C = 11.52$ kPa

2.116 $p = p_0 \exp(-xa_x\rho_0/p_0)$

2.118 $a_x = 2.04$ m/s^2, $a_y = -1.178$ m/s^2; $\theta = 13.3°$ **2.122** 3.275 rad/s

2.124 5.672 rad/s   **2.126** $2\sqrt{gh_0}/r_0$ **2.128** $r^2\left(\dfrac{\omega^2}{2g}\right) - r\sin\theta = \dfrac{p - p_0}{\rho g}$

2.132 $p = \left[p_0^{(n-1)/n} + \dfrac{n-1}{n}\dfrac{\rho_0\omega^2 r^2}{2p_0^{1/n}}\right]^{n/(n-1)}$ **2.134** 220.5 rpm   **2.136** 15.21 lb

2.138 $F_H = 471.7$ lb

Chapter 3

3.2 (a) 68.1%; (b) 28.14 ft·lb/lb; (c) 88.14 ft·lb/lb

3.4 $x\sqrt{y} = 2 = (5-z)/x$ **3.6** (a) 13.33 ft·lb/lb$_m$; (b) 30 ft·lb/lb$_m$

3.8 1.851 power   **3.10** 32 m/s, 62.83   **3.12** $V = 1.273/(0.8 - 0.6\,x/L)^2$ m/s

3.14 Yes   **3.22** 34.3×10^6 ft·lb   **3.24** 103.75 m·N/s

3.26 12.42 ft, 28.28 ft/s   **3.28** $\alpha = 70.1°$, $V_0 = 26.02$ m/s   **3.30** 1.037

3.32 -250ρ W   **3.34** 0.755 m, 2.74 m   **3.36** 0.82 m

3.38 0.51 m, 2.518 m   **3.40** 53.21 cfs/ft at A; 50.73 cfs/ft at B

3.42 1.751 cfs   **3.44** 3.52 cfs   **3.46** 6.11 velocity heads

3.48 0.0136 m³/s, 2.702 m   **3.50** $r = 1/[4(1 + y/H)^{0.25}]$

3.52 -39.21 kPa   **3.54** $Q = 2.436$ cfs, $p_2 = -372$ psf, $p_3 = 377$ psf

3.56 $Q = 1.436$ cfs, $H_p = 61.37$ ft   **3.58** $Q = 32.65$ cfs, $T = 21.884$ lb·ft

3.62 $V = 3.59$ ft/s   **3.64** $R = 0$, all H **3.66** 0.04 for both

3.68 0.1684 ft·lb/slug·°R   **3.70** $\frac{4}{3}$ **3.74** 0.0   **3.76** $F_y = 62.4$ lb

3.78 651.9 N   **3.80** Tension   **3.82** No change

3.84 $F_x = 2983$ lb, $F_y = 2983$ lb   **3.86** $F_x = 18,397$ N, $F_y = 37,642$ N

3.88 $F_x = -682.7$, $F_y = -1450.4$ lb   **3.90** 13,974 N downward

3.92 (a) 9324 lb;   (b) 957,000 lb;   (c) 6962 hp;   (d) 75%;   (e) 185.5 psf

3.94 46,500 lb, 4133 hp   **3.96** 2540 ft/s   **3.98** $Q = 1.859$ m³/s, $\eta = 0.356$

3.100 $D = 19.75$ cm, power $= 0.327$ MW   **3.102** 53,018 lb (any speed)

3.104 Yes, 0.923, yes   **3.106** 1372 m/s   **3.108** 131.8 km   **3.112** 84.26°

3.114 $-V_0/3$ **3.116** $F_x = -281$ lb, $F_y = 54.07$ lb

3.122 (a) 997 hp;   (b) 1495 hp   **3.124** 152.77°   **3.126** 0.50

3.128 23.79 N·m/N, 7.3 cm H₂O   **3.134** $y_2 = 11.04$ ft, losses $= 8.36$ ft

3.136 30.83 cfs/ft   **3.138** 0.161 ft/s²   **3.140** 242.11 lb

3.142 $T = 17.24$ lb·ft, hp $= 3.94$, $E = 38.97$ ft·lb/lb   **3.144** 463 rpm

3.146 285 rpm

Chapter 4

4.2 (a) $\dfrac{\rho V^2}{\Delta p}$;   (b) $\dfrac{Fg^2}{\rho V^6}$;   (c) $\dfrac{t\,\Delta p}{\mu}$ **4.4** 86.4×10^6 slugs

4.6 Dimensionless; T^{-1}; FLT^{-1}; FL; FL; FLT

4.8 $f\left(\dfrac{\Delta h}{l}, \dfrac{\mu D}{Q\rho}, \dfrac{Q^3\rho^5 g}{\mu^5}\right) = 0$ **4.10** $p = f(\gamma\,\Delta z)$ **4.12** $F_B = f(\rho \mathbb{U} g)$

4.16 $M = f\left(\dfrac{V}{\sqrt{p/\rho}}, k\right)$ **4.18** 10.436 ft/s   **4.20** $\gamma H^4 f\left(\dfrac{\omega H^3}{Q}, e\right)$

4.22 $\rho V^2 D^2 f(\mathbf{R},\mathbf{M})$ **4.24** 0.185

4.26 Choose model size one-seventy-fifth or less of prototype size;

$$\text{loss}_p = \text{loss}_m \left(\frac{D_m v_p}{D_p v_m}\right)^2 \qquad \textbf{4.28} \quad c = \sqrt{gd},\ c = \text{speed},\ d = \text{depth}$$

4.30 $\dfrac{\omega D^{3/2}\rho^{1/2}}{\sigma^{1/2}}$

4.32 36.98 m/s; 18.59 m³/s; losses the same when expressed in velocity heads

Chapter 5

5.2 $\dfrac{d}{dl}(p + \gamma h) = 2\mu U/a^2,\ Q = Ua/3$ **5.4** 0.126 lb, 1.064×10^{-6} cfs

5.6 $Q = \dfrac{(U - V)a}{2} - \dfrac{a^3}{12\mu}\dfrac{dp}{dl}$ **5.8** $\alpha = 1.543,\ \beta = 1.20$

5.12 0.694 psf to right   **5.14** eff $= 33.33\%$

5.16 $\dfrac{\rho a^5}{270\mu^2}\left[\dfrac{d}{dl}(p + \gamma h)\right]^2, \dfrac{0.00569 a^7}{\mu^3}\left[\dfrac{d}{dl}(p + \gamma h)\right]^3$, both per unit width

5.18 $\frac{4}{3}$ **5.20** $0.707 r_0$ **5.24** 0.0275 psf   **5.26** 0.00169 cfs, 22.9

5.28 0.00152 cfs   **5.30** 240 m   **5.32** 27,036   **5.34** 11 m/s

5.38 18.66 per liter   **5.40** 0.223   **5.42** 1549 lb

5.44 $\delta = 4.80 x/\sqrt{\mathbf{R}_x},\ \tau_0 = 0.327\sqrt{\mu U^3 \rho/x}$ **5.46** $\delta = 0.287 x/\mathbf{R}^{1/6}$

5.48 0.93°   **5.50** 38.24 m/s   **5.52** 1   **5.54** 3576 N   **5.56** 214 lb

5.62 2.85 mm/s   **5.64** 54.1 μm, 0.248 m/s   **5.66** 2.31 m/s   **5.68** 0.00244

5.70 1055 cfs   **5.72** 0.000482   **5.74** 310 cfs   **5.76** 9.24 ft/s   **5.78** $y^{8/3}$

5.80 1.754 ft   **5.82** $0.438 a$ **5.84** 1.594 hp   **5.86** 2.911 gpm

5.88 $\mathbf{R} = 1.3 \times 10^6$ **5.90** $D = 17.4$ cm   **5.94** 8.75 kW   **5.96** 1.155 MW

5.98 13 mi   **5.100** 0.539 cfs   **5.102** 0.054 l/s   **5.104** 1512 m³/min

5.106 86.2 kW   **5.108** 26.2 N/s   **5.110** 0.565 m   **5.112** $15,500

5.114   0.193 m   **5.116**   0.70 ft   **5.118**   0.0021 cfs   **5.120** $K = 9.2, L_{eq} = 145$ m

5.122   23.5 cfs   **5.124**   (a) 7.6 ft;   (b) 4.26 ft;   (c) 89.4 ft   **5.126**   21.7

5.132   120,000 psf, $x = 0.375$ ft

Chapter 6

6.2   0.331 kcal/kg·K   **6.4**   20   **6.6**   0.0573 kcal/K   **6.8**   4022 Btu

6.10 $\rho_1/\rho_2 = (T_1/T_2)^{1/(n-1)}$ **6.12**   2.276   **6.14**   16%   **6.16** $V = \sqrt{gy}$

6.18   114°F, 46.45 psia   **6.20**   Same

6.22   1.79 slugs/s, 0.583, 79.9 psia, 487°R

6.24   184.8 psia, 0.0248 slugs/ft³, 624.8°R   **6.26**   0.106 kg/s   **6.28**   0.312 in

6.30   0.119 m, 0.126 m, 0.158 m   **6.32**   321.3 m/s, 1.832 kg/s

6.34   0.095, 0.94   **6.36**   0.577, 90 kPa, 213°C, 254.9 m/s

6.38   1.552, 0.682, 15.9 psia, 299.4°F   **6.50**   68.2%   **6.52** $D = 0.98$ ft

6.54   0.215 kg/s   **6.56**   11.17 kcal/kg   **6.58**   0.168 slug/s

6.60   14.76 kcal/kg   **6.62** $q_H = (V_2^2 - V_1^2)/2$ **6.64**   18.62 m

6.66   3.172,   0.366 psi   **6.68**   0.128 ft

Chapter 7

7.2   0, 0, 0   **7.4** $\omega_x = \frac{3}{2}, \omega_y = -2, \omega_z = -\frac{1}{2}$ **7.6** $w = -2z(x + y)$

7.8 $\varphi = -4x + \frac{7}{2}(x^2 - y^2) - 6y + C$ **7.12** $\psi = \theta + \text{const}$

7.14 $\varphi = 18x + \text{const}$ **7.16** $\dfrac{\partial \varphi}{\partial r}\bigg]_{r=a} = 0, \dfrac{\partial \varphi}{\partial r}\bigg]_{r \to \infty} = 0, \dfrac{\partial \varphi}{\partial \theta}\bigg]_{r \to \infty} = 0$

7.18 $u = -2.785, v = 0.0, w = 0.0; u = 0.459, v = w = 1.022$

7.20 $\varphi = 5/r + 5r \cos \theta, \psi = 5 \cos \theta + 2.5r^2 \sin^2 \theta$

7.22 $\varphi = 12x + 3.641 \left[\dfrac{1}{\sqrt{(x - 1.456)^2 + \bar{\omega}^2}} - \dfrac{1}{\sqrt{(x + 1.456)^2 + \bar{\omega}^2}} \right]$

   $\psi = 6\bar{\omega}^2 + 3.641(\cos \theta_1 - \cos \theta_2)$

7.24 $p = 4787 - 40,500 \sin^2 \theta$ Pa   **7.32** $p = -2\rho\mu^2 y^2/(1 + y^2)^2; \pi\rho\mu^2$

7.36   At (0,1) $q = 4.15$ ft/s, $p = 121$ psf; at (1,1) $q = 5.132$ ft/s, $p = 112.8$ psf

Chapter 8

8.2 4.267 cm   **8.4** 5.67 m/s   **8.6** 68.1 ft/s   **8.8** 1.204   **8.10** 40.54 cfs

8.12 40.52°F; 859.44 ft/s   **8.14** 0.648 gpm   **8.16** 28.42 gpm

8.18 $y = 0.0181x^2$ **8.20** $Y = H \cos^2 \alpha$

8.24 $C_d = 0.773$; $C_v = 0.977$; $C_c = 0.791$ **8.26** 0.864 ft·lb/lb; 47.95 ft·lb/s

8.28 0.273 J/N; 112.35 W   **8.30** 3.61 cm   **8.32** 259.53 mm

8.34 $D = 1.402y^{1/4}$ (in meters)   **8.36** $D = 0.2253y^{3/4}$ (in meters)

8.38 248.95 s   **8.40** 200.05 cfs   **8.42** 2.95 psi

8.44 0.00419 slug/s; 540.05 ft/s   **8.46** 5.75 in   **8.48** 0.00784 slug/s

8.50 602.8 gpm **8.52** 24.7 cfs   **8.54** 0.435 m

8.56 (*a*) 3.055 ft; (*b*) 2.092 ft   **8.58** 3.51 N·m   **8.60** 8.42 mP

Chapter 9

9.2 $Q_c = (Q/N)n$; $H_c = (H/N^2)n^2$; c = corrected, n = const speed

9.4 Synchronization not exact   **9.6** $Q = 0.125Q_1$, $H = 4H_1$

9.8 $N_s = 17.93$ **9.10** 89 in; 300 rpm   **9.12** 3.0 m

9.14 (*a*) 1.78 ft; (*b*) 1200 rpm; (*c*) 201.18 hp; (*d*) 1.475 hp

9.16 14.78°   **9.18** 60.32 ft/s; 180.96 ft/s   **9.20** 117.52 ft   **9.22** 93.24%

9.24 $H = 16.10 - 176.63Q$ (units: m and m³/s)

9.26 (*a*) 515 rpm; (*b*) 3.89 m; (*c*) 38.9 N·m (*d*) 2098 W;

(*e*) 29,727 Pa

9.30 12.48 m   **9.32** 16.74 cm   **9.36** 0.153

Chapter 10

10.2 172.86   **10.4** 14.36 m   **10.6** 4.035 cfs; 8.13 psia   **10.10** 19.655 m

10.12 0.0744 m³/s   **10.14** 2.97 in   **10.16** 2.83 in   **10.18** 15.5 m

10.20 4.46 cfs   **10.24** 2.82 cfs   **10.26** 8.48 cfs

10.28 $Q_1 = 0.00343$ m³/s; $Q_2 = 0.00923$ m³/s; $Q_{tot} = 0.01266$ m³/s

10.30 1315 ft; 6.85 cfs   **10.32** $Q_{AJ} = 1.27$ cfs; $Q_{BJ} = 1.24$ cfs; $Q_{JC} = 2.51$ cfs

10.34 $Q_{JA} = 0.216$ cfs; $Q_{JB} = 1.765$ cfs   **10.36** 1.947 cfs; 103.2 ft

10.38 $Q_{J_1A} = 0.516$; $Q_{BJ_1} = 0.267$; $Q_{J_2J_1} = 0.250$; $Q_{CJ_2} = 0.454$;

      $Q_{J_2D} = 0.204$ (in m³/s)

10.40 58.51; 41.49; 2.36; 31.15; 43.85

10.42 $Q_A = 0.069$ cfs; $Q_B = 1.778$ cfs; $El_J = 100.1$ ft   **10.44** 0.392 ft/ft

10.46 0.004385 m³/s   **10.48** 4.43 ft

Chapter 11

11.2 0.00219 ft   **11.4** 175.1 m² per 100 m   **11.6** $m = \sqrt{3}/3$; 7.11 ft

11.8 $m = \sqrt{3}/3$; $b = 4.111$ m; $y = 3.561$ m   **11.10** 0.000166

11.16 5.32 ft; 2.28 ft   **11.18** 0.560 m   **11.20** 7.39 ft

11.22 2.58 ft; 82.96 cfs/ft   **11.26** 420 ft   **11.34** 0.7224 m rise

11.36 0.404 m

Chapter 12

12.2 1.831 ft/s, 1.429 s   **12.4** $z = V_0 t \exp(-mt)$

12.6 $z_1 = 14.6$ ft, $z_2 = 14.2$ ft   **12.8** 40.12 s   **12.10** 1044 m/s

12.12 1.78 s   **12.14** 500 m

12.16 At $t = 3$ s, $H = 274.9$ m at gate, $H = 92.3$ m at midpoint

12.18 146.2 psi   **12.28** $c = 24.1$ ft/s, $y_1 - y_2 = 0.83$ ft

12.30 2.55 m, $c = 3.564$ m/s

INDEX

INDEX

Accumulator, 670–672
Adiabatic flow, 117, 359–366
Aging of pipes, 580–581
Airfoil lift and drag, 281–282
Analogy:
  electric, 412
  of shock waves to open-channel
    waves, 375–377
Anemometer:
  air, 455
  hot-wire, 455
Angular momentum, 173–177,
  504–508
Artificially roughened pipes, 295
Atmosphere, 32, 33
  effect on plane areas, 50
  local, 33–35
  standard, 35
Available energy, 136–142
Avogadro's law, 16
Axial-flow pumps, 523, 524
Axially symmetric flow, 408–411,
  414–428

Bakhmeteff, B. A., 262, 600, 628

Barometer:
  aneroid, 35
  mercury, 36
Bearing:
  drag coefficients, 312
  journal, 310
  sliding, 309
Bends, forces on, 148
Bergeron, L., 700
Bernoulli equation, 134–139,
  403–405
  assumptions in, modification of,
  137, 138
Best hydraulic cross section, 592–595
Bisection method, 726–727
Blasius, H., 295
Blasius formula, 295
Blowers, 523–530
Borda mouthpiece, 462
Boundary conditions, 4, 410,
  658–673
Boundary layer, 117, 266–280
  critical Reynolds number, 274
  definition of, 267, 268
  laminar, 270–272
  momentum equation of, 268–270

Boundary layer:
  rough plates, 276
  smooth plates, 270–276
  turbulent, 272–276
Boundary layer flow, 117, 266–280
Bourdon gage, 33–35
Boyle's law, 15
Branching pipes, 563–565
Brater, E. F., 545n., 620n.
Bridgman, P. W., 238
Broad-crested weirs, 478
Buckingham, E., 211
Bulk modulus of elasticity, 17, 342,
  343, 653, 711, 712
Buoyant force, 60–64
Buzz bomb, 159

Cámbel, A. B., 371, 388
Capacitance gage, 451
Capillarity, 18
Capillary-tube viscometer, 486
Cascade theory, 504–506
Cavitation, 534–538
Cavitation index, 536
Cavitation parameter, 534
Center of pressure, 46–49
Centipoise, 11
Centrifugal compressor, 530–534
Centrifugal pumps, 511, 523–530,
  667–670
Centroids, 702–705
Characteristics solution, 654–661,
  682–685
Charles' law, 15
Chézy formula, 287
Chow, V. T., 628
Church, A. H., 543
Circular cylinder, flow around, 435
Circular disk, drag coefficients,
  279

Circulation, 431, 437
Classification:
  of open-channel flow, 591, 592
  of surface profiles, 611–615
Clausius inequality, 132, 133
Closed-conduit flow, 213, 216,
  291–309, 544–589
Closed system, 110
Colebrook, C. F., 293, 584n.
Colebrook formula, 291–293, 581n.
Column separation, 672, 673
Compressibility:
  of gases, 13–17
  of liquids, 17
Compressible flow, 333–388
  measurement of, 454–456, 467–472
    velocity, 454
  in pipes, 359–375
Compressor, centrifugal, 530–534
Computer programming aids,
  723–731
Concentric-cylinder viscometer, 484
Conduits, noncircular, 578–580
Conical expansion, 306
Conjugate depth, 171, 596–598
Conservation of energy, 114
Continuity equation, 114, 121–126,
  651–654, 681, 689, 690
Continuum, 13
Control section, 614, 615
Control surface, 110
Control volume, 110
Converging-diverging flow, 343–359
Conversion of energy, 244–246
Crane Company, 306
Critical conditions, 346
Critical depth, 600–604
Critical-depth meter, 620–622
Cross, Hardy, 566
Curl, 392
Current meter, 454

Curved surfaces, force components
    on, 53–60
    horizontal, 53–55
    vertical, 55–59
Cylinder:
    circular, 435–439
    drag coefficients, 280, 281

Daily, J. W., 514, 543
Dam, gravity, 51–53
Dam-break profile, 679, 680
Darcy-Weisbach formula, 287,
    294–302, 359–366, 371–375
Daugherty, R. L., 20
Deformation drag, 278
Del, 30, 125, 126, 390–396
Density, 13
Derivatives, partial, 706–710
Differentials, total, 706–710
Diffusion, 265, 266
Dimensional analysis, 207–238
Dimensionless parameters, 208–210
Dimensions, 210, 211
Discharge coefficient, 459
Disk:
    drag on, 279
    torque on, 484–486
Disk meter, 456–457
Divergence, 126, 392
Doublet:
    three-dimensional, 415
    two-dimensional, 432
Dowden, R. R., 497
Drag:
    airfoil, 281–282
    bearing, 312
    circular disk, 279
    compressibility effect on, 282–284
    cylinder, 280
    deformation, 278

Drag:
    flat plate, 268–276
    pressure, 276
    projectile, 284
    skin friction, 278
    sphere, 279, 284
    wave, 229
Dryden, H. L., 280
Dynamic pressure, 452
Dynamic similitude, 207–238

Eddy viscosity, 260
Efficiency:
    of centrifugal compressor, 530–534
    of centrifugal pump, 526
    hydraulic, 509, 510
    overall, 510
Eisenberg, P., 543
Elasticity, bulk modulus of, 17, 342,
    343, 653, 711, 712
Elbow meter, 472
Elbows, forces on, 148, 149
Electric analogy, 412
Electric strain gage, 451
Electromagnetic flow device, 482
Elementary wave, 375, 676–679
Elrod, H. G., Jr., 482n.
Energy:
    available, 136, 142
    conservation of, 114
    conversion of, 244–246
    flow, 136
    internal, 115, 334, 335
    kinetic, 136
    potential, 135
    pressure, 136
    specific, 600–604
Energy equation, 115, 130
Energy grade line, 142, 547–553
Energy gradient, 547–553
Entropy, 133, 335–340

Epp, R., 568n.
Equations:
  Bernoulli, 134–139, 403–405
  continuity, 114, 121–126, 651–654,
    681, 689, 690
  energy, 115, 130
  Euler's, 127, 128, 134, 396–401
  Hagen-Poiseuille, 213, 249–254,
    298, 486
  Laplace, 402
  momentum, 115, 144–177
    of boundary layer, 268–270
    of motion, 650, 651
    (*See also* Euler's *above*)
  Navier-Stokes, 239–241
  of state, 14–17
Equilibrium (*see* Relative
    equilibrium)
Equipotential lines, 411
Equivalent length, 307, 558, 559
Establishment of flow, 238, 252, 644,
    645
Euler's equation of motion, 127, 128,
    134, 396–401
Expansion factors, 468
Expansion losses:
  conical, 306
  sudden, 169, 304, 305
Exponential pipe friction, 544–547

$F + M$ curve, 596, 597
Falling head, 463, 464
Fanno lines, 355, 356
Fittings, losses for, 306, 307
Flat plate:
  drag coefficients, 268–276
  flow along, 266–280
Flettner rotor ship, 438
Flood routing, 680–687

Floodway, flow in, 595, 596
Flow:
  adiabatic, 117, 359–366
  through annulus, 249–254
  axially symmetric, 408–411,
    414–428
  boundary layer, 117, 266–280
  around circular cylinder, 435
  through circular tubes, 249–254
  with circulation, 437
  classification of, 591, 592
  through closed conduit, 213, 216,
    291–309, 544–589
  compressible (*see* Compressible
    flow)
  establishment of, 238, 252, 644,
    645
  along flat plate, 266–280
  in floodway, 595, 596
  frictionless, 117, 343–350, 389–445
    with heat transfer, 366–371
  gradually varied, 604–611
  ideal, 389–445
  irrotational, 389–445
  isentropic (*see* Isentropic flow)
  isothermal, 371–375
  laminar (*see* Laminar flow)
  measurement of, 449–497
  through noncircular section,
    578–580
  nonuniform, 118, 591
  normal, 287–290
    open-channel, 287–290, 591
  one-dimensional, 119
  open-channel (*see* Open-channel
    flow)
  between parallel plates, 241–248
  pipe, 227, 249–254, 291–309,
    544–589
  potential, 389–445
  rapid, 225, 591

Flow:
  reversible adiabatic (*see* Isentropic flow)
  separation, 276–280
  shooting, 591
  slip, 13
  steady, 116, 117
  supersonic, 343–376
  three-dimensional, 119, 414–424
  tranquil, 225, 591
  transition, 272, 590, 591
  turbulent, 116
  two-dimensional, 119, 428–439
  types of, 116–121
  uniform, 118, 417, 435, 591, 592
  unsteady (*see* Unsteady flow)
  varied, 604–611
Flow cases, 406–445
Flow energy, 136
Flow net, 411–414
Flow nozzle, 469, 470
Flow work, 136
Fluid:
  definition of, 4
  deformation of, 5–7
Fluid flow, ideal, 389–445
Fluid-flow concepts, 109–177
Fluid measurement, 449–497
Fluid meters, 457–481
Fluid properties, 4–19, 711–718
Fluid resistance, 239–313
Fluid statistics, 27–82
Force:
  buoyant, 60–64
  shear, 4, 5
  static pressure, 44–64
    units of, 14
Force systems, 701, 702
Forced vortex, 77
Forces:
  on curved surfaces, 53–62

Forces:
  on elbows, 148, 149
  on gravity dam, 51–53
  on plane areas, 44–52
FORTRAN IV, 302–304, 481, 547, 571–573, 616, 617, 659, 673, 686, 687, 723, 731
Fouse, R. R., 482n.
Fowler, A. G., 568n.
Francis turbine, 519–521
Free molecule flow, 13
Free vortex, 77, 431, 508
Friction factor, 287, 292–300, 359–366, 371–375, 544–547
Frictionless flow (*see* Flow, frictionless)
Froude number, 225, 229, 598–601, 678
Fuel injection system, 3, 4
Fuller, D. D., 313n.

Gage:
  Bourdon, 33–35
  capacitance, 451
  electric, 451
Gage height-discharge curve, 482
Gas constant, 14–16
  properties of, 712
  universal, 16
Gas dynamics, 13, 333–388
Gas law, perfect, 14–16, 333–340
Gas meter, 457
Gibson, A. H., 306
Gradient, 30, 391, 402
Gradually varied flow, 604–611
  computer calculation, 615–619
  integration method, 607–611
  standard step method, 605, 606
Gravity, specific, 14
Gravity dam, 51–53

Hagen, G. W., 250
Hagen-Poiseuille equation, 213,
  249–254, 298, 486
Half body, 420–423
Halliwell, A. R., 700
Hardy Cross method, 565–578
Hazen-Williams formula, 545
Head and energy relationships,
  509–511
Heat, specific, 333–335, 712
Heat transfer, 366–371
Hele-Shaw flow, 258
Henderson, F. M., 628, 687
Holt, M., 238
Homologous units, 498–503
Hot-wire anemometer, 455
Howard, C. D. D., 568n.
Hudson, W. D., 581n.
Hunsaker, J. C., 222, 238, 543
Hydraulic cross sections, best,
  592–595
Hydraulic efficiency, 509, 510
Hydraulic grade line, 142, 293,
  547–553, 650, 651
Hydraulic gradient, 547–553
Hydraulic jump, 170–173, 596–600,
  618
Hydraulic machinery, 230, 498–543
Hydraulic models, 226–231, 238
Hydraulic radius, 286
Hydraulic structures, 228
Hydrodynamic lubrication, 309–313
Hydrometer, 63
Hydrostatics, 27–71

Ideal fluid, 7, 117
Ideal-fluid flow, 389–445
Ideal plastic, 7
Imaginary free surface, 56
Impulse turbines, 511–517

Inertia:
  moment of, 704–705
  product of, 704–705
Internal energy, 115, 334
International System (SI) of units, 7
Ippen, A. T., 620n.
Ipsen, D. C., 238
Irreversibility, 129
Irrotational flow, 389–445
Isentropic flow, 117, 336, 343–350,
  713, 714
  through nozzles, 343–350, 467, 468
Isentropic process, 336
Isothermal flow, 371–375

Jennings, B. H., 371, 388
Jet propulsion, 155–162
Jets, fluid action of, 152–162
Journal bearing, 310

Kaplan turbine, 517–521
Kaye, J., 371, 388
Keenan, J. H., 364n., 371, 388
Kinematic eddy viscosity, 260
Kinematic viscosity, 11, 718
  of water, 711, 712, 718
Kinetic energy, 136
  correction factor, 140, 254
King, H. W., 545n., 620n.

Laminar flow, 116, 241–258
  through annulus, 249, 250
  losses in, 244–248
  between parallel plates, 241–248
  through tubes, 213, 249–255
Langhaar, H. L., 238, 252
Lansford, W. M., 473n.
Laplace equation, 402
Least squares, 480, 481, 546

Liepmann, H., 344, 350, 359, 360, 388
Lift, 281, 438
Lindsey, W. F., 281
Linear momentum, 115, 144–173
  unsteady, 115
Losses, 129
  conical expansion, 306
  fittings, 306
  laminar flow, 244–248
  minor, 304–309
  sudden contraction, 304, 305
  sudden expansion, 169
Lubrication mechanics, 309–313

Mach angle, 283
Mach number, 219, 226, 340, 375
Mach wave, 283
Magnus effect, 438
Manning formula, 287
Manning roughness factors, 288
Manometer:
  differential, 39–43
  inclined, 43, 44
  simple, 38–41
Mass, units of, 7–9
Mean free path, 13
Measurement:
  of compressible flow, 454–456, 467–472
  of flow, 449–497
  of river discharge, 482
  of static pressure, 449–451
  of turbulence, 482
  of velocity, 451–456
  of viscosity, 483–488
Metacenter, 65
Metacentric height, 66
Meters:
  critical-depth, 620–622
  current, 454

Meters:
  disk, 456–457
  elbow, 472
  fluid, 457–481
  gas, 457
  orifice, 458–464, 471, 472
  positive-displacement, 456–457
  rate, 457–481
  venturi, 139, 464–468
  wobble, 456
Method of characteristics, 654–673, 682–687
Micromanometer, 41–43
Milne-Thompson, L. M., 416n.
Minor losses, 304–309
  equivalent length for, 307–309
Mixed-flow pumps, 523–530
Mixing-length theory, 258–265
Model studies, 226–232
Moment:
  of inertia, 704, 705
  of momentum, 173–177, 504–511
Momentum:
  angular, 173–177, 504–508
  correction factor, 145, 146
  linear, 115, 144–173
    unsteady, 115
  molecular interchange of, 9
  moment of, 173–177, 504–511
Momentum equation, 115, 144–173
  of boundary layer, 268–270
Momentum theory for propellers, 152–155
Moody, L. F., 297, 522, 543
Moody diagram, 297
Moody formula, 522
Motion, equation of, 127, 128, 650, 651
  Euler's, 127, 128, 396–401

Natural coordinates, 399
Navier-Stokes equations, 239–241

Net positive suction head (NPSH), 537

Networks of pipes, 565–578

Neumann, E. P., 364

Newton (unit), 7, 9

Newton-Raphson method, 569, 727, 728

Newtonian fluid, 6

Newton's law of viscosity, 5–7

Nikuradse, J., 263, 292, 295

Noncircular conduits, 578–580

Non-Newtonian fluid, 6, 7

Nonuniform flow, 118, 591

Normal depth, 591

Normal flow, 287–290
    open-channel, 287–290, 591

Notation, 719–722

Nozzle:
  forces on, 151
  VDI flow, 469, 470

Nozzle flow, 343–359

NPSH (new positive suction head), 537

Numerical integration, 723, 724

One-dimensional flow, 119

One-seventh-power law, 141, 263, 272

Open-channel flow, 287–290, 590–628
  classification of, 591, 592
  gradually varied, 604–611
  steady uniform, 287–290

Open system, 110

Organick, E. I., 723n.

Orifice:
  determination of coefficients, 458–461
  falling head, 463, 464
  losses, 461, 462
  pipe, 471, 472
  in reservoir, 138, 458–463

Orifice:
  VDI (Verein-Deutscher-Ingenieure), 469, 470

Oscillation:
  of liquid in U-tube: frictionless, 630–632
    laminar resistance, 632–638
    turbulent resistance, 638–641
  of reservoirs, 641–645

Oswald-Cannon-Fenske viscometer, 488

Owczarek, J. A., 388

Page, L., 125n.

Parabolic interpolation, 724–726

Parallel pipes, 560–563

Parallel plates, 241–248

Parameters:
  cavitation, 534
  dimensionless, 208–210

Parmakian, J., 700

Partial derivatives, 706–710

Pascal (unit), 7, 36

Path of particle, 120

Pelton turbine, 511–517

Perfect gas, 14–17
  laws of, 14
  relationships, 333–340

Physical properties:
  of fluids, 4–19, 711–718
  of water, 711, 712

II theorem, 211–223

Piezometer opening, 449

Piezometer ring, 450

Pipe flow, 227, 249–254, 291–309, 544–589

Pipes, L. A., 727n.

Pipes:
  aging of, 580, 581
  branching, 563–565
  compressible flow in, 359–375
    (See also Pipe flow)
  equivalent, 558–560

Pipes:
  frictional resistance in, 249–254,
    291–309, 359–366, 371–375
  networks of, 565–578
  in parallel, 560–563
  in series, 556–560
  tensile stress in, 58, 59
Pitot-static tube, 437, 453, 454
Pitot tube, 143, 453, 454
Poise (unit), 11
Poiseuille (*see* Hagen-Poiseuille
    equation)
Polar vector diagram, 165, 507
Polytropic process, 336–340
Posey, C. J., 630
Positive-displacement meter,
    456, 457
Potential energy, 135
Potential flow, 389–445
Potential velocity, 401–403
Prandtl, L., 259, 266, 270, 272, 274,
    276, 284
Prandtl hypothesis, 266, 389
Prandtl mixing length, 258–265
Prandtl one-seventh-power law, 141
    272
Prandtl tube, 453
Pressure:
  dynamic, 452
  stagnation, 452
  static, 14, 27–29, 449
  total, 452
  vapor, 18, 537, 711, 712
Pressure center, 46, 50
Pressure coefficient, 223–225
Pressure measurement, 449–451
  units and scales of, 33–37
Pressure prism, 49–51
Pressure variation:
  compressible, 32, 33
  incompressible, 29–32
Price current meter, 454
Process, 129

Product of inertia, 704, 705
Programming aids, 723–731
Propeller turbine, 508, 519–521
Propellers:
  momentum theory, 152–155
  thrust, 220
Properties:
  fluids, 4–19, 711–718
  water, 711, 712
Propulsion, rocket, 159–162
Pseudo loop, 569–578
Pumps:
  axial-flow, 523, 524
  centrifugal, 511, 523–530, 667–670
  characteristic curves for, 527
  mixed-flow, 523, 524
  radial-flow, 523–530
  selection chart for, 526
  theoretical head-discharge curve,
    527
  theory of, 506–511

Quadratures, 723, 724

Radial-flow pumps, 523–530
Rainfall-runoff relations, 687–694
Rainville, E. D., 639*n.*
Ram jet, 159
Rankine bodies, 421–424
Rankine degrees, 15
Rapid flow, 225, 591
Rate meters, 457–481
Rate processes, 265, 266
Rayleigh lines, 357, 367
Reaction turbines, 517–523
Relative equilibrium, 71–82
  pressure forces in, 81
  uniform linear acceleration, 72–77
  uniform rotation, 77–81
Relative roughness, 295–303

Reservoirs:
 oscillation in, 641–645
 unsteady flow in, 463
Reversibility, 129, 130
Reversible adiabatic flow, 117
Reynolds, Osborne, 255
Reynolds apparatus, 255
Reynolds number, 219, 225,
 254–258
 critical, 256
 open-channel, 590, 591
Rheingans, W. J., 535n.
Rheological diagram, 6
Rightmire, B. G., 222, 238
River flow measurement, 482
Rocket propulsion, 159–162
Roshko, A., 344, 350, 359, 360, 388
Rotameter, 473
Rotation:
 in fluid, 392
 uniform, 77–81
Rotor ship, Flettner, 438
Runge-Kutta solution, 639, 728–731

Saybolt viscometer, 486–488
Scalar components of vectors,
 394–396
Schlichting, H., 276
Secondary flow, 276
Sedov, L. I., 238
Separation, 276–280, 506
Series pipes, 556–560
Shames, I., 713, 715
Shamir, U., 568n.
Shapiro, A. H., 371, 388
Sharp-crested weirs, 473–477
Shear stress, 4–7
 distribution of, 246
 turbulent, 256–259
Ship's resistance, 229
Shock waves, 350–359, 715, 716

SI (International System) units,
 7–9
Silt distribution, 266
Similitude, 226, 232
 dynamic, 207–238
Simpson's rule, 723, 724
Sink, 414–424, 430
Siphon, 142, 553–556
Skin friction, 278
Sliding bearing, 309
Slip flow, 13
Slipper bearing, 309
Sound wave, 340–343
Source:
 three-dimensional, 414–424
 two-dimensional, 429
Specific energy, 600–604
Specific gravity, 14
Specific heat, 333–335, 712
Specific-heat ratio, 335, 712
Specific speed, 502, 503, 538
Specific volume, 13
Specific weight, 14
Speed of sound, 340–343, 653
Speed factor, 514
Sphere:
 translation of, 424
 uniform flow around, 426
Stability, 64–71
 rotational, 67–71
Stagnation pressure, 452
Standing wave, 171
Stanton diagram, 297
State, equation of, 14–17
Static pressure, 14, 27, 449
 measurement of, 449–451
Static tube, 450
Steady flow, 116, 117
Stepanoff, A. J., 543
Stilling basins, 599, 600
Stoke (unit), 12
Stokes, G., 284, 424n.

Stokes' law, 284
  of viscosity, 240
Stokes' steam function, 408–411
Stoner, M. A., 568n.
Strain gage, electric, 451
Streak line, 120
Stream functions, 406–411
Stream surface, 408–411
Stream tube, 121
Streamline, 119, 120
Streamlined body, 278
Streeter, V. L., 141n., 438n., 700
Sudden expansion, 169
Supersonic flow, 343–376
Surface profiles, 611–619
Surface tension, 18–20
  water, 711, 712
Surge control, 645–647
Surge tank:
  differential, 646
  orifice, 646
  simple, 646
Surge waves:
  negative, 676–679
  positive, 674–676
Surroundings, 110
Sutton, G. W., 535n.
System:
  closed, 110
  open, 110

Teledeltos paper, 413
Tensile stress:
  in pipe, 58, 59
  in spherical shell, 59
Thermodynamics:
  first law, 114, 130–132
  second law, 132
Thixotropic substance, 7
Three-dimensional flow, 119,
    414–424

Time of emptying, 463, 464
Torque on disk, 484–486
Torricelli's theorem, 139
Trajectory method, 460
Tranquil flow, 225, 591
Transitions, 619–622
Tulin, M. P., 543
Turbines:
  Francis, 517–521
  impulse, 511–517
  Kaplan, 517–521
  Pelton, 511–517
  propeller, 508, 519–521
  reaction, 517–523
Turbocompressor, 530–534
Turbojet, 159
Turbomachinery, 498–543
Turbomachines, theory of, 504–511
Turboprop, 159
Turbulence, 255
  level of, 280
  measurement of, 482
Turbulent flow, 116
Two-dimensional flow, 119, 428–439

Ubbelohde viscometer, 488
Uniform flow, 118, 417, 435, 591, 592
Units:
  force and mass, 7, 210
  International System (SI) of,
    7–9
Universal constant, 260
Unsteady flow:
  closed conduits, 118, 629–673
  open channels, 673–694
  reservoirs, 463, 464, 641, 644

V-notch weir, 214, 473–481
Valve in line, 666, 667

Vanes:
    fixed, 162–165
    moving, 165–169
    series of, 166–169
Vapor-column separation, 672, 673
Vapor pressure, 18
    of water, 711, 712
Varied flow, 604–611
VDI flow nozzle, 469–470
VDI orifice, 472, 473
Vector cross product, 174, 392
Vector diagrams, 507, 529
Vector field, 30
Vector operator ∇, 390–396
Velocity:
    of sound, 340–343
    temporal mean, 118
Velocity deficiency law, 265
Velocity distribution, 140, 261–265,
    592
Velocity measurement, 451–456
Velocity potential, 401–403
Vena contracta, 305, 458
Venturi meter, 139, 464–468
Viscometer:
    capillary-tube, 486
    concentric-cylinder, 484
    Saybolt, 486–488
    Ubbelohde, 488
Viscosity, 9–12, 711–712, 717
    eddy, 260
    kinematic, 11, 711–712, 718
    kinematic eddy, 260
    measurement of, 483–488
    Newton's law of, 5–7
    units and conversions, 11, 12
Viscous effects, 239–313
von Kármán, T., 260, 268

Vortex, 77, 431, 437, 520
Vorticity, 392–394

Wake, 276–280
Water, physical properties of, 711,
    712
Water-tunnel tests, 227
Waterhammer, 647–673
    boundary conditions, 658–673
    characteristic solution, 654–673
    differential equations, 649–654
    program, 659
    valve closure: rapid, 649
        slow, 649
Waves:
    elementary, 375, 676–679
    surge, 674–680
Weber number, 219, 225
Weirs:
    broad-crested, 487
    sharp-crested, 473–477
    V-notch, 214, 473–481
Weisbach, J., 305
White, C. M., 581n.
Wiedemann, G., 252
Wind tunnel, 348
Wind-tunnel tests, 227
Windmill, 154
Wislicenus, G. F., 543
Wobble meter, 456
Wood, D. J., 302, 547
Wooding, R. A., 687n.
Wylie, E. B., 700

Yih, C.-S. 413n.